Advances in Compositional Data Analysis

Peter Filzmoser · Karel Hron ·
Josep Antoni Martín-Fernández ·
Javier Palarea-Albaladejo
Editors

Advances in Compositional Data Analysis

Festschrift in Honour of Vera Pawlowsky-Glahn

Springer

Editors
Peter Filzmoser
Institute of Statistics and Mathematical Methods in Economics
TU Wien, Vienna, Austria

Karel Hron
Department of Mathematical Analysis and Applications of Mathematics
Palacký University
Olomouc, Czech Republic

Josep Antoni Martín-Fernández
Department of Computer Science, Applied Mathematics and Statistics
University of Girona
Girona, Spain

Javier Palarea-Albaladejo
Biomathematics and Statistics Scotland
Edinburgh, UK

ISBN 978-3-030-71174-0 ISBN 978-3-030-71175-7 (eBook)
https://doi.org/10.1007/978-3-030-71175-7

This Springer imprint is published by the registered company Springer Nature Switzerland AG
The registered company address is: Gewerbestrasse 11, 6330 Cham, Switzerland

Published with permission of Juan José Egozcue

Preface

This book is conceived as a tribute to Vera Pawlowsky-Glahn for her extraordinary contributions and leadership in the field of the analysis of Compositional Data (CoDa) in the last decades marking the celebration of her 70th birthday anniversary in 2021. The book provides a snapshot of the state-of-the-art and recent advances in the theoretical fundamentals for CoDa and its applications across a wide range of disciplines such as, among others, geology, physical activities and microbiome.

Methods for the analysis and modelling of CoDa are the result of more than a century of efforts. The renowned statistician Karl Pearson already identified the difficulties in the statistical analysis of CoDa. In particular, he identified the problem of spurious correlation between ratios of variables caused by the scaling of the data. The issue was revisited in the 60s and 70s, explicitly connecting Pearson's work with CoDa and showing that some correlations between components must be negative because of the usual sum constraint. However, it was not until the 80s that John Aitchison presented a principled methodological approach to deal with the problems. He stated that the relative information contained in this type of multivariate data can be represented through ratios between the constituent parts of a composition, in particular using log ratios, and that the simplex is the natural sample space for CoDa. The simplex is formed by vectors of positive components with a constant sum (1 for proportions, 100 for percentages, one million for ppm). Despite the efforts of Aitchison and some of his co-authors, the methods for the analysis of CoDa did initially not have a strong impact in the applied sciences.

Vera started working in the field of regionalised CoDa in 1982, extending the log-ratio approach to spatially distributed compositions. Over the years, she has contributed to identify some problems which could not easily be explained in the framework of the log-ratio approach introduced by Aitchison. She has instigated a deepening in the mathematical foundations which has led to a better understanding of the algebraic-geometric structure of the simplex as sample space of CoDa and to establish the current mainstream approach to CoDa analysis; building on previous work and extending ideas and translating this into insights in diverse domains. Two key contributions were the characterisation of the simplex as Euclidean space and the definition of CoDa as equivalence classes of vectors of proportional positive components. Establishing the Euclidean structure of the simplex led to the general idea of mapping data defined on constrained spaces onto the ordinary real space through the use of real coordinates with respect to an orthonormal basis. This has had a great impact in practice by facilitating the use of ordinary statistical methods (typically designed for real data) to analyse data subject to constraints by expressing them on coordinates. An initial drawback of working with coordinates was in some cases their interpretability, particularly as the number of variables involved increased. Vera and her collaborators led the introduction and use of the nowadays popular sequential binary partition for creating *balances*. These are particular orthonormal coordinates that can be customised by the user to represent comparisons of interest between subsets of parts of a composition. In essence, the key methodological idea in the field and in statistical analysis in general is to put the sample space first as the matter of interest to be considered when dealing with a statistical problem. Paying attention to the structure of the sample space gives a deeper insight into which tools are appropriate to deal with it.

The impact of Vera's work in CoDa analysis cannot be overestimated. She has been an inspiration and guiding light to many others that have engaged with CoDa analysis in one way or another, mentoring with infinite patience both senior and early career researchers, and has greatly contributed to its popularisation through a wide range of collaborations across varied scientific disciplines. Although CoDa methods were initially mainly known in the geosciences, Vera was the 2008–2012 President of the International Association for Mathematical Geosciences, they have pervaded throughout diverse areas across the spectrum of the natural, health and social sciences over the last decades; a quick glance at the table of contents of this monograph provides an excellent illustration of this fact. Anyone who has had the opportunity to discuss with Vera about CoDa analysis, even if just for a few minutes, will have been captivated by her dedication and enthusiasm. Vera's work includes a significant amount of debate and advocacy of CoDa analysis, but also self-reflection when something could have been done better. It has been in her very own nature to challenge mainstream thinking and daring to question well-established conventions to advance the field of statistical analysis. Vera has had a significant task in both strengthening the mathematical underpinnings and raising awareness and convincing others to abandon ordinary approaches in favour of a well-founded compositional framework. Her support and encouragement to others have been commendable, fostering the currently vibrant and expanding community of CoDa researchers and practitioners.

We are personally deeply in debt with Vera for the reasons cited above, both on the scientific side for her outstanding contributions to the field as well as on the personal side for her inspiring personality, support and encouragement over these years. It has been a pleasure to organise the preparation of this monograph, and we sincerely thank all the authors for their willingness, the high quality of their contributions and the hard work to have them completed on time. Equally, we very much thank the timely and constructive feedback of the reviewers without whom this could not have been possible. Finally, we thank Springer that kindly supported this project and agreed to have it published.

Vienna, Austria Peter Filzmoser
Olomouc, Czech Republic Karel Hron
Girona, Spain Josep Antoni Martín-Fernández
Edinburgh, UK Javier Palarea-Albaladejo

Contents

Contributors

Núria Arimany–Serrat Department of Economics and Business, University of Vic-UCC, Vic, Spain

Peter M. Atkinson School of Natural and Built Environment, Queen's University Belfast, Belfast, Northern Ireland, UK;
Lancaster Environment Centre, Lancaster University, Lancaster, UK

Nihat Ay Max-Planck Institute for Mathematics in the Sciences, Leipzig, Germany;
Department of Mathematics and Computer Science, Leipzig University, Leipzig, Germany;
Santa Fe Institute, Santa Fe, NM, USA

Carles Barceló-Vidal Department IMAE, Campus Montilivi, Edif. P4, University of Girona, Girona, Spain

Antonella Buccianti Department of Earth Sciences, University of Florence, Florence, Italy

Raja Chakir INRAE, AgroParisTech, Economie Publique, Université Paris-Saclay, Thiverval-Grignon, France

Julien A. Chaput Department of Geological Sciences, University of Texas at El Paso, El Paso, TX, USA

Sebastien F. M. Chastin School of Health and Life Sciences, Glasgow Caledonian University, Glasgow, UK;
Department of Movement and Sport Sciences, Ghent University, Ghent, Belgium

Germà Coenders Department of Economics, University of Girona, Spain

Siobhan F. Cox School of Natural and Built Environment, Queen's University Belfast, Belfast, Northern Ireland, UK

William H. Craddock U.S. Geological Survey, Geology, Energy and Minerals Science Center, Reston, VA, USA

Philippa M. Dall School of Health and Life Sciences, Glasgow Caledonian University, Glasgow, UK

Rory Doherty School of Natural and Built Environment, Queen's University Belfast, Belfast, Northern Ireland, UK

Dorothea Dumuid Allied Health & Human Performance, Alliance for Research in Exercise, Nutrition and Activity (ARENA), University of South Australia, Adelaide, Australia

J. J. Egozcue Department of Civil and Environmental Engineering, Technical University of Catalonia (UPC), Barcelona, Spain;
Department of Applied Mathematics III, Technical University of Catalonia, Barcelona, Spain

Mark A. Engle Department of Geological Sciences, University of Texas at El Paso, El Paso, TX, USA

Ionas Erb Centre for Genomic Regulation (CRG), The Barcelona Institute of Science and Technology, Barcelona, Spain

Àngels Farreras–Noguer Department of Business Studies, University of Girona, Girona, Spain

Kamila Fačevicová Department of Mathematical Analysis and Applications of Mathematics, Palacký University Olomouc, Olomouc, Czech Republic

Damian Fogarty Belfast Health Trust, Belfast, Northern Ireland, UK

Michele Gallo University of Naples "L'Orientale" - DISUS, Naples, Italy

Briana L. Genge Department of Biochemistry, London, ON, Canada

Gregory B. Gloor Department of Biochemistry, London, ON, Canada

Caterina Gozzi Department of Earth Sciences, University of Florence, Florence, Italy

Jan Graffelman Department of Statistics and Operations Research, Universitat Politècnica de Catalunya, Barcelona, Spain;
Department of Biostatistics, University of Washington, Seattle, WA, USA

Karel Hron Department of Mathematical Analysis and Applications of Mathematics, Palacký University, Olomouc, Czech Republic

Lu Huang University of Pennsylvania, Philadelphia, PA, USA

Petra Kynčlová Statistics Division, United Nation Industrial Development Organization (UNIDO), Vienna, Austria

Thibault Laurent Toulouse School of Economics, CNRS, University of Toulouse Capitole, Toulouse, France

Hongzhe Li University of Pennsylvania, Philadelphia, PA, USA

David R. Lovell School of Computer Science, Centre for Data Science, Queensland University of Technology, Brisbane, Australia

Anna Lungarska INRAE, AgroParisTech, Economie Publique, Université Paris-Saclay, Thiverval-Grignon, France

Jean M. Macklaim Department of Biochemistry, London, ON, Canada

Karel Macků Department of Geoinformatics, Palacký University Olomouc, Olomouc, Czech Republic

Wilfredo L. Maldonado Faculty of Economics, Management and Accounting, University of São Paulo, Butantã, São Paulo, Brazil

Josep Antoni Martín-Fernández Department of Computer Science, Applied Mathematics and Statistics, University of Girona, Girona, Spain

Gloria Mateu-Figueras Department of Computer Science, Applied Mathematics and Statistics, University of Girona, Girona, Spain

Duncan E. McGregor School of Health and Life Sciences, Glasgow Caledonian University, Glasgow, UK;
Biomathematics and Statistics Scotland, Edinburgh, UK

Jennifer M. McKinley School of Natural and Built Environment, Queen's University Belfast, Belfast, Northern Ireland, UK

Alessandra Menafoglio MOX, Department of Mathematics, Politecnico di Milano, Milano, Italy

Gianna S. Monti Department of Economics, Management and Statistics, University of Milan-Bicocca, Milan, Italy

Ute Mueller School of Science, Edith Cowan University, Perth, WA, Australia

Ulrich Ofterdinger School of Natural and Built Environment, Queen's University Belfast, Belfast, Northern Ireland, UK

Timothy Olds Allied Health & Human Performance, Alliance for Research in Exercise, Nutrition and Activity (ARENA), University of South Australia, Adelaide, Australia

Ricardo A. Olea U.S. Geological Survey, Geology, Energy and Minerals Science Center, Reston, VA, USA

Javier Palarea-Albaladejo Biomathematics and Statistics Scotland, Edinburgh, Scotland, UK

Željko Pedišić Institute for Health and Sport, Victoria University, Melbourne, Australia

Anne Ruiz-Gazen Toulouse School of Economics, University of Toulouse Capitole, Toulouse, France

Elisabet Saus–Sala Department of Business Studies, University of Girona, Girona, Spain

Violetta Simonacci University of Naples "L'Orientale" - DISUS, Naples, Italy

Matthias Templ Institute of Data Analysis and Process Design, Zurich University of Applied Sciences, Winterthur, Switzerland

Nguyen Thi Huong An Toulouse School of Economics, University of Toulouse Capitole, Toulouse, France;
Danang University of Architecture, Da Nang, Vietnam

Christine Thomas-Agnan Toulouse School of Economics, University of Toulouse Capitole, Toulouse, France

Valentin Todorov United Nations Industrial Development Organization (UNIDO), VIC, Vienna, Austria

Jia R. Wu Department of Computer Science, University of Waterloo, Waterloo, Canada

An Interpretable Orthogonal Decomposition of Positive Square Matrices

J. J. Egozcue and Wilfredo L. Maldonado

Abstract This study of square matrices with positive entries is motivated by a previous contribution on exchange rates matrices. The sample space of these matrices is endowed with a group operation, the componentwise product or Hadamard product. Also an inner product, identified with the ordinary inner product of the componentwise logarithm of the matrices, completes the sample space to be a Euclidean space. This situation allows to introduce two orthogonal decompositions: the first one inspired on the independence of probability tables, and the second related to the reciprocal symmetry matrices whose transpose is the componentwise inverse. The combination of them results in an orthogonal decomposition into easily computable four parts. The merit of this decomposition is that, applied to exchange rate matrices, the four matrices of the decomposition admit an intuitive interpretation.

Dedication

Juan José Egozcue: Vera Pawlowsky-Glahn introduced me to the compositional world in 1998. Since then, we have had many discussions on compositions and their applications. One of Vera's characteristics is to dare to challenge the well-established main stream of thinking. I remember her sentence "we have to redefine variance for positive random variables; the scale should be logarithmic". My prudent answer was: "we could be burnt in a fire immediately; statisticians are very conservative!". To us the way of compositional analysis was and still is an adventure.

Wilfredo L. Maldonado: I met Vera in 2012, during a scientific visit to the Australian National University. Firstly, my family and I met her on a personal level: a

J. J. Egozcue (✉)
Department of Civil and Environmental Engineering, Technical University of Catalonia, Campus Nord c/Jordi Girona 1-3, C-2, E-08034 Barcelona, Spain
e-mail: juan.jose.egozcue@upc.edu

W. L. Maldonado
Faculty of Economics, Management and Accounting, University of São Paulo, Av. Professor Luciano Gualberto 908, Butantã, São Paulo CEP 05508-010, Brazil
e-mail: wilfredo.maldonado@usp.br

P. Filzmoser et al. (eds.), *Advances in Compositional Data Analysis*,
https://doi.org/10.1007/978-3-030-71175-7_1

charming person! Then, research questions arose in which she suggested the use of compositional calculus for the analysis. I, as usual, very skeptical regarding new techniques to be used, criticized the scope of compositional analysis in economics matters. She, with much patience and clarity, explained to me the details of this new technique. The results of this interaction were joint studies with Vera and Juanjo (Juan José Egozcue) in the dynamics of capital intermediation, compositional bubbles in the exchange rates and, more recently, studies of analytic hierarchy processes. I consider myself very fortunate to have met such a distinguished co-author and great friend.

1 Introduction

Positive square matrices have been studied for more than a century due to their many applications. Maybe, the most important result is that known as the 1907 Perron–Frobenius theorem (Horn and Johnson 1985). Today, the main motivations for these studies come from applications of square matrices with positive entries as transition matrices between states in discrete Markov Chains or input–output matrices introduced by Leontief (1966), Peterson and Olinick (1982). These applications use positive matrices in the context of linear systems of equations in which the main operations involved are standard addition and products of matrices. Recently, some attention has been paid to compositional tables (rectangular or square). The matrices are considered as a composition arranged within a matrix. The first contribution on these compositional tables was (Egozcue and Pawlowsky-Glahn 2010) followed by a more complete contribution (Egozcue et al. 2015). Compositional tables were interpreted as a joint probability distribution between two discrete variables. The main result was the decomposition of these tables into an independent table and an interaction table. Among the subsequent articles, the coordinate expression of these tables in orthonormal coordinates (Fačevicová et al. 2016) or their estimation (Ortego and Egozcue 2016) is remarkable. Also, some applications have been published in different fields (Todorov et al. 2013; Pawlowsky-Glahn et al. 2019; Maldonado et al. 2019). The characteristic of compositional tables or matrices is that they are not multiplied in the standard way of matrix multiplication but in a compositional perturbation way, that is, using multiplication componentwise or Hadamard–Shur product (e.g. Horn and Johnson 1985). These results complement both the Hadamard–Shur matrix theory and the traditional theory on positive matrices as they are using a different group operation.

The main motivation of this contribution is the study on no-arbitrage matrices of exchange rates (Maldonado et al. 2019). Exchange rates between currencies of different countries are presented in a (square) matrix. Arbitrage is the possibility of an instantaneous chain of exchanges producing a benefit or a loss. The conditions and properties of these matrices to be free of arbitrage are there studied. One of the remarkable properties is that, necessarily, the diagonal of the exchange rate matrix should be unitary and that symmetrical entries need to be reciprocal (inverse), so that binary arbitrage is avoided. This kind of reciprocal (square) matrices is here used to define an orthogonal decomposition of positive square matrices. However, the ones in the diagonal and the reciprocal characteristic are not easily treated in the strict framework of compositional data analysis as scale invariance is lost. As a consequence, the sample space for square positive matrices should be extended to include reciprocal matrices.

Section 2 presents the exchange rates matrices as an example motivating the proposed decomposition. Section 3 proposes a sample space for positive matrices and its structure. In two subsections, different orthogonal decompositions of the space are specified. Their combination gives an orthogonal decomposition in four components. Section 4 gives some intuitive ideas on the interpretation of the proposed decomposition.

2 Motivating Example

Consider a group of n commodities that can be exchanged and denote them C_i, $i = 1, 2, \ldots, n$. For instance, C_i can be the currency of a country, some raw material or general tradable good. The trade between C_i and C_j is made using the rate a_{ij}, namely, one unit of C_j can be trade with a_{ij} units of C_i. In terms of currencies, one unit of currency C_j can exchange a_{ij} units of currency C_i. For the n commodities, the rates a_{ij} can be arranged into a $(n \times n)$-matrix $A = [a_{ij}]$. Several questions can be asked on the properties of these matrices: are the exchanges charged with taxes or commissions? is there a possibility of changing from commodity to commodity to end at the initial commodity with a benefit or loss? Some observations are quite immediate. All rates a_{ij} can be assumed positive; changing from C_i to C_j and, then, from C_j to C_i, denoted as $C_i \to C_j \to C_i$, there is no benefit or loss whenever $a_{ij} = a_{ji}^{-1}$ and, particularly, $a_{ii} = 1$. If we are dealing with currencies, there is no binary arbitrage. However, alternative sequences of exchanges, larger than two, can still produce benefit or loss; these possibilities are called triangular arbitrages. This question was studied extensively in Maldonado et al. (2019). In these mentioned conditions, when there is no arbitrage, the matrix A is the product of two vectors, with positive components, one reciprocal to the other, that is $A = \mathbf{u}(\mathbf{u}^{-1})^\top$. Consequently, A has a rank equal to 1.

In practice, the exchange of currencies or other commodities is charged with taxes or commissions that can vary from commodity to commodity. If there are taxes, the two-step exchange $C_i \to C_j \to C_i$ is no longer neutral. Frequently, when arriving at the end of the exchange chain, you get less than you invested initially. In other frameworks, the situation can be reversed, in place of paying taxes, you can get some revenue or subsidy from changing from one commodity to another one. Even the rates a_{ii} can be different from 1 if any transaction pays taxes or receives a subsidy for staying as they are. Then, in a general scenario, the rates in $A = [a_{ij}]$ are positive values, $a_{ij} > 0$. For these general exchange rates, can we decompose A into matrices corresponding to no-arbitrage matrices, arbitrage matrices and matrices related to taxes, commissions or subsidies. The orthogonal decomposition proposed in the following sections studies the shape of these matrices and some of their properties.

3 Sample Space Operations and Orthogonality

Square matrices with real positive entries are elements of $\mathbb{R}_+^D$, $D = n^2$, where n is the number of rows or columns of the matrix. For the present purposes, all entries of such matrices are in a ratio scale and, therefore, their differences are measured by log-ratios between them.

Definition 1 (*Sample space for positive square matrices*) Positive $(n \times n)$-matrices are elements of $\mathbb{R}_+^D$, $D = n^2$. This set is taken as the sample space and it is denoted $\mathbb{P}^D$. The internal group operation is the componentwise multiplication: let $A = [a_{ij}]$ and $B = [b_{ij}]$ be in $\mathbb{P}^D$, then $C = A \oplus B$ has entries $c_{ij} = a_{ij} \cdot b_{ij}$. The external multiplication by scalars is $C = \alpha \odot A$, $\alpha \in \mathbb{R}$ where $c_{ij} = a_{ij}^{\alpha}$. Moreover,

$$\langle A, B \rangle_+ = \sum_{i=1}^{n} \sum_{j=1}^{n} \log a_{ij} \cdot \log b_{ij} \tag{1}$$

is taken as an inner product for positive square matrices.

The group operation $\oplus$ between square positive matrices is known as the Hadamard or Shur matrix product, although these terms are generally applied to rectangular matrices as well (Horn and Johnson 1985). Here the Hadamard product, frequently denoted by $\circ$, is denoted $\oplus$ so that the role of addition in $\mathbb{P}^D$ is enhanced. From the inner product in $\mathbb{P}^D$, the norm and distance in $\mathbb{P}^D$ are readily defined as $\|A\|_+^2 = \langle A, A \rangle_+$ and $\mathrm{d}_+(A, B) = \|A \ominus B\|_+$ where $A \ominus B = [a_{ij} b_{ij}^{-1}]$, that is $\ominus \equiv (-1) \odot \oplus$.

This kind of structure for $\mathbb{P}^D$ has been studied in Pawlowsky-Glahn et al. (2015a). The componentwise logarithmic transformation $\tilde{A} = \log A$, $\tilde{A} = [\tilde{a}_{ij}] = [\log a_{ij}]$, provides a representation in real coordinates of A. Similarly we use the componentwise exponential $\exp(\tilde{A}) = [\exp(a_{ij})]$. In fact, $\exp[(\alpha \cdot \tilde{A}) + \tilde{B}] = (\alpha \odot A) \oplus B$, i.e. it is a vector space isomorphism. Also the inner product (1) can be expressed using the coordinates: $\langle A, B \rangle_+ = \langle \tilde{A}, \tilde{B} \rangle$ where $\langle \cdot, \cdot \rangle$ is the ordinary inner product in $\mathbb{R}^D$ when the D-tuples are arranged in a $(n \times n)$-matrix. All this facts can be summarized in the following theorem.

Theorem 1 (Isometry to $\mathbb{R}^D$ Euclidean space) *The spaces $\mathbb{P}^D$ and $\mathbb{R}^D$ are isometric Euclidean spaces through logarithmic function $\log(\mathbb{P}^D) = \mathbb{R}^D$ where $\log$ applies to each entry of matrices. The dimension of these spaces is $D = n^2$ when the matrices $\mathbb{P}^D$ have n rows and n columns.*

Additionally, the space $\mathbb{R}_+^D$ is also isometric to $\mathbb{S}^D \times \mathbb{R}_+$ where $\mathbb{S}^D$ is the D-part simplex. This isometry $\mathcal{T} : \mathbb{P}^D \to \mathbb{S}^D \times \mathbb{R}_+$ is studied in Pawlowsky-Glahn et al. (2015a) except that the elements are arranged as vectors in place of square matrices. The isometry $\mathcal{T}$ consists in closing $A \in \mathbb{P}^D$ to a constant sum, and considering a magnitude, called total, related to the geometric mean of the elements in A, this is

$$\mathcal{C}A = \frac{1}{\sum_{ij} a_{ij}} A \,, \quad t = \left(\prod_{ij} a_{ij} \right)^{1/\sqrt{D}} ,$$

where $\mathcal{C}$ is the closure operator. Therefore, $\mathcal{T}(A) = (\mathcal{C}A, t)$. In some instances, the matrix A is scale invariant, that is, the information conveyed by A and its positive multiples are compositionally equivalent (Egozcue and Pawlowsky-Glahn 2019; Barceló-Vidal and Martín-Fernández 2016; Pawlowsky-Glahn et al. 2015b). Then, A can be treated as composition and the total t is then irrelevant. When A is considered as a composition, the isometric log-ratio (ilr) coordinates in the simplex can be used for its representation. The isometry $\mathcal{T}$ allows $A \in \mathbb{P}^D$ to be represented by $D-1$ real ilr coordinates, $\mathbf{a}^* = \text{ilr}(A) = (a_1^*, a_2^*, \ldots, a_{D-1}^*)^\top$, complemented with a coordinate $t^* = \log t$, thus forming a set of D orthogonal coordinates $(\mathbf{a}^*, t^*)$. The transformation ilr$(\cdot)$ assigns a real $(D-1)$-vector of coordinates to the composition arranged in the $(n \times n)$-matrix A. The ilr-coordinates $\mathbf{a}^*$ (Egozcue et al. 2003) can be selected in a variety of ways, and the most appealing being the one proposed in Fačevicová et al. (2016).

One interesting property, related to the isometry $\mathcal{T}$ is the computation of the square norm of a matrix in $\mathbb{P}^D$ in two different ways corresponding to the two different coordinate representations.

Property 1 (*Square norm of a positive matrix*) Let $A = [a_{ij}]$ be a positive matrix in $\mathbb{P}^D$, with logarithmic coordinates $\tilde{a}_{ij} = \log a_{ij}$. The compositional coordinates of A considered as a composition are $\text{ilr}(A) = (a_1^*, a_2^*, \ldots, a_{D-1}^*)$ for some choice of an orthonormal basis in the simplex; the coordinate corresponding to the total t is the logarithmic total $t^* = \log t = (1/\sqrt{D}) \sum_{ij} \log a_{ij}$. Then,

$$\|A\|_+^2 = \sum_{i,j=1}^{n} (\tilde{a}_{ij})^2 = \sum_{k=1}^{D-1} (a_k^*)^2 + (t^*)^2 \,.$$

Considering A or its closure $\mathcal{C}A$ as a composition, the term $\|A\|_a^2 = \sum_k (a_k^*)^2$ is the square Aitchison norm of A since it is the sum of squares of its ilr coordinates. The proof of Property 1 is based on Theorem 1 which was proven in Pawlowsky-Glahn et al. (2015a). The selection of orthonormal bases in the simplex can be found in Pawlowsky-Glahn et al. (2015b) and the concept of Aitchison norm in Pawlowsky-Glahn and Egozcue (2001) and Billheimer et al. (2001).

3.1 Independent and Interaction Matrices

In Egozcue et al. (2015), $\mathbb{P}^D$ matrices, square or not, were considered as tables representing joint probabilities of two discrete random variables and, accordingly, as D-part compositions arranged in a matrix. There, the normalization was avoided as proportional matrices were considered equivalent according to compositional equivalence (Barceló-Vidal and Martín-Fernández 2016; Pawlowsky-Glahn et al. 2015b). The main result was the orthogonal decomposition of a probability matrix into its independent part and the orthogonal complement called interaction part. In the context of positive matrices, the sample space is no longer $\mathbb{S}^D$ but $\mathbb{P}^D$. These two spaces share a lot of features as, in fact, it was shown (Pawlowsky-Glahn et al. 2015a) that $\mathbb{P}^D$ is isometric with $\mathbb{S}^D \times \mathbb{R}_+$.

Independent matrices inherit their name from the probability theory. A $(n \times n)$-matrix, $A \in \mathbb{P}^D$, is independent if it is the outer product of two n-vectors, with positive components. These vectors are called marginals in probability theory. The difference with probability is that here neither normalization of marginals nor the joint distribution is imposed.

Definition 2 (*Independent matrices*) A square matrix $A \in \mathbb{P}^D$, $D = n^2$, is an independent matrix if there exist two n-vectors, $\mathbf{u}_c$, $\mathbf{u}_r$, with positive components such that $A = \mathbf{u}_c \mathbf{u}_r^\top$. The set of independent $(n \times n)$-matrices is denoted $\mathbb{P}^D_{\text{ind}}$.

An independent matrix can be expressed as a function of its marginals using the perturbation in $\mathbb{P}^D$, that is

$$A = (\mathbf{u}_c \mathbf{1}^\top) \oplus (\mathbf{1}\mathbf{u}_r^\top) = U_c \oplus U_r^\top\ , \tag{2}$$

where $\mathbf{1} = (1, 1, \dots, 1)^\top$ and U_c, U_r are matrices in $\mathbb{P}^D$ whose n columns are equal to the column marginal $\mathbf{u}_c$ and row marginal $\mathbf{u}_r$, respectively. The marginals $\mathbf{u}_c$, $\mathbf{u}_r$ are defined up to a positive constant since $A = (a\mathbf{u}_c)(a^{-1}\mathbf{u}_r^\top)$, $a > 0$, is also a valid decomposition into marginals. This is not the case when a normalization of A, $\mathbf{u}_c$ and $\mathbf{u}_r$ is required, for instance, when they are normalized as probability table and vectors.

The proofs of the following theorems and properties in this Section are in the Appendix. The main result is the following theorem.

Theorem 2 (Subspace of independent matrices) *The set $\mathbb{P}^D_{\text{ind}}$ of square matrices with positive entries is a $2n$-dimensional subspace of $\mathbb{P}^D$.*

The next theorem gives a procedure of projection of $A \in \mathbb{P}^D$ onto $\mathbb{P}^D_{\text{ind}}$.

Theorem 3 (Closest independent positive matrix) *Let A be a matrix with positive entries, $A \in \mathbb{P}^D$, $D = n^2$. The orthogonal projection of A onto the subspace of independent matrices $\mathbb{P}^D_{\text{ind}}$ is A_{ind} such that its components are*

$$[A_{\text{ind}}]_{ij} = \frac{\mathrm{g_m}(\mathbf{a}^j)\mathrm{g_m}(\mathbf{a}_i)}{\mathrm{g_m}(A)}\ ,$$

where $\mathbf{a}^j$, $\mathbf{a}_i$ *are the rows and columns of* A, *respectively, and* g_m *denotes the geometric mean of the argument.*

Consider the orthogonal complement of $\mathbb{P}^D_{\text{ind}}$ with respect to $\mathbb{P}^D$, and denote it by $\mathbb{P}^D_{\text{int}} = (\mathbb{P}^D_{\text{ind}})^{\perp}$.

Corollary 1 (Subspace of interaction matrices) *The subspace* $\mathbb{P}^D_{\text{int}}$ *is a* $n(n-2)$-*dimensional subspace of* $\mathbb{P}^D$.

Some properties of matrices in $\mathbb{P}^D_{\text{ind}}$ and $\mathbb{P}^D_{\text{int}}$ also characterize these kinds of matrices.

Property 2 Let $A \in \mathbb{P}^D$ be a positive matrix with orthogonal decomposition $A = A_{\text{ind}} \oplus A_{\text{int}}$ with $A_{\text{ind}} \in \mathbb{P}^D_{\text{ind}}$, $A_{\text{int}} \in \mathbb{P}^D_{\text{int}}$. Then, the rank of A_{ind} is one and the rank of A_{int} is equal to the rank of A.

In general, the properties in Euclidean spaces hold. The two next properties are interesting.

Property 3 Let $A = A_{\text{ind}} \oplus A_{\text{int}}$ be the orthogonal decomposition of A and denote by $\|\cdot\|_+$ the norm in $\mathbb{P}^D$. Then,
(a) The decomposition is unique;
(b) $\|A\|_+^2 = \|A_{\text{ind}}\|_+^2 + \|A_{\text{int}}\|_+^2$.

The characteristics of the logarithmic coordinates of $\mathbb{P}^D_{\text{ind}}$ and $\mathbb{P}^D_{\text{int}}$ are remarkable.

Property 4 (*Logarithmic coordinates in* $\mathbb{P}^D_{\text{ind}}$ *and* $\mathbb{P}^D_{\text{int}}$) Let $A_{\text{ind}} \in \mathbb{P}^D_{\text{ind}}$ and $A_{\text{int}} \in \mathbb{P}^D_{\text{int}}$ such that $A = A_{\text{ind}} \oplus A_{\text{int}}$. Then, the respective logarithmic coordinates $\tilde{A}_{\text{ind}} = \log(A_{\text{ind}})$, $\tilde{A}_{\text{int}} = \log(A_{\text{int}})$ satisfy the following:
(a) $\tilde{A}_{\text{ind}}$ is obtained from the average coordinates of rows and columns (marginals) of $\tilde{A}$,

$$\begin{aligned}
\tilde{A}_{\text{ind}} &= \tilde{\mathbf{u}}_c \mathbf{1}^\top + \mathbf{1}\tilde{\mathbf{u}}_r^\top \,, \\
\tilde{\mathbf{u}}_r &= \frac{1}{n}\tilde{A}^\top \mathbf{1} - \frac{1}{2n^2}\mathbf{1}\mathbf{1}^\top \tilde{A}\mathbf{1} \,, \\
\tilde{\mathbf{u}}_c &= \frac{1}{n}\tilde{A}\mathbf{1} - \frac{1}{2n^2}\mathbf{1}\mathbf{1}^\top \tilde{A}\mathbf{1} \,.
\end{aligned}$$

(b) The (arithmetic) marginals of $\tilde{A}_{\text{int}}$ are null,

$$\frac{1}{n}\tilde{A}_{\text{int}}\mathbf{1} = \mathbf{0} \,, \quad \frac{1}{n}\tilde{A}_{\text{int}}^\top \mathbf{1} = \mathbf{0} \,.$$

(c) The logarithmic coordinates of the interaction part are obtained by double centering of $\tilde{A}$,

$$\tilde{A}_{\text{int}} = \left(I_n - \frac{1}{n}\mathbf{1}\mathbf{1}^\top\right) \tilde{A} \left(I_n - \frac{1}{n}\mathbf{1}\mathbf{1}^\top\right) ,$$

where I_n denotes the identity matrix of n rows.

Computation of the orthogonal projection of $A \in \mathbb{P}^D$ onto $\mathbb{P}^D_{\text{ind}}$ and $\mathbb{P}^D_{\text{int}}$ becomes simple in logarithmic coordinates, as it is a double centering of $\tilde{A}$; the matrix subtracted from $\tilde{A}$ is $\tilde{A}_{\text{ind}}$ and the remaining is $\tilde{A}_{\text{int}}$.

3.2 *Symmetric and Reciprocal Square Matrices*

When studying exchange matrices (Maldonado et al. 2019), the fact that the rate $A_i \to A_j$, a_{ij} and the rate $A_j \to A_i$, a_{ji} must be reciprocal became clear in order to avoid binary arbitrages. In fact, if $a_{ij} \cdot a_{ji} \neq 1$ points out binary arbitrage. Consequently, the condition $A = (-1) \odot A^\top$ becomes a necessary condition for no-arbitrage exchange matrices.

Definition 3 (*Reciprocal square matrices*) A square matrix $A \in \mathbb{P}^D$, $D = n^2$, is a reciprocal matrix if

$$A = (-1) \odot A^\top ,$$

equivalent to $a_{ij} = a_{ji}^{-1}$, $a_{ii} = 1$ for all elements in A.

Taking logarithms componentwise, the following property holds.

Property 5 (*Logarithmic coordinates of reciprocal matrices*) The matrix of logarithmic coordinates of a reciprocal matrix is anti-symmetric, that is $\tilde{A} = -\tilde{A}^\top$.

Theorem 4 (Subspace of reciprocal matrices) *The set of reciprocal matrices in* $\mathbb{P}^D$ *is a* $(n(n-1)/2)$*-dimensional subspace of* $\mathbb{P}^D$, *denoted* $\mathbb{P}^D_{\text{rec}}$.

Similarly to the case of independent matrices, consider the orthogonal complement of the reciprocal matrices. This space is constituted by all symmetrical matrices and is denoted $\mathbb{P}^D_{\text{sym}}$. In fact, the logarithmic coordinates of reciprocal matrices in $\mathbb{P}^D_{\text{rec}}$ are anti-symmetric thus implying that matrices in their orthogonal in $\mathbb{R}^D$ are symmetric. Taking exponential of a symmetric matrix yields a symmetric matrix. The dimension of $\mathbb{P}^D_{\text{sym}}$ is the difference between the respective dimensions of $\mathbb{P}^D$ and $\mathbb{P}^D_{\text{rec}}$, that is $n^2 - n(n-1)/2 = n(n+1)/2$.

3.3 *Independent-Reciprocal Orthogonal Decomposition*

Two different orthogonal decompositions of $\mathbb{P}^D$ into subspaces have been proposed in the previous subsections. Now considering the two decompositions simultaneously, a decomposition of $\mathbb{P}^D$ into four orthogonal subspaces is easily obtained by the intersection of the two decompositions.

Definition 4 (*Four subspaces of* $\mathbb{P}^D$) The intersection of subspaces $\mathbb{P}^D_{\text{ind}}$, $\mathbb{P}^D_{\text{int}}$, $\mathbb{P}^D_{\text{rec}}$, $\mathbb{P}^D_{\text{sym}}$, are denoted:

$\mathbb{P}^D_{\text{rind}} = \mathbb{P}^D_{\text{ind}} \cap \mathbb{P}^D_{\text{rec}}$ includes independent-reciprocal matrices (rind),
$\mathbb{P}^D_{\text{rint}} = \mathbb{P}^D_{\text{int}} \cap \mathbb{P}^D_{\text{rec}}$ includes interaction reciprocal matrices (rint),
$\mathbb{P}^D_{\text{sind}} = \mathbb{P}^D_{\text{ind}} \cap \mathbb{P}^D_{\text{sym}}$ includes independent symmetric matrices (sind), and
$\mathbb{P}^D_{\text{sint}} = \mathbb{P}^D_{\text{int}} \cap \mathbb{P}^D_{\text{sym}}$ includes interaction symmetric matrices (sint).

Theorem 5 (orthogonal decomposition) *The space of square positive matrices* $\mathbb{P}^D$, $D = n^2$, *is uniquely decomposed into the four mutually orthogonal subspaces* $\mathbb{P}^D_{\text{rind}}$, $\mathbb{P}^D_{\text{rint}}$, $\mathbb{P}^D_{\text{sind}}$, $\mathbb{P}^D_{\text{sint}}$ *whose dimensions are* $\dim(\mathbb{P}^D_{\text{rind}}) = n$, $\dim(\mathbb{P}^D_{\text{sind}}) = n$, $\dim(\mathbb{P}^D_{\text{rint}}) = n(n-3)/2$, $\dim(\mathbb{P}^D_{\text{sint}}) = n(n-1)/2$. *Any matrix* $A \in \mathbb{P}^D$ *is expressed in a unique way as*

$$A = A_{\text{rind}} \oplus A_{\text{rint}} \oplus A_{\text{sind}} \oplus A_{\text{sint}} \,. \tag{3}$$

In logarithmic coordinates, the decomposition of $A \in \mathbb{P}^D$ *is*

$$\tilde{A}_{\text{rec}} = \frac{1}{2}(\tilde{A} - \tilde{A}^\top)\,, \quad \tilde{A}_{\text{sym}} = \frac{1}{2}(\tilde{A} + \tilde{A}^\top)$$

and

$$\begin{aligned}
\tilde{A}_{\text{rint}} &= \left(I_n - \frac{1}{n}\mathbf{1}\mathbf{1}^\top\right) \tilde{A}_{\text{rec}} \left(I_n - \frac{1}{n}\mathbf{1}\mathbf{1}^\top\right) \,, \\
\tilde{A}_{\text{sint}} &= \left(I_n - \frac{1}{n}\mathbf{1}\mathbf{1}^\top\right) \tilde{A}_{\text{sym}} \left(I_n - \frac{1}{n}\mathbf{1}\mathbf{1}^\top\right) \,, \\
\tilde{A}_{\text{rind}} &= \tilde{A}_{\text{rec}} - \tilde{A}_{\text{rint}} \,, \\
\tilde{A}_{\text{sind}} &= \tilde{A}_{\text{sym}} - \tilde{A}_{\text{sint}} \,.
\end{aligned} \tag{4}$$

4 Interpretation in Exchange Scenarios

Consider a fictitious scenario, a market for simplicity, in which n different divisible commodities are traded among them. The rules of this exchange are specified in a positive matrix $A \in \mathbb{P}^D$, $D = n^2$, which in this first view is a constant. One unit of commodity C_j can be exchanged by a_{ij} units of commodity C_i. Availability of commodities is unlimited, independently of their demand. The target of operators in this market is to increase the value of their controlled commodities, called here portfolio. However, the operators cannot check the effective value of their portfolios except using the rates specified in A, that is, when they perform one or several instantaneous exchanges. The most elementary model for checking the value of the quantity of the C_i commodity is to perform a closed chain of exchanges like $C_i \to C_j \to \cdots \to C_i$ and then, compare the initial amount with the final amount of C_i. As explained in Maldonado et al. (2019), this evaluation corresponds to the product $v = a_{ij} \cdot a_{jk} \cdots a_{m\ell} \cdot a_{\ell i}$; if $v = 1$ there is no increase or decrease of the

Table 1 Fictitious exchange rate matrix in $\mathbb{P}^D$ between four commodities (left) and its logarithmic coordinates (right). Note that the diagonal values are not unitary and there are slight departures from reciprocity between symmetric entries (left) or from anti-symmetry for log-coordinates (right). The values in the diagonal appear equal when rounded to three decimal places, but they are slightly different

	$\mathbf{a}_1$	$\mathbf{a}_2$	$\mathbf{a}_3$	$\mathbf{a}_4$	;		$\log \mathbf{a}_1$	$\log \mathbf{a}_2$	$\log \mathbf{a}_3$	$\log \mathbf{a}_4$
$\mathbf{a}_1$	1.003	0.502	0.249	0.099	;	$\log \mathbf{a}_1$	0.003	−0.688	−1.390	−2.311
$\mathbf{a}_2$	2.000	1.003	0.497	0.202	;	$\log \mathbf{a}_2$	0.693	0.003	−0.699	−1.601
$\mathbf{a}_3$	3.914	2.003	1.003	0.403	;	$\log \mathbf{a}_3$	1.364	0.695	0.003	−0.909
$\mathbf{a}_4$	10.140	4.982	2.494	1.003	;	$\log \mathbf{a}_4$	2.317	1.606	0.914	0.003

value by performing the chain of exchanges; $v > 1$ and $v < 1$ represent increases of value (benefit for short) or decrease of value (loss for short), respectively.

Table 1 shows a exchange rate matrix A of four commodities (left) in a fictitious market. The logarithmic coordinates are shown in the right part of the Table. The main feature of A is that it seems close to be reciprocal (or its logarithmic coordinates are near to be anti-symmetric). In fact, the diagonal terms are not 1 but larger than 1 thus indicating that the market has some stimulus against exchanging all considered commodities. Paying attention to arbitrages, the exchange chains $C_1 \to C_2 \to C_1$ reveals some binary arbitrage $a_{12} \cdot a_{21} = 1.0048$ (a benefit in this case). Also order three chains produce arbitrage. For instance, $C_1 \to C_2 \to C_3 \to C_1$ results in a loss since $a_{12} \cdot a_{23} \cdot a_{31} = 0.977$. Up to present, we cannot distinguish which are the reasons for these arbitrages; they can be due to taxes in each transaction or to the structure of the whole matrix of exchange rates. This simple example reveals that the examination of these matrices is quite difficult. The orthogonal decomposition allows us interpreting separately different characteristics of these exchange rate matrices.

The exchange rate matrix A can be decomposed into its four orthogonal components, as in Eq. (3), that is

$$A = A_{\mathrm{rind}} \oplus A_{\mathrm{rint}} \oplus A_{\mathrm{sind}} \oplus A_{\mathrm{sint}} \ ,$$

where each component admits an interpretation in this scenario.

The matrix A_{rind} corresponds to a situation in which any circular chain of exchanges provides any benefit or loss, that is, there is no arbitrage. This was discussed in Maldonado et al. (2019) (Proposition 2.5) and these matrices were called NAMER (no-arbitrage matrices of exchange rates). As pointed in Property 4, NAMER, here denoted A_{rind}, are the outer product of a vector and its reciprocal $A = \mathbf{u}(\mathbf{u}^{-1})^\top$. The vector $\mathbf{u}$ is defined up to a multiplication by a positive scalar and consequently, it should be considered a composition arbitrarily scalable, i.e. $\mathbf{u} \in \mathbb{S}^n$. In Maldonado et al. (2019), $\mathbf{u}$ was identified to exchange rates for a (virtual) kind of currency representing neutral exchange rates.

Table 2 shows the component A_{rind} (left side), the vector $\mathbf{u}$ (first column) that generates it and the corresponding logarithmic coordinates (right side). The diagonal

Table 2 The A_{rind} matrix corresponding to example in Table 1. The first column is the vector $\mathbf{u}$ that generates the whole matrix. The columns of A_{rind} (left) are multiples of $\mathbf{u}$ and the columns of the logarithmic coordinates (right) are shifts of any of the columns

	$\mathbf{u}$	$\mathbf{a}_{\text{rind},1}$	$\mathbf{a}_{\text{rind},2}$	$\mathbf{a}_{\text{rind},3}$	$\mathbf{a}_{\text{rind},4}$	;		log $\mathbf{a}_{\text{rind},1}$	log $\mathbf{a}_{\text{rind},2}$	log $\mathbf{a}_{\text{rind},3}$	log $\mathbf{a}_{\text{rind},4}$
$\mathbf{a}_{\text{rind},1}$	0.091	1.000	0.500	0.250	0.100	;	log $\mathbf{a}_{\text{rind},1}$	0.000	0.002	0.009	−0.011
$\mathbf{a}_{\text{rind},2}$	0.182	2.000	1.000	0.500	0.200	;	log $\mathbf{a}_{\text{rind},2}$	−0.002	0.000	−0.004	0.006
$\mathbf{a}_{\text{rind},3}$	0.364	4.000	2.000	1.000	0.400	;	log $\mathbf{a}_{\text{rind},3}$	−0.009	0.004	0.000	0.005
$\mathbf{a}_{\text{rind},4}$	0.909	10.000	5.000	2.500	1.000	;	log $\mathbf{a}_{\text{rind},4}$	0.011	−0.006	−0.005	0.000

Table 3 The A_{rint} matrix (left) corresponding to example in Table 1 and its logarithmic coordinates (right). Note that diagonal elements of A_{rint} are exactly one and that log A_{rint} is anti-symmetric

	$\mathbf{a}_{\text{rint},1}$	$\mathbf{a}_{\text{rint},2}$	$\mathbf{a}_{\text{rint},3}$	$\mathbf{a}_{\text{rint},4}$	;		log $\mathbf{a}_{\text{rint},1}$	log $\mathbf{a}_{\text{rint},2}$	log $\mathbf{a}_{\text{rint},3}$	log $\mathbf{a}_{\text{rint},4}$
$\mathbf{a}_{\text{rint},1}$	1.000	1.002	1.009	0.989	;	log $\mathbf{a}_{\text{rint},1}$	0.000	0.002	0.009	−0.011
$\mathbf{a}_{\text{rint},2}$	0.998	1.000	0.996	1.006	;	log $\mathbf{a}_{\text{rint},2}$	−0.002	0.000	−0.004	0.006
$\mathbf{a}_{\text{rint},3}$	0.991	1.004	1.000	1.005	;	log $\mathbf{a}_{\text{rint},3}$	−0.009	0.004	0.000	0.005
$\mathbf{a}_{\text{rint},4}$	1.011	0.994	0.995	1.000	;	log $\mathbf{a}_{\text{rint},4}$	0.011	−0.006	−0.005	0.000

of A_{rind} only contains ones (zeros in the coordinate matrix). The matrix is reciprocal as can be easily seen in the coordinates which are anti-symmetric. Paying a bit more attention, one can realize that columns of A_{rind} are multiples of $\mathbf{u}$. The composition $\mathbf{u}$ can be thought of as the exchange rates of commodities with a fictitious fair currency which facilitates the exchange of commodities and it is defined up to a multiplicative constant.

One can ask for the meaning of the square Aitchison norm $\|\mathbf{u}\|_a^2$. This is simply a measure of inequality of the exchange rates between commodities (Egozcue and Pawlowsky-Glahn 2019) and the relative value of the respective units. An exchange market ruled by a constant A_{rind} is triangular arbitrage-free. For instance, the exchange chains mentioned above using the whole matrix A give neither loss nor benefit when using A_{rind}. In fact, $a_{\text{rind}12} \cdot a_{\text{rind}21} = 1$ and $a_{\text{rind}12} \cdot a_{\text{rind}23} \cdot a_{\text{rind}31} = 1$. These kinds of exchange rate matrices propitiate a kind of inertial exchange: the exchange itself does not produce neither loss nor benefit and does not generate any stress to change A. The entries of A_{rind} are true exchange rates and they have units, for instance, $[A_{\text{rind}}]_{ij}$ has units (units of C_j/units of C_i). The other three components of the decomposition, namely A_{rint}, A_{sind}, A_{sint}, are perturbations of the exchange rates in A_{rind} and they are dimensionless.

The component A_{rint} describes how ternary arbitrages are implicit in A. Binary arbitrages are excluded as A_{rint} is a reciprocal matrix. Table 3 shows A_{rint} from the example. It is a reciprocal matrix (anti-symmetric logarithmic coordinates) but contains ternary arbitrages, for instance, $a_{\text{rint}12} \cdot a_{\text{rint}23} \cdot a_{\text{rint}31} = 0.990$ thus producing a loss through the chain of exchanges $C_1 \to C_2 \to C_3 \to C_1$.

Table 4 The $A_{\text{sym}} = A_{\text{sind}} \oplus A_{\text{sint}}$ matrix (left) corresponding to example in Table 1 and its logarithmic coordinates (right)

	$\mathbf{a}_{\text{sym},1}$	$\mathbf{a}_{\text{sym},2}$	$\mathbf{a}_{\text{sym},3}$	$\mathbf{a}_{\text{sym},4}$	;		log $\mathbf{a}_{\text{sym},1}$	log $\mathbf{a}_{\text{sym},2}$	log $\mathbf{a}_{\text{sym},3}$	log $\mathbf{a}_{\text{sym},4}$
$\mathbf{a}_{\text{sym},1}$	1.003	1.002	0.987	1.003	;	log $\mathbf{a}_{\text{sym},1}$	0.003	0.002	−0.013	0.003
$\mathbf{a}_{\text{sym},2}$	1.002	1.003	0.998	1.003	;	log $\mathbf{a}_{\text{sym},2}$	0.002	0.003	−0.002	0.003
$\mathbf{a}_{\text{sym},3}$	0.987	0.998	1.003	1.003	;	log $\mathbf{a}_{\text{sym},3}$	−0.013	−0.002	0.003	0.003
$\mathbf{a}_{\text{sym},4}$	1.003	1.003	1.003	1.003	;	log $\mathbf{a}_{\text{sym},4}$	0.003	0.003	0.003	0.003

The exchange rates in $A_{\text{rind}} \oplus A_{\text{rint}}$ are exchange rates able to produce losses or benefits in closed instantaneous chains of exchanges (triangular arbitrage for currencies). The maximum size of these losses or benefits is related to the norm $\|A_{\text{rint}}\|_+$ (Definition 1). This allows us to interpret $\|A_{\text{rint}}\|_+$ as a measure of the stress on the market to achieve benefits or to avoid losses. It also points out the origin of forces trying to modify the exchange rates A to reduce the size of $\|A_{\text{rint}}\|_+$. Certainly, a large $\|A_{\text{rint}}\|_+$ is likely to produce excesses of demand or supply on certain commodities. Then, A_{rint} produces stresses tending to moderate or even eliminate this component of A. There are several potential origins of non-neutral A_{rint}. For instance, the lack or delay of information on prices can cause these movements of A_{int}.

When the exchange rate matrix A has a non-neutral symmetric component A_{sint} and/or A_{sind}, it means that each operation of exchange is charged or subsided in a proportion to its value. An entry in A_{sym} (symmetrical contribution to A) is smaller than 1, say $1+\epsilon$, $\epsilon < 0$, then the exchange is paying $100\epsilon\%$. If $\epsilon > 0$, the exchange is subsided in an $100\epsilon\%$. This charge of taxes, $\epsilon < 0$ are a kind of dumper in the general movement of the market, meanwhile subsidies encourage the movement of exchanged commodities in the direction of the subsidy. The contribution A_{sind} represents symmetric taxes or subsidies. One pays the same ratio of taxes when you buy or when you sell, a kind of reciprocity between currencies of different countries. Therefore, A_{sind} represents dumpers/amplifiers between two currencies or commodities. More general taxes/subsidies are represented in A_{sint} as they are able to concentrate the demand or supply in differential ways. Both A_{rint}, A_{sint} favour some commodities in front of others and they are sources of inequality between commodities.

Table 4 shows $A_{\text{sym}} = A_{\text{sind}} \oplus A_{\text{sint}}$ (left side) and its logarithmic coordinates (right side). The symmetry of the matrix is here easily checked. It is remarkable that departures from unit diagonal point out external actions on the marked. In this case, subsiding the exchange of commodities.

When A is able to change in time, the orthogonal decomposition proposed gives a hint on how the market is changing and the dynamics of the exchange. They can be summarized by the norm of the different elements. The $\|A_{\text{rind}}\|_+$ only describes differences between commodities and $\mathbf{u}$ contains the fair exchange rates with a kind of universal currency; $\|A_{\text{rint}}\|_+$ measures differential movements in the exchange of commodities potentially modifying supply and demand. Both $\|A_{\text{sind}}\|_+$, $\|A_{\text{sint}}\|_+$ measure the strength of dumping or stimulus on the exchange flow, the first one maintaining reciprocity or symmetry in selling and buying, and the second favouring certain directions of exchange in detriment of others.

Changing an exchange matrix, A consists of carrying out a perturbation of the initial matrix, for instance, $A' = A \oplus M$. An interesting question is how should be M in order to maintain $A'_{\text{rec}} = A_{\text{rec}}$ (no binary arbitrage, maintaining the level of triangular arbitrage, average exchange rates constant), but increasing taxes on buying one commodity. The orthogonal decomposition tells us that this is not possible. The proposed change in taxes is $M = (\mathbf{v}, \mathbf{1}, \ldots, \mathbf{1})$ where $\mathbf{v}$ is a column of numbers between 0 and 1. However, M is not symmetric and necessarily will introduce changes in the reciprocal part of A. Carrying out the orthogonal decomposition of such M, with $\mathbf{v} = (0.95, 0.95, 0.95, 0.95)^\top$, it yields M_{rint} neutral (all entries equal to 1) but the independent part M_{rind} is

$$M_{\text{rind}} = \begin{pmatrix} 1.00 & 1.03 & 1.03 & 1.03 \\ 0.97 & 1.00 & 1.00 & 1.00 \\ 0.97 & 1.00 & 1.00 & 1.00 \\ 0.97 & 1.00 & 1.00 & 1.00 \end{pmatrix}, \quad \mathbf{u}_M = \begin{pmatrix} 0.51 \\ 0.50 \\ 0.50 \\ 0.50 \end{pmatrix},$$

what represents a revaluation of the first commodity in front of the rest, as pointed out by the first component of $\mathbf{u}_M$. In this way, part of the taxes on the first commodity is incorporated into the fair exchange rate represented in A'_{rind}.

5 Conclusion

Square matrices with positive entries are considered within $\mathbb{R}_+^D$, $D = n^2$ as their sample space. The group operation is the componentwise product of matrices or Hadamard product. The vector space of these matrices, endowed with the inner product of the componentwise logarithms, constitutes a Euclidean space. A theoretical orthogonal decomposition into four matrices, combining independence and reciprocal properties, has been presented. Some connections of the theory with the compositional analysis are succinctly pointed out. This study was motivated by a previous contribution on exchange rates matrices (Maldonado et al. 2019). In that context, the orthogonal decomposition results in interpretable matrices potentially driving the dynamics of a market.

Acknowledgements Research by J.J. Egozcue was supported by the project Ministerio de Ciencia, Innovación y Universidades (Spain) (Ref: RTI2018-095518-B-C22, 2019-2021).
Wilfredo L. Maldonado would like to thank the CNPq of Brazil for financial support 306473/2018-6 and FAPDF for financial support 00193-00001833/2019-50.

Appendix: Proofs of Theorems and Properties

Theorem 2 (Subspace of independent matrices) *The set $\mathbb{P}^D_{\text{ind}}$ of square matrices with positive entries is a $2n$-dimensional subspace of $\mathbb{P}^D$.*

Proof For any A and B in $\mathbb{P}^D_{\text{ind}}$ and $\alpha \in \mathbb{R}$ it should be shown that $C = (\alpha \odot A) \oplus B$ is in $\mathbb{P}^D_{\text{ind}}$, that is, $\mathbb{P}^D_{\text{ind}}$ is closed under the vector space operations $\oplus$, $\odot$. Let us assume that $A = U_c \oplus U_r^\top$ and $B = W_c \oplus W_r^\top$. Then

$$C = (U_c^\alpha) \oplus (U_r^\alpha)^\top \oplus W_r^\top \oplus W_c \ ,$$

where the α powers operate componentwise and maintain positiveness as well as the product operations indicated by $\oplus$. Since $\oplus$ is commutative, reordering of terms yields

$$C = [(U_c^\alpha) \oplus W_c] \oplus [(U_r^\alpha) \oplus W_r]^\top \ ,$$

where each of two terms in brackets have all columns and rows, respectively, equal, thus showing that $C \in \mathbb{P}^D_{\text{ind}}$. The space of marginal column matrices like U_c is n; they are generated by vectors $\mathbf{u}_c$ or their logarithmic coordinates $\tilde{\mathbf{u}}_c = \log \mathbf{u}_c$ which form an n-dimensional subspace of $\mathbb{R}^n$. The same is valid for U_r. Since U_c and $U_r^\top$ are linearly independent in $\mathbb{P}^D$, the generated space of matrices in $\mathbb{P}^D_{\text{ind}}$ is $2n$. □

Theorem 3 (Closest independent positive matrix) *Let A be a matrix with positive entries, $A \in \mathbb{P}^D$, $D = n^2$. The orthogonal projection of A onto the subspace of independent matrices $\mathbb{P}^D_{\text{ind}}$ is A_{ind} such that its components are*

$$[A_{\text{ind}}]_{ij} = \frac{\mathrm{g_m}(\mathbf{a}^j)\mathrm{g_m}(\mathbf{a}_i)}{\mathrm{g_m}(A)} \ ,$$

where $\mathbf{a}^j$, $\mathbf{a}_i$ are the columns and rows of A, respectively.

Proof This proof follows the way used in Maldonado et al. (2019). The statement is equivalent to find $A_{\text{ind}} \in \mathbb{P}^D_{\text{ind}}$ that minimizes $\mathrm{d}^2_+(A, A_{\text{ind}})$. In order to carry out such a minimization, let $\mathbf{u}_c = (u_{c1}, u_{c2}, \ldots, u_{cn})^\top$, $\mathbf{u}_r = (u_{r1}, u_{r2}, \ldots, u_{rn})^\top$ vectors with positive entries such that $A_{\text{ind}} = \mathbf{u}_c \mathbf{u}_r^\top$. Then, they minimize

$$\mathrm{d}^2_+(A, A_{\text{ind}}) = \| \log A - \log A_{\text{ind}} \|^2 = \sum_{ij} [\log a_{ij} - (\log u_{ci} + \log u_{rj})]^2 \ . \quad (5)$$

Taking derivatives with respect to u_{ci} and u_{rj}, $i, j = 1, 2, \ldots, n$ and equating to zero, we find

$$\sum_{i=1}^{n} [\log a_{ij} - (\log u_{ci} + \log u_{rj})] = 0 \ , \quad \sum_{j=1}^{n} [\log a_{ij} - (\log u_{ci} + \log u_{rj})] = 0 \ . \quad (6)$$

Rearranging these $2n$ equations and taking exponentials yields

$$u_{ci} = \frac{\mathrm{g_m}(\mathbf{a}_i)}{\mathrm{g_m}(\mathbf{u}_r)}, \quad u_{rj} = \frac{\mathrm{g_m}(\mathbf{a}^j)}{\mathrm{g_m}(\mathbf{u}_c)},$$

where $\mathrm{g_m}$ denotes the geometric mean over the arguments; $\mathbf{a}_i$, $\mathbf{a}^j$ the rows and columns of A, respectively. Moreover, adding all equations in (6), we conclude that $\mathrm{g_m}(\mathbf{u}_c)\mathrm{g_m}(\mathbf{u}_r) = \mathrm{g_m}(A)$. Combining the two results,

$$[A_{\mathrm{ind}}]_{ij} = u_{ci} \cdot u_{rj} = \frac{\mathrm{g_m}(\mathbf{a}^j)\mathrm{g_m}(\mathbf{a}_i)}{\mathrm{g_m}(A)},$$

what proves the statement. Note that the factorization $A_{\mathrm{ind}} = \mathbf{u}_c\mathbf{u}_r^\top$ is not unique. □

Property 2 Let $A \in \mathbb{P}^D$ be a positive matrix with orthogonal decomposition $A = A_{\mathrm{ind}} \oplus A_{\mathrm{int}}$ with $A_{\mathrm{ind}} \in \mathbb{P}^D_{\mathrm{ind}}$, $A_{\mathrm{int}} \in \mathbb{P}^D_{\mathrm{int}}$. Then, the rank of A_{ind} is one and the rank of A_{int} is equal to the rank of A, that is $\mathrm{rank}(A_{\mathrm{ind}}) = 1$ and $\mathrm{rank}(A_{\mathrm{int}}) = \mathrm{rank}(A)$.

Proof The rank of A_{ind} is one since it is the outer product of two vectors, $A_{\mathrm{ind}} = \mathbf{u}_c\mathbf{u}_r^\top$ due to Definition 2. Let $\mathbf{a}^j$, $j = 1, 2, \ldots, n$, be the columns of A_{int}. If there is a linear combination $\sum_j \alpha_j \mathbf{a}^j = \mathbf{0}$ for some non null real α_i's, after multiplication of $\mathbf{a}^j$ and $\mathbf{u}_r$ componentwise, the ith-row is $\sum_j \alpha_j a_i^j u_{ri} = 0$ thus maintaining the combination null. The existence of the null linear combination of columns implies there exist another null linear combination of rows, which remains null after multiplication componentwise by $\mathbf{u}_c$. This means that after the matrix perturbation $A = A_{\mathrm{ind}} \oplus A_{\mathrm{int}}$, the possible null linear combinations of rows or columns remain as they were, null or non null; therefore, the rank is maintained after perturbation by A_{ind}. This property is a strong version of the inequality $\mathrm{rank}(A_{\mathrm{ind}} \oplus A_{\mathrm{int}}) \leq \mathrm{rank}(A_{\mathrm{ind}}) \cdot \mathrm{rank}(A_{\mathrm{int}})$ (Horn and Johnson 1985), in which $\mathrm{rank}(A_{\mathrm{ind}}) = 1$, and the equal sign is valid. □

Property 3 Let $A = A_{\mathrm{ind}} \oplus A_{\mathrm{int}}$ be the orthogonal decompositions of A and denote by $\|\cdot\|_+$ the norm in $\mathbb{P}^D$. Then,
(a) The decomposition is unique;
(b) $\|A\|_+^2 = \|A_{\mathrm{ind}}\|_+^2 + \|A_{\mathrm{int}}\|_+^2$.

Proof The uniqueness of orthogonal decomposition into a subspace and its orthogonal complement and the Pythagorean theorem are standard properties of Hilbert spaces and, particularly, Euclidean spaces. □

Property 4 (*Logarithmic coordinates in $\mathbb{P}^D_{\mathrm{ind}}$ and $\mathbb{P}^D_{\mathrm{int}}$*) Let $A_{\mathrm{ind}} \in \mathbb{P}^D_{\mathrm{ind}}$ and $A_{\mathrm{int}} \in \mathbb{P}^D_{\mathrm{int}}$ such that $A = A_{\mathrm{ind}} \oplus A_{\mathrm{int}}$. Then, the respective logarithmic coordinates $\tilde{A}_{\mathrm{ind}} = \log(A_{\mathrm{ind}})$, $\tilde{A}_{\mathrm{int}} = \log(A_{\mathrm{int}})$ satisfy the following:
(a) $\tilde{A}_{\mathrm{ind}}$ is obtained from the average coordinates of rows and columns (marginals) of $\tilde{A}$,

$$\tilde{A}_{\text{ind}} = \tilde{\mathbf{u}}_c \mathbf{1}^\top + \mathbf{1}\tilde{\mathbf{u}}_r^\top ,$$
$$\tilde{\mathbf{u}}_r = \frac{1}{n}\tilde{A}^\top \mathbf{1} - \frac{1}{2n^2}\mathbf{1}\mathbf{1}^\top \tilde{A}\mathbf{1} ,$$
$$\tilde{\mathbf{u}}_c = \frac{1}{n}\tilde{A}\mathbf{1} - \frac{1}{2n^2}\mathbf{1}\mathbf{1}^\top \tilde{A}\mathbf{1} .$$

(b) The (arithmetic) marginals of $\tilde{A}_{\text{int}}$ are null,

$$\frac{1}{n}\tilde{A}_{\text{int}}\mathbf{1} = \mathbf{0} , \quad \frac{1}{n}\tilde{A}_{\text{int}}^\top \mathbf{1} = \mathbf{0} .$$

(c) The logarithmic coordinates of the interaction part are obtained by double centering of $\tilde{A}$,

$$\tilde{A}_{\text{int}} = \left(I_n - \frac{1}{n}\mathbf{1}\mathbf{1}^\top\right) \tilde{A} \left(I_n - \frac{1}{n}\mathbf{1}\mathbf{1}^\top\right) ,$$

where I_n denotes the identity matrix of n rows.

Proof (a) $A \in \mathbb{P}^D_{\text{ind}}$, $A = (\mathbf{u}_c\mathbf{1}^\top) \oplus (\mathbf{u}_r^\top\mathbf{1})$ due to Definition 2. Taking log componentwise, $\log(A) = \tilde{A}$, products in $\oplus$ become sums and $\log(\mathbf{u}_c\mathbf{1}^\top) = \tilde{\mathbf{u}}_c\mathbf{1}^\top$ and $\log(\mathbf{u}_r\mathbf{1}^\top) = \tilde{\mathbf{u}}_r\mathbf{1}^\top$. The expressions of $\tilde{\mathbf{u}}_r$ and $\tilde{\mathbf{u}}_c$ are obtained from Theorem 3 and taking logarithms.

(b) From decomposition $A = A_{\text{ind}} \oplus A_{\text{int}}$, taking logarithmic coordinates $\tilde{A}_{\text{int}} = \tilde{A} - \tilde{A}_{\text{ind}}$. Computing the arithmetic marginals of both sides of the latter equality and taking into account the expressions of $\mathbf{u}_c$, $\mathbf{u}_r$ in (a), the desired result holds. Note that this means that the arithmetic marginals of $\tilde{A}$ and $\tilde{A}_{\text{ind}}$ are equal.

(c) The decomposition $A = A_{\text{ind}} \oplus A_{\text{int}}$ in logarithmic coordinates implies $\tilde{A}_{\text{int}} = \tilde{A} - \tilde{A}_{\text{ind}}$. Substituting the expressions of $\tilde{A}_{\text{ind}}$, $\tilde{\mathbf{u}}_c$ and $\tilde{\mathbf{u}}_r$ given in part (a) of this property, and after some tedious computation, the double centering expression is obtained. □

Theorem 4 (Subspace of reciprocal matrices) *The set of reciprocal matrices in* $\mathbb{P}^D$ *is a* $(n(n-1)/2)$*-dimensional subspace of* $\mathbb{P}^D$*, denoted* $\mathbb{P}^D_{\text{rec}}$.

Proof The statement is equivalent to prove that anti-symmetric matrices are a $(n(n-1)/2)$-dimensional subspace of square matrices equipped with the ordinary sum of matrices and the multiplication by real scalars. This is, if $\tilde{A}$ and $\tilde{B}$ are anti-symmetric and $\alpha \in \mathbb{R}$ then $\tilde{C} = \alpha\tilde{A} + \tilde{B}$ is also anti-symmetric. In fact, for any entry of $\tilde{C}$, $\tilde{c}_{ij} = \alpha \cdot \tilde{a}_{ij} + \tilde{b}_{ij}$. Using the anti-symmetry properties of $\tilde{A}$ and $\tilde{B}$, it yields

$$\tilde{c}_{ij} = -\alpha \cdot \tilde{a}_{ji} - \tilde{b}_{ji} = -\tilde{c}_{ji} , \quad \tilde{C} = -\tilde{C}^\top .$$

Therefore, $\tilde{C}$ is anti-symmetric and $C \in \mathbb{P}^D_{\text{rec}}$. The dimension of the subspace is equal to the number of free logarithmic coordinates, for instance those in the lower triangle of the matrix, excluding the diagonal, that is $n(n-1)/2$. □

Theorem 5 (orthogonal decomposition) *The space of square positive matrices* $\mathbb{P}^D$, $D = n^2$, *is uniquely decomposed into the four mutually orthogonal subspaces* $\mathbb{P}^D_{\text{rind}}$, $\mathbb{P}^D_{\text{rint}}$, $\mathbb{P}^D_{\text{sind}}$, $\mathbb{P}^D_{\text{sint}}$ *whose dimensions are* $\dim(\mathbb{P}^D_{\text{rind}}) = n$, $\dim(\mathbb{P}^D_{\text{sind}}) = n$, $\dim(\mathbb{P}^D_{\text{rint}}) = n(n-3)/2$, $\dim(\mathbb{P}^D_{\text{sint}}) = n(n-1)/2$. *Any matrix* $A \in \mathbb{P}^D$ *is expressed in a unique way as*

$$A = A_{\text{rind}} \oplus A_{\text{rint}} \oplus A_{\text{sind}} \oplus A_{\text{sint}} \ . \tag{7}$$

In logarithmic coordinates, the decomposition of $A \in \mathbb{P}^D$ *is*

$$\tilde{A}_{\text{rec}} = \frac{1}{2}(\tilde{A} - \tilde{A}^\top) \ , \quad \tilde{A}_{\text{sym}} = \frac{1}{2}(\tilde{A} + \tilde{A}^\top)$$

and

$$\begin{aligned}
\tilde{A}_{\text{rint}} &= \left(I_n - \frac{1}{n}\mathbf{1}\mathbf{1}^\top\right) \tilde{A}_{\text{rec}} \left(I_n - \frac{1}{n}\mathbf{1}\mathbf{1}^\top\right) \ , \\
\tilde{A}_{\text{sint}} &= \left(I_n - \frac{1}{n}\mathbf{1}\mathbf{1}^\top\right) \tilde{A}_{\text{sym}} \left(I_n - \frac{1}{n}\mathbf{1}\mathbf{1}^\top\right) \ , \\
\tilde{A}_{\text{rind}} &= \tilde{A}_{\text{rec}} - \tilde{A}_{\text{rint}} \ , \\
\tilde{A}_{\text{sind}} &= \tilde{A}_{\text{sym}} - \tilde{A}_{\text{sint}} \ .
\end{aligned} \tag{8}$$

Proof The orthogonality of the four subspaces is a consequence of the two orthogonal decompositions. Consider $A \in \mathbb{P}^D$ and obtain its decomposition as in Eq. (3). Take A_{rind}; it is orthogonal to A_{rint} due to orthogonal decomposition in Corollary 1. Additionally, it is orthogonal to A_{sind} and A_{sint} due to orthogonality with the subspace $\mathbb{P}^D_{\text{sym}}$. Similarly, this reasoning can be repeated for the four components of the decomposition. The uniqueness statement comes from the uniqueness of orthogonal decompositions.

The dimension of the spaces can be studied using logarithmic coordinates. The $\mathbb{P}^D_{\text{rec}} \cup \mathbb{P}^D_{\text{sym}} = \mathbb{P}^D$ whose dimension is $D = n^2$. Logarithmic (orthogonal) coordinates that determine $\mathbb{P}^D_{\text{rec}}$ can be those in the lower triangle, excluding the diagonal, as they determine the upper triangle, which are the anti-symmetric values, and the coordinates in the diagonal are null. Therefore, the dimension of $\mathbb{P}^D_{\text{rec}}$ is $n(n-1)/2$. Since matrices in $\mathbb{P}^D_{\text{rind}}$ can be constructed from a single n-vector, the dimension is n and the dimension of $\mathbb{P}^D_{\text{rint}}$ is $n(n-1)/2 - n$. The dimension of $\mathbb{P}^D_{\text{sym}}$ is $n(n+1)/2$ since the logarithmic coordinates needed to reconstruct the symmetric matrix is the lower triangle including the diagonal. The dimension of $\mathbb{P}^D_{\text{sind}}$ is n as these matrices are represented as two equal n-vectors and its marginals. Then, the dimension of $\mathbb{P}^D_{\text{sint}}$ is $n(n+1)/2 - n$.

The fact that logarithmic coordinates of $A_{\text{rec}} \in \mathbb{P}^D_{\text{rec}}$ are anti-symmetric leads to decompose $\tilde{A}$ into symmetric and anti-symmetric components, that is a unique decomposition, and $\tilde{A}_{\text{rec}}$, $\tilde{A}_{\text{sym}}$ are identified to the anti-symmetric and symmetric parts, respectively. As separation of interaction components corresponds to a double centering, $\tilde{A}_{\text{rint}}$, $\tilde{A}_{\text{sint}}$ are so obtained and the subtracted matrices are $\tilde{A}_{\text{rind}}$, $\tilde{A}_{\text{sind}}$. $\square$

References

C. Barceló-Vidal, J.A. Martín-Fernández, The mathematics of compositional analysis. Austrian J. Stat. **45**, 57–71 (2016)

D. Billheimer, P. Guttorp, W. Fagan, Statistical interpretation of species composition. J. Amer. Stat. Assoc. **96**(456), 1205–1214 (2001)

J.J. Egozcue, V. Pawlowsky-Glahn, A compositional approach to contingency tables, in *Proceedings of the Ninth International Conference on Computer Data Analysis and Modelling*, vol. 1 (Publishing center of BSU, Minsk, 2010), pp. 101–107

J.J. Egozcue, V. Pawlowsky-Glahn, Compositional data: the sample space and its structure (with discussion). TEST **28**(3), 599–638 (2019). https://doi.org/10.1007/s11749-019-00670-6

J.J. Egozcue, V. Pawlowsky-Glahn, G. Mateu-Figueras, C. Barceló-Vidal, Isometric logratio transformations for compositional data analysis. Math. Geol. **35**(3), 279–300 (2003)

J.J. Egozcue, V. Pawlowsky-Glahn, M. Templ, K. Hron, Independence in contingency tables using simplicial geometry. Commun. Stats. A-Theor. **44**(18), 3978–3996 (2015)

K. Fačevicová, K. Hron, V. Todorov, M. Templ, General approach to coordinate representation of compositional tables. Scand. J. Stat. **43**(4), 962–977 (2016)

R.A. Horn, C.R. Johnson, *Matrix Analysis* (Reprint 1988 ed.) (Cambridge University Press, New York, 1985), 561 p

W.W. Leontief, *Input-Output Economics* (Oxford University Press, New York, 1966), p. 359

W.L. Maldonado, J.J. Egozcue, V. Pawlowsky-Glahn, No-arbitrage matrices of exchange rates: some characterizations. Int. J. Econ. Theory (2019). https://doi.org/10.1111/ijet.12249

M.I. Ortego, J.J. Egozcue, Bayesian estimation of the orthogonal decomposition of a contingency table. Austrian J. Stat. **45**(4), 45–56 (2016)

V. Pawlowsky-Glahn, J.J. Egozcue, D. Lovell, Tools for compositional data with a total. Stat. Model **15**(2), 175–190 (2015a)

V. Pawlowsky-Glahn, J.J. Egozcue, R. Tolosana-Delgado, *Modeling and Analysis of Compositional Data*, Statistics in Practice (Wiley, Chichester, 2015b), p. 272

V. Pawlowsky-Glahn, J.J. Egozcue, Geometric approach to statistical analysis on the simplex. Stoch. Env. Res. Risk A **15**(5), 384–398 (2001)

V. Pawlowsky-Glahn, J.J. Egozcue, M. Planes-Pedra, Survey data on perceptions of contraceptive measures as compositional tables. Rev. Lat. Amer. Psicol. **50**(3), 179–186 (2019)

B. Peterson, M. Olinick, Markov chains, substochastic matrices and positive solutions of matrix equations. Math. Modell. **3**, 221–239 (1982)

V. Todorov, K. Fačevicová, K. Hron, D. Guo, M. Templ, Statistical analysis of compositional 2x2 tables with an economic application, in *Proceedings of the 5th Workshop on Compositional Data Analysis, CoDaWork 2013*, ISBN: 978-3-200-03103-6 (2013), pp. 123–130

Fundamentals

The Information-Geometric Perspective of Compositional Data Analysis

Ionas Erb and Nihat Ay

Abstract Information geometry uses the formal tools of differential geometry to describe the space of probability distributions as a Riemannian manifold with an additional dual structure. The formal equivalence of compositional data with discrete probability distributions makes it possible to apply the same description to the sample space of Compositional Data Analysis (CoDA). The latter has been formally described as a Euclidean space with an orthonormal basis featuring components that are suitable combinations of the original parts. In contrast to the Euclidean metric, the information-geometric description singles out the Fisher information metric as the only one keeping the manifold's geometric structure invariant under equivalent representations of the underlying random variables. Well-known concepts that are valid in Euclidean coordinates, e.g., the Pythagorean theorem, are generalized by information geometry to corresponding notions that hold for more general coordinates. In briefly reviewing Euclidean CoDA and, in more detail, the information-geometric approach, we show how the latter justifies the use of distance measures and divergences that so far have received little attention in CoDA as they do not fit the Euclidean geometry favored by current thinking. We also show how Shannon entropy and relative entropy can describe amalgamations in a simple way, while Aitchison distance requires the use of geometric means to obtain more succinct relationships. We proceed to prove the information monotonicity property for Aitchison distance. We close with some thoughts about new directions in CoDA where the rich structure that is provided by information geometry could be exploited.

I. Erb (✉)
Centre for Genomic Regulation (CRG), The Barcelona Institute of Science and Technology, Barcelona, Spain
e-mail: ionas.erb@crg.eu

N. Ay
Max-Planck Institute for Mathematics in the Sciences, Leipzig, Germany
e-mail: nihat.ay@mis.mpg.de

Department of Mathematics and Computer Science, Leipzig University, Leipzig, Germany

Santa Fe Institute, Santa Fe, NM, USA

P. Filzmoser et al. (eds.), *Advances in Compositional Data Analysis*,
https://doi.org/10.1007/978-3-030-71175-7_2

1 Introduction

Information geometry and Compositional Data Analysis (CoDA) are fields that have ignored each other so far. Independently, both have found powerful descriptions that led to a deeper understanding of the geometric relationships between their respective objects of interest: probability distributions and compositional data. Although both of these live on the same mathematical space (the simplex), and some of the mathematical structures are identically described, surprisingly, both fields have come to focus on quite different geometric aspects. On the one hand, the tools of differential geometry have revealed the underlying duality of the manifold of probability distributions, with the Fisher information metric playing a central role. On the other hand, the classical log-ratio approach led to the modern description of the compositional sample space as Euclidean and affine. We think it is time that CoDA starts to profit from the rich structures that information geometry has to offer. This paper intends to build some bridges from information geometry to CoDA. In the first section, we will give a brief description of the Euclidean CoDA perspective. The second and main part of the paper describes in some detail the approach centred around the Fisher metric, with a description of the dual coordinates, exponential families, and of how dual divergence functions generalize the notion of Euclidean distance. To ease understanding, throughout this section, we link those concepts to the ones familiar in CoDA. In the third part of the paper, we show how information-based measures can lead to simpler expressions when amalgamations of parts are involved, and an important monotonicity result that holds for relative entropy is derived for Aitchison distance. We conclude with a short discussion about where we could go from here.

2 The Euclidean CoDA Perspective

Compositional data analysis is now unthinkable without the log-ratio approach pioneered by Aitchison (1986). It has both led to a variety of data-analytic developments (for the most recent review, see Greenacre 2021), and more formal mathematical descriptions (see Egozcue and Pawlowsky-Glahn 2019). Following these, compositions can be described as equivalence classes whose representatives are points in a Euclidean space. We will give a brief recount here for completeness of exposition. Compositional data are defined as vectors of strictly positive numbers describing the parts of a whole for which the relevant information is only relative. As such, the absolute size of a D-part composition $\boldsymbol{x} \in \mathbb{R}^D_+$ is irrelevant, and all the information is conveyed by the ratios of its components. This can further be formalized by considering equivalent compositions $\boldsymbol{x}$, $\boldsymbol{y}$ such that $\boldsymbol{y} = c\boldsymbol{x}$ for a positive constant c. A composition is then an equivalence class of such proportional vectors (Barceló-Vidal and Martín-Fernández 2016). A closed composition is the simplicial representative $C\boldsymbol{x} := \boldsymbol{x} / \sum_i x_i$, where the symbol C denotes the closure operation (i.e., the division by the sum over the components). Closed compositions are elements of the simplex

$$\mathcal{S}^D := \left\{ (x_1, \ldots, x_D)^T \in \mathbb{R}^D : x_i > 0, i = 1, \ldots, D, \sum_i^D x_i = 1 \right\}, \tag{1}$$

where T denotes transposition. The simplex $\mathcal{S}^D$ can be equipped with a Euclidean structure by the vector space operations of perturbation and powering (playing the role of vector addition and scalar multiplication in real vector spaces), defined by

$$\boldsymbol{x} \oplus \boldsymbol{y} := C(x_1 y_1, \ldots, x_D y_D)^T, \tag{2}$$

$$\alpha \odot \boldsymbol{x} := C(x_1^\alpha, \ldots, x_D^\alpha)^T. \tag{3}$$

An inverse perturbation is given by $\ominus \boldsymbol{x} := (-1) \odot \boldsymbol{x}$.

As a vector space, $\mathcal{S}^D$ also carries the structure of an affine space, and we can study affine subspaces, which are referred to as *linear manifolds* in Egozcue et al. (2011). In order to do so, we require a set of vectors $\boldsymbol{x}_1, \ldots, \boldsymbol{x}_m$, which we assume to be linearly independent, and an origin point $\boldsymbol{x}_0$. Here, independence means the following: Let $\boldsymbol{n} = C(1, \ldots, 1)$ be the neutral element. A set of m compositions $\boldsymbol{x}_1, \ldots, \boldsymbol{x}_m \in \mathcal{S}^D$ is called *perturbation-independent* if the fact that $\boldsymbol{n} = \bigoplus_{i=1}^m (\alpha_i \odot \boldsymbol{x}_i)$ implies $\alpha_1 = \cdots = \alpha_m = 0$. With this, an affine subspace is given as the set of compositions $\boldsymbol{y}$ such that

$$\boldsymbol{y} = \boldsymbol{x}_0 \oplus \bigoplus_{i=1}^m (\alpha_i \odot \boldsymbol{x}_i) \tag{4}$$

for any real constants $\alpha_i, i = 1, \ldots, m$. Due to the linear independence of the vectors $\boldsymbol{x}_1, \ldots, \boldsymbol{x}_m$, this is an m-dimensional space.

It is convenient to define the inner product for our Euclidean space via the so-called centred log-ratio transformation. Its definition (Aitchison 1982) and inverse operation are

$$\boldsymbol{v} = \mathrm{clr}(\boldsymbol{x}) := \left(\log \frac{x_1}{g(\boldsymbol{x})}, \ldots, \log \frac{x_D}{g(\boldsymbol{x})} \right)^T, \tag{5}$$

$$\boldsymbol{x} = \mathrm{clr}^{-1}(\boldsymbol{v}) = C\exp(\boldsymbol{v}), \tag{6}$$

where g denotes the geometric mean $g(\boldsymbol{x}) = \left(\prod_{i=1}^D x_i \right)^{\frac{1}{D}}$. Note that the sum over the components clr_i of clr-transformed vectors is 0. An inner product can then be defined by

$$\langle \boldsymbol{x}, \boldsymbol{y} \rangle_A := \sum_{i=1}^D \mathrm{clr}_i(\boldsymbol{x}) \mathrm{clr}_i(\boldsymbol{y}). \tag{7}$$

The corresponding (squared) norm and distance are

$$\|\boldsymbol{x}\|_A^2 = \langle \boldsymbol{x}, \boldsymbol{x} \rangle_A, \quad d_A^2(\boldsymbol{x}, \boldsymbol{y}) = \|\boldsymbol{x} \ominus \boldsymbol{y}\|_A^2 . \tag{8}$$

This distance is known as Aitchison distance (denoted by the A subscript). Note that the clr transformation is an isometry $\mathcal{S}^D \to \mathcal{T}^D \subset \mathbb{R}^D$ between $(D-1)$-dimensional Euclidean spaces, where

$$\mathcal{T}^D := \left\{ \boldsymbol{v} \in \mathbb{R}^D : \sum_{i=1}^{D} v_i = 0 \right\} . \tag{9}$$

We can thus obtain orthonormal bases in $\mathcal{S}^D$ from orthonormal bases in $\mathcal{T}^D$. Such orthonormal basis vectors are defined by the columns $\boldsymbol{v}_i$, $i = 1, \ldots, D-1$, of $\mathbf{V}$, a matrix of order $D \times (D-1)$ obeying

$$\mathbf{V}^T\mathbf{V} = \mathbf{I}_{D-1}, \tag{10}$$

$$\mathbf{V}\mathbf{V}^T = \mathbf{I}_D - \frac{1}{D}\mathbf{1}\mathbf{1}^T, \tag{11}$$

with $\mathbf{I}_D$ the $D \times D$ identity matrix and $\mathbf{1}$ a D-dimensional vector where each component is 1. The first equation ensures orthonormality, the second equation makes sure components sum to zero. Now the vectors

$$\boldsymbol{e}_i = \mathrm{clr}^{-1}(\boldsymbol{v}_i) = C\exp(\boldsymbol{v}_i) \tag{12}$$

constitute an orthonormal basis in $\mathcal{S}^D$. Euclidean coordinates $\boldsymbol{z}$ of $\boldsymbol{x} \in \mathcal{S}^D$ with respect to the basis $\boldsymbol{e}_i$, $i = 1, \ldots, D-1$, then follow from the so-called isometric log-ratio transformation (Egozcue et al. 2003). Its definition and inverse operation are

$$\boldsymbol{z} = \mathrm{ilr}(\boldsymbol{x}) := \mathbf{V}^T \log(\boldsymbol{x}), \tag{13}$$

$$\boldsymbol{x} = \mathrm{ilr}^{-1}(\boldsymbol{z}) = C\exp(\mathbf{V}\boldsymbol{z}). \tag{14}$$

The second equation shows the composition as generated from the coordinates $\boldsymbol{z}$. This can also be written in the usual component form using the basis vectors of Eq. (12):

$$\boldsymbol{x} = \bigoplus_{i=1}^{D-1} (z_i \odot \boldsymbol{e}_i). \tag{15}$$

Equations (10) and (11) characterize any orthonormal basis of $\mathcal{S}^D$. $\mathbf{V}$ is known under the name of contrast matrix. Many choices for this matrix are possible. Balance coordinates (Egozcue and Pawlowsky-Glahn 2005) are popular for their relative simplicity and interpretability, where pivot coordinates (Hron et al. 2017) are sometimes preferred.

3 The Information-Geometric Perspective

Information geometry started out as a study of the geometry of statistical estimation. The set of probability distributions that constitute a statistical model is seen as a manifold whose invariant geometric structures (Chentsov 1982) are studied in relation to the statistical estimation using this model. A Riemannian metric and a family of affine connections are naturally introduced on such a manifold (Rao 1945). It turns out that a fundamental duality underlies these structures (Amari 2000). While the first ideas about the geometry of statistical estimation are from the first half of the twentieth century and go back to a variety of authors, the first attempt at a unified exposition of the topic by Amari (1985) was published around a similar time as Aitchison's book.

Here we try to highlight some of the main concepts, borrowing from chapter two in Ay et al. (2017), but mainly following the treatment in Amari (2016). The latter also showcases the many applications information geometry has found during the last decades. While the best-known book on the topic may be (Amari 2000), the most formal and complete exposition can currently be found in Ay et al. (2017).

The ideas that require more advanced notions from differential geometry will not be touched upon in our short outline.

3.1 Dual Coordinates and Fisher Metric in Finite Information Geometry

To exploit the equivalence of probability distributions with compositional data, we only consider the manifold of discrete distributions $\mathcal{S}^D$. To emphasize the equivalence, we will denote the probabilities by the same symbol as our compositions, i.e., $\boldsymbol{x} \in \mathcal{S}^D$ is a vector of D probabilities. To complete the probabilistic picture, we need a random variable R which can take values $r \in \{1, \dots, D\}$ with the respective probabilities $x_r = \mathbb{P}\{R = r\}$, the coordinates of $\boldsymbol{x}$. The distribution of this random variable can now be written as

$$p_R = \boldsymbol{x} = (x_1, \dots, x_D)^T \in \mathcal{S}^D. \tag{16}$$

From the information-geometric perspective, there are two natural ways to parametrize the set $\mathcal{S}^D$ of all (strictly positive) distributions of R. The first possibility is quite obvious. The first $D - 1$ probabilities in Eq. (16) that are free to be specified, that is $x_1, \dots, x_{D-1}$, can be considered parameters, which we denote by $\boldsymbol{\eta} = (x_1, \dots, x_{D-1})^T$. In these coordinates, probability distributions are written as

$$p(r; \boldsymbol{\eta}) = \begin{cases} \eta_r, & \text{if } r \leq D-1 \\ 1 - \sum_{i=1}^{D-1} \eta_i, & \text{if } r = D \end{cases}, \qquad r = 1, \dots, D. \tag{17}$$

Alternatively, our distribution can be parametrized using what is known as the alr-transformation (Aitchison 1982) in CoDA:

$$\theta^i = \log \frac{x_i}{x_D}, \quad i = 1, \ldots, D-1. \tag{18}$$

With this, we can write our distribution in the form

$$p(r;\boldsymbol{\theta}) = \exp\left(\sum_{k=1}^{D-1} \theta^k \mathbb{1}_k(r) - \psi(\boldsymbol{\theta})\right), \qquad r = 1, \ldots, D, \tag{19}$$

where $\mathbb{1}_k(r) = 1$ if $r = k$, and $\mathbb{1}_k(r) = 0$ otherwise. The function ψ ensures normalization, that is

$$\psi(\boldsymbol{\theta}) = \log\left(1 + \sum_{i=1}^{D-1} e^{\theta^i}\right) = -\log x_D. \tag{20}$$

The parametrization of Eq. (19) in terms of $\boldsymbol{\theta} \in \mathbb{R}^{D-1}$ can be used in order to define a linear structure on $\mathcal{S}^D$. The addition of two distributions $p(\cdot;\boldsymbol{\theta}_1)$ and $p(\cdot;\boldsymbol{\theta}_2)$ can simply be defined by taking their product and then normalizing. With this vector addition, denoted by $\oplus$, we obviously have

$$p(r;\boldsymbol{\theta}_1) \oplus p(r;\boldsymbol{\theta}_2) = p(r;\boldsymbol{\theta}_1 + \boldsymbol{\theta}_2). \tag{21}$$

Thus, the operation $\oplus$ is consistent with the usual addition in the parameter space $\mathbb{R}^{D-1}$. The multiplication of a distribution $p(\cdot;\boldsymbol{\theta})$ with a scalar $\alpha \in \mathbb{R}$ can be correspondingly defined by potentiating and then normalizing. This defines a scalar multiplication $\odot$, and we have

$$\alpha \odot p(r;\boldsymbol{\theta}) = p(r;\alpha \cdot \boldsymbol{\theta}). \tag{22}$$

Obviously, the scalar multiplication $\odot$ is consistent with the usual multiplication in the parameter space. Note that the vector space structure defined here, which is well known in information geometry, coincides with the structure defined by Eqs. (2) and (3). Given the linear structure, we can consider affine subspaces of $\mathcal{S}^D$. These are well-known and fundamental families in statistics, statistical physics, and information geometry, the so-called *exponential families*. We basically obtain them from the representation of Eq. (19) if we replace the functions $\mathbb{1}_k$ by d ($d \leq D-1$) functions $X_k : \{1, \ldots, D\} \to \mathbb{R}$, and shift the whole family by some reference measure p_0:

$$p(r;\boldsymbol{\theta}) = p_0(r) \exp\left(\sum_{k=1}^{d} \theta^k X_k(r) - \psi(\boldsymbol{\theta})\right), \qquad r = 1, \ldots, D. \tag{23}$$

Here, $\psi(\boldsymbol{\theta})$ ensures normalization, but it does not reduce to the simple structure of Eq. (20) in general. In statistical physics, the function ψ is known as the free energy (in other contexts it is also known as the cumulant-generating function). Note that, given the same linear structure on $\mathcal{S}^D$, the exponential families coincide with the linear manifolds of Eq. (4), which were introduced into the field of Compositional Data Analysis more recently.

In what follows we restrict attention to the parametrizations of Eqs. (17) and (19) of the full simplex $\mathcal{S}^D$ as one instance of the general structure that underlies information geometry. The function ψ, given by Eq. (20), is a convex function, and we can get back the $\boldsymbol{\eta}$ coordinates from it via a Legendre transformation (Amari 2000)

$$\boldsymbol{\eta} = \nabla\psi(\boldsymbol{\theta}), \tag{24}$$

where ∇ denotes the partial derivatives $(\partial\psi/\partial\theta^i)_{i=1}^D$. The Legendre dual of $\psi(\boldsymbol{\theta})$ is another convex function defined by

$$\phi(\boldsymbol{\eta}) = \max_{\boldsymbol{\theta}} \{\boldsymbol{\theta}\cdot\boldsymbol{\eta} - \psi(\boldsymbol{\theta})\}, \tag{25}$$

which is given by the negative Shannon entropy

$$\phi(\boldsymbol{\eta}) = \sum_{i=1}^{D-1} \eta_i \log \eta_i + \left(1 - \sum_{i=1}^{D-1} \eta_i\right) \log\left(1 - \sum_{i=1}^{D-1} \eta_i\right). \tag{26}$$

In equivalence to Eq. (24), from $\phi(\boldsymbol{\eta})$ we can get back the $\boldsymbol{\theta}$ coordinates:

$$\boldsymbol{\theta} = \nabla\phi(\boldsymbol{\eta}). \tag{27}$$

There is thus a fundamental duality mediated by the Legendre transformation which links the two types of parameters $\boldsymbol{\eta}$ and $\boldsymbol{\theta}$ as well as the convex functions ψ and ϕ. Legendre transformations are well known to play a fundamental role in phenomenological thermodynamics. Their importance for information geometry was established by Amari and Nagaoka (2000).

In what follows, we use the parameters $\boldsymbol{\theta}$. Like $\boldsymbol{\eta}$, they define a (local) coordinate system of the manifold $\mathcal{S}^D$. From each point $\boldsymbol{\theta}$, $D-1$ coordinate curves $\gamma_i : t \mapsto \gamma_i(t)$ in $\mathcal{S}^D$, $\gamma_i(0) = p_{\boldsymbol{\theta}}$, emerge when holding $D-2$ of the θ_i constant. Consider their velocities

$$\mathrm{e}_i := \left.\frac{d}{dt}\gamma_i(t)\right|_{t=0}, \qquad i = 1, \ldots, D-1. \tag{28}$$

In the point $\boldsymbol{\theta}$ itself, the $D-1$ vectors e_i pointing in the direction of each coordinate curve form the basis of the so-called *tangent space* $T_{\boldsymbol{\theta}}\mathcal{S}^D = \mathcal{T}^D$ of this point, see Fig. 1a. Similarly, we can define vectors e^i with respect to the coordinates $\boldsymbol{\eta}$, which also span the tangent space.

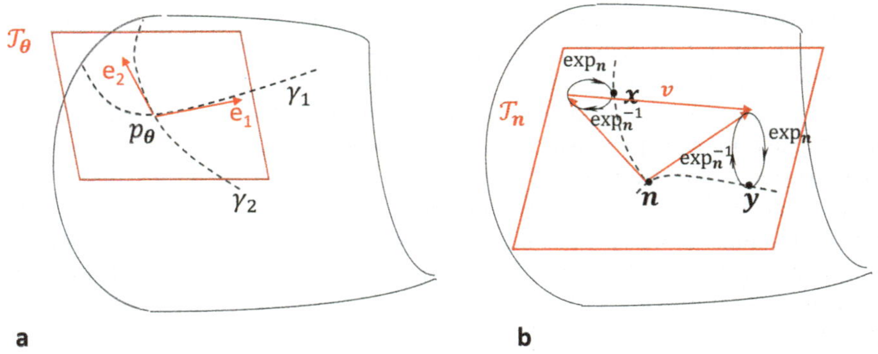

Fig. 1 **a** A manifold (in black) and its tangent space in point p_θ (in red) where two coordinate curves (dashed lines) are crossing. The velocities of the coordinate curves are the basis vectors e_1 and e_2. **b** Exponential map and its inverse via the tangent space anchored in a reference point $\boldsymbol{n}$. The points $\boldsymbol{x}$ and $\boldsymbol{y}$ are mapped via $\exp_{\boldsymbol{n}}^{-1}$ to the tangent space. The difference vector between the two mapped points is $\boldsymbol{v} = \mathrm{vec}(\boldsymbol{x}, \boldsymbol{y})$

For a Riemannian manifold, a metric tensor **G** is defined. This metric is obtained via an inner product on the tangent space:

$$g_{ij} = \langle \mathrm{e}_i, \mathrm{e}_j \rangle, \tag{29}$$

which depends on the coordinates chosen. The coordinate system is Euclidean if $g_{ij} = \delta_{ij}$ (this is the case for the parametrization achieved by Eq. 13). For the Riemannian metric of exponential families, the basis vectors can be identified with the so-called score[1] function $\partial \log p(r, \boldsymbol{\theta})/\partial \theta^i$. The resulting Riemannian metric is known as the Fisher information matrix:

$$g_{ij}(\boldsymbol{\theta}) = E_{\boldsymbol{\theta}} \left[\frac{\partial}{\partial \theta^i} \log p(r; \boldsymbol{\theta}) \frac{\partial}{\partial \theta^j} \log p(r; \boldsymbol{\theta}) \right], \tag{30}$$

$$= E_{\boldsymbol{\theta}} \left[(\mathbb{1}_i - E(\mathbb{1}_i))(\mathbb{1}_j - E(\mathbb{1}_j)) \right] \tag{31}$$

$$= \begin{cases} -\eta_i \eta_j, & \text{if } i \neq j \\ \eta_i(1-\eta_i), & \text{if } i = j \end{cases}. \tag{32}$$

with E denoting the expectation value, and

$$\eta_i = \frac{e^{\theta^i}}{1 + \sum_{k=1}^{D-1} e^{\theta^k}}, \qquad i = 1, \ldots, D-1.$$

Note that this is the covariance matrix of the random vector $(\mathbb{1}_1, \ldots, \mathbb{1}_{D-1})$, as expressed by Eq. (31). Convex functions have positive definite Hessian matrices.

[1]The score function plays an important role in maximum-likelihood estimation.

Here, their elements are given by the Fisher metric itself:

$$g_{ij}(\boldsymbol{\theta}) = \frac{\partial}{\partial \theta^i} \frac{\partial}{\partial \theta^j} \psi(\boldsymbol{\theta}), \tag{33}$$

$$g^{ij}(\boldsymbol{\eta}) = \frac{\partial}{\partial \eta_i} \frac{\partial}{\partial \eta_j} \phi(\boldsymbol{\eta}). \tag{34}$$

The second matrix is the inverse of the first, which follows from their Legendre duality. This means the second matrix is an inverse covariance, an important object in the theory of graphical models (Whittaker 1990). Although the Fisher metric is not Euclidean, i.e., $\langle \mathrm{e}_i, \mathrm{e}_j \rangle \neq \delta_{ij}$, we do have a generalization of this when mixing the dual coordinates: $\langle \mathrm{e}_i, \mathrm{e}^j \rangle = \delta_{ij}$.

Note that for a general exponential family of the form of Eq. (23), the Fisher information matrix can be expressed as a covariance matrix, which generalizes Eq. (31):

$$g_{ij}(\boldsymbol{\theta}) = E_{\boldsymbol{\theta}} \left[\frac{\partial}{\partial \theta^i} \log p(r; \boldsymbol{\theta}) \frac{\partial}{\partial \theta^j} \log p(r; \boldsymbol{\theta}) \right], \tag{35}$$

$$= E_{\boldsymbol{\theta}} \left[(X_i - E(X_i))(X_j - E(X_j)) \right]. \tag{36}$$

3.2 Distance Measures and Divergences

Our affine structure on $\mathcal{S}^D$ can be reformulated additively via an exponential map $\mathcal{S}^D \times \mathcal{T}^D \to \mathcal{S}^D$, $(\boldsymbol{x}, \boldsymbol{v}) \mapsto \boldsymbol{y}$:

$$\boldsymbol{y} = \exp_{\boldsymbol{x}}(\boldsymbol{v}) := \boldsymbol{x} \oplus e^{\boldsymbol{v}}, \tag{37}$$

$$\boldsymbol{v} = \mathrm{vec}(\boldsymbol{x}, \boldsymbol{y}) := \exp_{\boldsymbol{x}}^{-1}(\boldsymbol{y}) = \mathrm{clr}(\boldsymbol{y}) - \mathrm{clr}(\boldsymbol{x}), \tag{38}$$

using the notation introduced in Sect. 2, and shown here together with its inverse. This map is used in Ay and Erb (2005), where ordinary linear differential equations with the time derivative defined for $\mathrm{vec}(\boldsymbol{x}(t_0), \boldsymbol{x}(t))$ letting $t \to t_0$ are considered. These turn out to be replicator equations with special properties that are known from population dynamics. Note that, e.g., with the center of the simplex $\boldsymbol{n}$ as defined before, we have $\mathrm{vec}(\boldsymbol{n}, \boldsymbol{x}) = \mathrm{clr}(\boldsymbol{x})$. With the exponential map, we can interpret $\mathrm{vec}(\boldsymbol{x}, \boldsymbol{y})$ as the difference vector between two compositions, see Fig. 1b. The set of all such difference vectors for a given point can be interpreted as the gradient field of a convex function (a.k.a. a potential). In order to highlight the generality of this concept in information geometry, we consider a general convex function U with respect to some parameters $\boldsymbol{\eta}$ from a convex domain. In the following, we will denote the compositions for which the parameters are to be evaluated by subscripts. The linearization of U in $\boldsymbol{\eta}_y$ is given by

$$\overline{U}(\boldsymbol{\eta}_x) := U(\boldsymbol{\eta}_y) + \nabla U(\boldsymbol{\eta}_y) \cdot (\boldsymbol{\eta}_x - \boldsymbol{\eta}_y).$$

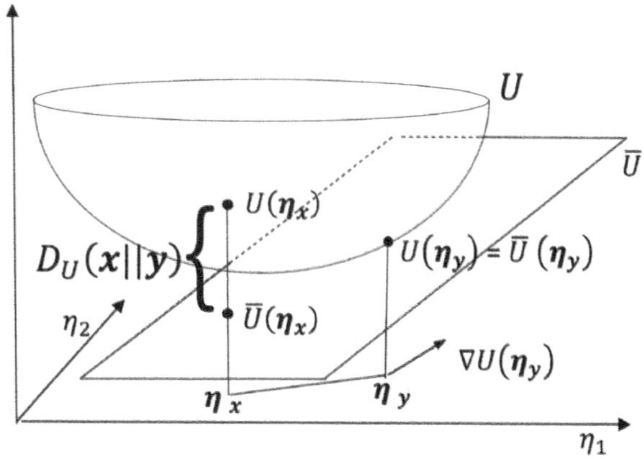

Fig. 2 The graph of the potential U, of its linearization $\overline{U}$, and a visualization of the corresponding Bregman divergence D_U. The gradient vector ∇U points in the direction of greatest change of the function U in the space of the coordinates η

The graph of this linearization is a hyperplane of dimension $D-1$ touching the graph of U in the point $(\eta_y, U(\eta_y))$, see Fig. 2. The difference between U and its linearization $\overline{U}$ in η_y, evaluated at η_x, defines a so-called *Bregman divergence*, a class of divergences that plays an important role in information geometry. More precisely,

$$D_U(\boldsymbol{x}||\boldsymbol{y}) := U(\eta_x) - \overline{U}(\eta_x) = U(\eta_x) - U(\eta_y) - \nabla U(\eta_y) \cdot (\eta_x - \eta_y). \quad (39)$$

Divergences are similar to distance functions but they are not necessarily symmetric and need not fulfill the triangle inequality. As an example, let us consider the potential naturally associated with the structure of Eqs. (37) and (38), the squared Aitchison norm

$$A(z_x) = \sum_{i=1}^{D} \text{clr}_i(\boldsymbol{x})^2 = \sum_{i=1}^{D-1} \text{ilr}_i(\boldsymbol{x})^2, \quad (40)$$

with ilr_i the i-th ilr-coordinate z_i, see Eq. (13), and z_x the vector of coordinates z_i. We then have

$$D_A(\boldsymbol{x}||\boldsymbol{y}) = \sum_{i=1}^{D-1} (\text{ilr}_i(\boldsymbol{x}) - \text{ilr}_i(\boldsymbol{y}))^2 = d_A^2(\boldsymbol{x}, \boldsymbol{y}), \quad (41)$$

coinciding with the squared Aitchison distance. This is the special case of a Euclidean divergence, which is also a (squared) distance function.

Let us now come to the divergences that arise when replacing $U(\eta)$ in Eq. (39) by our dually convex functions $\psi(\boldsymbol{\theta})$ and $\phi(\eta)$. They turn out to be the relative entropies (a.k.a. Kullback–Leibler divergences)

$$D_\phi(\boldsymbol{x}||\boldsymbol{y}) = \sum_{i=1}^{D} x_i \log \frac{x_i}{y_i}, \tag{42}$$

$$D_\psi(\boldsymbol{x}||\boldsymbol{y}) = \sum_{i=1}^{D} y_i \log \frac{y_i}{x_i}. \tag{43}$$

Thus, the symmetry we had in the Euclidean case finds its generalization for our dual case in

$$D_\psi(\boldsymbol{x}||\boldsymbol{y}) = D_\phi(\boldsymbol{y}||\boldsymbol{x}). \tag{44}$$

Moreover, one can show that we can "complete the square" via

$$D_\psi(\boldsymbol{x}||\boldsymbol{y}) = \psi(\boldsymbol{\theta}_x) + \phi(\boldsymbol{\eta}_y) - \boldsymbol{\theta}_x \cdot \boldsymbol{\eta}_y. \tag{45}$$

Of course, symmetrizations of relative entropy exist, with the Jensen–Shannon divergence perhaps the most prominent example. Also, a symmetric "compositional" version of relative entropy has been proposed because of its additional properties that are often regarded as indispensable[2] in CoDA (Martín-Fernández et al. 1999). While such measures have some interesting properties, they do not make use of the duality of our parametrizations and are therefore less suitable for our approach. Indeed, although a dual divergence is not a distance measure in the strict sense, it can quantify the distance between points along a curve in a similar way, and generalizations of well-known relationships from Euclidean geometry are available. Geodesic lines connecting two compositions can be constructed via convex combinations of the parameters $\boldsymbol{\theta}$ and the corresponding dual geodesics from convex combinations of the dual parameters $\boldsymbol{\eta}$. Such geodesics are orthogonal to each other when the inner product of their tangent vectors, with respect to the Fisher metric, vanishes at the point of intersection. In this case, a generalized Pythagorean theorem holds for the corresponding dual divergence, e.g.,

$$D_\psi(\boldsymbol{x}||\boldsymbol{y}) = D_\psi(\boldsymbol{x}||\boldsymbol{z}) + D_\psi(\boldsymbol{z}||\boldsymbol{y}), \tag{46}$$

where the geodesics $t\boldsymbol{\eta}_x + (1-t)\boldsymbol{\eta}_z$ and $t\boldsymbol{\theta}_z + (1-t)\boldsymbol{\theta}_y$ intersect orthogonally.

3.3 *Distances Obtained from the Fisher Metric Compared with Those Used in CoDA*

We have seen that the potential of Eq. (40) led to a divergence that is also a Euclidean distance. Let us now consider a generalization of our dual divergences that includes

[2] As an example, the translation invariance of distance measures, known under the name of perturbation invariance in CoDA, has its information-geometric analog in the invariance of the inner product of tangent vectors $\mathbf{u}, \mathbf{v}$ under parallel transport P and its dual P^*: $\langle \mathbf{u}, \mathbf{v} \rangle = \langle P(\mathbf{u}), P^*(\mathbf{v}) \rangle$.

as a special case a Euclidean distance that is related to the Fisher metric. The so-called α-divergence (closely related to the Renyi 1961 and Tsallis 1988 entropies) is defined as

$$D_\alpha(\boldsymbol{x}||\boldsymbol{y}) = \frac{4}{1-\alpha^2}\left(1 - \sum_{i=1}^{D} y_i^{\frac{1+\alpha}{2}} x_i^{\frac{1-\alpha}{2}}\right). \tag{47}$$

In the limit, the cases $\alpha = \pm 1$ correspond to D_ψ and D_ϕ. The case $\alpha = 0$ (where the divergence is self-dual, i.e., symmetric) corresponds to the Euclidean distance of the points $(\sqrt{x_1}, \ldots, \sqrt{x_D})$, and $(\sqrt{y_1}, \ldots, \sqrt{y_D})$:

$$\begin{aligned} d_H^2(\boldsymbol{x}, \boldsymbol{y}) &= \sum_{i=1}^{D}(\sqrt{x_i} - \sqrt{y_i})^2 \\ &= \sum_{i=1}^{D}(x_i - 2\sqrt{x_i y_i} + y_i) \qquad (48) \\ &= 2\left(1 - \sum_{i=1}^{D}\sqrt{x_i y_i}\right) \\ &= \frac{1}{2} D_0(\boldsymbol{x}\|\boldsymbol{y}). \qquad (49) \end{aligned}$$

This is the so-called Hellinger distance. It is closely related to the Riemannian distance[3] between two compositions with respect to the Fisher metric. This so-called Fisher distance can be expressed explicitly (Ay et al. 2017) as

$$d_F^2(\boldsymbol{x}, \boldsymbol{y}) = 4 \arccos^2\left(\sum_{i=1}^{D}\sqrt{x_i y_i}\right), \tag{50}$$

and its relation to the Hellinger distance is given by

$$d_H^2(\boldsymbol{x}, \boldsymbol{y}) = 2\left(1 - \cos\frac{d_F(\boldsymbol{x}, \boldsymbol{y})}{2}\right). \tag{51}$$

These distances can be better understood when noting the role that the angle plays between the two points on the sphere, i.e., when comparing Eq. (50) with the cosine of the angle φ between the rays going from the origin through the transformed compositions (see Fig. 3), also referred to as Bhattacharyya coefficient (Bhattacharyya 1943):

$$\cos\varphi = \sum_{i=1}^{D}\sqrt{x_i y_i}. \tag{52}$$

[3]The Riemannian distance between two points on a manifold is the minimum of the lengths of all the piecewise smooth paths joining the two points.

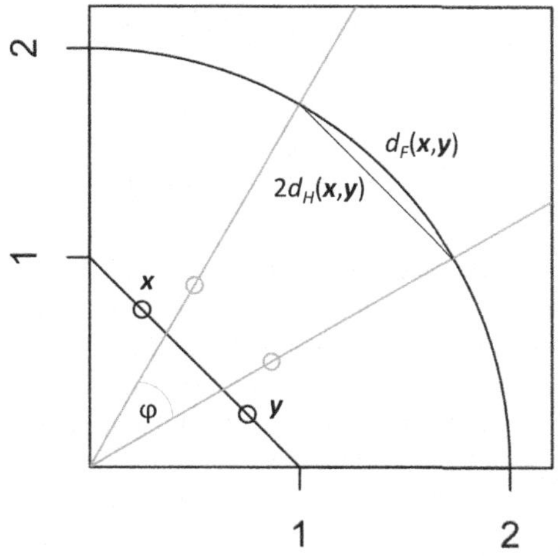

Fig. 3 Hellinger and Fisher distances between compositions $\boldsymbol{x}$ and $\boldsymbol{y}$. Shown are the 2-simplex and the positive orthant of a sphere with radius 2. The gray points indicate the square-root-transformed compositions. They lie on a sphere with a radius one, and their Euclidean distance is the Hellinger distance. The lines from the origin going through these points cross the sphere with a radius two. The length of the arc between the resulting points is the Fisher distance, while their Euclidean distance is twice the Hellinger distance

Let us now come back to Aitchison distance and discuss in which structural aspects it differs from divergences obtained from the Fisher metric. In fact, in data analysis, parametrized classes of distance measures are common (Greenacre 2009). In the same way as in Eq. (47), they are mediated by the Box-Cox transformation, which has the limit

$$\lim_{\beta \to 0} \frac{x^{\beta} - 1}{\beta} = \log(x). \tag{53}$$

This has been applied in CoDA to obtain log-ratio analysis as a Correspondence Analysis of power-transformed data (Greenacre 2010). There, we have the following family of distance measures:

$$d_{\beta}^{2}(\boldsymbol{x}, \boldsymbol{y}) = \frac{1}{\beta^{2}} \sum_{i=1}^{D} \omega_i \left(C(\boldsymbol{x}^{\beta})_i - C(\boldsymbol{y}^{\beta})_i \right)^2, \tag{54}$$

where ω_i are suitable weights. For the case $\beta = 1$, this is the (square of the) symmetric χ^2-distance used in Correspondence Analysis, while $\beta = 0$ gives[4] Aitchison distance (when $\omega_i = D^2$ for all i), see the Appendix for a proof. Although the case $\beta = 1/2$ has a direct relationship with it, the Hellinger distance does not form part of this family because of the closure operation that makes us stay inside the simplex. Similarly, the Aitchison distance cannot be obtained from the alpha divergences of Eq. (47).

[4]This can be seen as the high-temperature limit in statistical physics.

Alpha divergences are included in a more general class of divergences known under the name of f-divergence. They have the form

$$D_f(\boldsymbol{x}, \boldsymbol{y}) = \sum_{i=1}^{D} x_i f\left(\frac{y_i}{x_i}\right), \tag{55}$$

where f is a convex function. It is a well-established result in information geometry that f-divergences are the only decomposable[5] divergences that behave monotonically under coarse graining of information (Amari 2016), i.e., when compositional parts are amalgamated into higher-level parts. This important invariance property is called information monotonicity. The Aitchison distance is not decomposable, as in each summand we use information from all compositional parts when evaluating their geometric mean. Nevertheless, in the next section, we will show that it fulfills information monotonicity.

It is interesting to note that the two Euclidean distances (Hellinger and Aitchison) are each related to different isometries of the simplex. In the case of the Hellinger distance, compositions are isometrically mapped into the positive orthant of the sphere.[6] This isometry also holds for the Fisher metric itself (Ay et al. 2017) (which makes it possible to view the Fisher metric as a Euclidean metric on the $\mathbb{R}^D$ where the sphere is embedded). In the case of the Aitchison distance, the isometry in question is of central interest in CoDA. It is the clr transformation, i.e., the map between $\mathcal{S}^D$ and $\mathcal{T}^D$. Here, however, there is no corresponding isometry of the Fisher metric. Although a Euclidean metric may appear convenient, it does not have the same desirable properties as the Fisher metric. In fact, a central result in information geometry states that the Fisher metric is the only metric (Chentsov 1978) that stays invariant under coarse graining of information.

4 Information Monotonicity of Aitchison Distance

4.1 Amalgamations Lead to Coarse-Grained Information

Let us denote by $\mathcal{A}$ a subset of $\mathcal{D} := \{1, \ldots, D\}$, and let $\boldsymbol{x}_{\mathcal{A}}$ denote the corresponding subcomposition, i.e., the vector where parts with indices not belonging to $\mathcal{A}$ have been removed. Let $H(\boldsymbol{x})$ denote the Shannon entropy of composition $\boldsymbol{x}$, i.e., the potential $-\phi(\eta_x)$. Consider its decomposition

$$H(\boldsymbol{x}) = (1 - s(\boldsymbol{x}_{\mathcal{A}}))H(C\boldsymbol{x}_{\mathcal{D}\setminus\mathcal{A}}) + s(\boldsymbol{x}_{\mathcal{A}})H(C\boldsymbol{x}_{\mathcal{A}}) + H(s(\boldsymbol{x}_{\mathcal{A}}), 1 - s(\boldsymbol{x}_{\mathcal{A}})), \tag{56}$$

[5] A divergence is decomposable if it can be written as a sum of terms that only depend on individual components.

[6] But note that this mapping does not obtain the spherical representative of the composition in the sense of the definition of an equivalence class.

where $s(\boldsymbol{x}_{\mathcal{A}}) = \sum_{i\in\mathcal{A}} x_i$. We see that this is a convex combination of the entropies of the two subcompositions plus a binary entropy, where all terms involve the amalgamation $s(\boldsymbol{x}_{\mathcal{A}})$. Probabilistically speaking, this particular amalgamation corresponds to the probability that an event occurs for any of the events $\mathcal{A}$ that are left out to obtain the subcomposition $\boldsymbol{x}_{\mathcal{D}\setminus\mathcal{A}}$. Generally, amalgamation is nothing else but a coarse graining of the events and their probabilities. The corresponding coarse graining of information is described by this alternative decomposition of Shannon entropy:

$$H(\boldsymbol{x}) = H(\boldsymbol{x}_{\mathcal{D}\setminus\mathcal{A}}, s(\boldsymbol{x}_{\mathcal{A}})) + s(\boldsymbol{x}_{\mathcal{A}})H(C\boldsymbol{x}_{\mathcal{A}}). \tag{57}$$

As the second summand is greater than/equal to zero, this also shows that information cannot grow after coarse graining.

4.2 *The Notion of Monotonicity for Divergences and Distance Measures*

These considerations lead us to an important result that concerns the divergence associated with the potential $\phi(\eta_x)$, i.e., the relative entropy. This result is the information monotonicity under coarse graining, where the notion of monotonicity is somewhat related to the notion of subcompositional dominance. The latter refers to the property that a measure of distance does not increase when evaluating it on a subset of parts only. This is often seen as a desirable property of distances in CoDA (and not fulfilled by distances like Hellinger and Batthacharyya, see Palarea-Albaladejo (2012) for a discussion of distance measures with respect to compositions). A similar—but perhaps more natural—requirement that has not received attention yet in the CoDA community is the one that a distance between compositions should not increase when comparing it with the one after amalgamating over a subset of parts.[7] As we have seen in the previous subsection, we cannot gain information when amalgamating parts, so we should lose resolution when comparing the amalgamated compositions. This is also related to the notion of sufficient statistic, see Amari (2016). The information monotonicity property of relative entropy can thus be expressed as

$$D_\phi(\boldsymbol{x}||\boldsymbol{y}) \geq D_\phi\left((\boldsymbol{x}_{\mathcal{D}\setminus\mathcal{A}}, s(\boldsymbol{x}_{\mathcal{A}})) \,||\, (\boldsymbol{y}_{\mathcal{D}\setminus\mathcal{A}}, s(\boldsymbol{y}_{\mathcal{A}}))\right). \tag{58}$$

This result can be shown for the more general case of f-divergences and continuous distributions using Jensen's inequality (Amari 2000).

Note that in Egozcue and Pawlowsky-Glahn (2005), when discussing amalgamations of parts, the notion of "monotonicity" is used differently. There, the authors argue against amalgamations referring to the observation that the Aitchison dis-

[7]Subcompositional coherence, i.e., the fundamental requirement that quantities remain identical on a renormalized subcomposition, is not an issue for amalgamation: after amalgamation, there is no need for renormalization.

tances between amalgamated compositions and the amalgamated center of the simplex show a non-monotonic behavior along an ilr-coordinate axis defined before amalgamation. We will show below that information monotonicity does hold for the Aitchison distance. We see the lack of distance monotonicity as discussed in Egozcue and Pawlowsky-Glahn (2005) rather as an argument against the use of a Euclidean coordinate system here.

4.3 Monotonicity of Aitchison Distance

A symmetrized version of relative entropy has recently been used in the context of data-driven amalgamation (Quinn and Erb 2021), where it was shown to be better preserved between samples than the Aitchison distance. While the information-theoretic meaning and mathematical properties reflected in the decompositions shown in Sect. 4.1 make Shannon entropy an ideal measure of information, alternative indices that can sometimes be evaluated more easily on real-world data (e.g., making use of sums of squares) have also been considered. In our context, it is interesting that $A(z_x)$, the potential of Eq. (40), has been proposed as an alternative measure of information within an attempt to reformulate information theory from a CoDA point-of-view (Egozcue and Pawlowsky-Glahn 2018). More recently, it has also been proposed as an inequality index (when divided by the number of parts D) (Egozcue and Pawlowsky-Glahn 2019). Here, the following decomposition was shown:

$$||\boldsymbol{x}||_A^2 = ||(\boldsymbol{x}_{\mathcal{D}\setminus\mathcal{A}}, \boldsymbol{x}_{\mathcal{A}})||_A^2 = ||\boldsymbol{x}_{\mathcal{D}\setminus\mathcal{A}}||_A^2 + ||\boldsymbol{x}_{\mathcal{A}}||_A^2 + \frac{a(D-a)}{D}\left(\log\frac{g(\boldsymbol{x}_{\mathcal{D}\setminus\mathcal{A}})}{g(\boldsymbol{x}_{\mathcal{A}})}\right)^2, \tag{59}$$

where we denoted the set size of $\mathcal{A}$ by a. If we now want to decompose with respect to a composition that was partly amalgamated, we find a corresponding relationship that is more complicated[8]:

$$\begin{aligned} ||\boldsymbol{x}||_A^2 &= ||(\boldsymbol{x}_{\mathcal{D}\setminus\mathcal{A}}, s(\boldsymbol{x}_{\mathcal{A}}))||_A^2 + ||\boldsymbol{x}_{\mathcal{A}}||_A^2 \\ &+ \frac{a(D-a)}{D}\left(\log\frac{g(\boldsymbol{x}_{\mathcal{D}\setminus\mathcal{A}})}{g(\boldsymbol{x}_{\mathcal{A}})}\right)^2 - \frac{D-a}{D-a+1}\left(\log\frac{g(\boldsymbol{x}_{\mathcal{D}\setminus\mathcal{A}})}{s(\boldsymbol{x}_{\mathcal{A}})}\right)^2. \end{aligned} \tag{60}$$

Clearly, if we replace the amalgamation $s(\boldsymbol{x}_{\mathcal{A}})$ by the geometric mean $g(\boldsymbol{x}_{\mathcal{A}})$, we get a simpler equality:

[8] It is interesting to note that the two interaction terms have the form of squares of the balance and pivot coordinates mentioned in Sect. 2.

$$||\boldsymbol{x}||_A^2 = ||(\boldsymbol{x}_{\mathcal{D}\setminus\mathcal{A}}, g(\boldsymbol{x}_{\mathcal{A}}))||_A^2 + ||\boldsymbol{x}_{\mathcal{A}}||_A^2 + \left(\frac{a(D-a)}{D} - \frac{D-a}{D-a+1}\right)\left(\log\frac{g(\boldsymbol{x}_{\mathcal{D}\setminus\mathcal{A}})}{g(\boldsymbol{x}_{\mathcal{A}})}\right)^2. \quad (61)$$

Aggregating by geometric means or by amalgamations has been a subject of debate in the CoDA community (Egozcue and Pawlowsky-Glahn 2019; Greenacre 2020). As we can see, measures like the Aitchison norm lend themselves much better to taking geometric means rather than amalgamations. There is, however, no straightforward probabilistic interpretation of geometric means,[9] and the more elegant formal expressions that result often suffer from reduced interpretability.

To the best of our knowledge, the information monotonicity property in its general form has not been considered yet for the Aitchison distance. We here exploit the various decompositions stated above for proving it. Results are summarized in the following propositions. All proofs can be found in the Appendix.

Proposition 1 *Let $\mathcal{D}$ and $\mathcal{A} \subset \mathcal{D}$ be two index sets with sizes D and a, respectively. Further, let $\boldsymbol{x}$ and $\boldsymbol{y}$ be the simplicial representatives of two compositions in $\mathcal{S}^D$. The amalgamation of $\boldsymbol{x}$ over the subset $\mathcal{A}$ of parts be denoted by $s(\boldsymbol{x}_{\mathcal{A}}) = \sum_{i\in\mathcal{A}} x_i$. Then, the following decomposition of the Aitchison distance holds:*

$$||\boldsymbol{x} \ominus \boldsymbol{y}||_A^2 = ||(\boldsymbol{x}_{\mathcal{D}\setminus\mathcal{A}}, s(\boldsymbol{x}_{\mathcal{A}})) \ominus (\boldsymbol{y}_{\mathcal{D}\setminus\mathcal{A}}, s(\boldsymbol{y}_{\mathcal{A}}))||_A^2 + ||\boldsymbol{x}_{\mathcal{A}} \ominus \boldsymbol{y}_{\mathcal{A}}||_A^2 + \frac{a(D-a)}{D}\left(\log\frac{g(\boldsymbol{x}_{\mathcal{D}\setminus\mathcal{A}})}{g(\boldsymbol{x}_{\mathcal{A}})} - \log\frac{g(\boldsymbol{y}_{\mathcal{D}\setminus\mathcal{A}})}{g(\boldsymbol{y}_{\mathcal{A}})}\right)^2 - \frac{D-a}{D-a+1}\left(\log\frac{g(\boldsymbol{x}_{\mathcal{D}\setminus\mathcal{A}})}{s(\boldsymbol{x}_{\mathcal{A}})} - \log\frac{g(\boldsymbol{y}_{\mathcal{D}\setminus\mathcal{A}})}{s(\boldsymbol{y}_{\mathcal{A}})}\right)^2.$$

Corollary 1 *When aggregating a subset of parts in form of their geometric mean, we have the following decomposition of the Aitchison distance*

$$||\boldsymbol{x} \ominus \boldsymbol{y}||_A^2 = ||(\boldsymbol{x}_{\mathcal{D}\setminus\mathcal{A}}, g(\boldsymbol{x}_{\mathcal{A}})) \ominus (\boldsymbol{x}_{\mathcal{D}\setminus\mathcal{A}}, g(\boldsymbol{y}_{\mathcal{A}}))||_A^2 + ||\boldsymbol{x}_{\mathcal{A}} \ominus \boldsymbol{y}_{\mathcal{A}}||_A^2 + \left(\frac{a(D-a)}{D} - \frac{D-a}{D-a+1}\right)\left(\log\frac{g(\boldsymbol{x}_{\mathcal{D}\setminus\mathcal{A}})}{g(\boldsymbol{x}_{\mathcal{A}})} - \log\frac{g(\boldsymbol{y}_{\mathcal{D}\setminus\mathcal{A}})}{g(\boldsymbol{y}_{\mathcal{A}})}\right)^2. \quad (62)$$

From this decomposition, we get the following monotonicity result:

Corollary 2 *With parts aggregated by geometric means, the following inequality holds:*

$$||\boldsymbol{x} \ominus \boldsymbol{y}||_A^2 \geq ||(\boldsymbol{x}_{\mathcal{D}\setminus\mathcal{A}}, g(\boldsymbol{x}_{\mathcal{A}})) \ominus (\boldsymbol{x}_{\mathcal{D}\setminus\mathcal{A}}, g(\boldsymbol{y}_{\mathcal{A}}))||_A^2 + ||\boldsymbol{x}_{\mathcal{A}} \ominus \boldsymbol{y}_{\mathcal{A}}||_A^2.$$

---

[9]The product over parts specifies the probability that *all* events in the subset occur, but this is then re-scaled by the exponent to the probability of a single event.

As we can see, for geometric-mean summaries, the sum of the interaction terms (i.e., of the terms not involving norms) will remain greater than/equal to zero. This is no longer true for amalgamation of parts, and it is less straightforward to show the corresponding inequality:

Proposition 2 *Aitchison distance fulfills the information monotonicity*

$$||\boldsymbol{x} \ominus \boldsymbol{y}||_A^2 \geq ||(\boldsymbol{x}_{\mathcal{D}\setminus\mathcal{A}}, s(\boldsymbol{x}_{\mathcal{A}})) \ominus (\boldsymbol{x}_{\mathcal{D}\setminus\mathcal{A}}, s(\boldsymbol{y}_{\mathcal{A}}))||_A^2.$$

5 Discussion and Outlook

In our little outline of finite information geometry, we could but scratch the surface of the formal apparatus that is at our disposal. We are certain it can serve to advance the field of Compositional Data Analysis in various ways. Differential geometry provides a universally valid framework for the problems occurring with constrained data. Considering the simplex a differentiable manifold enables a general approach from which specific problems like the compositional differential calculus (Barceló-Vidal et al. 2011) follow naturally. Clearly, there is (and has to be) overlap in methodology between the information-geometric perspective and the CoDA approach. An example is the use of the exponential map anchored at the center of the simplex discussed in Sect. 3.2, which allows to identify the simplex with a linear space that is central to the Euclidean CoDA approach. Another example is the fundamental role that exponential families play in information geometry and which have been studied in the CoDA context in the so-called Bayes spaces (Egozcue et al. 2013). But we also think that some of the current limitations of CoDA can be overcome using the additional structures that information geometry can provide. The ease with which amalgamations of parts can be handled by Kullback–Leibler divergence might partly resolve the debate surrounding this issue in the CoDA community. Further, maximum-entropy projections, where Kullback–Leibler divergences are the central tool, seem an especially promising avenue to pursue in the context of data that are only partially available or subject to constraints. Also, our description has focused on the equivalence of compositions with discrete probability distributions, but information geometry can of course be used to describe the distributions themselves. These are no longer finite but continuous and contain a constraint that introduces dependencies among their random variables, calling for the use of more general versions of the concepts presented here.

Appendix

Proof that Eq. 54 tends to Aitchison distance for $\beta \to 0$

$$d_\beta^2(\boldsymbol{x}, \boldsymbol{y}) = \frac{1}{\beta^2} \sum_{i=1}^{D} \omega_i \left(C(\boldsymbol{x}^\beta)_i - C(\boldsymbol{y}^\beta)_i \right)^2 = \sum_{i=1}^{D} \omega_i \left(\frac{x_i^\beta}{\beta \sum_{k=1}^D x_k^\beta} - \frac{y_i^\beta}{\beta \sum_{k=1}^D y_k^\beta} \right)^2.$$

The terms inside the bracket can be written as

$$\frac{x_i^\beta}{\beta \sum_{k=1}^D x_k^\beta} - \frac{y_i^\beta}{\beta \sum_{k=1}^D y_k^\beta} = \frac{x_i^\beta - \frac{1}{D}\sum_{k=1}^D x_k^\beta}{\beta \sum_{k=1}^D x_k^\beta} - \frac{y_i^\beta - \frac{1}{D}\sum_{k=1}^D y_k^\beta}{\beta \sum_{k=1}^D y_k^\beta} \tag{63}$$

when subtracting $1/(\beta D)$ and adding it back. Similarly,

$$= \frac{x_i^\beta - 1 - \frac{1}{D}\sum_{k=1}^D (x_k^\beta - 1)}{\beta \sum_{k=1}^D x_k^\beta} - \frac{y_i^\beta - 1 - \frac{1}{D}\sum_{k=1}^D (y_k^\beta - 1)}{\beta \sum_{k=1}^D y_k^\beta}. \tag{64}$$

In this, we recognize the Box-Cox transform in the numerators and can make use of the limit in Eq. (53). The sums in the denominators clearly tend to D for $\beta \to 0$. We can thus evaluate the limit as a quotient of finite limits and conclude

$$\lim_{\beta \to 0} d_\beta^2(\boldsymbol{x}, \boldsymbol{y}) = \sum_{i=1}^{D} \frac{\omega_i}{D^2} \left(\log \frac{x_i}{g(\boldsymbol{x})} - \log \frac{y_i}{g(\boldsymbol{y})} \right)^2. \tag{65}$$

Proof of Proposition 1 We start by defining $\boldsymbol{z} = (x_i/y_i)_{i \in \mathcal{D}}$. We can now use the decomposition

$$||\boldsymbol{z}||_A^2 = ||\boldsymbol{z}_{\mathcal{D}\setminus\mathcal{A}}||_A^2 + ||\boldsymbol{z}_{\mathcal{A}}||_A^2 + \frac{a(D-a)}{D} \left(\log \frac{g(\boldsymbol{z}_{\mathcal{D}\setminus\mathcal{A}})}{g(\boldsymbol{z}_{\mathcal{A}})} \right)^2, \tag{66}$$

which can be derived using equalities like

$$g(\boldsymbol{x}) = \left(g(\boldsymbol{x}_{\mathcal{A}})^a g(\boldsymbol{x}_{\mathcal{D}\setminus\mathcal{A}})^{D-a} \right)^{\frac{1}{D}} = g(\boldsymbol{x}_{\mathcal{A}}) g(\boldsymbol{x}_{\mathcal{A}})^{\frac{a}{D}-1} g(\boldsymbol{x}_{\mathcal{D}\setminus\mathcal{A}})^{\frac{D-a}{D}},$$

which can be used to expand

$$\sum_{i \in \mathcal{A}} \left(\log \frac{z_i}{g(\boldsymbol{z})} \right)^2 = \sum_{i \in \mathcal{A}} \left(\log \frac{z_i}{g(\boldsymbol{z}_{\mathcal{A}})} + \frac{D-a}{D} \log \frac{g(\boldsymbol{z}_{\mathcal{A}})}{g(\boldsymbol{z}_{\mathcal{D}\setminus\mathcal{A}})} \right)^2$$

and doing the square, cross terms vanish after summation.

Next, we observe that, for an arbitrary s_z which we join as an additional component with the vector $z_{\mathcal{D}\setminus\mathcal{A}}$, we have

$$||(z_{\mathcal{D}\setminus\mathcal{A}}, s_z)||_A^2 = ||z_{\mathcal{D}\setminus\mathcal{A}}||_A^2 + \frac{D-a}{D-a+1}\left(\log\frac{g(z_{\mathcal{D}\setminus\mathcal{A}})}{s_z}\right)^2 \tag{67}$$

because, similarly as before, we have

$$\sum_{i\in\mathcal{D}\setminus\mathcal{A}}\left(\log\frac{z_i}{g((z_{\mathcal{D}\setminus\mathcal{A}}, s_z))}\right)^2 + \left(\log\frac{s_z}{g((z_{\mathcal{D}\setminus\mathcal{A}}, s_z))}\right)^2 =$$
$$\sum_{i\in\mathcal{D}\setminus\mathcal{A}}\left(\log\frac{z_i}{g(z_{\mathcal{D}\setminus\mathcal{A}})} + \frac{1}{D-a+1}\log\frac{g(z_{\mathcal{D}\setminus\mathcal{A}})}{s_z}\right)^2 + \left(\frac{D-a}{D-a+1}\log\frac{s_z}{g(z_{\mathcal{D}\setminus\mathcal{A}})}\right)^2.$$

We can now choose for s_z the expression $s_z = (\sum_{i\in\mathcal{A}} x_i)/(\sum_{i\in\mathcal{A}} y_i)$. With this, we express $||z_{\mathcal{D}\setminus\mathcal{A}}||_A^2$ using the expression following from Eq. (67). We then substitute the corresponding term in Eq. (66), proving the proposition:

$$||z||_A^2 = ||(z_{\mathcal{D}\setminus\mathcal{A}}, s)||_A^2 + ||z_{\mathcal{A}}||_A^2$$
$$+\frac{a(D-a)}{D}\left(\log\frac{g(z_{\mathcal{D}\setminus\mathcal{A}})}{g(z_{\mathcal{A}})}\right)^2 - \frac{D-a}{D-a+1}\left(\log\frac{g(z_{\mathcal{D}\setminus\mathcal{A}})}{(\sum_{i\in\mathcal{A}} x_i)/(\sum_{i\in\mathcal{A}} y_i)}\right)^2. \tag{68}$$

Proof of Corollary 1 This follows from the fact that we can insert $g(z_{\mathcal{A}})$ for s_z in Eq. (67). This gets us an expression for $||z_{\mathcal{D}\setminus\mathcal{A}}||_A^2$, which is inserted in Eq. (66).

Proof of Corollary 2 To prove the corollary, we have to show that the last term in the decomposition of Corollary 1 is greater than or equal to zero. We thus need to show

$$(D-a)\frac{a(D-a+1)-D}{D(D-a+1)} \geq 0. \tag{69}$$

Since $D > a > 1$, and the quadratic equation $a^2 - a(D+1) + D = 0$ has solutions $a = 1$ and $a = D$, between these values we are either above or below zero. We are above because the first derivative of Eq. (69) in $a = 1$ is $(D-1)^2/D^2$, which is bigger zero.

Proof of Proposition 2 Let z denote the vector with components $x_i/y_i, i = 1, \ldots, D$. To prove the proposition, we have to bound the terms after the first plus sign in Proposition 1 from below, i.e.,

$$R(\boldsymbol{x}, \boldsymbol{y}) := \sum_{i \in \mathcal{A}} \left(\log \frac{z_i}{g(z_{\mathcal{A}})} \right)^2 + \frac{a(D-a)}{D} \left(\log \frac{g(z_{\mathcal{D} \setminus \mathcal{A}})}{g(z_{\mathcal{A}})} \right)^2 - \frac{D-a}{D-a+1} \left(\log \frac{g(z_{\mathcal{D} \setminus \mathcal{A}})}{(\sum_{i \in \mathcal{A}} x_i)/(\sum_{i \in \mathcal{A}} y_i)} \right)^2 \geq 0. \quad (70)$$

Let us start with the second summand. We rewrite it as

$$\left(\frac{a(D-a)}{D} - \frac{D-a}{D-a+1} \right) \left(\log \frac{g(z_{\mathcal{D} \setminus \mathcal{A}})}{g(z_{\mathcal{A}})} \right)^2 + \frac{D-a}{D-a+1} \left(\log \frac{g(z_{\mathcal{D} \setminus \mathcal{A}})}{g(z_{\mathcal{A}})} \right)^2$$

Since we showed Eq. (69), the summand on the left is greater than/equal to zero. Thus we have

$$\begin{aligned} R(\boldsymbol{x}, \boldsymbol{y}) &\geq \sum_{i \in \mathcal{A}} \left(\log \frac{z_i}{g(z_{\mathcal{A}})} \right)^2 \\ &\quad + \frac{D-a}{D-a+1} \left(\left(\log \frac{g(z_{\mathcal{D} \setminus \mathcal{A}})}{g(z_{\mathcal{A}})} \right)^2 - \left(\log \frac{g(z_{\mathcal{D} \setminus \mathcal{A}})}{(\sum_{i \in \mathcal{A}} x_i)/(\sum_{i \in \mathcal{A}} y_i)} \right)^2 \right) \\ &\geq \sum_{i \in \mathcal{A}} \left(\log \frac{z_i}{g(z_{\mathcal{A}})} \right)^2 - \left(\log \frac{(\sum_{i \in \mathcal{A}} x_i)/(\sum_{i \in \mathcal{A}} y_i)}{g(z_{\mathcal{A}})} \right)^2, \end{aligned} \quad (71)$$

where the second inequality follows because the big bracket (with a prefactor smaller one) has a structure that can be bounded like

$$|(A-B)^2 - (A-C)^2| \leq (B-C)^2,$$

with log $g(z_{\mathcal{D} \setminus \mathcal{A}})$ playing the role of A. Finally, the last term in Eq. (71) can be bounded from above as follows. Since $x_i \leq y_i \cdot \max x_i/y_i$, we also have

$$\sum_{i \in \mathcal{A}} x_i \leq \max_{k \in \mathcal{A}} \frac{x_k}{y_k} \sum_{i \in \mathcal{A}} y_i, \quad (72)$$

so the ratio of sums is smaller than the maximum over the ratios. Without restricting generality, let us assume that $g(\boldsymbol{x}_{\mathcal{A}}) \geq g(\boldsymbol{y}_{\mathcal{A}})$, i.e., $g(z_{\mathcal{A}}) \geq 1$. The bound on the sum ratio implied by Eq. (72) is then sufficient for proving the proposition:

$$R(\boldsymbol{x}, \boldsymbol{y}) \geq \max_{i \in \mathcal{A}} \left(\log \frac{x_i/y_i}{g(z_{\mathcal{A}})} \right)^2 - \left(\log \frac{(\sum_{i \in \mathcal{A}} x_i)/(\sum_{i \in \mathcal{A}} y_i)}{g(z_{\mathcal{A}})} \right)^2 \geq 0. \quad (73)$$

References

J. Aitchison, *The Statistical Analysis of Compositional Data* (Chapman and Hall, London, 1986)
J. Aitchison, The statistical analysis of compositional data. J. R. Stat. Soc. B **44**(2), 139–160 (1982)
S. Amari, *Differential-Geometric Methods in Statistics*. Lecture Notes in Statistics, vol. 28 (Springer, Berlin, 1985)
S. Amari, *Information Geometry and Its Applications*. Applied Mathematical Sciences, vol. 194 (Springer, Berlin, 2016)
S. Amari, H. Nagaoka, Methods of Information Geometry. Translations of Mathematical Monographs, vol. 191 (American Mathematical Society, Providence, 2000)
N. Ay, J. Jost, H.V. Le, L. Schwachhöfer, *Information Geometry*. A Series of Modern Surveys in Mathematics, vol. 64 (Springer, Berlin, 2017)
N. Ay, I. Erb, On a notion of linear replicator equations. J Dyn. Diff. Eqs. **17**(2), 427–451 (2005)
C. Barceló-Vidal, J.A. Martín-Fernández, G. Mateu-Figueras, Compositional differential calculus on the simplex, in *Compositional Data Analysis: Theory and Applications*, ed. by V. Pawlowsky-Glahn, A. Buccianti (Wiley, New York, 2011), pp. 176–190
C. Barceló-Vidal, J.A. Martín-Fernández, The mathematics of compositional analysis. Austrian J. Stat. **45**(4), 57–71 (2016)
A. Bhattacharyya, On a measure of divergence between two statistical populations defined by their probability distributions. Bull. Calcutta Math. Soc. **35**, 99–109 (1943)
N. Chentsov, Algebraic foundation of mathematical statistics. Math. Oper. Forsch. Stat., Ser. Stat. **9**, 267–276 (1978)
N. Chentsov, *Statistical Decision Rules and Optimal Inference*, vol. 53, Nauka (1972) (in Russian); English translation in: Monographs in Mathematics, vol. 53 (American Mathematical Society, Providence, 1982)
J.J. Egozcue, C. Barceló-Vidal, J.A. Martín-Fernández, E. Jarauta-Bragulat, J.L. Díaz-Barrero, G. Mateu-Figueras, V. Pawlowsky-Glahn, A. Buccianti, Elements of simplicial linear algebra and geometry, in *Compositional Data Analysis: Theory and Applications*, ed. by V. Pawlowsky-Glahn, A. Buccianti (Wiley, New York, 2011), pp. 141–157
J.J. Egozcue, V. Pawlowsky-Glahn, R. Tolosana-Delgado, M.I. Ortego, K.G. van den Boogaart, Bayes spaces: use of improper distributions and exponential families. Revista de la Real Academia de Ciencias Exactas, Físicas y Naturales. Serie A. Matematicas, **107**(2), 475–486 (2013)
J.J. Egozcue, V. Pawlowsky-Glahn, Groups of parts and their balances in compositional data analysis. Math. Geol. **37**(7), 795–828 (2005)
J.J. Egozcue, V. Pawlowsky-Glahn, Evidence functions: a compositional approach to information. SORT **42**(2), 101–124 (2018)
J.J. Egozcue, V. Pawlowsky-Glahn, Compositional data: the sample space and its structure. TEST **28**(3), 599–638 (2019)
J.J. Egozcue, V. Pawlowsky-Glahn, G. Mateu-Figueras, C. Barceló-Vidal, Isometric Logratio transformations for compositional data analysis. Math. Geol. **35**(3), 279–300 (2003)
M. Greenacre, Compositional data analysis. Annu. Rev. Stat. Appl. **8**(1), 271–299 (2021)
M. Greenacre, Power transformations in correspondence analysis. Comput. Stat. & Data Anal. **53**(8), 3107–3116 (2009)
M. Greenacre, Log-ratio analysis is a limiting case of correspondence analysis. Math. Geosci. **42**, 129 (2010)
M. Greenacre, Amalgamations are valid in compositional data analysis, can be used in agglomerative clustering, and their logratios have an inverse transformation. Appl. Comp. Geosc. **5**, 100017 (2020)
K. Hron, P. Filzmoser, P. de Caritat, E. Fišerová, A. Gardlo, Weighted pivot coordinates for compositional data and their application to geochemical mapping. Math. Geosci. **49**, 797–814 (2017)
J.A. Martín-Fernández, M. Bren, C. Barceló-Vidal, V. Pawlowsky-Glahn, A measure of difference for compositional data based on measures of divergence, in *Proceedings of the Fifth Annual*

Conference of the International Association for Mathematical Geology, Trondheim (Norway), vol. 1, ed. by S.J. Lippard, A. Naess, R. Sinding-Larsen (1999), pp. 211–215

J. Palarea-Albaladejo, J.A. Martín-Fernández, J.A. Soto, Dealing with distances and transformations for fuzzy c-means clustering of compositional data. J. Classif. **29**(2), 144–169 (2012)

T.P. Quinn, I. Erb, Amalgams: data-driven amalgamation for the dimensionality reduction of compositional data. NAR genom. bioinform. **2**(4) lqaa076 (2021)

C.R. Rao, Information and the accuracy attainable in the estimation of statistical parameters. Bull. Calcutta Math. Soc. **37**, 81–89 (1945)

A. Rényi, On measures of entropy and information, in *Proceedings of the 4th Berkeley Symposium on Mathematical Statistics and Probability*, vol. 1 (University of California Press, Berkeley, 1961), pp. 547–561

C. Tsallis, Possible generalization of Boltzmann-Gibbs statistics. J. Stat. Phys. **52**, 479–487 (1988)

J. Whittaker, *Graphical Models in Applied Multivariate Statistics* (Wiley, New York, 1990)

Log-Ratio Analysis of Finite Precision Data: Caveats, and Connections to Digital Lines and Number Theory

David R. Lovell

Abstract Log-ratio analysis (LRA) is a popular and theoretically coherent framework for investigating and modelling compositional data. Most empirical compositional data will be measured and recorded with finite precision; count data is a special instance of this in which the fundamental quantity of interest is discrete, but it is also common and practical to round continuous variables to the nearest convenient multiple of the unit of measurement. LRA is often applied to such finite precision measurements without considering the underlying discrete nature of the data (with the exception of the special case of zero values). Here we examine how the characteristics of finite precision data can manifest in LRA so that theoreticians and practitioners can be mindful of situations in which finite precision might affect their conclusions. We focus in particular on log-ratio variance—a fundamental measure of pairwise association between components—and demonstrate situations in which finite precision can have a profound effect on this statistic and related measures of proportionality. We also make connections to computer science concepts about digital lines and to mathematical concepts in number theory, including Farey sequences, to understand how finite precision approximations can affect the value of log-ratio variance even when the underlying continuous variables are perfectly proportional.

1 Introduction

Compositional data analysis is motivated by the desire to make sense of experimental data in which the *relative* abundances of components are of interest. Log-ratio analysis (LRA) does this by using the ratios of components (or combinations of components), treating components that have the same ratio in different observations as equivalent. Pioneered by Aitchison (1986) and advanced by a range of exponents including Pawlowsky-Glahn and Buccianti (2011), the theory of log-ratio analysis has expanded, and its breadth of application has increased as an awareness of the

D. R. Lovell (✉)
School of Computer Science, Centre for Data Science, Queensland University of Technology, Brisbane, Australia
e-mail: David.Lovell@qut.edu.au

P. Filzmoser et al. (eds.), *Advances in Compositional Data Analysis*,
https://doi.org/10.1007/978-3-030-71175-7_3

compositional nature of many experimental datasets has grown. Log-ratio analysis respects that compositional data with D non-negative components "live" in the (constrained) sample space of the D-dimensional unit simplex, $\mathcal{S}^D$. LRA uses the logarithm of ratios of components to transform these data into the (unconstrained) space of real numbers $\mathbb{R}^D$ where an arsenal of statistical methods can be applied without fear of creating results that are compositionally invalid.

In practice, empirical compositional data is typically measured and recorded with finite precision (Bacon Shone 2008). This can happen when the underlying quantity of interest is continuous (e.g., time, mass and parts per million) or inherently discrete (e.g., counts). We recently began to explore the implications of this for log-ratio analysis of count data in the molecular biosciences, motivated by the fact that metagenomic measurements often yield many low or zero counts (Lovell et al. 2020). We coined the term "lattice composition" to describe compositions of finite precision observations (including counts) and showed that the discrete nature of these data can have a profound effect on log-ratio analysis, including statistics that measure pairwise association.

While our focus is on LRA, it is important to be aware of other approaches that respect the underlying discrete nature of lattice compositions, such as methods based on the log-normal Poisson distribution (Aitchison and Ho 1989; Bacon Shone 2008), correspondence analysis (Greenacre 2017) (which can be related to LRA by the Box-Cox transformation (Greenacre 2009)), multinomial logistic-normal modelling (Cargnoni et al. 1997; Billheimer et al. 2001; Silverman et al. 2018), Dirichlet-multinomial models (Holmes et al. 2012; Chen and Li 2013; Gloor et al. 2016) and log-linear modelling with generalised estimating equations (Warton and Guttorp 2011). We also want to note that data recorded with finite precision can be treated as interval censored data (Lindsey and Ryan 1998) and that Bayesian approaches to its analysis have been investigated (Leung 2017). Here, however, our aim is to highlight how finite precision can affect log-ratio analysis.

In the following, we recap key points about the log-ratio analysis of finite precision data (Lovell et al. 2020) before extending our exploration. We begin by taking advantage of the opportunities that low-dimensional lattice compositions present for visualisation, and then show how their finite precision affects measures of pairwise association. Next, we characterise the lattice approximations of pairwise proportional relationships on a finite range of data. We show how these approximations are related to digital lines and number theory and use that understanding to compute the number of possible distinct lattice approximations to proportional lines on a finite range of data.

Finally, we explain how the discrete nature of finite precision data makes log-ratio variance take on a discrete set of positive values for the lowest error approximations to lines of perfect proportionality. This is an important caveat for practitioners familiar with the fact that the log-ratio variance of perfectly proportional continuous components is zero.

2 Visualising Lattice Compositions and Lattice Line Approximations

Low-dimensional compositions of discrete values lend themselves to visualisations that illustrate fundamental concepts in compositional data analysis as well as highlight the characteristics of compositions with discrete components.

We use the term *"lattice composition"* to describe a composition whose components take on discrete numerical values. This includes count data, whose components are from the set of natural numbers $\mathbb{N}$, as well as data recorded with finite precision, which covers the vast majority of experimentally measured data.

"When we speak of lattice systems, we are imagining grids of points in space connected like monkey bars on a playground" (Olds et al. 2000): this is clearly illustrated in the integer lattice of 1000 (x, y, z) triples shown in Fig. 1, where x, y and z are integers from 1 to 10. This figure also contains three coloured line segments that represent the proportional relationships

$$\left(x, \frac{1}{3}x, 1\right) \quad \text{in red,} \tag{1}$$

$$\left(1, \frac{1}{10}z, z\right) \quad \text{in green,} \tag{2}$$

$$\left(\frac{9}{10}z, 1, z\right) \quad \text{in blue} \tag{3}$$

where x, y and z are positive real numbers. Three sets of coloured points show the corresponding integer lattice approximations to these lines with zeros replaced by 1 to enable log-ratio analysis, i.e.,

$$\left(x, \max\left(\left\lfloor \frac{1}{3}x \right\rceil, 1\right), 1\right) \quad \text{in red, for } x \in \{1, 2, \ldots 10\} \tag{4}$$

$$\left(1, \max\left(\left\lfloor \frac{1}{10}z \right\rceil, 1\right), z\right) \quad \text{in green, for } z \in \{1, 2, \ldots 10\} \tag{5}$$

$$\left(\max\left(\left\lfloor \frac{9}{10}z \right\rceil, 1\right), 1, z\right) \quad \text{in blue, for } z \in \{1, 2, \ldots 10\}. \tag{6}$$

The notation $\lfloor x \rceil$ indicates the value of x rounded to the nearest integer. There are several different approaches to rounding which differ primarily in their treatment of values halfway between integers. In this paper, halves will be rounded in a consistent direction (upwards, i.e., $\lfloor x \rceil \triangleq \lfloor x + 0.5 \rfloor$) to ensure that our approximations obey the properties of digital straight line segments (Rosenfeld 1974; Bresenham 1965).

The lattice approximations of these lines are *digital arcs* (Rosenfeld 1974): each approximation is a set of lattice points in which all but the end points have exactly two neighbours and the end points have exactly one. The neighbourhood of a point encompasses all $3^D - 1$ points adjacent to it in each of the D lattice dimensions, e.g.,

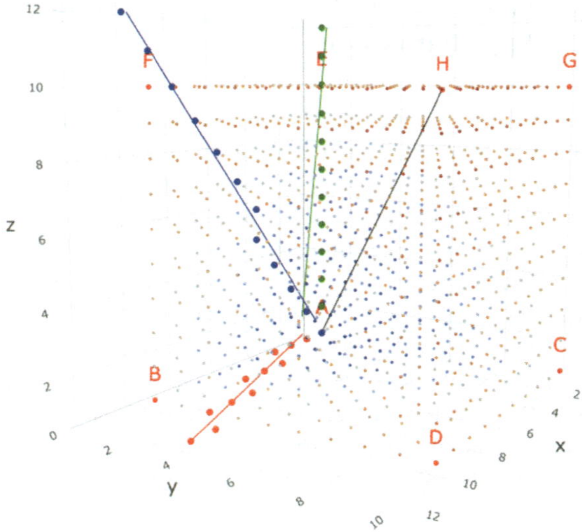

Fig. 1 The three-dimensional lattice of 1000 (x, y, z) triples, where x, y and z are integers from 1 to 10. Points are coloured from blue to red as distance to the origin increases. The black line from (1, 1, 1) to (10, 10, 10) is shown for reference, and the extreme points are labelled A–H. The red, green and blue points and lines are lines of proportionality and their lattice approximations as described in Eqs. 1–6

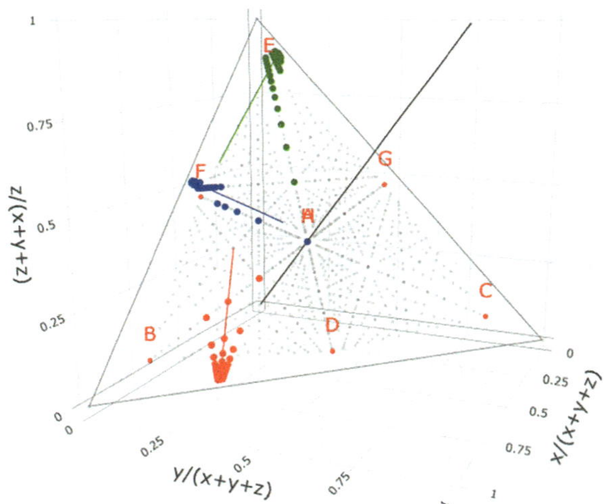

Fig. 2 The red, green and blue points and lines in Fig. 1 after *closure*, a many-to-one transformation from $\mathbb{N}^3$ to the triangular simplex $\mathcal{S}^3$ where $x + y + z = 1$. The transformed lattice points are shown in grey, the black line from (0, 0, 0) to (1, 1, 1) is shown for reference and the extreme points from Fig. 1 are labelled

in 2-dimensions, a lattice point has a neighbourhood of 8 points that are horizontally, vertically or diagonally adjacent (Freeman 1961). Each of the digital arcs in Fig. 1 involves one or more *digital straight line segments* (Rosenfeld 1974); approximations that involve no zero replacement (e.g., the blue points) have just one segment.

Figure 2 shows the points and lines of Fig. 1 after closure, i.e., dividing each component x, y and z by the total $x + y + z$. The closed data now lie in the triangular section of the plane $x + y + z = 1$, the *simplex* in 3-dimensions, $\mathcal{S}^3$.

The discrepancies between Fig. 1 and the closed data in Fig. 2 deserve explanation. For example, in Fig. 1, the green line appears close to point A (1, 1, 1) and distant from point F (10, 1, 10). After closure however, the green line is *distant* from point A (which has been transformed to $(1/3, 1/3, 1/3)$) but close to point F $(10/21, 1/21, 10/21)$. In fact, when $z = 1$, point F lies on the green line in the simplex. This is because, in Fig. 1, point F $(10, 1, 10)$ and the green line at $(1, 1/10, 1)$ lie on the same ray from the origin. Each point on the simplex corresponds to an equivalence class of vectors that lie on the ray from the origin through that point. In other words, closure maps (infinitely) many proportional vectors of D-dimensional data to a single point in the D-dimensional simplex $\mathcal{S}^D$.

But what about lattice points? How many of the 1000 points in Fig. 1 will map to the same points in Fig. 2? This is the first of our connections to number theory which tells us that if a lattice point is picked at random in D dimensions, the probability that it is visible from the origin is $1/\zeta(D)$, where $\zeta(D)$ is the Riemann zeta function (Weisstein 2021). This suggests there will be around $1000/\zeta(3) \approx 832$ points visible from the origin in Fig. 1; there are actually 841 unique points in the closed data.

One final remark about the geometry of the red, green and blue lines of proportionality on the simplex in Fig. 2: one end of each line segment will be at a vertex of the simplex (e.g., the red line has an end at $(0, 0, 1)$); the other will be at the point on the opposite edge which divides that edge into sections proportional to the closure of the variable parts of Eqs. 1–3 (e.g., the red line has an end at $(3/4, 1/4, 0)$).

Log-ratio analysis uses logarithmic transformation to map the (constrained) simplex $\mathcal{S}^D$ to the (unconstrained) space of real numbers $\mathbb{R}^D$ in which statistical methods can be applied without fear of creating compositionally invalid results. These transformations include the *centred log-ratio* (clr), *arithmetic log-ratio* (alr) and *isometric log-ratio* (ilr) transforms, as explained in Pawlowsky-Glahn and Buccianti (2011).

Figure 3 shows the 1000 (x, y, z) triples on the natural number lattice after clr-transformation onto the centred log-ratio plane in $\mathbb{R}^3$. As with closure, this is a many-to-one mapping and there are 841 unique points in the clr-transformed data,

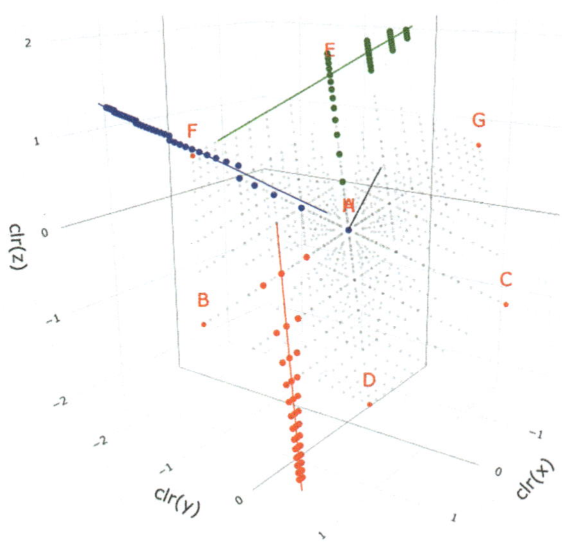

Fig. 3 The red, green and blue points and lines in Fig. 1 after the *centred log-ratio* (clr) transformation. These points and lines lie on the clr-plane $\log x + \log y + \log z = 0$. The black line from $(0, 0, 0)$ to $(1, 1, 1)$ is shown for reference and the extreme points from Fig. 1 are labelled. The axis labels are abbreviated for presentation, e.g., $\text{clr}(x)$ stands for $\log x - \log(xyz)^{1/3}$

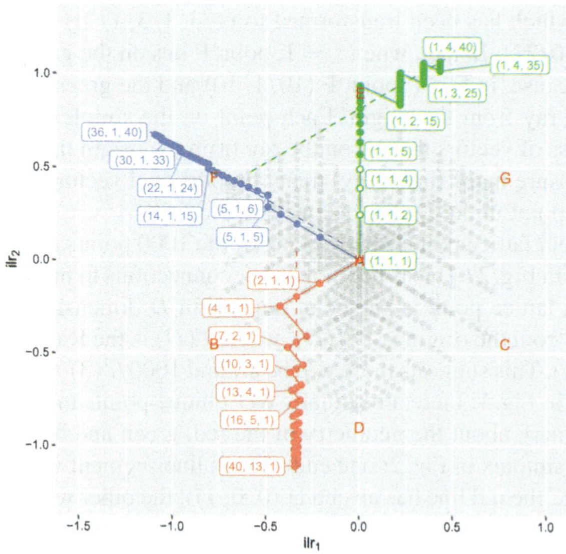

Fig. 4 An isometric log-ratio transformation of the 1000 (x, y, z) triples in Fig. 1. Isometric log-ratios define a family of transformations rather than a single transform, and we have used the specific ilr transformation $\text{ilr}_1 = -\sqrt{1/2}\log_{10}(x/y)$, $\text{ilr}_2 = -\sqrt{2/3}\log_{10}(\sqrt{xy}/z)$ which corresponds to the view of the clr-plane in Fig. 3 along the ray $(1, 1, 1)$. Some points have been labelled with their original (x, y, z) values to show correspondences to Fig. 1. Hollow points indicate where zeros have been replaced by 1 in the lattice approximations of the dashed lines

e.g., the corner points A (1, 1, 1) and H (10, 10, 10) in Fig. 1 are mapped to (0, 0, 0) by clr-transformation, as is every other point where $x = y = z$. The perspective view in Fig. 3 aims to show how the clr-transformation maps compositions to a plane in $\mathbb{R}^3$, specifically the plane where $\log x + \log y + \log z = 0$.

Isometric log-ratio transformation into *ilr-coordinates* provides a $D - 1$-dimensional view of compositions of D parts (Egozcue et al. 2003), and we use that to visualise our data in Fig. 4. In this figure, we use dashed lines to represent the three exact lines of proportionality in Eqs. 1–3 and solid lines to connect the points of their lattice approximations (Eqs. 4–6). Hollow points indicate where zeros have been replaced by 1, e.g., for $z \in \{1, \ldots, 4\}$ in green (Eq. 5). The ilr projection illustrates that lattice approximations of lines become more precise as the values of the underlying finite precision variables increase. It also shows how lattice point approximations zig-zag around the exact lines they approximate; this affects log-ratio variance and other measures of pairwise association as we shall now discuss.

3 The Impact of Lattice Approximation on Pairwise Association

The statistical assessment of pairwise association is a key method in making sense of data. It needs to be approached with particular care for compositional data because change in one component necessarily affects the relative abundance of others. Hence, Pearson's correlation is not a valid measure of association in compositions, as Pearson himself showed in 1897 (Pearson 1897).

Proportionality is a valid measure of association for data that carry relative information (Lovell et al. 2015), and here we consider how finite precision affects measures of proportionality. Figure 5 shows proportional lines with the same slopes as Eqs. 1–3 but in components x and y (with $z = 1$) to simplify our exposition. Each line is labelled with the values of three statistics which measure the extent to which pairs of components in the lattice approximation (i.e., the coloured points) are proportional: variance of log-ratios (vlr) (Aitchison 1986), and two scaled versions of that, ϕ_s and ρ_p (Erb and Notredame 2016; Lovell et al. 2015). (All logarithms are base 10 in the calculations of this paper.)

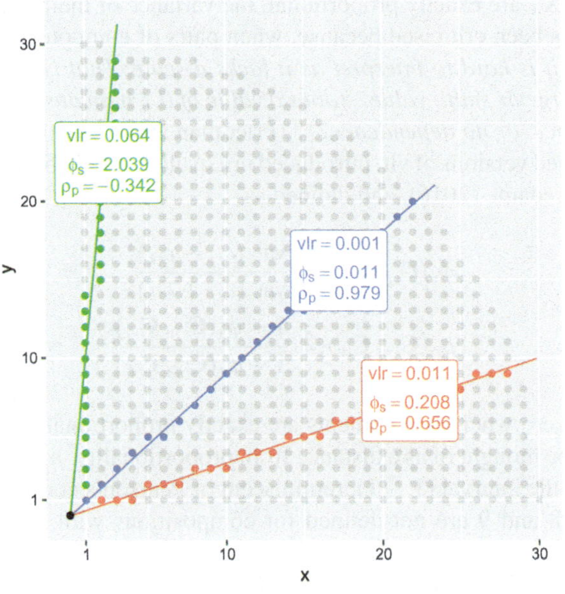

Fig. 5 Lattice compositions cannot, in general, exactly represent proportional relationships between pairs of components, and this affects statistical measures of proportionality. The coloured lines of exact proportionality between x and y have slopes 1/3 (red), 9/10 (blue) and 10/1 (green); coloured points show their discrete lattice approximation. The labels show corresponding proportionality statistics of the lattice approximations where each coloured point is treated as a distinct sample. We use a third component $z = 1$ (not shown) to ensure that ϕ_s and ρ_p are defined and again replace zeros with 1 in the lattice approximations

To introduce these statistics, we first define $\mathbf{X}$ as an $N \times D$ matrix of N observations where the ith observation is the D-part composition $\mathbf{x}_i = (x_{i1}, \ldots, x_{ij}, \ldots, x_{iD})$. The clr representation of composition $\mathbf{x}_i$ is the logarithm of the components after dividing by the geometric mean of $\mathbf{x}_i$:

$$\operatorname{clr}(\mathbf{x}_i) = \left(\log \frac{x_{i1}}{\operatorname{g_m}(\mathbf{x_i})}, \ldots, \log \frac{x_{ij}}{\operatorname{g_m}(\mathbf{x_i})}, \ldots, \log \frac{x_{iD}}{\operatorname{g_m}(\mathbf{x_i})}, \right)$$

hence the sum of the elements of $\operatorname{clr}(\mathbf{x}_i)$ is zero. The row-wise clr-transformed version of $\mathbf{X}$ is written as $\mathbf{C}$, its jth column is written as $\mathbf{C}_{\cdot j}$ and element i, j is denoted as c_{ij}.

The *variance of the log-ratios* of parts j and k is defined as (Aitchison 1986)

$$\begin{aligned} \operatorname{vlr}(\mathbf{X}_{\cdot j}, \mathbf{X}_{\cdot k}) &\triangleq \operatorname*{var}_i \left(\log \frac{x_{ij}}{x_{ik}} \right) \\ &= \operatorname*{var}_i (\log x_{ij} - \log x_{ik}) \\ &= \operatorname*{var}_i (c_{ij} - c_{ik}). \end{aligned} \tag{7}$$

When $\mathbf{X}_{\cdot j}$ and $\mathbf{X}_{\cdot k}$ are exactly proportional, the variance of their log-ratios is 0. But this statistic has been criticised because, when pairs of components are not exactly proportional, *"it is hard to interpret as it lacks a scale. That is, it is unclear what constitutes a large or small value...(does a value of 0.1 indicate strong dependence, weak dependence, or no dependence?)"* (Friedman and Alm 2012). This led to the proposal of scaled versions of vlr, introduced in Lovell et al. (2015), developed further in Erb and Notredame (2016), and defined as

$$\phi_s(\mathbf{C}_{\cdot j}, \mathbf{C}_{\cdot k}) \triangleq \frac{\operatorname{var}_i(c_{ij} - c_{ik})}{\operatorname{var}_i(c_{ij} + c_{ik})} \tag{8}$$

$$\rho_p(\mathbf{C}_{\cdot j}, \mathbf{C}_{\cdot k}) \triangleq \frac{1 - \phi_s(\mathbf{C}_{\cdot j}, \mathbf{C}_{\cdot k})}{1 + \phi_s(\mathbf{C}_{\cdot j}, \mathbf{C}_{\cdot k})}. \tag{9}$$

Like vlr, ϕ_s is zero when its arguments are exactly proportional and positive otherwise, so it can be thought of as a distance from proportionality. ρ_p maps this distance from $[0, \infty)$ to the interval $[1, -1)$, reminiscent of Pearson's correlation coefficient. Note that Eqs. 8 and 9 are not defined for compositions with two parts since, by definition, $c_{i1} + c_{i2} = 0$.

As Fig. 5 suggests, lattice compositions cannot exactly represent proportional relationships ($y = kx$) between pairs of components, except when $k = 1$. We showed in Lovell et al. (2020) how proportionality statistics behave across a range of lattice approximations by computing them on an equiangular sample of lines of positive slopes between zero and infinity. We stated (but did not show in Lovell et al. 2020) that the proportionality statistics should be piecewise constant functions of a slope, with discontinuities where different lattice approximations incorporate different points.

Since then, we have come to understand the exact nature of the different lattice approximations to lines of proportionality on a finite range of points. To explain that, we must first consider some important characteristics of digital lines.

4 Characteristics of Digital Lines

In 1965, Bresenham published an algorithm to compute the set of points on the integer lattice with minimum total vertical distance to a line defined by rational coefficients (Bresenham 1965; Harris and Reingold 2004). In theory, it is not hard to find the points on the integer lattice closest to a straight line; Bresenham's practical contribution lay in facilitating this computation. His algorithm computed digital straight lines (Freeman 1961) using only integer arithmetic and could thus be implemented to run fast, an important consideration for the computers and display devices of the 1960s. Bresenham's algorithm remains fundamental to computer graphics to this day. In this section, we will show how the digital lines produced by this algorithm relate to certain compositions of finite precision data.

Harris and Reingold (2004) show the relationship between Bresenham's algorithm, leap years and Euclid's algorithm for the greatest common divisor of two integers. We benefit greatly from their lucid exposition and adopt their notation here. We will also (as they do) refer to the *sets of points* generated by Bresenham's algorithm as *Bresenham lines*.

Bresenham's algorithm calculates the coordinates (x_n, y_n), of the digital straight line from the integer starting point (x_S, y_S) to the integer ending point (x_E, y_E) as depicted in Fig. 6. Since we are interested in lines of proportionality, we set $(x_S, y_S) = (0, 0)$. For simplicity, we restrict our attention to the octant $0 \leq y \leq x$ (as in Harris and Reingold 2004). The ideas we present can be extended to the octant $y \geq x \geq 0$ by interchanging x and y axes.

Bresenham lines give the least-error discrete approximation of continuous lines. There are a finite number of different Bresenham lines that can be formed on a finite region of the integer lattice, each one corresponding to a range of continuous line slopes. Consider, for example, the red line of slope 1/3 in Fig. 5: lines with slightly different slopes, say 100/301 or 100/299, will have the same Bresenham line—and therefore the same proportionality statistics—on that region of the lattice. We will call Bresenham lines that approximate proportional relationships *Bresenham lines of proportionality*.

So how many different Bresenham lines of proportionality can there be on a finite range of integer lattice? Each point (x_n, y_n) in a Bresenham line is associated with slopes ranging from $[k_{n_{\min}}, k_{n_{\max}})$ where

Fig. 6 The coloured regions show the range of slopes $[k_{n_{\min}}, k_{n_{\max}})$ in which the coloured points (10, 1) green, (3, 1) red and blue (10, 9) will appear in Bresenham lines of proportionality. The red point (3, 1) will appear in Bresenham lines approximating slopes from $k_{n_{\min}} = (1 - \frac{1}{2})/3$ to $k_{n_{\max}} = (1 + \frac{1}{2})/3$, such as the Bresenham line from the black points $(x_S, y_S) = (0, 0)$ to $(x_E, y_E) = (10, 3)$

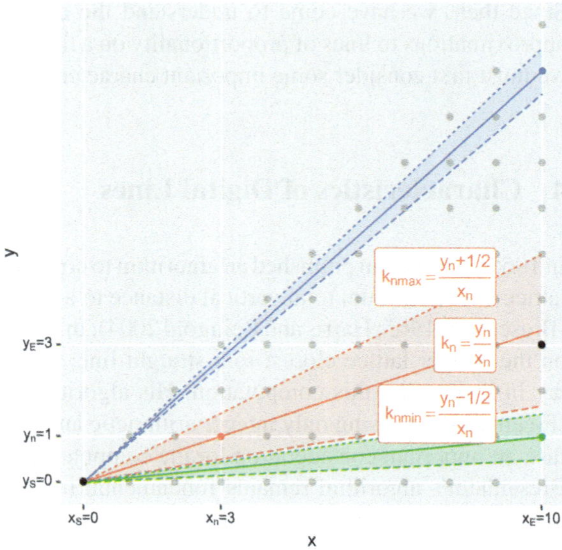

$$k_{n_{\max}}(x_n, y_n) = \frac{y_n + \frac{1}{2}}{x_n} = \frac{2y_n + 1}{2x_n} \tag{10}$$

$$k_{n_{\min}}(x_n, y_n) = \frac{y_n - \frac{1}{2}}{x_n} = \frac{2y_n - 1}{2x_n}. \tag{11}$$

The number of different Bresenham lines of proportionality will be the number of different slope ranges possible with points in the lattice region. Figure 6 shades in red the range of slopes that will include the point $(x_n, y_n) = (3, 1)$ in their corresponding Bresenham lines of proportionality. Since $k_{n_{\max}}(x_n, y_n) = k_{n_{\min}}(x_n, y_n + 1)$, this is the same as the number of distinct $k_{n_{\min}}(x_n, y_n)$ values possible in the lattice region. For the region $x_S = 0$ to x_E, this is equal to 1 plus the number of terms with even denominators in $\mathcal{F}_{2x_E}$, the Farey sequence of order $2x_E$.

The Farey sequence of order N, denoted by $\mathcal{F}_N$, is the set of all reduced fractions between 0 and 1 whose denominators are N or less, arranged in increasing order (Graham et al. 1994). Table 1 shows Farey sequences to order 8, with terms that have even denominators in bold. Thus, for $x_S = 0$ to $x_E = 1$, there are two distinct Bresenham lines of proportionality: one consisting of the points (0, 0), (1, 0), the other consisting of points (0, 0), (1, 1). For $x_S = 0$ to $x_E = 4$, there are 9 terms with even denominators in $\mathcal{F}_{2x_E} = \mathcal{F}_8$, and the corresponding $9 + 1 = 10$ Bresenham lines of proportionality are shown in Fig. 7. Broader ranges of $x_S = 0$ to x_E consume too much space to plot the Bresenham lines themselves but we can show compactly where the y values of each different line are incremented. We do this in Fig. 8 for $x_S = 0$ to $x_E = 30$. This representation also reveals the patterns that arise as Bresenham's algorithm spaces y increment as evenly as possible (Harris and Reingold 2004).

Table 1 Farey sequences $\mathcal{F}_N$ up to order $N = 8$ laid out in a *Stern-Brocot* tree (Graham et al. 1994). Terms with even denominators are shown in bold. The terms with even denominators in the sequences of even order $N = 2x_E$ define the range of slopes represented by each Bresenham line of proportionality from $x_s = 0$ to x_E. For example, the 10 different Bresenham lines of proportionality that can be formed on $x_s = 0$ to $x_E = 4$ represent the slope ranges defined by the terms of $\mathcal{F}_8$ with even denominators: $\left[\frac{0}{1}, \frac{1}{8}\right), \left[\frac{1}{8}, \frac{1}{6}\right), \ldots, \left[\frac{7}{8}, \frac{1}{1}\right)$ as shown in Fig. 7

$\mathcal{F}_1$:	$\frac{0}{1}$																						$\frac{1}{1}$
$\mathcal{F}_2$:	$\frac{0}{1}$											$\mathbf{\frac{1}{2}}$											$\frac{1}{1}$
$\mathcal{F}_3$:	$\frac{0}{1}$							$\frac{1}{3}$				$\mathbf{\frac{1}{2}}$				$\frac{2}{3}$							$\frac{1}{1}$
$\mathcal{F}_4$:	$\frac{0}{1}$					$\mathbf{\frac{1}{4}}$		$\frac{1}{3}$				$\mathbf{\frac{1}{2}}$				$\frac{2}{3}$		$\mathbf{\frac{3}{4}}$					$\frac{1}{1}$
$\mathcal{F}_5$:	$\frac{0}{1}$				$\frac{1}{5}$	$\mathbf{\frac{1}{4}}$		$\frac{1}{3}$		$\frac{2}{5}$		$\mathbf{\frac{1}{2}}$		$\frac{3}{5}$		$\frac{2}{3}$		$\mathbf{\frac{3}{4}}$	$\frac{4}{5}$				$\frac{1}{1}$
$\mathcal{F}_6$:	$\frac{0}{1}$			$\mathbf{\frac{1}{6}}$	$\frac{1}{5}$	$\mathbf{\frac{1}{4}}$		$\frac{1}{3}$		$\frac{2}{5}$		$\mathbf{\frac{1}{2}}$		$\frac{3}{5}$		$\frac{2}{3}$		$\mathbf{\frac{3}{4}}$	$\frac{4}{5}$	$\mathbf{\frac{5}{6}}$			$\frac{1}{1}$
$\mathcal{F}_7$:	$\frac{0}{1}$		$\frac{1}{7}$	$\mathbf{\frac{1}{6}}$	$\frac{1}{5}$	$\mathbf{\frac{1}{4}}$	$\frac{2}{7}$	$\frac{1}{3}$		$\frac{2}{5}$	$\frac{3}{7}$	$\mathbf{\frac{1}{2}}$	$\frac{4}{7}$	$\frac{3}{5}$		$\frac{2}{3}$	$\frac{5}{7}$	$\mathbf{\frac{3}{4}}$	$\frac{4}{5}$	$\mathbf{\frac{5}{6}}$	$\frac{6}{7}$		$\frac{1}{1}$
$\mathcal{F}_8$:	$\frac{0}{1}$	$\mathbf{\frac{1}{8}}$	$\frac{1}{7}$	$\mathbf{\frac{1}{6}}$	$\frac{1}{5}$	$\mathbf{\frac{1}{4}}$	$\frac{2}{7}$	$\frac{1}{3}$	$\mathbf{\frac{3}{8}}$	$\frac{2}{5}$	$\frac{3}{7}$	$\mathbf{\frac{1}{2}}$	$\frac{4}{7}$	$\frac{3}{5}$	$\mathbf{\frac{5}{8}}$	$\frac{2}{3}$	$\frac{5}{7}$	$\mathbf{\frac{3}{4}}$	$\frac{4}{5}$	$\mathbf{\frac{5}{6}}$	$\frac{6}{7}$	$\mathbf{\frac{7}{8}}$	$\frac{1}{1}$

Figure 8 highlights the Bresenham lines of proportionality that correspond to the continuous lines of slopes 1/10, 1/3 and 9/10 in Fig. 6. The line of slope $1/3 = 0.3\dot{3}$ is approximated by the 127th Bresenham line of proportionality which represents continuous lines of proportionality with slopes k from $19/58 = 0.327586$ to $19/56 = 0.339285$.

What about the number of distinct Bresenham lines of proportionality possible on the range x_S to x_E where x_S is greater than 0? At this stage, we have not found a neat relationship between this number and the terms of Farey sequences, though we know it must again involve terms with even denominators because of Eqs. 10 and 11. We can however compute all unique values of these equations for $x_S \leq x_n \leq x_E$ and $y_n \leq x_n$, and we use this approach in the next section to explore the behaviour of log-ratio variance of Bresenham lines of proportionality.

5 The Impact of Finite Precision on Log-Ratio Variance

In Sect. 3, we showed the impact that lattice approximation has on lines of proportionality and log-ratio measures of proportionality (Eqs. 7–9). Now that we know the ranges of slope represented by each distinct Bresenham line of proportionality, we can show properly the discrete, piecewise constant values of vlr, ϕ_s and ρ_p for lattice approximations of lines of proportionality with slopes from 0 to 1. Figure 9 shows

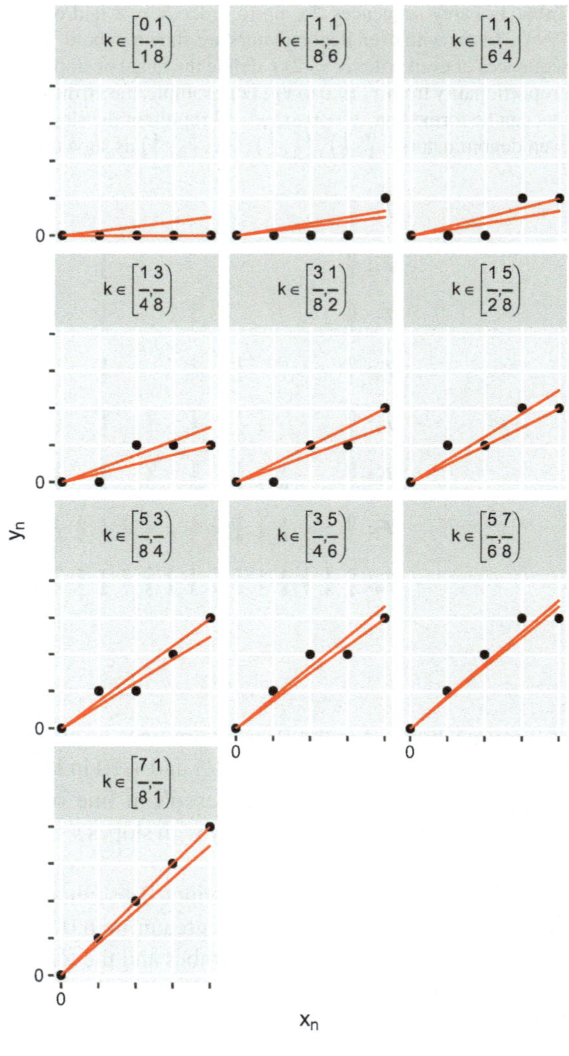

Fig. 7 The black points in each panel show one of the 10 different Bresenham lines of proportionality that can be formed on $x_s = 0$ to $x_E = 4$. The red lines and the panel labels show the range of slopes that each Bresenham line of proportionality represents. These correspond to the terms of the Farey sequence $\mathcal{F}_8$ with even denominators as described in Table 1

that these statistics demonstrate intricate variation for lattice approximations of the continuous proportional lines whose values would be vlr $= 0$, $\phi_s = 0$ and $\rho_p = 1$.

As the slope of the lines being approximated nears 0, vlr attains its maximum when $y_n = 1$ at all values of $x_n = 1$ from x_S to x_E, i.e., when all values of $\lfloor kx_n \rceil$ are 0 and therefore replaced by $y_n = 1$. Note that replacing zero values of $\lfloor kx_n \rceil$ with positive values less than 1 introduces more fluctuations in log-ratio variance at low angles. Even before any zero replacement occurs (i.e., for slopes above 1/2 (about 27°), lattice approximation has an impact on the three proportionality statistics, most noticeably on ρ_p.

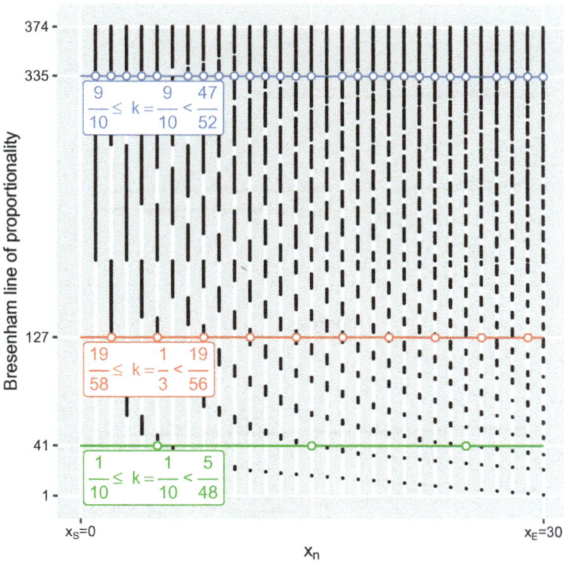

Fig. 8 In total, 374 different Bresenham lines of proportionality can be formed from $x_S = 0$ to $x_E = 30$, approximating lines $y = kx$ with slopes $0 <= k <= 1$. Each row represents a different Bresenham line. Each column corresponds to the x value of each Bresenham line's lattice points. In each row, a black dot indicates that the Bresenham line's y value has been incremented by 1; a white dot means no increment. The coloured lines indicate the Bresenham lines that approximate the continuous lines with slopes 1/10, 1/3 and 9/10. In these lines, hollow points show increments in y

The effect of the finite precision lattice approximation is more prominent the smaller the region of lattice involved because there are fewer distinct Bresenham lines of proportionality available. Compare the proportionality statistics obtained when x_n ranges from 1 to 10 (which has 39 distinct Bresenham lines of proportionality after zero replacement) to those where x_n ranges from 1 to 100 (on which there are 3898 distinct Bresenham lines of proportionality after zero replacement).

6 Conclusions

Our investigation into the impact of finite precision data was initially motivated by a desire to understand compositional data analysis of count data in the molecular biosciences, especially in situations where those counts can be low or zero (Lovell et al. 2020). As we have explored this topic, we have come to realise that not just counts, but finite precision data in general can also impact log-ratio analysis. The information carried by count and other finite precision data depends on their scale and distribution relative to their smallest quantum of change. For counts, this quantum

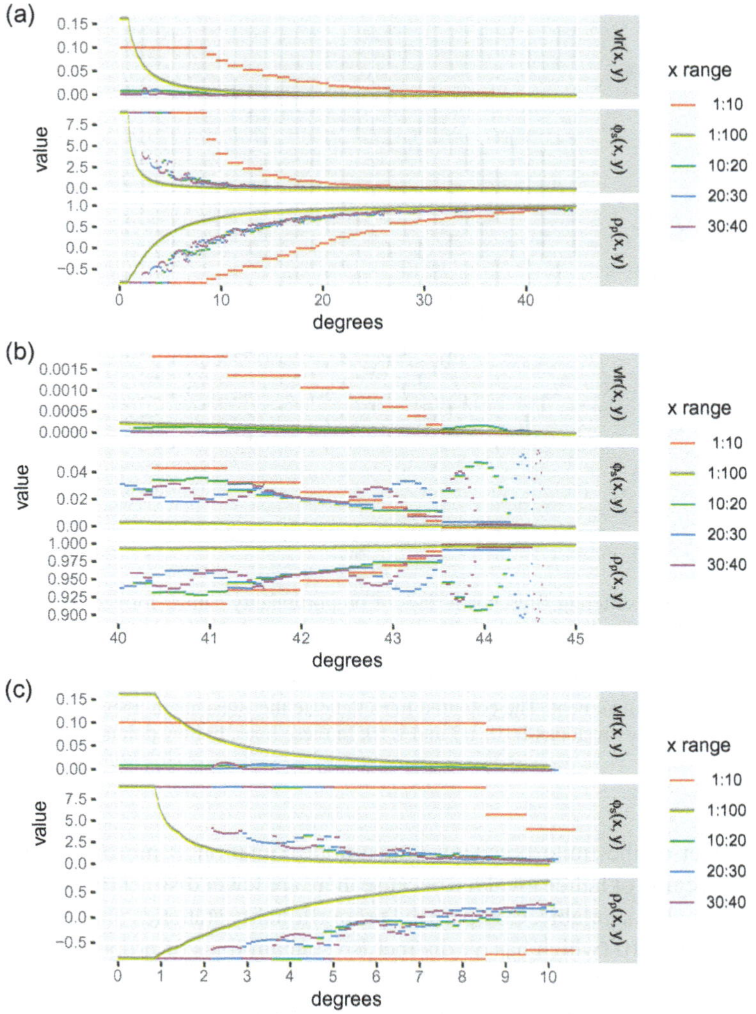

Fig. 9 **a** The exact values of vlr, ϕ_s and ρ_p for lattice approximations to a line of proportionality as its slope k sweeps from 0 to 1 (angles 0–45°). Different coloured lines indicate these statistics for different ranges of x values ranging from x_S to x_E. **b** Shows detail of vlr, ϕ_s and ρ_p at slopes near 1 (40–45°). **c** Shows detail of vlr, ϕ_s and ρ_p at slopes near 0 (0–10°). Note that as in Eqs. 4–6, we have replaced all zeros in the lattice approximations with 1 to permit log-ratio analysis. Also, these statistics are symmetric about slope 1: their values 45–90° will be the mirror image of these plots

is 1; for other finite precision data, the quantum relates to the least significant digit of measurement.

Lattice compositions have fundamental limits to the amount of information they can represent, and these limits become apparent to us when components have values close to the finite precision with which they are measured. We have used relationships between finite precision compositions, digital lines and number theory to characterise the impact of quantisation on log-ratio proportionality statistics, showing that the conclusions we draw from log-ratio analysis depend on the scale of data involved. We have seen that variation comes from the finite precision of the data itself as well as the variation of interest in the components being studied. Our hope is that this exposition will add to the wisdom of those undertaking log-ratio analysis of finite precision data.

Acknowledgements The author thanks the two anonymous reviewers of this work for their valuable insights. The author also gratefully acknowledges the support, encouragement and enlightenment he has received from Profs. Vera Pawlowsky-Glahn and Juan José Egozcue who graciously responded when he was introduced to them by Prof. John Aitchison in 2008.

References

J. Aitchison, *The Statistical Analysis of Compositional Data* (Chapman and Hall Ltd, London, 1986). ISBN 0-412-28060-4

J. Aitchison, C.H. Ho, The Multivariate poisson-log normal distribution. Biometrika **76**(4), 643–653 (1989). ISSN 0006-3444. https://doi.org/10.2307/2336624

J. Bacon Shone, Discrete and continuous compositions, in *Proceedings of CoDAWork'08*, ed. by J. Daunis-I-Estadella, J.-A. Martín-Fernández (University of Girona, Girona, 2008)

D. Billheimer, P. Guttorp, W.F. Fagan, Statistical interpretation of species composition. J. Amer. Stat. Ass. **96**(456), 1205–1214 (2001). ISSN 0162-1459. https://doi.org/10.1198/016214501753381850

J.E. Bresenham, Algorithm for computer control of a digital plotter. IBM Syst. J. **4**(1), 25–30 (1965). ISSN 0018-8670. https://doi.org/10.1147/sj.41.0025

C. Cargnoni, P. Muller, M. West, Bayesian forecasting of multinomial time series through conditionally Gaussian dynamic models. J. Amer. Stat. Ass. **92**(438), 640–47 (Analysis, Statistical Information Theory, and Other Special) (1997). ISSN 0162-1459

J. Chen, H. Li, Variable selection for sparse Dirichlet-multinomial regression with an application to microbiome data analysis. Ann. Appl. Stat. **7**(1), 418–442 (2013). ISSN 1932-6157

J.J. Egozcue, V. Pawlowsky-Glahn, G. Mateu-Figueras, C. Barceló-Vidal, Isometric logratio transformations for compositional data analysis. Math. Geol. **35**(3), 279–300 (2003). ISSN 1573-8868. https://doi.org/10.1023/A:1023818214614

I. Erb, C. Notredame, How should we measure proportionality on relative gene expression data? Theory Biosci. **135**(1), 21–36 (2016). ISSN 1611-7530. https://doi.org/10.1007/s12064-015-0220-8

H. Freeman, On the encoding of arbitrary geometric configurations. IRE Trans. Electron. Comput. **EC-10**(2), 260–268 (1961). ISSN 0367-9950. https://doi.org/10.1109/TEC.1961.5219197

J. Friedman, E.J. Alm, Inferring correlation networks from genomic survey data. PLoS Comput. Biol. **8**(9) (2012). https://doi.org/10.1371/journal.pcbi.1002687

G.B. Gloor, J.M. Macklaim, M. Vu, A.D. Fernandes, Compositional uncertainty should not be ignored in high-throughput sequencing data analysis. Austrian J. Stat. **45**(4), 73 (2016). ISSN 1026597X. https://doi.org/10.17713/ajs.v45i4.122

R.L. Graham, D.E. Knuth, O. Patashnik, *Concrete Mathematics: A Foundation for Computer Science*, 2nd edn (Addison-Wesley, Reading, 1994). ISBN 978-0-201-55802-9

M. Greenacre, Power transformations in correspondence analysis. Comput. Stat. & Data Anal. **53**(8), 3107–3116 (2009). ISSN 01679473. https://doi.org/10.1016/j.csda.2008.09.001

M. Greenacre, *Correspondence Analysis in Practice*, 3rd edn. (Chapman and Hall/CRC, London, 2017)

M.A. Harris, E.M. Reingold, Line drawing, leap years, and Euclid. ACM Comput. Sur. (CSUR) **36**(1), 68–80 (2004). ISSN 0360-0300, 1557-7341. https://doi.org/10.1145/1013208.1013211

I. Holmes, K. Harris, C. Quince, Dirichlet multinomial mixtures: generative models for microbial metagenomics. PLoS ONE **7**(2), e30126 (2012). ISSN 1932-6203. https://doi.org/10.1371/journal.pone.0030126

T.-C. Leung, Bayesian censoring approach to rounded zeros in compositional data. Ph.D. Thesis, The University of Hong Kong (Pokfulam, Hong Kong, 2017)

J.C. Lindsey, L.M. Ryan, Methods for interval-censored data. Stat. Med. **17**(2), 219–238 (1998). ISSN 1097-0258. https://doi.org/10.1002/(SICI)1097-0258(19980130)17:2<219::AID-SIM735>3.0.CO;2-O

D.R. Lovell, V. Pawlowsky-Glahn, J.J. Egozcue, S. Marguerat, J. Bähler, Proportionality: a valid alternative to correlation for relative data. PLOS Comput. Biol. **11**(3), e1004075 (2015). ISSN 1553-7358. https://doi.org/10.1371/journal.pcbi.1004075

D.R. Lovell, X.-Y. Chua, A. McGrath, Counts: an outstanding challenge for log-ratio analysis of compositional data in the molecular biosciences. NAR Genom. Bioinf. **2**(2) (2020). https://doi.org/10.1093/nargab/lqaa040

C.D. Olds, A. Lax, G.P. Davidoff, *The Geometry of Numbers*. Number v. #41 in The Anneli Lax New Mathematical Library (Mathematical Association of America, Washington, 2000). ISBN 978-0-88385-643-7

V. Pawlowsky-Glahn, A. Buccianti (eds.) *Compositional Data Analysis: Theory and Applications* (2011). ISBN 978-0-470-71135-4 978-1-119-97646-2

K. Pearson, Mathematical contributions to the theory of evolution.–On a form of spurious correlation which may arise when indices are used in the measurement of organs. Proc. R. Soc. Lond. **60**, 489–498 (1897). ISSN 03701662

A. Rosenfeld, Digital straight line segments. IEEE Trans. Comput. **C-23**(12), 1264–1269 (1974). ISSN 1557–9956. https://doi.org/10.1109/T-C.1974.223845

J.D. Silverman, H.K. Durand, R.J. Bloom, S. Mukherjee, L.A. David, Dynamic linear models guide design and analysis of microbiota studies within artificial human guts. Microbiome **6**(1), 1–20 (2018). ISSN 2049-2618. https://doi.org/10.1186/s40168-018-0584-3

D.I. Warton, P. Guttorp, Compositional analysis of overdispersed counts using generalized estimating equations. Env. Ecol. Stat. **18**(3), 427–446 (2011). ISSN 1573-3009. https://doi.org/10.1007/s10651-010-0145-9

E.W. Weisstein, Visible Point. http://mathworld.wolfram.com/VisiblePoint.html

Distributions on the Simplex Revisited

Gloria Mateu-Figueras, Gianna S. Monti, and J. J. Egozcue

Abstract A large number of families of distributions are available to model multivariate real vectors. On the contrary, for the simplex sample space, we have only a limited number of families arising through the generalization of the Dirichlet family or the logratio normal family. This chapter tries to summarize those models and some generalizations with a special emphasis on the algebraic-geometric structure of the simplex and on the measure which is considered compatible. In particular, the shifted-scaled Dirichlet distribution is studied and the logratio t distribution is rewritten and studied with respect to the Aitchison measure.

1 Introduction

In this work we revisit some distributions on the simplex, the sample space for compositional data. We focus our attention to the Dirichlet and to the logratio normal models because they have a relevant role in the analysis of compositional data and they are frequently used in practice. Nevertheless, both models were introduced before the Euclidean vector space structure of the simplex was known. In particular the Dirichlet distribution is a historical model that derives its name by the Dirichlet's integral formula (Dirichlet 1839). The logratio normal models appear in the eighties when John Aitchison reformulates the statistical analysis of compositional data (Aitchison

G. Mateu-Figueras (✉)
Department of Computer Science, Applied Mathematics and Statistics, University of Girona, Girona, Spain
e-mail: gloria.mateu@udg.edu

G. S. Monti
Department of Economics, Management and Statistics, University of Milan-Bicocca, Milan, Italy
e-mail: gianna.monti@unimib.it

J. J. Egozcue
Department of Applied Mathematics III, Technical University of Catalonia, Barcelona, Spain
e-mail: juan.jose.egozcue@upc.es

P. Filzmoser et al. (eds.), *Advances in Compositional Data Analysis*,
https://doi.org/10.1007/978-3-030-71175-7_4

1986, 1982). Although they are models restricted on the simplex, both distributions are defined using the standard definitions and the standard structure of the real space.

Nowadays we know that the vector space structure of the simplex is different from the classical structure of real space. Further, V. Pawlowsky-Glahn defined a new measure on the simplex compatible with this new space structure. This measure represents an alternative to the classical Lebesgue measure and it opens the door to work with statistical models considering not only the assumed distribution, but also taking into account the measure which is considered as appropriate or natural in this specific sample space.

The purpose of this work is to revise some models and to study how they have been rewritten or adapted to the simplex space structure. Some of the important contributions made by V. Pawlowsky-Glahn in this field are included. This paper is organized as follows. In Sect. 2 we recall some fundamental tools for the statistical analysis of compositional data, in particular the operations on the simplex and the Aitchison measure. In Sect. 3 the Dirichlet and the Scaled-Dirichlet distribution are revisited. Then, a power transformation is added to obtain the shifted-scaled Dirichlet model first presented by Monti et al (2011). This model was early presented by Craiu and Craiu (1969) and called *Generalized Dirichlet.* Sect. 4 is focused on the logratio normal distribution, Sect. 5 presents the logratio t distribution and finally, Sect. 6 lists some other distributions on the simplex with special attention to generalizations of the Dirichlet distribution and generalizations of the logratio normal.

2 Algebra for Compositions

In this section we present the algebra for compositions that includes perturbation and powering operations and the Aitchison measure on the simplex.

2.1 The Vector Space Structure of the Simplex

An appropriate sample space for compositional data, i.e. non-negative bounded-sum data, is the simplex, denoted

$$\mathcal{S}^D = \left\{ \mathbf{x} = (x_1, \dots, x_D),\ x_i > 0,\ \sum_{i=1}^{D} x_i = \kappa \right\},$$

where κ is a constant, usually taken to be one or one hundred.

Given two compositions $\mathbf{x}, \mathbf{y} \in \mathcal{S}^D$ and a scalar $\alpha \in \mathbb{R}$, the internal operation, $\oplus$, called *perturbation*, and the external operation, $\odot$, called *powering*, are defined in Aitchison (1986) as

$$\mathbf{x} \oplus \mathbf{y} = \mathcal{C}(x_1 y_1, \ldots, x_D y_D), \qquad \alpha \odot \mathbf{x} = \mathcal{C}(x_1^{\alpha}, \ldots, x_D^{\alpha}),$$

where the *closure* operator $\mathcal{C}$ normalizes the contained vector by dividing each component by the sum of its components and multiplying them by the constant κ so that, after closure, the components sum to κ. The symbol $\oplus$ emphasizes the analogy with vector addition in real space. Perturbation defines a commutative group structure on the simplex with identity element $(1/D, \ldots, 1/D)$ and inverse $\mathbf{x}^{-1} = \mathcal{C}(1/x_1, \ldots, 1/x_D)$. Powering is analogous to multiplication by scalars in real space. The symbol $\odot$ emphasizes the analogy with multiplication by a scalar in real space.

Furthermore, an inner product $\langle \cdot, \cdot \rangle_a$ can be defined as

$$\langle \mathbf{x}, \mathbf{y} \rangle_a = \sum_{i=1}^{D} \ln \frac{x_i}{\mathrm{g_m}(\mathbf{x})} \ln \frac{y_i}{\mathrm{g_m}(\mathbf{y})},$$

where $\mathrm{g_m}(\mathbf{x})$ denotes the geometric mean of the components of $\mathbf{x}$ (Billheimer et al. 2001; Pawlowsky-Glahn and Egozcue 2001).

With these operations $(\mathcal{S}^D, \oplus, \odot, \langle \cdot, \cdot \rangle_a)$ has a $(D-1)$-dimensional real Euclidean vector space structure. The geometry on the simplex is called *simplicial* or *Aitchison geometry* (Pawlowsky-Glahn and Egozcue 2001), for this reason we use the subindex "a". To complete the metric vector space structure of the simplex we report the expression of the Aitchison distance between two compositions $\mathbf{x}$ and $\mathbf{y}$ in $\mathcal{S}^D$

$$\mathrm{d}_a(\mathbf{x}, \mathbf{y}) = \sqrt{\frac{1}{D} \sum_{i<j} \left(\ln \frac{x_i}{x_j} - \ln \frac{y_i}{y_j} \right)^2}, \tag{1}$$

which is compatible with the above vector space structure of the simplex.

In our discussion we use an orthonormal basis in $\mathcal{S}^D$ for deriving coordinate representations of compositional vectors. Given an orthonormal basis $\{\mathbf{e}_1, \ldots, \mathbf{e}_{D-1}\}$ of $\mathcal{S}^D$, which elements satisfy the conditions $\langle \mathbf{e}_i, \mathbf{e}_i \rangle_a = 1$ and $\langle \mathbf{e}_i, \mathbf{e}_j \rangle_a = 0$, $(i = 1, \ldots, D-1,\ i \neq j)$, a composition $\mathbf{x} \in \mathcal{S}^D$ can be expressed as a linear-combination of the elements of the basis,

$$\mathbf{x} = (\langle \mathbf{x}, \mathbf{e}_1 \rangle_a \odot \mathbf{e}_1) \oplus (\langle \mathbf{x}, \mathbf{e}_2 \rangle_a \odot \mathbf{e}_2) \oplus \ldots \oplus (\langle \mathbf{x}, \mathbf{e}_{D-1} \rangle_a \odot \mathbf{e}_{D-1}).$$

A composition $\mathbf{x}$ can always be represented in a unique way by its coordinates with respect to an orthonormal basis, namely the vector

$$h(\mathbf{x}) = (\langle \mathbf{x}, \mathbf{e}_1 \rangle_a, \cdots, \langle \mathbf{x}, \mathbf{e}_{D-1} \rangle_a) \in \mathbb{R}^{D-1} \tag{2}$$

which is uniquely determined. Importantly, once an orthonormal basis has been chosen, all standard statistical methods can be applied to coordinates and transferred to the simplex preserving their properties (Mateu-Figueras et al. 2011). The vector of coordinates $h(\mathbf{x})$ is often called *isometric logratio* transformed vector and denoted $\operatorname{ilr}(\mathbf{x})$ (Egozcue et al. 2003). In general, one component of $\operatorname{ilr}(\mathbf{x})$ is a log-contrast , the simplex analogue of a linear combination, like

$$\sum_{i=1}^{D} a_i \ln(x_i)\,, \qquad \sum_{i=1}^{D} a_i = 0\,.$$

When the orthonormal basis of the simplex is built using a sequential binary partition (Egozcue and Pawlowsky-Glahn 2005), the ilr coordinates take the form of balances. The term balance is used as the obtained ilr coordinates are normalized logratios of the geometric mean of two groups of different parts. Recently, ilr coordinates, assigning coordinates have been termed orthogonal log-ratio coordinates (olr) after the discussion by Egozcue and Pawlowsky-Glahn (2019) and Martín-Fernández (2019). In this paper, we adopt the notation $h(\mathbf{x})$ just to emphasize that all possible orthonormal coordinates can be used.

Other frequent representations of compositions involve transformations based on logratios, such as the additive logratio (alr) transformation and the centered logratio (clr) introduced by Aitchison (1986, 1982). The alr transformation assigns a vector of coordinates

$$\operatorname{alr}(\mathbf{x}) = \left(\ln\frac{x_1}{x_D}, \ldots, \ln\frac{x_{D-1}}{x_D}\right) \in \mathbb{R}^{D-1}\,, \tag{3}$$

to the composition $\mathbf{x}$, but the coordinates obtained correspond to an oblique basis (Egozcue et al. 2011; Mateu-Figueras et al. 2011).

The clr transformation is defined as

$$\operatorname{clr}(\mathbf{x}) = \left(\ln\frac{x_1}{g_{\mathrm{m}}(\mathbf{x})}, \ldots, \ln\frac{x_D}{g_{\mathrm{m}}(\mathbf{x})}\right) \in \mathbb{R}^{D}\,, \tag{4}$$

but its coefficients are coordinates in a generating system, not coordinates with respect to a basis (Mateu-Figueras et al. 2011).

2.2 The Aitchison Measure on the Simplex

Traditionally but not conveniently, a Lebesgue measure has been used in the simplex. A composition $\mathbf{x}$ in $\mathcal{S}^D$ is represented by $(D-1)$ parts $\mathbf{x}_- = (x_1, x_2, \ldots, x_{D-1})$ living in a $(D-1)$-dimensional space, because the last component of $\mathbf{x}$, x_D, is implicitly equal to κ minus the sum of the others, so that all D components sum to κ. The vector $\mathbf{x}_-$ is considered as an element of $\mathbb{R}^{D-1}$ and the measure in this space is then taken as the Lebesgue measure λ.

An alternative measure on the simplex, denoted as λ_a and called Aitchison measure, which is compatible with the inner vector space structure of the simplex is defined by Pawlowsky-Glahn (2003). The same strategy can be used to define a

Lebesgue type measure on any Euclidean space (Eaton 1983). A simple way to define the Aitchison measure is to translate the Lebesgue measure in the space of orthonormal coordinates into the simplex. Let $\mathbf{e}_i$, $(i = 1, 2, \ldots, D-1)$ be an orthonormal basis of $\mathcal{S}^D$. In the space of coordinates, $\mathbb{R}^{D-1}$, a hyper-parallelepiped R is characterized by two points, say $\mathbf{a} = (a_1, a_2, \ldots, a_{D-1})$ and $\mathbf{b} = (b_1, b_2, \ldots, b_{D-1})$. The hyper-volume (Lebesgue measure) of the hyper-parallelepiped R is the product of the lengths of the edges in the direction of each coordinate, that is

$$\lambda_{D-1}(R) = \prod_{i=1}^{D-1} |b_i - a_i| \,,$$

where the Lebesgue measure on $\mathbb{R}^{D-1}$ has been subscripted with $D-1$ in order to distinguish it from the previous λ. The difference between λ and λ_{D-1} is not the space where they are defined but on how the elements of $\mathbb{R}^{D-1}$ are interpreted: $(D-1)$ parts of a composition, $\mathbf{x}_-$, for λ; and $(D-1)$ coordinates for λ_{D-1}. Using the inverse h^{-1}, the Aitchison measure of a subset $S = h^{-1}(R) \subset \mathcal{S}^D$, with $R \subset \mathbb{R}^{D-1}$, is defined as

$$\lambda_a(S) = \lambda_a(h^{-1}(R)) = \lambda_{D-1}(R) \,.$$

The measure for hyper-parallelepipeds is extended to the whole sigma-algebra following the general measure theory (Ash and Doléans-Dade 2000; Athreya and Lahiri 2006). Invariance with respect to the choice of the orthonormal basis derives from the same property in $\mathbb{R}^{D-1}$.

Consider a probability measure P defined on the simplex. If P is absolutely continuous with respect to λ and λ_a, the Radon-Nikodym derivative of P with respect to each of the two measures is a probability density, and it is denoted $dP/d\lambda$, respectively $dP/d\lambda_a$. Their integrals on a measurable set, $S \subseteq \mathcal{S}^D$, are the corresponding probabilities:

$$\mathrm{P}(S) = \int_S \frac{dP}{d\lambda}\, d\lambda = \int_S \frac{dP}{d\lambda_a}\, d\lambda_a \,.$$

The relationship between the two probability densities, $dP/d\lambda_a$ and $dP/d\lambda$, is given by the chain rule for measures

$$\frac{dP}{d\lambda} = \frac{dP}{d\lambda_a} \cdot \frac{d\lambda_a}{d\lambda} \,,$$

where the Jacobian $d\lambda_a/d\lambda$ describes the relationship between the Lebesgue measure for parts of the simplex and the Aitchison measure. As shown by Van Den Boogart et al. (2010) and Mateu-Figueras et al. (2013), $d\lambda_a/d\lambda$ equals to the Jacobian of h, the one-to-one mapping from $\mathcal{S}^D$ to $\mathbb{R}^{D-1}$. Using the Jacobian of the alr transformation (Aitchison 1986, p. 115) and the matrix relationship between $h(\mathbf{x})$ and $\mathrm{alr}(\mathbf{x})$ (Egozcue et al. 2003) the expression

$$\frac{d\lambda_a}{d\lambda} = \frac{1}{\sqrt{D}\prod_{i=1}^{D} x_i}, \tag{5}$$

is obtained by Mateu-Figueras (2003, p. 53)

Thus, the Aitchison measure λ_a is absolutely continuous with respect to λ. In the next sections the Dirichlet, the logratio normal and the logratio t distributions are analyzed changing the measure from λ to λ_a. The Lebesgue measure on the simplex helps the visualization of a sample on a ternary diagram, as the points are visually distributed following λ. Alternatively, the natural way of plotting isolines of a probability density is to draw $dP/d\lambda_a$. The way to homogenize the visualization of samples and probability densities is using the space of orthogonal coordinates where Lebesgue measure is appropriate for both ends.

2.3 Center and Variability

The representation of the distribution of random compositions by densities defined with respect to λ or λ_a requires a review of the measures of location and variability.

The expected value is defined as the integral of the random vector with respect to its probability measure. We can proceed using the standard way and compute the expected value using the Lebesgue measure, i.e.

$$\mathrm{E}(\mathbf{X}) = \int_S \mathbf{X}\frac{dP}{d\lambda}\,d\lambda\,.$$

We also could use the Aitchison measure

$$\mathrm{E}_a(\mathbf{X}) = \int_S \mathbf{X}\frac{dP}{d\lambda_a}\,d\lambda_a\,.$$

Obviously we will have difficulties to compute this second integral but we can apply the principle of working on coordinates (Mateu-Figueras et al. 2011) and we obtain $\mathrm{E}_a(\mathbf{X}) = h^{-1}(\mathrm{E}(h(\mathbf{X})))$ where $\mathrm{E}(h(\mathbf{X}))$ is an expected value on $\mathbb{R}^{D-1}$ and consequently with respect to the Lebesgue measure λ_{D-1}. Note that to obtain this expected value we only need the probability measure of the vector $h(\mathbf{X})$. Mateu-Figueras et al. (2013), following the work by Eaton (1983) defines the expectation of a random vector $\mathbf{X}$ supported in any constrained sample space as $h^{-1}(\mathrm{E}(h(\mathbf{X})))$, where h is a one-to-one differentiable mapping from the sample space to the real space. In our case, h is the coordinate function.

Importantly, given a random composition $\mathbf{X}$, the values $\mathrm{E}_a(\mathbf{X})$ and $\mathrm{E}(\mathbf{X})$ will be, in general, different. This means that depending on the measure which we consider, our measure of location will change. As noted by Mateu-Figueras et al. (2011) the value $\mathrm{E}_a(\mathbf{X})$ equals to *the so called center*, an alternative measure of location defined by Aitchison (1997) as a minimisation of Kullback-Leibler divergence from the center to the random composition. Aitchison introduced the center because he perceived that the usual expected value E(**X**) does not provide a meaningful measure of central location. The geometric interpretation of the expected value and the philosophy introduced by Fréchet (1948) were then used to define it as the *metric center* (Pawlowsky-Glahn and Egozcue 2001, 2002). Importantly, the metric center or center equals to the expected value with respect to the Aitchison distance.

A global measure of variability with respect to the Aitchison measure λ_a is the *metric variance* (Pawlowsky-Glahn and Egozcue 2001, 2002), equivalent to the concept of *total variance* (Aitchison 1997). Given $\mathbf{X} \in \mathcal{S}^D$, the dispersion or metric variance around $\mathrm{E}_a(\mathbf{X})$ is the expected value of the squared distance between $\mathbf{X}$ and $\mathrm{E}_a(\mathbf{X})$

$$\mathrm{Mvar}(\mathbf{X}) = \mathrm{E}\left(\mathrm{d}_a^2\left(\mathbf{X}, \mathrm{E}_a(\mathbf{X})\right)\right) ,$$

where d_a is the Aitchison distance defined in (Eq. 1).

The metric variance can also be computed using orthonormal coordinates, $h(\mathbf{X})$, and the Euclidean distance in $\mathbb{R}^{D-1}$, (Pawlowsky-Glahn and Egozcue 2001) as

$$\mathrm{Mvar}(\mathbf{X}) = \mathrm{E}\left(\mathrm{d}^2\left(\mathbf{X}, \mathrm{E}(h(\mathbf{X}))\right)\right) .$$

This equality also holds using the clr coordinates but not using the alr coordinates because they are coordinates with respect to an oblique basis and the distances are not preserved.

Aitchison (1997) defines the total variance as the trace of the matrix $\boldsymbol{\Gamma} = \mathrm{Cov}(\mathrm{clr}(\mathbf{X}))$. As the total variance equals to the metric variance (Pawlowsky-Glahn and Egozcue 2001) we can see that

$$\mathrm{Mvar}(\mathbf{X}) = \mathrm{trace}\,(\boldsymbol{\Gamma}) = \mathrm{trace}\,(\boldsymbol{\Upsilon}) = \mathrm{trace}\,(\boldsymbol{\Sigma}) - D^{-1}\mathbf{j}'\boldsymbol{\Sigma}\mathbf{j} . \qquad (6)$$

where $\boldsymbol{\Gamma} = \mathrm{Cov}(\mathrm{clr}(\mathbf{X}))$, $\boldsymbol{\Upsilon} = \mathrm{Cov}(\mathrm{ilr}(\mathbf{X}))$ and $\boldsymbol{\Sigma} = \mathrm{Cov}(\mathrm{alr}(\mathbf{X}))$ and $\mathbf{j}$ represents a column vector of units. The equalities in (6) can be easily proved using the following matrix relationship between $\boldsymbol{\Gamma}, \boldsymbol{\Upsilon}$ and $\boldsymbol{\Sigma}$.

$$\boldsymbol{\Gamma} = \mathbf{F}^{\star}\boldsymbol{\Sigma}\mathbf{F}^{\star\prime} ,$$
$$\boldsymbol{\Upsilon} = \left(\mathbf{U}'\mathbf{F}^{\star}\right)\boldsymbol{\Sigma}\left(\mathbf{U}'\mathbf{F}^{\star}\right)' = \mathbf{U}'\boldsymbol{\Gamma}\mathbf{U} ,$$

where $\mathbf{U}$ is a $(D, D-1)$-matrix with column vectors $\mathbf{u}_i = \mathrm{clr}\,(\mathbf{e}_i)$, $i = 1, \ldots, (D-1)$, and $\mathbf{F}^{\star}$ is the $(D, D-1)$-matrix

$$\mathbf{F}^{\star} = \frac{1}{D}\begin{bmatrix} D-1 & -1 & \ldots & -1 \\ -1 & D-1 & \ldots & -1 \\ \vdots & \vdots & \ddots & \vdots \\ -1 & -1 & \ldots & D-1 \\ -1 & -1 & \ldots & -1 \end{bmatrix} .$$

3 The Dirichlet Distribution and Some Generalizations Using $\oplus$ and $\odot$

Historically the prevalent model for compositional data was the Dirichlet distribution (Gupta and Richards 2001). As the multivariate version of the Beta distribution, the Dirichlet model is defined for random vectors restricted to the $\mathcal{S}^D$. It has been widely used in geology, biology, and chemistry for handling compositional data such as rates, percentages and proportions. For example, one may be interested in modeling the proportions of a household budget, the proportions of nucleotides in a DNA sequence, the proportions of time people spend a day in different activities among work, sport and extra work, or the impact of the market share distribution on industry performance. In Bayesian statistics, the Dirichlet distribution is the conjugate prior of the probabilities of the multinomial distribution for counts (Kotz et al. 2000; Wong 1998).

Unfortunately, the Dirichlet family is not rich enough to describe the variability of compositional data in many applications. The drawbacks of this distribution are widely described by Aitchison (1986). They include, among others, a completely negative correlation structure and a very strong independence structure. This last drawback has prompted a number of authors to search for generalizations with less independence structure (Connor and Mosimann 1969; Darroch and James 1974; James 1981; James and Mosimann 1980; Patil 1975; Rayens and Srinivasan 1994) but, as it is stated by Aitchison (1986), "Their efforts have been met with only limited success ...".

One of the generalizations was the *scaled Dirichlet* distribution, which is just a perturbation (shift) in the simplex and thus belongs to the same class of distributions as the Dirichlet one (Monti et al. 2011b). According to the vector-space structure of the simplex (Aitchison et al. 2002; Billheimer et al. 2001; Pawlowsky-Glahn and Egozcue 2001), a power transformation is added by Monti et al (2011) and a further generalization called *shifted-scaled* Dirichlet distribution is obtained.

3.1 Dirichlet Distribution

The definition of the Dirichlet distribution can be found in Kotz et al. (2000); Wilks (1962), and Ng et al. (2011).

Definition 1 (*Dirichlet*) A random composition $\mathbf{X} \in \mathcal{S}^D$ has a D-variate Dirichlet distribution with parameter $\boldsymbol{\alpha} = (\alpha_1, \ldots, \alpha_D) \in \mathbb{R}_+^D$, and denoted as $\mathbf{X} \sim \mathcal{D}^D(\boldsymbol{\alpha})$, if it has density function

$$f(\mathbf{x}) = \frac{dP}{d\lambda}(\mathbf{x}) = \frac{\Gamma(\alpha_+)}{\prod_{i=1}^D \Gamma(\alpha_i)} \prod_{i=1}^D x_i^{\alpha_i - 1}, \quad \mathbf{x} \in \mathcal{S}^D, \tag{7}$$

where P is the Dirichlet probability measure, $\alpha_+ = \sum_{i=1}^{D} \alpha_i$, and Γ denotes the Euler gamma function.

Property 1 (Genesis) *The Dirichlet distribution can be obtained by normalizing a set of independent, equally scaled gamma random variables. Formally, if $W_i \sim Ga(\alpha_i, \beta)$, $i = 1, \ldots, D$, then $\mathbf{X} = \mathcal{C}(\mathbf{W}) \sim \mathcal{D}^D(\boldsymbol{\alpha})$ where $\mathcal{C}$ means the closure operation.*

It is thus clear that the variables in a Dirichlet random composition exhibit strong conditional independence relationships. Moreover, as pointed out by Rayens and Srinivasan (1994), the Dirichlet distribution admits a strong form of independence structure called complete subcompositional independence (Aitchison 1982). The subcompositions formed from any partition of the composition form an independent set. One implication of this property is that each ratio X_i/X_j is independent of any other ratio X_k/X_l for any $i \neq j \neq k \neq l$ (Aitchison 1986).

Equation (7) corresponds to a classical density, i.e. it is a Radon-Nikodym derivative with respect to the Lebesgue measure λ in $\mathcal{S}^D$ considering $(D-1)$ parts. Using (5) we can change the measure and express this density with respect to the measure λ_a (Mateu-Figueras and Pawlowsky-Glahn 2005). The resulting expression for the Dirichlet density function is

$$f_a(\mathbf{x}) = \frac{dP}{d\lambda_a}(\mathbf{x}) = \frac{\Gamma(\alpha_+)\sqrt{D}}{\prod_{i=1}^{D} \Gamma(\alpha_i)} \prod_{i=1}^{D} x_i^{\alpha_i} .$$

This change of measure gives us exactly the same law of probability but it produces changes in some characteristic values of the distribution.

Property 2 (Location) *Let $\mathbf{X} \sim \mathcal{D}^D(\boldsymbol{\alpha})$. The mode and the mean of a Dirichlet r.v., with respect to the measures λ and λ_a, are:*

$$\text{mode}(\mathbf{X}) = \left(\frac{\alpha_1 - 1}{\alpha_+ - D}, \ldots, \frac{\alpha_D - 1}{\alpha_+ - D} \right) ,$$
$$\text{mode}_a(\mathbf{X}) = \left(\frac{\alpha_1}{\alpha_+}, \ldots, \frac{\alpha_D}{\alpha_+} \right) ,$$

$$\mathrm{E}(\mathbf{X}) = \left(\frac{\alpha_1}{\alpha_+}, \ldots, \frac{\alpha_D}{\alpha_+} \right) ,$$
$$\mathrm{E}_a(\mathbf{X}) = \mathcal{C}\left(e^{\psi(\alpha_1)}, \ldots, e^{\psi(\alpha_D)} \right) ,$$

where $\psi(t) = d \ln \Gamma(t)/dt$ is the digamma function (Abramovitz and Stegun 1965) and $\mathcal{C}$ is the closure operator.

See Monti et al. (2011b) for a complete proof.

Note that the mean $\mathrm{E}(\mathbf{X})$ coincides with the normalized parameter vector $\boldsymbol{\alpha}$, the same as the mode of $\mathbf{X}$ with respect to the Aitchison measure. Also, α_+ may be regarded as a precision parameter, as when it increases, the distributions become more tightly concentrated around the mean. With respect to the Lebesgue measure, when the components of $\boldsymbol{\alpha}$ are all greater than 1, the density has a single mode; otherwise when the components of $\boldsymbol{\alpha}$ are all less than 1, the density is unbounded when approaching the boundary of the simplex. As $\alpha_i \to 1$ $(i = 1, \ldots, D)$ their sum tends to D and the Dirichlet probability function tends to a constant function on the simplex.

In the literature we can find the expression of the covariance and correlation between components of the Dirichlet composition computed using the density with respect to the Lebesgue measure (Ng et al. 2011). But in compositional data analysis the metric or total variance deserves special attention (Aitchison 1986; Pawlowsky-Glahn and Egozcue 2001).

Property 3 (Metric variance) *Given a random composition* $\mathbf{X} \sim \mathcal{D}^D(\boldsymbol{\alpha})$ *defined in the unit simplex* $\mathcal{S}^D$ *the metric variance of* $\mathbf{X}$ *is*

$$\mathrm{Mvar}(\mathbf{X}) = \frac{D-1}{D}\left(\psi'(\alpha_1) + \ldots + \psi'(\alpha_D)\right), \tag{8}$$

where $\psi'(t)$, $(t > 0)$ *is the trigamma function (Abramovitz and Stegun 1965).*

Proof To prove this result, we use that $\mathrm{Mvar}(\mathbf{X}) = \mathrm{trace}\,(\boldsymbol{\Sigma}) - D^{-1}\mathbf{j}'\boldsymbol{\Sigma}\mathbf{j}$. As provided by Aitchison (1986, p. 60)

$$\begin{aligned} \mathrm{Var}\left(\ln\frac{X_i}{X_j}\right) &= \psi'(\alpha_i) + \psi'(\alpha_j), \\ \mathrm{Cov}\left(\ln\frac{X_i}{X_k}, \ln\frac{X_j}{X_k}\right) &= \psi'(\alpha_k). \end{aligned} \tag{9}$$

Using (9) we see that the diagonal elements of matrix $\boldsymbol{\Sigma}$ equals to $\psi'(\alpha_i) + \psi'(\alpha_D)$ for $i = 1, \ldots, D-1$ and all off-diagonal elements equals to $\psi'(\alpha_D)$. Then, it is direct to obtain that $\mathrm{trace}\,(\boldsymbol{\Sigma}) = \psi'(\alpha_1) + \cdots + \psi'(\alpha_{D-1}) + (D-1)\psi'(\alpha_D)$ and $D^{-1}\mathbf{j}'\boldsymbol{\Sigma}\mathbf{j} = D^{-1}\left(\psi'(\alpha_1) + \cdots + \psi'(\alpha_{D-1}) + (D-1)^2\psi'(\alpha_D)\right)$. □

Densities of the multidimensional exponential families take the form

$$f(\mathbf{x}) = b(\mathbf{x}) \exp\left(\sum_{i=1}^{l} \theta_i(\boldsymbol{\omega}) t_i(\mathbf{x}) - c(\boldsymbol{\omega})\right) \tag{10}$$

where $\theta(\cdot)$ and $t(\cdot)$ are functions of $\boldsymbol{\omega}$ and $\mathbf{x}$ respectively and $\boldsymbol{\omega}$ is the vector of parameters. Many of the commonly used distributions, the Dirichlet among them, belong to an exponential family. We recall here the well known result about the expression of the Dirichlet density in terms of an exponential density

$$f(\mathbf{x}) = \exp\left(\sum_{i=1}^{D}(\alpha_i - 1)\ln x_i - c(\boldsymbol{\alpha})\right),$$

where $c(\boldsymbol{\alpha}) = \ln\Gamma(\alpha_+) - \sum_{i=1}^{D}\ln\Gamma(\alpha_i)$. An equivalent expression can be obtained using the density with respect to λ_a measure.

3.2 *Scaled Dirichlet Distribution*

In a Bayesian analysis of a multinomial outcomes it is necessary to choose a suitable prior from the set of all probability measures on the simplex. Conjugate analysis leads to the Dirichlet family as a natural choice. But while this way is mathematically convenient, it is not the most appropriate solution; e.g. it does not take into account relative positions between categories or multinomial cells (Lochner 1975). In this framework Savage, in a personal and unpublished work, presented a transformed Dirichlet density (Dichey 1968) as an alternative to the Dirichlet density, taking

$$\tilde{x}_i = \frac{p_i x_i}{\sum_{i=1}^{D} p_i x_i}, \qquad x_i = \frac{\tilde{x}_i / p_i}{\sum_{i=1}^{D}(\tilde{x}_i / p_i)},$$

where $\mathbf{X} \sim \mathcal{D}^D(\boldsymbol{\alpha})$ and each $p_i > 0$. This kind of generalization allows some more flexibility in the choice of a prior in multinomial problems. The probability density function of $\tilde{\mathbf{X}}$ is the scaled Dirichlet distribution (Monti et al. 2011b).

Definition 2 (*Scaled Dirichlet distribution*) A random composition $\mathbf{X} \in \mathcal{S}^D$ has a scaled Dirichlet distribution with parameters $\boldsymbol{\alpha} = (\alpha_1, \ldots, \alpha_D)$ and $\boldsymbol{\beta} = (\beta_1, \ldots, \beta_D) \in \mathbb{R}_+^D$ if its density function is

$$f_s(\mathbf{x}) = \frac{dP_s}{d\lambda}(\mathbf{x}) = \frac{\Gamma(\alpha_+)}{\prod_{i=1}^{D}\Gamma(\alpha_i)} \frac{\prod_{i=1}^{D}\beta_i^{\alpha_i} x_i^{\alpha_i - 1}}{(\sum_{i=1}^{D}\beta_i x_i)^{\alpha_+}}, \quad \mathbf{x} \in \mathcal{S}^D, \tag{11}$$

where P_s is the scaled Dirichlet probability measure, $\alpha_+ = \sum_{i=1}^{D}\alpha_i$, and Γ denotes the gamma function. This distribution will be denoted as $\mathbf{X} \sim \mathcal{SD}^D(\boldsymbol{\alpha}, \boldsymbol{\beta})$.

Note that the number of parameters is $2D$. If $\beta_1 = \cdots = \beta_D$ then the scaled Dirichlet distribution is identical to a Dirichlet one with parameter $\boldsymbol{\alpha}$, thus the nature of this generalization is obvious.

Property 4 (Genesis) *The scaled Dirichlet distribution can be obtained by normalizing a composition of D independent, scaled, gamma r.v.s* $W_i \sim Ga(\alpha_i, \beta_i)$, $i = 1, 2, \ldots, D$. *Formally, if* $\mathbf{X} = \mathcal{C}(\mathbf{W})$ *then* $\mathbf{X} \sim \mathcal{SD}^D(\boldsymbol{\alpha}, \boldsymbol{\beta})$.

Note that the scaled Dirichlet distribution is obtained removing the requirement of equal scaled parameters for the gamma r.v.s. in Property 1. Thus, it has the same strong independence structure. At this point, if we take into account the algebraic geometric structure of $\mathcal{S}^D$, we can see that the scaled Dirichlet distribution can also be obtained starting from a perturbed random composition with a Dirichlet density (Monti et al. 2011b). In fact, let $\mathbf{X} \sim \mathcal{D}^D(\boldsymbol{\alpha})$ be a random composition defined in $\mathcal{S}^D$, and let $\mathbf{p} \in \mathcal{S}^D$ be a composition. The random composition $\widetilde{\mathbf{X}} = \mathbf{p} \oplus \mathbf{X}$ has distribution $\mathcal{SD}^D(\boldsymbol{\alpha}, \boldsymbol{\beta} = \boldsymbol{p}^{-1})$. Note that in this case the $\boldsymbol{\beta}$ parameter is a composition and the number of parameters of the model is $2D - 1$. The consequence is that the parameter space $\boldsymbol{\beta} \in \mathbb{R}_+^D$ specified in Definition 2 should be substituted by $\boldsymbol{\beta} \in \mathcal{S}^D$, because proportional $\boldsymbol{\beta}$'s lead to equal distributions. Thus, the scaled Dirichlet density could be viewed just as a translation (shift) of a Dirichlet density in the simplex. For this reason, the term shifted Dirichlet distribution is also used in the literature (Monti et al. 2011b). It justifies why the Dirichlet and the scaled Dirichlet belong to the same class of distributions and they are not from two different classes.

We can express Eq. (11) with respect to the measure λ_a in the same way we did for the Dirichlet distribution,

$$f_{sa}(\mathbf{x}) = \frac{dP_s}{d\lambda_a}(\mathbf{x}) = \frac{\Gamma(\alpha_+)\sqrt{D}}{\prod_{i=1}^D \Gamma(\alpha_i)} \frac{\prod_{i=1}^D (\beta_i x_i)^{\alpha_i}}{(\sum_{i=1}^D \beta_i x_i)^{\alpha_+}} .$$

Again, this change of measure does not modify the law of probability but produces changes in moments and characteristic measures.

Property 5 (Location) *The mode and the expected value of* $\widetilde{\mathbf{X}} \sim \mathcal{SD}^D(\boldsymbol{\alpha}, \boldsymbol{\beta})$ *with respect to the measure* λ_a *are, respectively,*

$$\begin{aligned} \text{mode}_a(\widetilde{\mathbf{X}}) &= (\ominus\boldsymbol{\beta}) \oplus \text{mode}_a(\mathbf{X}) , \\ \text{E}_a(\widetilde{\mathbf{X}}) &= (\ominus\boldsymbol{\beta}) \oplus \text{E}_a(\mathbf{X}) , \end{aligned} \tag{12}$$

where $\mathbf{X} \sim \mathcal{D}^D(\boldsymbol{\alpha})$ *and* $\ominus$ *is the inverse operation of the perturbation, by analogy with standard operations in real space.*

Proof Given that $\widetilde{\mathbf{X}} = (\ominus\boldsymbol{\beta}) \oplus \mathbf{X}$, where $\mathbf{X} \sim \mathcal{D}^D(\boldsymbol{\alpha})$, the expressions for the mode and the expected value of $\widetilde{\mathbf{X}}$ can be obtained using the linearity of these operators. □

For a composition $\widetilde{\mathbf{X}} \sim \mathcal{SD}^D(\boldsymbol{\alpha}, \boldsymbol{\beta})$, there is no closed form for $\text{E}(\widetilde{\mathbf{X}})$ with respect to the Lebesgue measure λ in real space. It is necessary to use numerical integration to compute it. The $\text{mode}(\widetilde{\mathbf{X}})$ can be computed solving a system of equations (Ng et al. 2011, p. 262)

Equation (12) shows how $\boldsymbol{\beta}$ operates in a simplicial way. The vector of parameters $\boldsymbol{\beta}$ acts on the location of $\widetilde{\mathbf{X}}$, and not on the scale; i.e. it is not related to the measure of scale. Moreover, for $\boldsymbol{\alpha}$ a vector with equal components ($\boldsymbol{\alpha} = (\alpha, \alpha, \dots, \alpha)$), the mean and the mode with respect to the Aitchison measure coincide and are the neutral element in the simplex.

Property 6 (Metric variance) *Given* $\widetilde{\mathbf{X}} \sim \mathcal{SD}^D(\boldsymbol{\alpha}, \boldsymbol{\beta})$, *then* $\text{Mvar}(\widetilde{\mathbf{X}}) = \text{Mvar}(\mathbf{X})$ *for* $\mathbf{X} \sim \mathcal{D}^D(\boldsymbol{\alpha})$.

Proof The metric variance is invariant under perturbation (Aitchison 2002; Pawlowsky-Glahn and Egozcue 2002). As $\widetilde{\mathbf{X}} = (\ominus\boldsymbol{\beta}) \oplus \mathbf{X}$, the value given in Eq. (8) is obtained. This property is equivalent to invariance under translation in real space. □

There is no closed form for the covariance matrix of a scaled Dirichlet with respect to the Lebesgue measure λ for $(D-1)$ parts of the composition. It is necessary to use numerical integration to compute it.

Monti et al. (2011b) show that only the subfamily obtained fixing the parameter vector $\boldsymbol{\beta} = (\beta, \ldots, \beta)$ belongs to the exponential family of the Dirichlet distribution. Note that in this case the scaled Dirichlet model reduces to a Dirichlet one.

3.3 *Shifted Scaled Dirichlet Distribution*

The shifted-scaled Dirichlet distribution is introduced by Monti et al (2011) as the distribution obtained after applying the perturbation and powering operations to a Dirichlet random composition, i.e, the density of the composition $\widetilde{\mathbf{X}} = \mathbf{p} \oplus (a \odot \mathbf{X})$, with $\boldsymbol{p} \in \mathcal{S}^D, a \in \mathbb{R}_+$ and $\mathbf{X} \sim \mathcal{D}^D(\boldsymbol{\alpha})$. The probability density function was derived by Craiu and Craiu (1969) and called *Generalized Dirichlet.* This model has been recently used by Alsuroji et al. (2018) for studying a finite mixture model. Recently and using a basis of independent generalized Gamma distributions, Graf (2020) obtains this model called *simplicial generalized Beta,* and uses it to present a novel regression method for compositions.

Definition 3 (*Shifted-scaled Dirichlet distribution*) A random composition $\mathbf{X} \in \mathcal{S}^D$ has a shifted-scaled Dirichlet distribution with parameters $\boldsymbol{\alpha} = (\alpha_1, \ldots, \alpha_D) \in \mathbb{R}_+^D$, $a \in \mathbb{R}_+$, $\boldsymbol{p} = (p_1, \ldots, p_D) \in \mathcal{S}^D$ if its density function is

$$f_{ss}(\mathbf{x}) = \frac{dP_{ss}}{d\lambda}(\mathbf{x}) = \frac{\Gamma(\alpha_+)}{\prod_{i=1}^D \Gamma(\alpha_i)} \frac{1}{a^{D-1}} \frac{\prod_{i=1}^D p_i^{-(\alpha_i/a)} x_i^{(\alpha_i/a)-1}}{\left(\sum_{i=1}^D (x_i/p_i)^{(1/a)}\right)^{\alpha_+}}, \tag{13}$$

where P_{ss} is the shifted-scaled Dirichlet probability measure, $\alpha_+ = \sum_{i=1}^D \alpha_i$, and Γ is the gamma function. This distribution will be denoted $\mathbf{X} \sim \mathcal{SSD}^D(\boldsymbol{\alpha}, \boldsymbol{p}, a)$.

The number of parameters is $2D$. Furthermore, for $a = 1$ a scaled Dirichlet model with parameters $\boldsymbol{\alpha}$ and $\boldsymbol{\beta} = \ominus\boldsymbol{p}$ is obtained. If $a = 1$ and the composition $\boldsymbol{p} = \mathcal{C}(p, p, \ldots, p)$ for some constant p, we obtain a Dirichlet model because they correspond to the neutral elements with respect to the corresponding operations.

We can express Eq. (13) with respect to the measure λ_a, in the same way we did for the scaled Dirichlet density,

$$f_{ssa}(\mathbf{x}) = \frac{dP_{ss}}{d\lambda_a}(\mathbf{x}) = \frac{\sqrt{D}\Gamma(\alpha_+)}{\prod_{i=1}^D \Gamma(\alpha_i)} \frac{1}{a^{D-1}} \frac{\prod_{i=1}^D (x_i/p_i)^{(\alpha_i/a)}}{\left(\sum_{i=1}^D (x_i/p_i)^{(1/a)}\right)^{\alpha_+}}. \tag{14}$$

Property 7 (Location) *The mode and the expected value of* $\widetilde{\mathbf{X}} \sim \mathcal{SSD}^D(\boldsymbol{\alpha}, \boldsymbol{p}, a)$ *with respect to the measure* λ_a *are, respectively,*

$$\text{mode}_a(\widetilde{\mathbf{X}}) = (\boldsymbol{p}) \oplus (a \odot \text{mode}_a(\mathbf{X})) ,$$
$$\text{E}_a(\widetilde{\mathbf{X}}) = (\boldsymbol{p}) \oplus (a \odot \text{E}_a(\mathbf{X})) ,$$

where $\mathbf{X} \sim \mathcal{D}^D(\boldsymbol{\alpha})$.

Proof Given that $\widetilde{\mathbf{X}} = \mathbf{p} \oplus (a \odot \mathbf{X})$, where $\mathbf{X} \sim \mathcal{D}^D(\boldsymbol{\alpha})$, the expressions for the mode and the expected value of $\widetilde{\mathbf{X}}$ can be directly obtained using the linearity of these operators. □

Property 8 (Metric variance) *Given* $\widetilde{\mathbf{X}} \sim \mathcal{SSD}^D(\boldsymbol{\alpha}, \boldsymbol{p}, a)$, *then* $\text{Mvar}(\widetilde{\mathbf{X}}) = a^2\text{Mvar}(\mathbf{X})$ *for* $\mathbf{X} \sim \mathcal{D}^D(\boldsymbol{\alpha})$.

Proof Given that $\widetilde{\mathbf{X}} = \mathbf{p} \oplus (a \odot \mathbf{X})$, where $\mathbf{X} \sim \mathcal{D}^D(\boldsymbol{\alpha})$, the expressions for the metric variance of $\widetilde{\mathbf{X}}$ can also be directly obtained using the linearity of these operators (Aitchison 2002; Pawlowsky-Glahn and Egozcue 2002). □

Property 9 *The shifted-scaled Dirichlet density has complete permutation symmetry.*

Proof Any X_i and X_j $(i, j = 1, \ldots, D)$ may be interchanged, as long as the corresponding α_i, α_j, and p_i, p_j, are interchanged at the same time. As can be seen in Eqs. (13) and (14), the density is the same, independently of the measure of reference. □

Note that this property also holds for the Dirichlet and Scaled Dirichlet because they are particular cases.

Property 10 (Exponential family representation) *The* $\widetilde{\mathbf{X}} \sim \mathcal{SSD}^D(\boldsymbol{\alpha}, \mathbf{p}, a)$ *does not have an exponential form.*

Proof The density function with respect to λ or λ_a, cannot take the form given in (10). The term in the denominator $\left(\sum_{i=1}^{D} (x_i/p_i)^{(1/a)}\right)^{\alpha_+}$ or, taking logarithm, $\alpha_+ \ln\left(\sum_{i=1}^{D} (x_i/p_i)^{(1/a)}\right)$ cannot be factorized as a linear function of $t(\mathbf{x})$ and $\omega(\boldsymbol{\alpha}, \mathbf{p}, a)$. □

Sometimes a family is exponential only if some of its parameters are held fixed. This is the case of the shifted-scaled Dirichlet distribution, when $a = 1$ and $\boldsymbol{p} = \mathcal{C}(p, p, \ldots, p)$ for some constant p, resulting in the Dirichlet distribution. This is not true for the scaled Dirichlet distribution as it is not exponential.

The influence of the perturbation and power operations on a Dirichlet distribution can be observed in Fig. 1 where isodensity curves of a Dirichlet, scaled Dirichlet and shifted-scaled Dirichlet distribution with parameter $\boldsymbol{\alpha} = (1.5, 1.5, 1.5)$, $\boldsymbol{\beta} = \mathcal{C}(\mathbf{p}^{-1}) = (0.75, 0.15, 0.12)$ and $a = 1.5$ on the simplex with respect Aitchison measure (top) and in ilr-coordinates (bottom) are plotted. Coordinate expression is much more helpful to understand the effect of the perturbation (a shift) and power (a scaling) operations on a Dirichlet distribution.

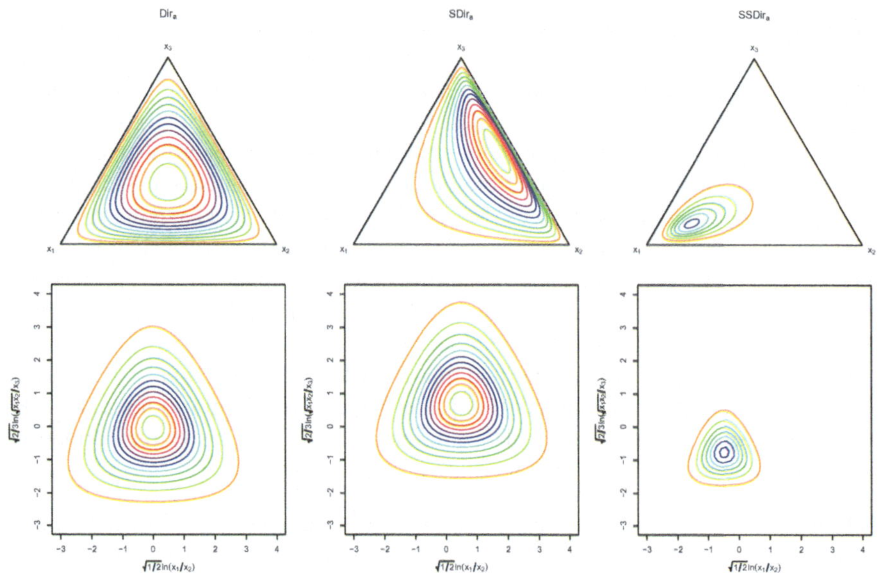

Fig. 1 Isodensity levels of a Dirichlet, scaled Dirichlet and shifted-scaled Dirichlet distribution with parameter $\boldsymbol{\alpha} = (1.5, 1.5, 1.5)$, $\boldsymbol{\beta} = (0.75, 0.15, 0.12)$ and $a = 1.5$ on the simplex with respect Aitchison measure (top) and in ilr-coordinates (bottom)

4 The Logratio Normal Distribution

The objective of this section is to review the Normal distribution on the simplex, also called the logratio normal or logistic normal distribution. Without any doubt, it is the most used distribution for random compositions but all this different terminology may be confusing and a clarification is needed. The objective of this section is not to review in detail all the properties of this model. We only want to focus our attention in the different definitions that had been appeared in the literature and on the measure with respect to which the model is expressed. For the reader interested in the properties of this model, we refer to Mateu-Figueras et al. (2013).

In light of the strong independence structure of the Dirichlet model and some generalizations, an alternative way to define distributions on $\mathcal{S}^D$ is proposed by Aitchison (1986, 1982). By analogy with the log-normal approach, a logratio transformation is applied to map the original D parts into $\mathbb{R}^{D-1}$. Then the multinormal distribution is used to model the transformed vector and, finally, the inverse logistic transformation is used to return to $\mathcal{S}^D$. As the additive logistic transformation (3) was first proposed, the model is called Additive logistic normal (Aitchison 1982). The expression of its density function with respect to λ, the Lebesgue measure in $\mathcal{S}^D$, is provided and some properties are studied by Aitchison (1986). Due to the matrix relationship between the alr and clr transformations, an equivalent model can be defined using the centred logratio vector (4) but a degenerate model is obtained, thus hindering its use. The isomeric logratio transformation (2) was proposed later and the term isometric logistic normal distribution is also adopted for the model. Its density function can be easily obtained using the matrix relationship between the logratio vectors (2)–(4) (Egozcue et al. 2003). From a probabilistic point of view, these three logistic transformations define the same law of probability on $\mathcal{S}^D$.

The Euclidean vector space structure of the simplex summarised in Sect. 2.1, the principle of working on coordinates (Mateu-Figueras et al. 2011) and the Aitchison measure described in Sect. 2.2 are used to define the Normal distribution on the simplex (Mateu-Figueras et al. 2013). In this case we stay in the simplex because the model is defined using the orthonormal coordinates of our random composition but a density with respect to λ_a is finally obtained. To emphasise the idea of staying on the simplex, the model is named as Normal model on $\mathcal{S}^D$ (Mateu-Figueras et al. 2013). Nowadays, there are some authors that prefer to refer it as logratio normal model (Comas-Cufí et al. 2016) to highlight that, instead of using the logistic transformation, some logratio coordinates and the measure λ_a are involved.

Definition 4 (*Logratio normal*) A random composition $\mathbf{X} \in \mathcal{S}^D$ has a logratio-normal distribution with parameters $\boldsymbol{\mu} \in \mathbb{R}^{D-1}$ and $\boldsymbol{\Upsilon}$ if it has density function

$$f_{na}(\mathbf{x}) = \frac{dP_n}{d\lambda_a}(\mathbf{x}) = (2\pi)^{-(D-1)/2}|\boldsymbol{\Upsilon}|^{-1/2} \exp\left(-\frac{1}{2}(h(\mathbf{x}) - \boldsymbol{\mu})'\boldsymbol{\Upsilon}^{-1}(h(\mathbf{x}) - \boldsymbol{\mu})\right), \tag{15}$$

where $h(\cdot)$ stands for the orthonormal coordinates and P_n is the logratio normal probability measure. This distribution will be denoted as $\mathbf{X} \sim \mathcal{N}_{\mathcal{S}}^D(\boldsymbol{\mu}, \boldsymbol{\Upsilon})$.

Note that this is the regular, or non-singular, normal model since the covariance matrix $\boldsymbol{\Upsilon}$ is assumed to be invertible.

Using the inverse of the Jacobian (5) we can change the measure and express this density function with respect to the Lebesgue measure λ in $\mathbb{R}^{D-1}$. The resulting expression is

$$f_n(\mathbf{x}) = \frac{dP_n}{d\lambda}(\mathbf{x}) = \frac{(2\pi)^{-(D-1)/2}|\boldsymbol{\Upsilon}|^{-1/2}}{\sqrt{D}\prod_{i=1}^{D} x_i} \exp\left(-\frac{1}{2}(h(\mathbf{x}) - \boldsymbol{\mu})'\boldsymbol{\Upsilon}^{-1}(h(\mathbf{x}) - \boldsymbol{\mu})\right) \tag{16}$$

We know that any orthonormal coordinates $h(\mathbf{x})$ correspond to a particular isometric logratio transformation, then, the logratio normal law of probability on $\mathcal{S}^D$ equals to the isometric logistic normal law.

Even though the model is widely studied and densities (15) and (16) compared by Mateu-Figueras et al. (2013), we want to emphasize here the main differences as a consequence of the measure.

The effect of change of measure can be observed in the isodensity curves in Fig. 2. The differences are obvious, in particular the bimodality when the Lebesgue measure is considered. On the contrary, considering the Aitchison measure, we observe a unique mode.

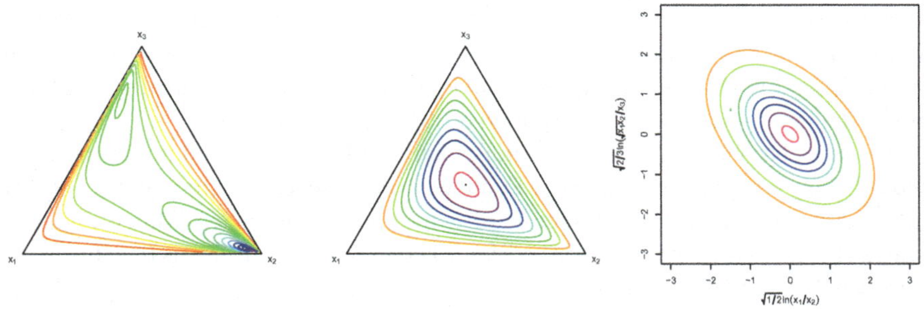

Fig. 2 Isodensity levels of a logratio normal distribution with $\boldsymbol{\mu} = (0, 0)$ and $\boldsymbol{\Upsilon} = \begin{pmatrix} 1 & -0.5 \\ -0.5 & 1 \end{pmatrix}$ with respect Lebesgue and Aitchison measure and in ilr-coordinates (from left to right)

Property 11 (Location) *The mode and the expected value of* $\mathbf{X} \sim \mathcal{N}_{\mathcal{S}}^{D}(\boldsymbol{\mu}, \boldsymbol{\Upsilon})$ *with respect to the measure* λ_a *coincide and are*

$$\text{mode}_a(\mathbf{X}) = \text{E}_a(\mathbf{X}) = h^{-1}(\boldsymbol{\mu}),$$

independently of the orthonormal coordinates $h(\mathbf{x})$.

For a composition $\mathbf{X} \sim \mathcal{N}_{\mathcal{S}}^{D}(\boldsymbol{\mu}, \boldsymbol{\Upsilon})$ there is no closed form for $\text{E}(\mathbf{X})$ with respect to the Lebesgue measure λ. Since the integral expression exists, it is not reducible to any simple form and it is necessary to use numerical integration to compute it (Aitchison 1986). Importantly here to note that, in compositional data analysis, the value $\text{E}(\mathbf{X})$ is not used in practice and the alternative center is recommended. This value coincides with $\text{E}_a(\mathbf{X})$ (Mateu-Figueras et al. 2013).

Property 12 (Metric variance) *Given* $\mathbf{X} \sim \mathcal{N}_{\mathcal{S}}^{D}(\boldsymbol{\mu}, \boldsymbol{\Upsilon})$, *then* $\text{Mvar}(\mathbf{X}) = \text{trace}\,(\boldsymbol{\Upsilon})$.

To finish this section we want to note that densities (15) and (16) are expressed in terms of $h(\mathbf{x})$, the coordinates of our composition $\mathbf{x}$ with respect to an orthonormal basis. Certainly, we could express density (15) in terms of clr coordinates (4) but a degenerate model with a singular covariance matrix is obtained thus, hindering its application. Remember that the clr coordinates have D parts instead of $D - 1$. We could also express our density using alr coordinates (3). In this case the covariance matrix is invertible but its interpretation is misleading. For instance, when the alr covariance matrix is diagonal we cannot interpret it as independent alr coordinates. Remember that vector $\text{alr}(\mathbf{x})$ are the coordinates of $\mathbf{x}$ with respect to an oblique basis.

5 The Logratio t Distribution

Katz and King (1999) proposed an Additive logistic multivariate t distribution to model multiparty electoral data. So the multivariate t distribution is used to model the transformed alr vector (3) and, finally, the inverse logistic transformation is used to return to $\mathcal{S}^D$. The additive logistic multivariate t distribution includes the additive logistic normal as a limiting special case. Here we define a multivariate t distribution on $\mathcal{S}^D$ using this idea and the orthonormal logratio coordinates. By analogy with the model defined in Sect. 4, it can be called as logratio t distribution.

The model is obtained using the orthonormal coordinates of a random composition and a density with respect to λ_a. The Euclidean vector space structure of the simplex summarised in Sect. 2.1, the principle of working on coordinates (Mateu-Figueras et al. 2011) and the Aitchison measure described in Sect. 2.2 allows us to define such isometric logistic t distribution.

Definition 5 (*Logratio t*) A random composition $\mathbf{X} \in \mathcal{S}^D$ has a logratio t distribution with location $\boldsymbol{\mu} \in \mathbb{R}^{D-1}$, shape matrix $\boldsymbol{\Upsilon}$ and ν degrees of freedom, if it has density function

$$f_{ta}(\mathbf{x}) = \frac{dP_{ta}}{d\lambda_a}(\mathbf{x}) = \frac{\Gamma(\frac{\nu+D}{2})|\boldsymbol{\Upsilon}|^{-1/2}}{\Gamma(\nu/2)(\nu\pi)^{D/2}} \left(1 + \frac{1}{\nu}(h(\mathbf{x}) - \boldsymbol{\mu})'\boldsymbol{\Upsilon}^{-1}(h(\mathbf{x}) - \boldsymbol{\mu})\right)^{-\left(\frac{\nu+D}{2}\right)}$$

where $h(\cdot)$ stands for a set of orthonormal coordinates and P_{ta} is the logratio t probability measure. This distribution will be denoted as $\mathbf{X} \sim t_{\mathcal{S}}^D(\boldsymbol{\mu}, \boldsymbol{\Upsilon}, \nu)$.

As in the case of the logratio normal, this is a regular, or non-singular, t model as the existence of the inverse covariance matrix $\boldsymbol{\Upsilon}^{-1}$ is assumed.

Using the inverse of the Jacobian (5) we can change the measure and express this density function with respect to the Lebesgue measure λ in $\mathbb{R}^{D-1}$. The resulting expression is

$$f_t(\mathbf{x}) = \frac{dP_t}{d\lambda}(\mathbf{x}) = \frac{\Gamma(\frac{\nu+D}{2})|\boldsymbol{\Upsilon}|^{-1/2}}{\Gamma(\nu/2)(\nu\pi)^{D/2}\sqrt{D}\prod_{i=1}^{D} x_i} \left(1 + \frac{1}{\nu}(h(\mathbf{x}) - \boldsymbol{\mu})'\boldsymbol{\Upsilon}^{-1}(h(\mathbf{x}) - \boldsymbol{\mu})\right)^{-\left(\frac{\nu+D}{2}\right)}. \tag{17}$$

Any orthonormal coordinates $h(\mathbf{x})$ correspond to a particular isometric logratio vector, then, using the matrix relationship between vectors (2) and (3), we can express density (17) in terms of the alr$(\mathbf{x})$ obtaining the expression of the density proposed by Katz and King (1999).

Property 13 (Location) *The mode and the expected value of* $\mathbf{X} \sim t_{\mathcal{S}}^D(\boldsymbol{\mu}, \boldsymbol{\Upsilon}, \nu)$, *if* $\nu > 1$, *else is undefined, with respect to the measure* λ_a *coincide and are*

$$\text{mode}_a(\mathbf{X}) = \text{E}_a(\mathbf{X}) = h^{-1}(\boldsymbol{\mu}),$$

independently of the orthonormal coordinates $h(\mathbf{x})$.

For a composition $\mathbf{X} \sim t_{\mathcal{S}}^D(\boldsymbol{\mu}, \boldsymbol{\Upsilon}, \nu)$ there is no closed form for $\text{E}(\mathbf{X})$ with respect to the Lebesgue measure λ. If $\text{E}(\mathbf{X})$ exists, numerical integration is required to compute its value.

Property 14 (Metric variance) *Given* $\mathbf{X} \sim t_{\mathcal{S}}^D(\boldsymbol{\mu}, \boldsymbol{\Upsilon}, \nu)$, *when* $\nu > 2$, *then* $\text{Mvar}(\mathbf{X}) = \text{trace}\,(\boldsymbol{\Upsilon})$.

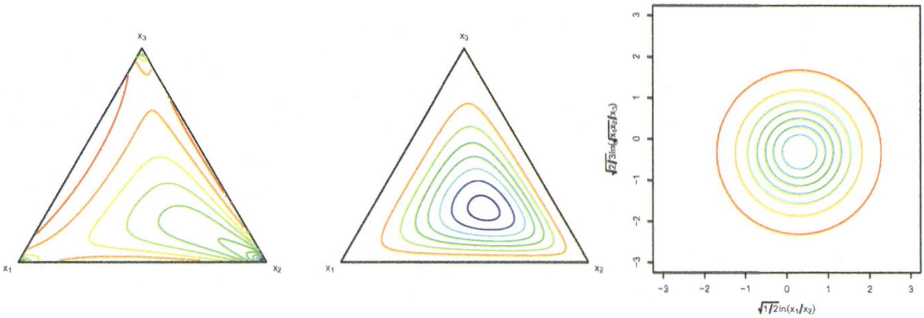

Fig. 3 Isodensity levels of logratio t distributions with $\boldsymbol{\mu} = (0.3, -0.3)$ and $\boldsymbol{\Upsilon} = \mathbf{I}$ and $\nu = 4$ with respect Lebesgue and Aitchison measure and in ilr-coordinates (from left to right)

The effect of change of measure can be observed in the isodensity curves in Fig. 3. The multi-mode is due to the Jacobian, which expresses the ratio of the Lebesgue and the Aitchison measures over the simplex, when the Lebesgue measure is considered. On the contrary, considering the Aitchison measure, we observe a unique mode. This behavior is not exclusive of the logratio t model. Indeed, beta and Dirichlet densities defined under Lebesgue measure are multimodal when their respective parameters tend to 0. Also the logratio normal model exhibits multimodality when the density with respect to the Lebesgue measure is used (see Fig. 2).

6 Other Distributions

Many other distributions on the simplex have been proposed in the literature. Rayens and Srinivasan (1994) indicate that essentially there are two different ways to define distributions for random compositions. To look for a family of distribution that generalizes the Dirichlet model or to look for a family of distribution that generalizes the logratio normal model.

Among others we recall some generalizations of the Dirichlet distribution: the generalized Dirichlet distribution (Dickey 1983; Dickey et al. 1987) useful to handle incomplete categorical data, misclassification data, and screening/diagnostic test data, the hyper-Dirichlet distribution (Hankin 2010) proposed as conjugate to the multinomial distribution when some observations are censored or grouped, the Flexible Dirichlet distribution (Ongaro et al. 2008), constructed normalizing an appropriate set of positive, but dependent, random variables, which allows various dependence structures, the Liouville distribution (Rayens and Srinivasan 1994), and the Connor and Mosimann's distribution (Connor and Mosimann 1969) which is a mixture of Dirichlet distributions.

To generalize the logratio normal model we can use different logratio transformations and the multivariate normal model in the transformed space. Aitchison (1986) proposes the multiplicative logistic transformation and defines the multiplicative logistic normal. Nevertheless this model has few nice properties compared with the logratio normal. The well-known Box-Cox transformation is applied to the ratios of parts in Barceló-Vidal et al. (1996) and has the additive logratio transformation as a limiting case. The resulting model is used to identify outliers. Another possibility to generalize the logratio normal model is to use other multivariate distributions instead of the normal one as for example the multivariate t distribution we saw in Sect. 5. The multivariate skew-normal distribution is also used to model the logratio vector and the skew-normal on the simplex is obtained, useful to fit asymmetrical distributions (Mateu-Figueras and Pawlowsky-Glahn 2007; Mateu-Figueras et al. 2005).

Furthermore the Aitchison distribution (Aitchison 1986), obtained as a posterior distribution in a multinomial sampling, contains both the Dirichlet and the logratio normal distribution as special cases.

Finally, Comas-Cufí et al. (2016) describe and compare the logratio normal and the Dirichlet models to define finite mixture models for compositions.

Acknowledgements This research was supported by the Ministerio de Ciencia, Innovacin y Universidades under the project METhods for COmpositional analysis of DAta (CODAMET) (Ref: RTI2018-095518-B-C21, 2019-2021). GSM acknowledges the financial support by the FAR (Fondo di Ateneo per la Ricerca - University of Milano-Bicocca).

References

M. Abramovitz, I. Stegun, *Handbook of Mathematical Functions* (Dover, New York, 1965), p. 1046

J. Aitchison, Simplicial inference, in *Algebraic Methods in Statistics and Probability*, volume 287 of *Contemporary Mathematics Series*, vol. 287, ed. by M.A.G. Viana, D.S.P. Richards (American Mathematical Society, Providence, 2002), pp. 1–22

J. Aitchison, The one-hour course in compositional data analysis or compositional data analysis is simple, in *Proceedings of the III Annual Conference of the International Association for Mathematical Geology (vol. I)*, ed. by V. Pawlowsky-Glahn (Barcelona, Spain, 1997), pp. 3–35, CIMNE

J. Aitchison, *The Statistical Analysis of Compositional Data*. Monographs on Statistics and Applied Probability (Chapman & Hall Ltd, London). (Reprinted in 2003 with additional material by The Blackburn Press), 1986. 416 p

J. Aitchison, C. Barceló-Vidal, J.J. Egozcue, V. Pawlowsky-Glahn, A concise guide for the algebraic-geometric structure of the simplex, the sample space for compositional data analysis, in *Proceedings of IAMG'02 — The Eight Annual Conference of the International Association for Mathematical Geology*, Vols. I and II (Selbstverlag der Alfred-Wegener-Stiftung, Berlin, 2002), pp. 387–392

J. Aitchison, The statistical analysis of compositional data (with discussion). J. R. Stat. Soc., Ser. B **44**, 139–177 (1982)

R. Alsuroji, N. Zamzami, N. Bouguila, Model selection and estimation of a finite shifted-scaled dirichlet mixture model, in *2018 17th IEEE International Conference on Machine Learning and Applications (ICMLA)* (2018), pp. 707–713

R.B. Ash, C.A. Doléans-Dade, *Probability and Measure Theory*, 2nd edn. (Academic, San Diego, 2000), p. 516

K.B. Athreya, S.N. Lahiri, *Measure Theory and Probability Theory* (Springer, New York, 2006), p. 625

C. Barceló-Vidal, V. Pawlowsky-Glahn, E. Grunsky, Some aspects of transformations of compositional data and the identification of outliers. Math. Geol. **28**, 501–518 (1996). ISSN 0882-8121
D. Billheimer, P. Guttorp, W.F. Fagan, Statistical interpretation of species composition. J. Am. Stat. Assoc. **96**, 1205–1214 (2001)
M. Comas-Cufí, J.A. Martín-Fernández, G. Mateu-Figueras, Log-ratio methods in mixture models for compositional data sets. SORT **40**, 349–374 (2016)
R.J. Connor, J.E. Mosimann, Concepts of independence for proportions with a generalization of the Dirichlet distribution. J. Am. Stat. Assoc. **64**(325), 194–206 (1969)
R.J. Connor, J.E. Mosimann, Concepts of independence for proportions with a generalization of the Dirichlet distribution. J. Am. Stat. Assoc. **64**, 194–206 (1969)
M. Craiu, V. Craiu, Repartitia dirichlet generalizatá. Analele Universitatii Bucuresti, Mathematicá-Mecanicá **18**, 9–11 (1969)
J.N. Darroch, I.R. James, F-independence and null correlations of bounded-sum, positive variables. J. R. Stat. Soc., Ser. B **36**, 467–483 (1974)
J.M. Dichey, Three multidimensional-integral identities with Bayesian applications. Ann. Math. Stat. **39**, 1615–1627 (1968)
J.M. Dickey, Multiple hypergeometric functions: probabilistic interpretations and statistical uses. J. Am. Stat. Assoc. **78**(383), 628–637 (1983)
J.M. Dickey, J.M. Jiang, J.B. Kadane, Bayesian methods for censored categorical data. J. Am. Stat. Assoc. **82**(399), 773–781 (1987)
P.G.L. Dirichlet, Sur une nouvelle méthode pour la détermination des intégrales multiples. Liouville, Journal de Mathématiques, Ser I **4**, 164–168 (1839). Reprinted in Dirichlet's Werke **1**, 377–380
M.L. Eaton, *Multivariate Statistics. A Vector Space Approach* (Wiley, New York, 1983). 512 p
J.J. Egozcue, C. Barceló-Vidal, J.A. Martín-Fernández, E. Jarauta-Bragulat, J.L. Díaz-Barrero, G. Mateu-Figueras, *Elements of Simplicial Linear Algebra and Geometry, Chapter 11* (Wiley, Ltd, New Jersey, 2011), pp. 139–157. ISBN 9781119976462
J.J. Egozcue, V. Pawlowsky-Glahn, Groups of parts and their balances in compositional data analysis. Math. Geol. **37**, 795–828 (2005). ISSN 0882-8121
J.J. Egozcue, V. Pawlowsky-Glahn, Rejoinder on: compositional data: the sample space and its structure. TEST **28**(3), 658–663 (2019). https://doi.org/10.1007/s11749-019-00674-2
J.J. Egozcue, V. Pawlowsky-Glahn, G. Mateu-Figueras, C. Barceló-Vidal, Isometric logratio transformations for compositional data analysis. Math. Geol. **35**, 279–300 (2003). ISSN 0882-8121
M. Fréchet, Les éléments aléatoires de nature quelconque dans un espace distancié. Annales de L'Institut Henri Poincaré **10**, 215–308 (1948)
M. Graf, Regression for compositions based on a generalization of the dirchlet distribution. Stat. Methods Appl. 658–663 (2020). https://doi.org/10.1007/s10260-020-00512-y
R.D. Gupta, D.S.P. Richards, The history of the Dirichlet and Liouville distributions. Int. Stat. Rev. **69**, 433–446 (2001)
R.K.S. Hankin, A generalization of the Dirichlet distribution. J. Stat. Softw. **33**(11), 1–13 (2010)
I.R. James, Distributions associated with neutrality properties for random proportions, in *Statistical Distributions in Scientific Work—Models, Structures, and Characterizations*, ed. by C. Taillie, G.P. Patil, B.A. Baldessari (D. Reidel Publishing Co., Dordrecht, 1981), pp. 125–136
I.R. James, I.R. Mosimann, A new characterisation of the Dirichlet distribution through neutrality. Ann. Stat. **8**, 183–189 (1980)
J.N. Katz, G. King, A statistical model for multiparty electoral data. Am. Polit. Sci. Rev. **93**(1), 15–32 (1999)
S. Kotz, N. Balakrishnan, N.L. Johnson, *Continuous Multivariate Distributions: Models and Applications*. Wiley Series in Probability and Statistics, vol. 1, 2nd edn. (2000)
R.H. Lochner, A Generalized Dirichlet distribution in Bayesian life testing. J. R. Stat. Soc. Ser. B (Methodol.) **37**, 103–113 (1975)
J.A. Martín-Fernández, Comments on: Compositional data: the sample space and its structure, by egozcue and pawlowsky-glahn. TEST **28**(3):(in press) (2019). ISSN 1133-0686. https://doi.org/10.1007/s11749-019-00670-4

G. Mateu-Figueras, *Models de distribució sobre el símplex* (Universitat Politècnica de Catalunya, Ph.D. Diss., 2003), p. 202
G. Mateu-Figueras, V. Pawlowsky-Glahn, The Dirichlet distribution with respect to the Aitchison measure on the simplex — a first approach, in *Compositional Data Analysis Workshop – CoDaWork'05, Proceedings*, ed. by G. Mateu-Figueras, C. Barceló-Vidal (Universitat de Girona, 2005)
G. Mateu-Figueras, V. Pawlowsky-Glahn, J.J. Egozcue, The principle of working on coordinates, in *Pawlowsky-Glahn and Buccianti* (2011)
G. Mateu-Figueras, V. Pawlowsky-Glahn, The skew-normal distribution on the simplex. Commun. Stat. - Theory Methods **36**(9), 1787–1802 (2007)
G. Mateu-Figueras, V. Pawlowsky-Glahn, C. Barceló-Vidal, The additive logistic skew-normal distribution on the simplex. Stoch. Environ. Res. Risk Assessment **18**, 205–214 (2005)
G. Mateu-Figueras, V. Pawlowsky-Glahn, J.J. Egozcue, The normal distribution in some constrained sample spaces. SORT **37**, 29–56 (2013)
G.S. Monti, G. Mateu-Figueras, V. Pawlowsky-Glahn, Notes on the scaled Dirichlet distribution, in *Pawlowsky-Glahn and Buccianti* (2011), pp. 128–138
G.S. Monti, G. Mateu-Figueras, V. Pawlowsky-Glahn, J.J. Egozcue, The shifted-scaled Dirichlet distribution in the simplex, in *Proceedings of the 4th International Workshop on Compositional Data Analysis*, ed. by J.J. Egozcue, R. Tolosa-Delgado, M.I.E. Ortego (Universitat de Girona, 2011)
K.W. Ng, G.L. Tian, M.L. Tang, *Dirichlet and Related Distributions. Theory, Methods and Applications* (Wiley Ltd, New York, 2011)
A. Ongaro, S. Migliorati, G.S. Monti, A new distribution on the simplex containing the Dirichlet family, in *Proceedings of CoDaWork'08, The 3rd Compositional Data Analysis Workshop*, ed. by J. Daunis-i Estadella, J.E. Martín-Fernández (University of Girona, Girona (Spain), 2008)
G.P. Patil, S.O.J.E. Kotz, *Statistical Problems of Size and Shape. II. Characterizations of the Lognormal, Gamma and Dirichlet Distributions* (1975)
V. Pawlowsky-Glahn, Statistical modelling on coordinates, in *Compositional Data Analysis Workshop – CoDaWork'03, Proceedings*, ed. by S. Thió-Henestrosa, J.A. Martín-Fernández (University of Girona, Girona (Spain), 2003)
V. Pawlowsky-Glahn, A. Buccianti (eds.), *Compositional Data Analysis: Theory and Applications* (Wiley Ltd, New York, 2011)
V. Pawlowsky-Glahn, J.J. Egozcue, Geometric approach to statistical analysis on the simplex. Stoch. Environ. Res. Risk Assessment **15**, 384–398 (2001)
V. Pawlowsky-Glahn, J.J. Egozcue, Blu estimators and compositional data. Math. Geol. **34**, 259–274 (2002)
W.S. Rayens, C. Srinivasan, Dependence properties of generalized Liouville distributions on the simplex. J. Am. Stat. Assoc. **89**(428), 1465–1470 (1994)
K. Van Den Boogart, J.J. Egozcue, V. Pawlowsky-Glahn, Bayes linear spaces. SORT **34**(3), 201–222 (2010)
S.S. Wilks, *Mathematical Statistics* (Wiley, New York, 1962), p. 644
T.T. Wong, Generalized Dirichlet distribution in Bayesian analysis. Appl. Math. Comput. **97**, 165–181 (1998)

Compositional Biplots: A Story of False Leads and Hidden Features Revealed by the Last Dimensions

Jan Graffelman

Abstract Logratio principal component analysis is often one of the first steps in exploring a compositional data set. Compositional biplots based on the first two principal components are frequently used to uncover proportionality between parts or to detect one-dimensional patterns of variability for larger subcompositions. This article argues that this approach is likely to produce false leads and proposes an alternative procedure based on condition indices and low-variance principal components. We advocate the calculation of condition indices, combined with biplots of the last few principal components and lists of subcompositions with large condition numbers, and these are shown to be useful for detecting proportionality and one-dimensional relationships. The detection of such patterns in compositional data sets is shown to be closely related to the analysis of multicollinearity as employed in linear regression. Two example data sets, amino acid compositions in calves and chemical components of coffee aroma, are used as illustrations.

1 Introduction

Compositional data sets are often first explored by logratio principal component analysis (Aitchison 1983). Logratio principal component analysis (LR-PCA) can easily be performed by applying a standard principal component analysis to compositions that are transformed with the centred logratio transformation (clr, Aitchison 1986). LR-PCA provides a low-rank approximation to the clr transformed data, from which a biplot of the compositional data is easily constructed. LR-PCA is typically covariance based, because all parts are registered in the same scale. Biplot properties and interpretation rules for compositional multivariate methods have been described in detail by several authors (Aitchison and Greenacre 2002; Pawlowsky-Glahn et al.

J. Graffelman (✉)
Department of Statistics and Operations Research, Universitat Politècnica de Catalunya, Carrer Jordi Girona, 1-3, 08034 Barcelona, Spain

Department of Biostatistics, University of Washington, 4333 Brooklyn Avenue NE, Seattle, WA 98105, USA
e-mail: jan.graffelman@upc.edu

P. Filzmoser et al. (eds.), *Advances in Compositional Data Analysis*,
https://doi.org/10.1007/978-3-030-71175-7_5

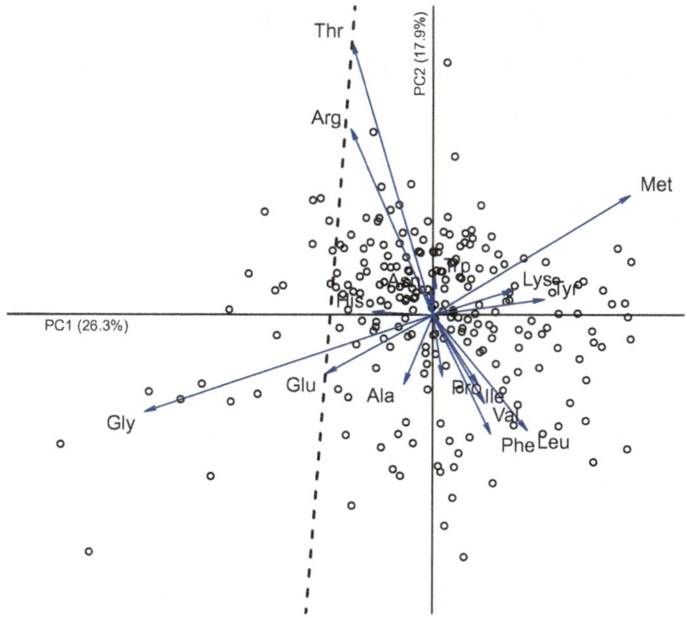

Fig. 1 Biplot of 16-part adjusted amino acid compositions. A set of approximately collinear biplot vectors (Glu-Arg-Thr) is connected by a dashed line

2015; Filzmoser et al. 2018; Graffelman et al. 2018) and are not exposed in full detail here. However, we draw attention to two principles used in the interpretation of compositional biplots: (a) close-to-coincident biplot vectors; (b) close-to-collinear biplot vectors. The first suggest proportionality between the two parts involved, whereas the second suggest a one-dimensional pattern of variation for the subcomposition of the involved parts. In this contribution, we will specifically focus on the detection of one-dimensional patterns in compositional data sets.

An example of a compositional biplot is given in Fig. 1. The data concerns 226 measurements of the 16-part amino acid composition in blood for a set of calves fed on different diets (milk replacers). The amino acid compositions were adjusted for diet and the calves' developmental stage. This data set is described in more detail by Ortuzar-Fernández (2019).

Figure 1 shows a close-to-perfect collinear relationship for the triple of amino acids (Glu-Arg-Thr), marked by a dashed line. This three-part subcomposition was extracted and plotted in a ternary diagram and also in logratio coordinates, as shown in Fig. 2A and B, respectively. Both representations show considerable variation around the respective curve and line that represent the first principal component. More than one-third of the total variance in Fig. 2B resides in the second dimension. Apparently, the biplot in Fig. 1 provided a false lead, as the suggested one-dimensional pattern is not really confirmed by this closer inspection of the data.

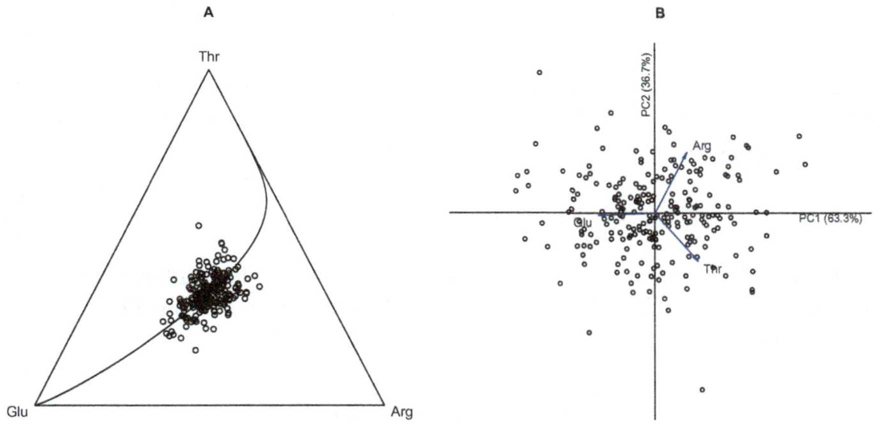

Fig. 2 A: Ternary diagram of the (Glu-Arg-Thr) subcomposition. The curve represents the first principal component. B: Compositional biplot of the (Glu-Arg-Thr) subcomposition obtained by LR-PCA

Proportionality and one-dimensional relationships are of paramount importance in the analysis of compositional data (Egozcue et al. 2018). Proportionality, if it is detected, can help to avoid redundancy by eliminating or combining proportional parts, and one-dimensional relationships often represent a gain in the understanding of the data. Ideally, one would want to safely detect truly one-dimensional relationships in the data without being distracted by suggested patterns that are not confirmed upon closer inspection. The goal of this paper is to search for and explore alternative procedures for uncovering one-dimensional patterns of variation in compositional data sets. The remainder of this article is organised as follows. In Sect. 2, we develop some theory for uncovering one-dimensional relationships and propose some procedures. In Sect. 3, we illustrate these procedures with the amino acid data set presented above and also discuss a second example using the chemical composition of coffee aroma. A discussion, in Sect. 4, completes the article.

2 The Detection of One-Dimensional Patterns

Aitchison (1999) showed that natural chemical, geological and genetic laws governing three-part compositions can be uncovered by looking for near-constant log-contrasts obtained by LR-PCA. LR-PCA is usually performed using the clr transformed data, and this is briefly summarised here. Let matrix $\mathbf{X}_\ell$ contain the log-transformed compositions, that is, $\mathbf{X}_\ell = \ln(\mathbf{X})$ with the natural logarithmic transformation applied element-wise. The clr transformed data can be obtained by row-centring this matrix, using the centring matrix $\mathbf{H}_r = \mathbf{I} - \frac{1}{D}\mathbf{1}\mathbf{1}'$, where D is the number of parts of the composition.

$$\mathbf{X}_{\text{clr}} = \mathbf{X}_\ell \mathbf{H}_r. \tag{1}$$

LR-PCA can be performed by extracting eigenvectors of the covariance matrix of the clr transformed data. This covariance matrix can be obtained by first column-centring the data matrix, such that a double-centred data matrix $\mathbf{X}_{c\text{clr}}$ is obtained

$$\mathbf{X}_{c\text{clr}} = \mathbf{H}_c \mathbf{X}_{\text{clr}} = \mathbf{H}_c \mathbf{X}_\ell \mathbf{H}_r, \tag{2}$$

where $\mathbf{H}_c$ is the idempotent centring matrix $\mathbf{H}_c = \mathbf{I} - (1/n)\mathbf{1}\mathbf{1}'$, with n representing the sample size. The sample covariance matrix of the clr transformed data is then obtained as

$$\mathbf{S}_{\text{clr}} = \frac{1}{n-1} \mathbf{X}_{c\text{clr}}' \mathbf{X}_{c\text{clr}} = \mathbf{H}_r \mathbf{S}_\ell \mathbf{H}_r = \mathbf{V} \mathbf{D}_\lambda \mathbf{V}', \tag{3}$$

where $\mathbf{S}_\ell$ represents the covariance matrix of the log-transformed data. Covariance matrix $\mathbf{S}_{\text{clr}}$ can thus be obtained by double-centring the covariance matrix of the log-transformed data. After the spectral decomposition in Eq. (3), with diagonal matrix $\mathbf{D}_\lambda$ containing the eigenvalues in non-increasing order, principal components are obtained as

$$\mathbf{F}_p = \mathbf{X}_{c\text{clr}} \mathbf{V}. \tag{4}$$

The covariance $\mathbf{S}_{\text{clr}}$ is structurally singular due to the double-centring operation, as $\mathbf{1}'\mathbf{H}_r = \mathbf{0}$, and matrix $\mathbf{S}_{\text{clr}}$ therefore has zero eigenvalue $\lambda_D = 0$. We will ignore this uninteresting dimension, considering $\mathbf{V}$ and $\mathbf{F}_p$ to have $D - 1$ columns, and will refer to column $(D - 1)$ of $\mathbf{F}_p$ as the *last* principal component. A biplot of the clr transformed data is obtained by defining $\mathbf{F}_p$ to be the matrix of row markers (in principal scaling) and $\mathbf{G}_s = \mathbf{V}$ to be the matrix of column markers (in standard scaling). We thus have the biplot factorisation

$$\mathbf{X}_{c\text{clr}} = \mathbf{F}_p \mathbf{G}_s', \tag{5}$$

which is known as the *form* biplot. Alternatively, a biplot based on the standardised principal components ($\mathbf{F}_s$) can be obtained as

$$\mathbf{X}_{c\text{clr}} = \mathbf{F}_s \mathbf{G}_p', \tag{6}$$

with $\mathbf{F}_s = \mathbf{F}_p \mathbf{D}_\lambda^{-\frac{1}{2}}$ and $\mathbf{G}_p = \mathbf{G}_s \mathbf{D}_\lambda^{\frac{1}{2}}$, and this is known as the *covariance* biplot. Usually, compositional biplots are conveniently obtained from the singular value decomposition of the clr transformed data (Pawlowsky-Glahn et al. 2015). Here we prefer the spectral decomposition of $\mathbf{S}_{\text{clr}}$, as this emphasises, by Eq. (4), that principal components obtained by LR-PCA are logcontrasts and therefore low-variance (near-constant) components with small eigenvalues signal the uni-dimensional relationships of interest. A logcontrast is defined by the equation (Aitchison 1986)

$$\sum_{i=1}^{D} a_i \ln(x_i) \text{ with } \sum_{i=1}^{D} a_i = 0, \tag{7}$$

which can be written in vector form as $\mathbf{a}'\mathbf{x}_{\ell i}$, where $\mathbf{x}_{\ell i}$ is the ith row of the matrix of log-transformed compositions $\mathbf{X}_\ell$. In matrix form, the full set of $D-1$ logcontrasts is given by $\mathbf{X}_\ell\mathbf{A}$ with $\mathbf{1}'\mathbf{A} = \mathbf{0}$. Correspondingly, in LR-PCA we have that the coefficients of the logcontrasts in $\mathbf{A}$ are given by

$$\mathbf{F}_p = \mathbf{X}_{\text{cclr}}\mathbf{V} = \mathbf{H}_c\mathbf{X}_\ell\mathbf{H}_r\mathbf{V} = \mathbf{H}_c\mathbf{X}_\ell\mathbf{A} \text{ with } \mathbf{A} = \mathbf{H}_r\mathbf{V} = \mathbf{V}, \tag{8}$$

where the last equality follows from the fact that Eq. (3) implies that the eigenvectors are centred. Consequently, the eigenvectors directly give the coefficients of the logcontrasts.

The issue arises as to how small a trailing eigenvalue needs to be in order to signal a close-to-constant logcontrast and provide evidence for the existence of some natural law relating the parts of the composition. The size of the eigenvalues depends on the total variation in the compositional data set, and if the total variation is lower, smaller eigenvalues will necessarily result. This problem is similar to the multicollinearity problem in regression modelling. In regression modelling, we would want to avoid the use of a predictor that is almost an exact linear combination of the other predictors, and there is a keen interest in the detection of such linear dependencies among the predictors. In the current compositional context, we wish to detect additional linear dependencies among the columns of $\mathbf{X}_{\text{cclr}}$, beyond the trivial dependency that the columns sum to zero. In regression, condition indices (κ_k) are often used to study multicollinearity. These are calculated from the spectral decomposition of the correlation matrix between the predictors ($\mathbf{R}_{xx} = \mathbf{V}\mathbf{D}_\lambda\mathbf{V}'$) and are defined as

$$\kappa_k = \sqrt{\lambda_1/\lambda_k}. \tag{9}$$

We suggest to calculate the condition indices κ_k of $\mathbf{S}_{\text{clr}}$ for $k = 1, 2, \ldots, D-1$. The largest of these, $\kappa_{(k)} = \kappa_{D-1}$, is known as the *condition number*, in this case of $\mathbf{S}_{\text{clr}}$. In the remainder, we will use κ_k to indicate the condition *index*—i.e., the ratio obtained using the kth eigenvalue, and $\kappa_{(k)}$ to represent the condition *number*—i.e., the maximum of all ratios considered. Values above 30 are used in practice as signalling (severe) multicollinearity (Belsley et al. 1980; Friendly and Kwan 2009). Principal components with a large condition index identify possible uni-dimensional relationships of interest, and those parts with large coefficients (in absolute value) for these components are identified as involved in that relationship. Principal components can be sorted on the basis of their condition index, which just amounts to reversing their order, as the last principal component will invariably have the smallest eigenvalue and therefore the largest κ_k. To investigate uni-dimensionality, besides computation of κ_k, biplots of the corresponding principal components can reveal the parts that are involved. Invariably, a biplot of the last two principal components will then be most informative for detecting one-dimensional relationships.

All of the foregoing concerns the study of the *full* D-part composition. Alternatively, subcompositions may be formed, and the same principles can be applied. Depending on the size of D, many subcompositions are possible, which may all be exhaustively enumerated and analysed. For a compositional data set with D parts, we have $\binom{D}{2} = \frac{1}{2}D(D-1)$ pairwise logratios. All their variances are usually calculated and reported in the variation matrix (Aitchison 1986). The variation matrix is typically screened for close-to-zero variances, which signal proportionality of parts. The computation of the variation matrix for this purpose is exhaustive, as all pairwise ratios are considered.

The all-combinations principle extends to logratios involving three (or more) parts in order to detect one-dimensional patterns. All combinations of three parts can be generated, and each three-part composition is analysed by logratio PCA. For each PCA, eigenvalues and eigenvectors are stored. A one-dimensional pattern is likely to be present if the first eigenvalue is large with respect to the second—i.e., when the condition index $\kappa_2 = \sqrt{\lambda_1/\lambda_2}$ is large. All three-part combinations are ranked (from high to low) according to the obtained condition index. By inspection of the eigenvectors of the second principal component, the nature of possible one-dimensional relationships can be studied in more detail. Obviously, the principle can be extended to larger subcompositions at an increased computational cost.

3 Examples

We explore the practical usefulness of the procedures outlined in the previous section with two examples. We first proceed with the analysis of the amino acid compositions presented in the introduction, and subsequently also analyse a compositional data set of compounds of coffee aroma.

3.1 Amino Acid Compositions

A scree plot of the eigenvalues of the LR-PCA of the amino acid data is shown in Fig. 3. This plot shows not only the structural zero eigenvalue, characteristic of clr transformed compositions, but also several small variance components with close-to-zero eigenvalues. Condition indices, eigenvalues and eigenvectors obtained in LR-PCA are shown in Table 1. The last two principal components (15 and 14) have large condition indices, well above the aforementioned cut-off of 30 and are clearly candidates for signalling uni-dimensional relationship.

A biplot of the last two principal components is shown in Fig. 4, where biplot vectors were conveniently scaled by dividing their coordinates by five. This biplot shows very long links for the essential amino acid pairs (Leu, Ile), (Ile, Val) and (Leu,

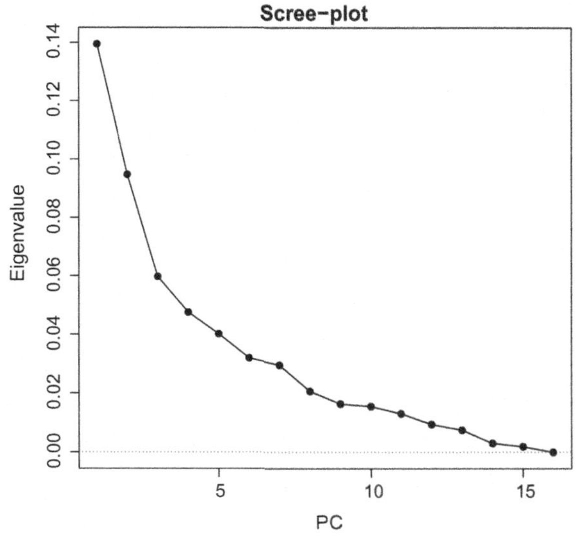

Fig. 3 Scree plot of the eigenvalues of the amino acid data

Val). These logratios have strong relationships with the last two principal components, and therefore have a low variance, and represent three pairs of approximately proportional amino acids. More generally, any direction in Fig. 4 represents a rotation of the principal components and defines a low-variance logcontrast of potential interest (see the Discussion section).

In order to further investigate low-variance principal components, a lower diagonal biplot matrix of the last four components was constructed and is shown in Fig. 5. This reveals that the 13th principal component reflects a proportionality between Asn and Pro, something that can also be glanced from Table 1.

The uncertainty in the first two condition indices has been assessed by using the bootstrap. Figure 8 shows the bootstrap distribution of the largest two condition indices corresponding to the last two principal components. Condition indices have positive skew and their distribution is more conveniently shown in a logarithmic scale. Both graphs show that the whole bootstrap distribution is above the threshold value of 30, indicating severe collinearity, and thus strong uni-dimensional relationships. We continue the analysis of the amino acid data by considering subcompositions. The first half of Table 2 shows the ten three-part compositions with the largest condition number. All these compositions have one element of the second eigenvector that is close to zero, meaning that the corresponding amino acid does not contribute to the second principal component; in fact, the full top ten of detected relationships completely concerns the previously detected proportionalities of (Val, Leu), (Ile, Leu) and (Val, Ile). The second half of Table 2 shows the same information for the top ten of four-part subcompositions. Again, nine of these represent the aforementioned proportionalities. Interestingly, if a subcomposition contains only two of these three amino acids, a proportionality between the two amino acids is observed. However, if

Table 1 Condition indices (κ_k), eigenvalues (λ_k) and eigenvectors of the last six principal components of the amino acid data set. Principal components are sorted decreasingly according to their condition index

PC	κ_k	λ_k	Asn	Glu	Gly	His	Arg	Thr	Ala	Pro	Tyr	Val	Met	Lys	Ile	Leu	Phe	Trp
15	74.11	0.0019	–0.24	–0.04	0.00	0.04	–0.13	–0.07	–0.15	0.19	0.13	**0.57**	0.05	0.09	**0.86**	**–1.35**	0.05	–0.00
14	46.37	0.0030	–0.23	–0.08	–0.09	0.24	0.04	0.05	–0.06	0.21	–0.14	**–1.30**	–0.12	0.16	**1.02**	0.18	0.22	–0.09
13	18.67	0.0075	**–1.34**	–0.10	0.06	0.04	0.16	0.03	0.09	**0.93**	–0.08	0.08	0.10	0.22	–0.39	0.13	–0.21	0.29
12	14.81	0.0094	0.44	–0.05	0.02	–0.72	0.43	–0.20	–0.20	0.76	–0.38	0.11	0.31	–0.92	0.42	0.21	–0.40	0.17
11	10.68	0.0131	–0.24	–0.25	0.16	0.68	0.38	–0.56	0.25	–0.48	0.58	0.11	0.15	–0.29	0.37	0.25	–1.02	–0.08
10	8.99	0.0155	0.41	0.11	0.02	–0.72	–0.06	–0.26	0.10	0.29	0.67	–0.33	–0.59	0.79	–0.08	–0.17	–0.65	0.46

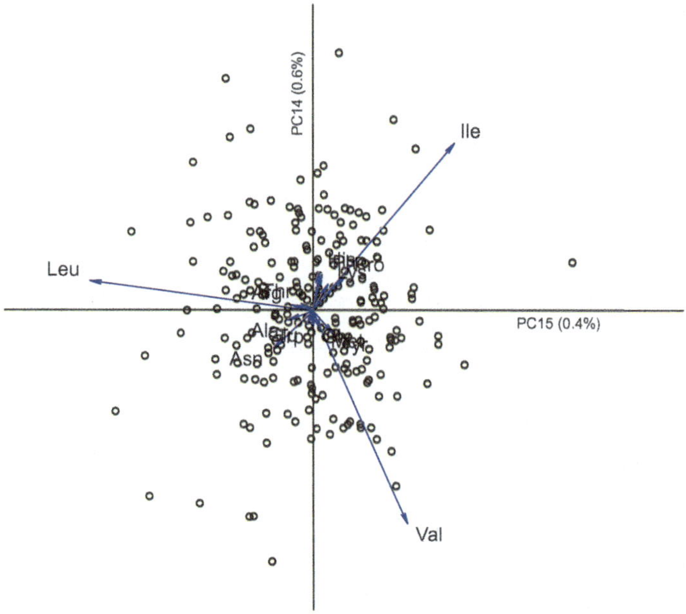

Fig. 4 Biplot of the last two principal components of the amino acid data

the three amino acids are present, a balance between the three is found, and roughly amounts to $\sqrt{\text{Val} \times \text{Ile}}/\text{Leu}$ being approximately constant.

The analysis of subcompositions was repeated for 5 through 15 part subcompositions, and ranking all subcompositions according to their condition number. In all cases, the last principal component had large coefficients for Val, Ile and Leu revealing a proportionality if two of these amino acids are present, or a three-part balance if the three are present (results not shown). The two-part proportionality and three-part balance of these three amino acids are apparently a very prominent feature of the data, emerging in the analysis of all subcompositions with large condition numbers where these parts are present.

3.2 Coffee Aroma Composition

We consider a data set of six components of coffee aroma, described by Korhoňová et al. (2009) and part of the R package `robCompositions` (Templ et al. 2011). The components are Acetic acid (AcA), Furfural (Fur), Methylpyrazine (MetP), Furfuryl alcohol (FurA), 2,6-Dimethylpyrazine (DimP) and 5-Methylfurfural (MetF) which form a subcomposition of a larger set of compounds (Korhoňová et al. 2009) measured for 30 coffees. For the sake of the homogeneity of the data, three outlying Robusta coffees were removed, retaining only Arabica and blended coffees.

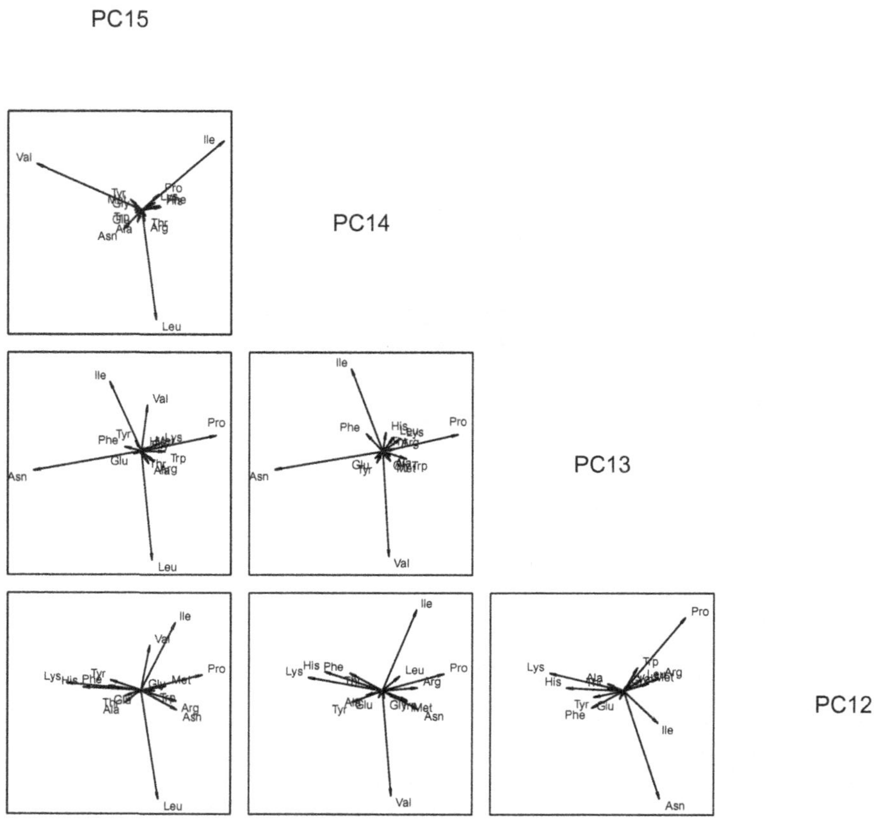

Fig. 5 Biplot matrix of the last four principal components of the amino acid data

Figure 6A shows a compositional biplot obtained by LR-PCA, and this plot explains 87.2% of the total variance of the compositions. Blended coffees tend towards negative values for the first component, whereas Arabica coffees tend to have positive values. Biplot Fig. 6A suggests approximate collinearity between MetP, MetF, and Fur but this is not confirmed by further inspection of the data. Table 3 shows the condition indices, eigenvalues and eigenvectors of LR-PCA. The last principal component has a large condition index, signalling multicollinearity, and the eigenvector suggests a one-dimensional relationship involving Fur, DimP and MetF. These parts do certainly not appear as collinear vectors in the standard compositional biplot in Fig. 6A. Figure 6B shows a compositional biplot of the last two principal components, and shows the last low-variance component is largely determined by Fur, DimP and MetF. The bootstrap distribution of the two largest condition indices is shown in Fig. 9. This shows that only the largest condition index is above 30 with large probability, confirming the one-dimensional relationship involving Fur, DimP and MetF to be the most relevant one.

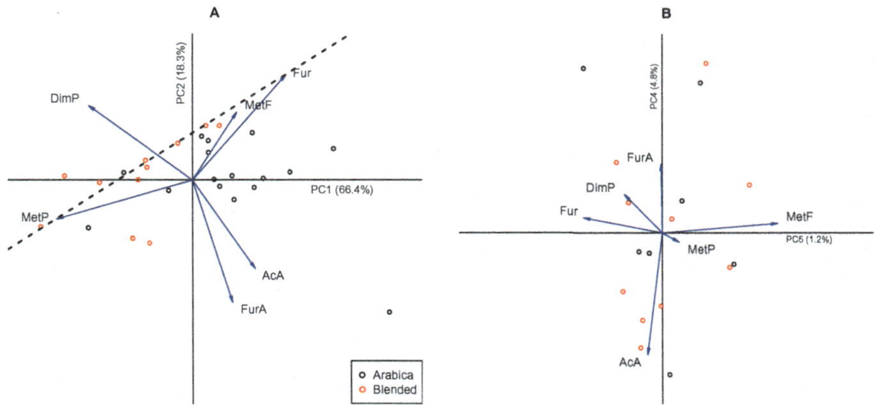

Fig. 6 Compositional biplots of the coffee data. A: first two principal components; B: last two principal components

Table 2 Condition numbers ($\kappa_{(k)}$), amino acid composition and corresponding second (or third, as corresponds) eigenvector for ten three-part subcompositions with the largest condition number, and for ten four-part subcompositions with the largest condition number. The bold entry identifies a balance (Val, Ile, Leu)

	Three-part top ten							Four-part top ten								
	$\kappa_{(k)}$	Amino acids			Eigenvector			$\kappa_{(k)}$	Amino acids				Eigenvector			
1	4.87	Gly	Val	Leu	–0.03	0.72	–0.69	5.64	Gly	Val	Met	Leu	0.03	–0.70	–0.05	0.72
2	4.82	Gly	Ile	Leu	–0.06	0.74	–0.67	5.48	Gly	Met	Ile	Leu	0.06	0.00	–0.74	0.67
3	4.39	Thr	Ile	Leu	–0.10	0.75	–0.65	5.35	Gly	Tyr	Val	Leu	–0.04	0.08	0.69	–0.72
4	4.38	Gly	Val	Ile	–0.03	–0.69	0.72	5.24	Gly	Thr	Ile	Leu	–0.03	–0.09	0.76	–0.64
5	4.35	Thr	Val	Leu	–0.07	0.74	–0.67	5.22	Gly	Val	Met	Ile	–0.03	–0.66	–0.06	0.75
6	4.00	Arg	Val	Leu	0.09	–0.75	0.66	5.21	Gly	Val	Ile	Leu	–0.06	**0.40**	**0.45**	**–0.80**
7	3.96	Val	Met	Leu	0.68	0.05	–0.73	5.19	Gly	Tyr	Ile	Leu	0.06	–0.03	–0.72	0.69
8	3.93	Arg	Ile	Leu	–0.10	0.75	–0.65	5.18	Gly	Ile	Leu	Phe	0.07	–0.68	0.72	–0.11
9	3.73	Val	Met	Ile	–0.68	–0.05	0.73	5.16	Gly	Arg	Val	Leu	–0.01	–0.08	0.75	–0.66
10	3.70	Glu	Val	Leu	–0.04	0.73	–0.69	5.12	Gly	Arg	Ile	Leu	–0.03	–0.08	0.76	–0.64

Table 3 Condition indices (κ_k), eigenvalues and eigenvectors of LR-PCA of the coffee data

PC	κ_k	λ_k	AcA	MetP	Fur	FurA	DimP	MetF
5	44.51	0.0124	–0.08	0.06	**–0.49**	–0.08	**–0.25**	**0.83**
4	16.96	0.0326	–0.70	0.29	0.29	0.46	–0.35	0.02
3	6.74	0.0820	–0.27	–0.55	–0.27	0.50	0.54	0.06
2	3.63	0.1521	–0.41	–0.18	0.50	–0.57	0.35	0.32
1	1.00	0.5524	0.29	–0.64	0.44	0.19	–0.49	0.21

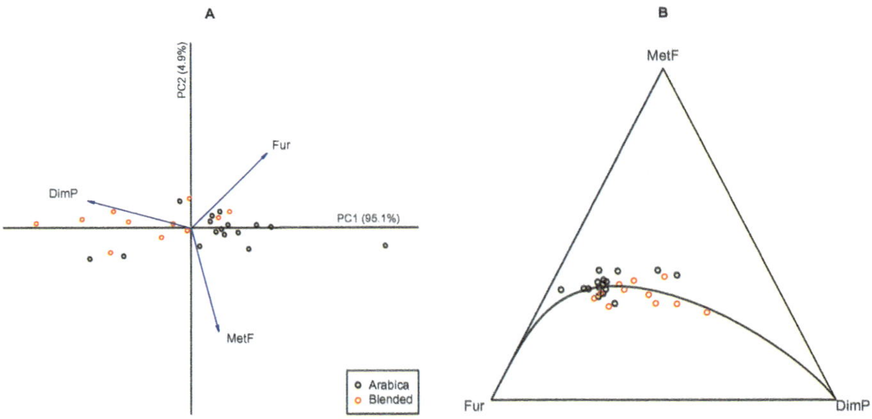

Fig. 7 A: compositional biplot of the (Fur, DimP, MetF) subcomposition; B: Ternary diagram with the first principal component

Table 4 Condition number ($\kappa_{(k)}$), coffee compounds and corresponding second eigenvector for ten three-part subcompositions with the largest condition index

	$\kappa_{(k)}$	Components			Eigenvector		
1	5.24	MetP	Fur	DimP	–0.61	–0.16	0.77
2	4.34	MetP	DimP	MetF	–0.52	0.80	–0.28
3	3.83	AcA	MetP	DimP	–0.15	–0.62	0.77
4	3.71	MetP	Fur	MetF	–0.28	–0.53	0.80
5	3.68	MetP	FurA	DimP	–0.62	–0.15	0.77
6	3.61	Fur	DimP	MetF	0.43	0.39	–0.82
7	3.01	Fur	FurA	MetF	0.56	0.23	–0.79
8	1.97	AcA	Fur	MetF	0.22	0.57	–0.79
9	1.70	FurA	DimP	MetF	0.07	–0.74	0.67
10	1.62	AcA	DimP	MetF	–0.18	–0.60	0.78

LR-PCA was repeated for the three-part (Fur, DimP, MetF) subcomposition. Figure 7 shows the compositional biplot as well as a ternary diagram of this subcomposition. These plots confirm a one-dimensional relationship between the three parts, and the dimension is seen to separate Arabica and blended coffees to some extent. Table 4 shows the top ten of all three-part subcompositions that have the largest condition index. The (Fur, DimP, MetF) subcomposition revealed by the previous LR-PCA is the sixth of this list. Several other compositions in this list suggest approximate proportionality between DimP and MetP. Assuming constancy of the second principal component, this amounts to the approximate constancy of $Fur^{0.52} \times DimP^{0.48}/MetF$.

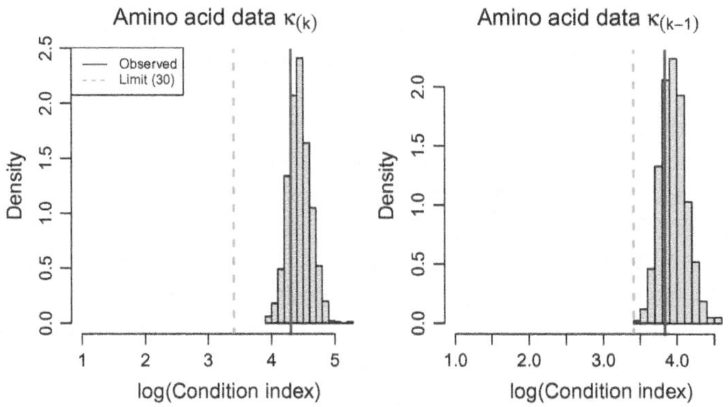

Fig. 8 Bootstrap distribution for the amino acid data obtained used 1,000 bootstrap samples for the condition indices $\kappa_{(k)}$ and $\kappa_{(k-1)}$. The solid vertical line indicates the observed value for the sample, and the dashed line the critical limit of 30

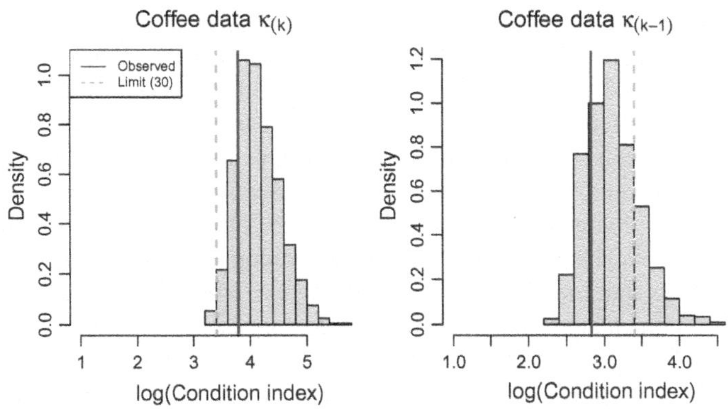

Fig. 9 Bootstrap distribution for the coffee data obtained used 1,000 bootstrap samples for the condition indices $\kappa_{(k)}$ and $\kappa_{(k-1)}$. The solid vertical line indicates the observed value for the sample, and the dashed line the critical limit of 30

4 Discussion

This article shows that compositional biplots that use the first two principal components are of limited value for the detection of one-dimensional relationships in a compositional data set. This is particularly true if the amount of explained variance is not high, and the studied compositions have a large number of parts. High-dimensional compositions generate many biplot vectors, and in a dense biplot, approximately collinear biplot vectors can easily arise by chance, suggesting one-dimensional patterns that are in fact not present in the data. Obviously, biplots with low goodness-of-fit—i.e., a small percentage of explained variance—are more prone to such misinterpretation, and reporting the goodness-of-fit of a biplot therefore remains crucial.

The discovery of one-dimensional patterns can be further complicated by outliers, which are common in compositional data, in particular, if combined with a small sample size.

The use of the condition indices for selecting principal components with low variance seems successful in the analysis of the examples discussed in the paper. It has often been stated that compositional *data* carries relative information (Pawlowsky-Glahn et al. 2015; Filzmoser et al. 2018), but this also holds true for some of the statistics derived from a compositional analysis. Natural laws can be inferred from "small" or "near-zero" trailing eigenvalues in LR-PCA. However, a trailing eigenvalue can only be considered small if the first eigenvalue is large. The proposed use of the condition index clearly recognises the relative nature of the size of the eigenvalues obtained in LR-PCA. Condition indices are random variables which can be expected to vary over repeated samples taken from the same population. In this article, we have used the bootstrap to assess variability in the conditions indices. Distributional results on condition indices (Edelman 1988) may be exploited for formal hypothesis testing with condition indices in a compositional context.

If several low-variance principal components are present, such as exemplified by the amino acid data in Figs. 3 and 4, then rotations of these components will also have low variance. Indeed, in the latter figure, the links (Leu, Ile), (Ile, Val) and (Leu, Val) have different directions but are all linear combinations of the last two principal components, and their pairwise ratios have low variance. Interestingly, in such a biplot the analyst can in fact choose a linear combination of the involved outstanding parts that is most interesting or interpretable in the light of the problem being investigated. The rotation of the principal components can therefore generate new logcontrasts that may be of theoretical interest. Rotation is sometimes used in principal component analysis (Jolliffe 1986), and its use for compositional data sets with multiple low-variance components could be further explored.

Finally, the examples in this article also show that interesting features of a compositional data set may remain hidden if only the first few components are inspected. Indeed, it may well be argued that a compositional biplot of the last few principal components is at least as interesting as a biplot of the first few.

5 Software

The type of analysis proposed in the paper is easily carried out in the R environment (R development core team 2004) and using the R packages `robCompositions` (Templ et al. 2011), `calibrate` (Graffelman and van Eeuwijk 2005) and `ToolsForCoDa` (Graffelman et al. 2018). The functions `lrpca` and `largest.kappas` of the latter package were written to generate tables with condition indices for LR-PCA and top ten lists of (sub)compositions with their condition numbers. Appendix A contains R code reproducing the example of the coffee aroma data in Sect. 3.2.

Acknowledgements This work was partially supported by grants RTI2018-095518-B-C22 of the Spanish Ministry of Science, Innovation and Universities and the European Regional Development Fund, and by grant R01 GM075091 from the United States National Institutes of Health. Àlex Bach and Marta Terré from the Institute of Agrifood Research and Technology (IRTA, Caldes de Montbui, Spain) are gratefully acknowledged for providing the amino acid data used in the article. I also thank two anonymous reviewers for their suggestions that helped to improve the article.

Appendix A: R Code for the Coffee Aroma Example

Program Code

```
#
# R code for reproducing the coffee aroma data example
#
#install.packages("robCompositions")
#install.packages("calibrate")
#install.packages("ToolsForCoDa")
library(robCompositions)
library(calibrate)
library(ToolsForCoDa)

#
# Data preparation
#

data(coffee)
X <- coffee
X <- X[X$sort!="robusta",]
so <- factor(X$sort)
X <- X[,-1]
colnames(X) <- c("AcA","MetP","Fur","FurA","DimP","MetF")
Xc <- X/rowSums(X)
n <- nrow(X)
p <- ncol(X)

#
# Logratio principal component analysis
#

out.pca <- lrpca(Xc)
Fp <- out.pca$Fp
Gs <- out.pca$Gs
decom <- out.pca$decom

colvec <- rep("black",n)
colvec[so=="blended"] <- "red"

#
# Biplot of first two principal components
#

bplot(Fp,2*Gs,colch=NA,rowch=1,rowlab="",rowcolor = colvec,
```

```
      collab=colnames(Xc))
legend("bottomright",c("Arabica","Blended"),col=c("black","red"),pch=1)
pc1 <- round(100*decom[2,1],digits=1)
text(1.4,-0.1,paste("PC1 (",toString(pc1),"%)",sep=""),cex=0.9)
pc2 <- round(100*decom[2,2],digits=1)
text(-0.1,1.0,paste("PC2 (",toString(pc2),"%)",sep=""),cex=0.9,srt=90)

#
# Biplot of last two principal components
#

dims <- c(5,4)

bplot(Fp[,dims],0.3*Gs[,dims],rowlab="",collab=colnames(X),colch=NA,
      rowch=1,rowcolor=colvec)
pc5 <- round(100*decom[2,5],digits=1)
text(0.3,-0.015,paste("PC5 (",toString(pc5),"%)",sep=""),cex=0.75)
pc4 <- round(100*decom[2,4],digits=1)
text(-0.015,0.30,paste("PC4 (",toString(pc4),"%)",sep=""),cex=0.75,srt=90)

#
# Condition index table
#

out.pca$kappalist

#
# Condition numbers for top ten 3-part subcompositions.
#

TopTenThreeParts <- largest.kappas(Xc,nparts=3,sizetoplist = 10)

#
# Ternary diagram for subcomposition
#

Xpro <- Xc[,c(3,5,6)]
Xpro <- Xpro/rowSums(Xpro)
ternaryDiag(Xpro,line="pca",grid=FALSE,col=colvec,main="B")

#
# Compositional biplot for subcomposition
#

out.pca.sub <- lrpca(Xpro)
Fp <- out.pca.sub$Fp
Gs <- out.pca.sub$Gs
decom <- out.pca.sub$decom

bplot(Fp,Gs,rowch=1,rowlab="",collab=colnames(Xpro),colch=NA,
      rowcolor=colvec,main="A")
legend("bottomright",c("Arabica","Blended"),col=c("black","red"),pch=1)
pc1 <- round(100*decom[2,1],digits=1)
text(1.35,0.06,paste("PC1 (",toString(pc1),"%)",sep=""),cex=0.9)
pc2 <- round(100*decom[2,2],digits=1)
text(-0.06,1.2,paste("PC2 (",toString(pc2),"%)",sep=""),cex=0.9,srt=90)
```

References

J. Aitchison, Principal component analysis of compositional data. Biometrika **70**(1), 57–65 (1983)

J. Aitchison. *The Statistical Analysis of Compositional Data* (The Blackburn Press, Caldwell, 1986). 2003 printing

J. Aitchison, Logratios and natural laws in compositional data analysis. Math. Geol. **31**(5), 563–580 (1999)

J. Aitchison, M. Greenacre, Biplots of compositional data. J. R. Stat. Soc., Ser. C (Appl. Stat.) **51**(4), 375–392 (2002)

D. Belsley, E. Kuh, R. Welsch, *Regression Diagnostics: Identifying Influential Data and Sources of Collinearity* (Wiley, New York, 1980)

A. Edelman, Eigenvalues and condition numbers of random matrices. SIAM J. Matrix Anal. Appl. **9**(4), 543–560 (1988)

J. Egozcue, V. Pawlowsky-Glahn, G. Gloor, Linear association in compositional data analysis. Aust. J. Stat. **47**, 3–31 (2018)

P. Filzmoser, K. Hron, M. Templ, *Applied Compositional Data Analysis: With Worked Examples in R* (Springer, Cham, 2018)

M. Friendly, W. Kwan, Where's Waldo? Visualizing collinearity diagnostics. Am. Stat. **63**(1), 56–65 (2009)

J. Graffelman, F.A. van Eeuwijk, Calibration of multivariate scatter plots for exploratory analysis of relations within and between sets of variables in genomic research. Biometr. J. **47**(6), 863–879 (2005)

J. Graffelman, V. Pawlowsky-Glahn, J. Egozcue, A. Buccianti, Exploration of geochemical data with compositional canonical biplots. J. Geochem. Explor. **194**, 120–133 (2018)

I. Jolliffe, *Principal Component Analysis* (Springer, New York, 1986)

M. Korhoňová, K. Hron, D. Klimčíková, L. Müller, P. Bednář, P. Barták, Coffee aroma' statistical analysis of compositional data. Talanta **80**(2), 710–715 (2009)

I. Ortuzar-Fernández, Analysis of calf growth with compositional mixed effects models. Master's thesis, Universitat Politècnica de Catalunya, Department of Statistics and Operations Research (2019)

V. Pawlowsky-Glahn, J. Egozcue, R. Tolosana-Delgado, *Modeling and Analysis of Compositional Data* (Wiley, Chichester, 2015)

R Development Core Team. *R: a Language and Environment for Statistical Computing*. R Foundation for Statistical Computing (Vienna, Austria, 2004). http://www.r-project.org. ISBN 3-900051-00-3

M. Templ, K. Hron, P. Filzmoser, robcompositions: an R-package for robust statistical analysis of compositional data, in *Compositional Data Analysis: Theory and Applications*, ed. by V. Pawlowsky-Glahn, A. Buccianti (Wiley Ltd, Chichester, 2011), pp. 341–355

Statistical Methodology

Geographically Weighted Regression Analysis for Two-Factorial Compositional Data

Kamila Fačevicová, Petra Kynčlová, and Karel Macků

Abstract The chapter focuses on the modelling and analysis of spatial dependent two-factorial compositional data. Spatial statistics provides a wide range of methods for the analysis of data with local variations but only a few of them are accommodated for the purposes of modelling relative structures. In this chapter, the geographically weighted regression model is introduced to analyse the relationship between the dependent variable and an explanatory variable reflecting a structure expressed in terms of a compositional table. The methodology is motivated by the problem of modelling local variations of the relationship between at-risk-of-poverty rates and the structure of the highest attained educational level in the German population aged 30–34. The real data study shows how information included in a compositional table and further expressed in real-valued coordinates can be highly valuable in selecting variables and prioritising them with respect to a research interest to facilitate the final interpretation of the model.

1 Introduction

Modelling of spatial dependent data represents one of the well-developed fields of statistical analysis. Methods of spatial statistics are based on the assumption that elements that are spatially close to one another are also more closely related. Thus,

---

K. Fačevicová (✉)
Department of Mathematical Analysis and Applications of Mathematics, Palacký University Olomouc, 17. listopadu 12, 77146 Olomouc, Czech Republic
e-mail: kamila.facevicova@gmail.com

P. Kynčlová
Statistics Division, United Nation Industrial Development Organization (UNIDO), Wagramerstr. 5, P.O. Box 300, 1400 Vienna, Austria
e-mail: p.kynclova@unido.org

K. Macků
Department of Geoinformatics, Palacký University Olomouc, 17. listopadu 50, 77146 Olomouc, Czech Republic
e-mail: karel.macku@upol.cz

P. Filzmoser et al. (eds.), *Advances in Compositional Data Analysis*,
https://doi.org/10.1007/978-3-030-71175-7_6

it is a direct link to Tobler's first law of geography saying that "Everything is related to everything else, but near things are more related than distant things." (Tobler 1970). Spatial data are often mistreated without taking into account their specific characteristics resulting from spatial relationships. These linkages may disrupt the assumptions of common statistical analyses, such as homogeneity, or absence of autocorrelation (Fotheringham 1997), making standard statistical methods unsuitable (Anselin 2013).

Many authors have been dealing with the design of spatial statistical methods. There are several well-developed methods and tools for solving spatial autocorrelation problems (Cliff and Ord 1970), applying regression models on spatial data (Anselin 2001; Brunsdon et al. 1998), estimating local spatial and geographically weighted correlations (Kalogirou 2012) or performing geographically weighted principal component analysis (Harris et al. 2011). Nevertheless, all these methods are primarily designed for real-valued data and, therefore, if the data at hand are of the relative (compositional) nature, a proper modification of the methods is needed.

Analysis of spatial dependent compositional data is quite frequent in practice. Let us mention here the Kola data set consisting of a geochemical structure of over 600 soil samples from different locations in Norway, Finland and Russia (Reimann et al. 1997) as a well-known example of spatial dependent vector compositional data. Relative-valued data also naturally appear in the econometric studies, when the demographic structure of countries (Gollini et al. 2015) or the election results in different electoral districts are of interest. Finally, not just vector but also more complex compositional structures appear in practice. The compositional approach to the analysis of manufacturing output distributed according to technology intensity and resource efficiency was introduced in Fačevicová et al. (2016). Furthermore, the educational structure reflecting the field of study and the highest reached degree was analysed in de Sousa et al. (2020). In both cases, the set of spatial dependent compositional tables was analysed. In the latter case, a more detailed structure can be considered when inhabitants' gender is taken into account resulting in a set of three factorial compositional data—compositional cubes.

Even though spatial dependency is quite common for compositional data, the methods for its analysis are still not well developed. One of the first steps towards this goal was taken by Pawlowsky-Glahn et al. (2015); Tolosana-Delgado et al. (2008) who focused on simplicial kriging and cokriging. The authors of Nordhausen et al. (2015) deal with the task of blind source separation. They propose the use of an independent component analysis method tailored for spatial dependent compositional data. Other authors have recently focused on methods for the analysis of spatial dependence (Andreano et al. 2019; Pirzamanbein et al. 2018), spatial regression modelling (Leininger et al. 2013) or multivariate spatial processes (Yoshida and Tsutsumi 2018). However, all the current methods are designed for vector compositional data and do not reflect the complexity of their more-factorial alternatives. This chapter represents the first step towards analysing more-factorial compositional data using spatial statistical methods. The main aim is to introduce a geographically weighted regression model for exploring spatial heterogeneity of the relationship between the dependent variable and an explanatory variable reflecting a structure expressed in terms of a compositional table.

The contribution itself is motivated by the problem of modelling a relationship between at-risk-of-poverty rates and the educational attainment structure of the population aged 30–34 in 36 German regions in 2017. As the at-risk-of-poverty rates vary significantly across different German regions, the relationship with the structure of the highest attained education level is explored assuming local variations over the space. The compositional aspect of the educational structure of the German population plays an important role in this study. The system of coordinates is thus constructed in such a way to focus primarily on the relative dominance of attained tertiary education as a key factor of at-risk-of-poverty rates.

The chapter is organised as follows. The theory of compositional tables and their coordinate representation is introduced in Sect. 2. Moreover, this section proposes an ordinary linear regression model with a compositional table in the role of the independent variable, and focuses on the interpretation of its coefficients. The conventional geographically weighted regression model is defined in Sect. 3 and further extended to its compositional version in Sect. 4. Section 5 demonstrates a practical application of the method followed by concluding remarks in Sect. 6.

2 Two-Factorial Compositional Regression Model

Analysis of more-factorial compositional data represents a natural extension of the traditional log-ratio methodology, which is described in detail, e.g. in Chaps. 1, 2, 3 and 11 of Pawlowsky-Glahn and Buccianti (2011). When the relative structure results from the combination of levels of two factors, it can be expressed in the form of an $I \times J$ compositional table

$$\mathbf{x} = \begin{pmatrix} x_{11} & \cdots & x_{1J} \\ \vdots & \ddots & \vdots \\ x_{I1} & \cdots & x_{IJ} \end{pmatrix}, \tag{1}$$

whose cells $x_{ij} > 0$, for $i = 1, 2, \ldots, I$ and $j = 1, 2, \ldots J$ convey relative contributions on the whole. An example of such a structure is shown in Table 1 where each cell represents a percentage of the given educational-gender category on the whole population aged 30–34 in Leipzig in 2017.

Similar to the case of vector compositional data, the sample space of compositional tables is a subset of an IJ-part simplex $\mathcal{S}^{IJ}$ where both basic operations of the Aitchison geometry, perturbation of tables $\mathbf{x}$ and $\mathbf{x}'$ and powering by a real constant α can be extended as follows:

$$\mathbf{x} \oplus \mathbf{x}' = \mathcal{C}\left(x_{ij}x'_{ij}\right)_{i,j=1}^{I,J} \quad \text{and} \quad \alpha \odot \mathbf{x} = \mathcal{C}\left(x_{ij}^{\alpha}\right)_{i,j=1}^{I,J}, \tag{2}$$

Table 1 The structure of attained education in German population aged 30–34 living in Leipzig in 2017 (in per cent). The educational classes are Less than primary, primary and lower secondary education (ED0-2), upper secondary and post-secondary non-tertiary education (ED3-4) and tertiary education (ED5-8)

	ED0-2	ED3-4	ED5-8
Women	1.63	25.35	20.94
Men	4.53	27.71	19.84

where $\mathcal{C}$ denotes closure and constitutes the equivalence classes of compositional tables, which differ only by the sum of their parts κ. The vector space structure is moreover extended with the Aitchison inner product defined for compositional tables as

$$\langle \mathbf{x}, \mathbf{x}' \rangle_A = \frac{1}{2IJ} \sum_{i,j} \sum_{k,l} \ln \frac{x_{ij}}{x_{kl}} \ln \frac{x'_{ij}}{x'_{kl}}. \tag{3}$$

Similar to compositional vectors, compositional tables are also defined as having strictly positive entries. The situation when some cells equal zero would disable the application of the log-ratio approach to further analysis, and therefore its imputation is needed. Various techniques for dealing with zeros or missing values have been introduced, e.g. in Filzmoser et al. (2018), ch. 13. Nevertheless, all of them are primarily designed for vector compositional data and not their more-factorial alternatives.

According to Egozcue et al. (2008, 2015), it is possible to decompose each compositional table into its independent and interactive parts

$$\mathbf{x} = \mathbf{x}_{ind} \oplus \mathbf{x}_{int} \tag{4}$$

and to analyse them separately. More specifically, both $\mathbf{x}_{ind}$ and $\mathbf{x}_{int}$ are also $I \times J$ compositional tables with cells

$$x_{ij}^{ind} = \left(\prod_{k=1}^{I} \prod_{l=1}^{J} x_{kj} x_{il} \right)^{\frac{1}{IJ}} \quad \text{and} \quad x_{ij}^{int} = \left(\prod_{k=1}^{I} \prod_{l=1}^{J} \frac{x_{ij}}{x_{kj} x_{il}} \right)^{\frac{1}{IJ}}, \tag{5}$$

respectively.

Before further analysing the compositional structures, it is convenient to express them in a system of $D - 1$ real-valued coordinates $\mathbf{z} = (z_1, \ldots, z_{D-1})'$. More specifically, a vector of isometric log-ratio (ilr) coordinates represents a mapping from $\mathcal{S}^D$ to $\mathbb{R}^{D-1}$. This real space representation of compositional data is defined as log-contrasts

$$z_i = \sum_{j=1}^{D} \xi_{ij} \ln x_j, \quad \sum_{j=1}^{D} \xi_{ij} = 0, \quad i = 1, \ldots, D - 1, \tag{6}$$

where $\mathbf{e}_i = \exp(\boldsymbol{\xi}_i) = \left[\exp(\xi_{i1}), \ldots, \exp(\xi_{iD})\right]'$, $i = 1, \ldots, D-1$ is a system of orthonormal vectors which forms an orthonormal basis of $\mathcal{S}^D$.

According to Hron et al. (2012), the linear regression model with a D-part compositional vector in the role of the explanatory variable can then be defined as

$$\mathrm{E}(Y|\mathbf{z}) = \beta_0 + \beta_1 z_1 + \cdots + \beta_{D-1} z_{D-1}. \tag{7}$$

Its alternative form will be mainly used in this chapter as follows:

$$\mathrm{E}\left[\ln(Y)|\mathbf{z}\right] = \beta_0 + \beta_1 z_1 + \cdots + \beta_{D-1} z_{D-1}. \tag{8}$$

Different ilr coordinate systems are orthogonal rotations of each other and therefore the general characteristics of the regression models (AIC, R^2, residuals, β_0, etc.) are invariant to the specific choice of the coordinate system. Despite this, the parameters β_1, ..., β_{D-1} vary according to the structure of coordinates, which is usually set with regards to a convenient interpretation. An ilr system, formed by three groups of coordinates, is defined in Fačevicová et al. (2018) to enable the real space representation of compositional tables. The construction is based on sequential binary partition (SBP) (Egozcue and Pawlowsky-Glahn 2005) applied separately to levels of each of the constituting factors. The first two groups of coordinates $\mathbf{z}^r$ and $\mathbf{z}^c$ of a compositional table $\mathbf{x}$ are thus simple balances between the whole rows and columns,

$$z_i^r = \sqrt{\frac{stJ}{s+t}} \ln \frac{\left[g(\mathbf{x}_{j_1\bullet}) \cdots g(\mathbf{x}_{j_s\bullet})\right]^{1/s}}{\left[g(\mathbf{x}_{k_1\bullet}) \cdots g(\mathbf{x}_{k_t\bullet})\right]^{1/t}}, \quad \text{for } i = 1, 2, \ldots, I-1 \tag{9}$$

and

$$z_j^c = \sqrt{\frac{uvI}{u+v}} \ln \frac{\left[g(\mathbf{x}_{\bullet l_1}) \cdots g(\mathbf{x}_{\bullet l_u})\right]^{1/u}}{\left[g(\mathbf{x}_{\bullet m_1}) \cdots g(\mathbf{x}_{\bullet m_v})\right]^{1/v}}, \quad \text{for } j = 1, 2, \ldots, J-1, \tag{10}$$

where s, t (u, v) are numbers of rows (columns) separated in the ith (jth) step of SBP, indices $(j_1\bullet, \ldots, j_s\bullet)$ and $(k_1\bullet, \ldots, k_t\bullet)$ or $(\bullet l_1, \ldots, \bullet l_u)$ and $(\bullet m_1, \ldots, \bullet m_v)$ denote rows/columns involved in this step and $g(.)$ stands for the geometric mean. These coordinates describe relations between levels of one factor when the influence of the second factor is suppressed. The coordinate system is complemented with a third group of $(I-1)(J-1)$ log-contrasts, which describe relations between factors. The interpretation of these coordinates is closely connected to the odds ratio between four groups of parts, which are uniquely determined by the initial row and column SBPs and represented by the geometric mean of the respective parts. When a, b, c, d are the numbers of parts within each group and $i_., j_., k_., l_.$ are indices of parts within each group, these coordinates can be in general defined as

Table 2 List of orthonormal coordinates and their values used for the practical example from Table 1

z_1^r	z_1^c	z_2^c	z_{11}^{OR}	z_{12}^{OR}
$\sqrt{\frac{3}{2}}\ln\frac{(x_{21}x_{22}x_{23})^{1/3}}{(x_{11}x_{12}x_{13})^{1/3}}$	$\sqrt{\frac{4}{3}}\ln\frac{(x_{13}x_{23})^{1/2}}{(x_{11}x_{12}x_{21}x_{22})^{1/4}}$	$\ln\frac{(x_{12}x_{22})^{1/2}}{(x_{11}x_{21})^{1/2}}$	$\sqrt{\frac{1}{3}}\ln\frac{(x_{11}x_{12})^{1/2}x_{23}}{(x_{21}x_{22})^{1/2}x_{13}}$	$\frac{1}{2}\ln\frac{x_{11}x_{22}}{x_{12}x_{21}}$
0.4317	1.0120	2.2777	−0.3520	−0.4769

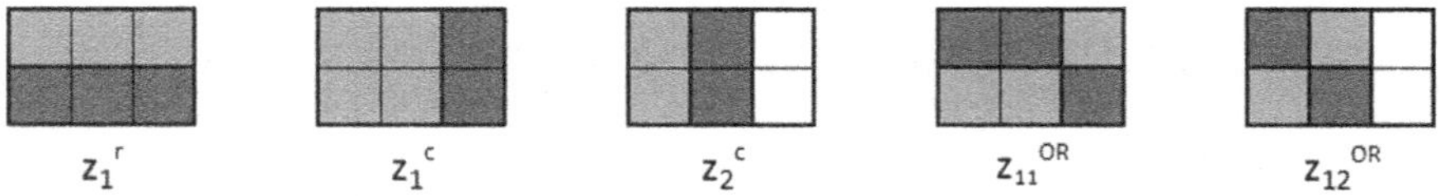

Fig. 1 Graphical representation of orthonormal coordinates defined in Table 2

$$z^{OR} = \sqrt{\frac{a \cdot d}{a+b+c+d}} \ln \frac{\left(x_{i_1} \cdots x_{i_a}\right)^{1/a} \left(x_{l_1} \cdots x_{l_d}\right)^{1/d}}{\left(x_{j_1} \cdots x_{j_b}\right)^{1/b} \left(x_{k_1} \cdots x_{k_c}\right)^{1/c}}. \tag{11}$$

The construction of coordinates $\mathbf{z}^{OR}$ is described in detail in Fačevicová et al. (2018), and the whole coordinate system can be computed with the function `tabCoord` from the R package `robCompositions` (Templ et al. 2011).

As an example, let us consider the 2 × 3 compositional table from Table 1. In order to represent the structure in orthonormal coordinates, the row and column SBPs need to be defined first. For the purpose of deeper analysis in Sect. 5, it seems to be convenient to focus on the relative dominance of men over women (which is quantified by the coordinate z_1^r). It furthermore seems convenient to concentrate on the overall population aged 30–34 with attained tertiary education (ED5-8) over the geometric mean of the remaining groups (ED0-2 and ED3-4)—z_1^c, complemented by the relative dominance of the upper secondary qualification (ED3-4) over those with the lowest achieved degree of education—z_2^c. The whole relative structure can therefore be represented with the system of coordinates that are listed in Table 2 and whose simple graphical description is presented in Fig. 1.

According to the values of the coordinate z_1^r, men in Leipzig dominated women across all attained educational classes in 2017. Considering the column balances z_1^c and z_2^c, we can conclude that there is (on average) a higher proportion of people with a university degree than those with primary or secondary educational qualification. Moreover, the coordinate z_2^c indicates that the population with attained secondary education prevails over those who attained only the lowest level of education.

An even deeper insight into the structure is provided by the odds ratio coordinates z_{11}^{OR} and z_{12}^{OR}. The negative value of z_{11}^{OR} highlights the fact that the proportion of women with tertiary education (in comparison to the geometric mean of those with the primary or secondary one) is slightly higher than the respective proportion of men with tertiary education. Furthermore, the relative predominance of secondary education over the primary one is slightly higher for women due to the negative value of z_{11}^{OR}. This difference is presumably caused by a small percentage of women with only a primary qualification.

The linear regression model with a compositional table as the explanatory variable can be written as

$$\begin{aligned} \mathrm{E}(Y|\mathbf{z}) = \beta_0 &+ \beta_1^r z_1^r + \cdots + \beta_{I-1}^r z_{I-1}^r + \beta_1^c z_1^c + \cdots \beta_{J-1}^c z_{J-1}^c \\ &+ \beta_{11}^{OR} z_{11}^{OR} + \cdots + \beta_{I-1,J-1}^{OR} z_{I-1,J-1}^{OR} \end{aligned} \tag{12}$$

or by incorporating a log-transformed response with respect to the application introduced in this chapter

$$\begin{aligned} \mathrm{E}\left[\ln(Y)|\mathbf{z}\right] = \beta_0 &+ \beta_1^r z_1^r + \cdots + \beta_{I-1}^r z_{I-1}^r + \beta_1^c z_1^c + \cdots \beta_{J-1}^c z_{J-1}^c \\ &+ \beta_{11}^{OR} z_{11}^{OR} + \cdots + \beta_{I-1,J-1}^{OR} z_{I-1,J-1}^{OR}. \end{aligned} \tag{13}$$

Similar to the case of classical regression, the intercept β_0 equals the predicted value of Y (or $\ln(Y)$) when all coordinates are zero. Zero values of all odds ratio coordinates correspond to the situation when the row and column factors are completely independent. Moreover, the zero values of all row and column balances reflect a situation when all parts of the compositional table are equal. The remaining parameters quantify the effect of a unit change of the respective coordinate (while fixing the remaining ones) on the response variable. A deeper insight into the model structure is needed to enable the interpretation of this effect in terms of the original compositional parts.

Consider a linear regression model (13) with regressors in the form of ilr coordinates of a 2×3 table, defined in Table 2. The model can be written as

$$\mathrm{E}\left[\ln(Y)|\mathbf{z}\right] = \beta_0 + \beta_1^r z_1^r + \beta_1^c z_1^c + \beta_2^c z_2^c + \beta_{11}^{OR} z_{11}^{OR} + \beta_{12}^{OR} z_{12}^{OR}. \tag{14}$$

The value of the parameter β_1^r quantifies the effect of a k-times growth of the relative dominance of the second level of the row factor. More specifically, this change, with simultaneously fixed values of the remaining coordinates, results in an $\exp\left(\sqrt{3/2}\beta_1^r \ln k\right)$-times growth of the response variable. Similar reasoning can also be done for the other coefficients. The effects of ratio changes forming the remaining regressors are summarised in Table 3. Note here that only finite changes of the coordinates are considered within this chapter and its infinitesimal alternatives are for the case of a vector compositional covariate discussed in Morais et al. (2018).

The condition of fixed regressors implies that the change of the coordinate of interest should be caused by a simultaneous increase of one constituting group and a decrease of the other. For instance, an increase of the coordinate z_2^c can reflect the situation when the proportion of ED3-4 men and women grows together with a decrease of proportions belonging to the group ED0-2 and an unchanged proportion of men and women with tertiary education. Another complex case can be derived when an increase of the coordinate of interest is not caused by simultaneous changes

Table 3 Interpretation of regression parameters of model (14). Column z denotes the coordinate corresponding to the given parameter β and z' its changed value. The original prediction y and its new value y', resulting from the model with z' and other coordinates fixed, are compared in $\ln(y'/y)$. Column $\mathbf{x}'$ introduces possible changes in the compositional table $\mathbf{x}$, which lead to a change in the coordinate z and simultaneously keep the remaining coordinates unchanged.

β	z	z'	$\ln(y'/y)$	$\mathbf{x}'$
β_1^r	$\sqrt{\frac{3}{2}} \ln \frac{(x_{21}x_{22}x_{23})^{1/3}}{(x_{11}x_{12}x_{13})^{1/3}}$	$\sqrt{\frac{3}{2}} \ln k \frac{(x_{21}x_{22}x_{23})^{1/3}}{(x_{11}x_{12}x_{13})^{1/3}}$	$\sqrt{\frac{3}{2}} \beta_1^r \ln k$	$x_{11} \quad x_{12} \quad x_{13}$ $kx_{21} \quad kx_{22} \quad kx_{23}$
β_1^c	$\sqrt{\frac{4}{3}} \ln \frac{(x_{13}x_{23})^{1/2}}{(x_{11}x_{12}x_{21}x_{22})^{1/4}}$	$\sqrt{\frac{4}{3}} \ln k \frac{(x_{13}x_{23})^{1/2}}{(x_{11}x_{12}x_{21}x_{22})^{1/4}}$	$\sqrt{\frac{4}{3}} \beta_1^c \ln k$	$x_{11} \quad x_{12} \quad kx_{13}$ $x_{21} \quad x_{22} \quad kx_{23}$
β_2^c	$\ln \frac{(x_{12}x_{22})^{1/2}}{(x_{11}x_{21})^{1/2}}$	$\ln k \frac{(x_{12}x_{22})^{1/2}}{(x_{11}x_{21})^{1/2}}$	$\beta_2^c \ln k$	$\frac{1}{\sqrt{k}}x_{11} \quad \sqrt{k}x_{12} \quad x_{13}$ $\frac{1}{\sqrt{k}}x_{21} \quad \sqrt{k}x_{22} \quad x_{23}$
β_{11}^{OR}	$\sqrt{\frac{1}{3}} \ln \frac{(x_{11}x_{12})^{1/2}x_{23}}{(x_{21}x_{22})^{1/2}x_{13}}$	$\sqrt{\frac{1}{3}} \ln k \frac{(x_{11}x_{12})^{1/2}x_{23}}{(x_{21}x_{22})^{1/2}x_{13}}$	$\sqrt{\frac{1}{3}} \beta_{11}^{OR} \ln k$	$\sqrt[6]{k}x_{11} \quad \sqrt[6]{k}x_{12} \quad \frac{1}{\sqrt[3]{k}}x_{13}$ $\frac{1}{\sqrt[6]{k}}x_{21} \quad \frac{1}{\sqrt[6]{k}}x_{22} \quad \sqrt[3]{k}x_{23}$
β_{12}^{OR}	$\frac{1}{2} \ln \frac{x_{11}x_{22}}{x_{12}x_{21}}$	$\frac{1}{2} \ln k \frac{x_{11}x_{22}}{x_{12}x_{21}}$	$\frac{1}{2} \beta_{12}^{OR} \ln k$	$\sqrt[4]{k}x_{11} \quad \frac{1}{\sqrt[4]{k}}x_{12} \quad x_{13}$ $\frac{1}{\sqrt[4]{k}}x_{21} \quad \sqrt[4]{k}x_{22} \quad x_{23}$

of all involved compositional parts (as described in Table 3) but, e.g. by an increase of only those parts, which form the nominator of the respective formula. Such changes of the relative structure would also affect (some of) the other coordinates. The resulting effect on the response will therefore be strengthened or weakened with these supplementary effects. Some of these scenarios are with respect to the model (14) described in Table 4.

3 Geographically Weighted Regression Model

There are situations where the analysed phenomena vary over space. That is when each observation corresponds to some geographic location. An approach to deal with the problem of spatial non-stationarity was given in Brunsdon et al. (1998) where authors define the geographically weighted regression (GWR) model. The main idea of this approach is that the regression parameters $\boldsymbol{\beta}$ can vary over space. Therefore a system of n *local models* is estimated. Each model corresponds to one geographical location i and is based on the observations which are geographically *close enough* to the given one. The local model can be formally written as

$$\mathrm{E}(Y_i|\mathbf{v}_i) = \beta_{i0} + \sum_{k=1}^{m} \beta_{ik} v_{ik}, \tag{15}$$

where Y_i and $\mathbf{v}_i = (v_{i1}, \ldots, v_{im})'$ are values of the dependent and independent variables measured at the location i and $\boldsymbol{\beta}_i = (\beta_{i0}, \ldots, \beta_{im})'$ are the respective regression coefficients. Within the GWR approach, the observation which is closer to the loca-

Table 4 List of possible changes in the relative structure of a compositional table **x** (column **x**′), their impact on ilr coordinates (z'), when the not-listed coordinates remain unchanged, and the final effect on the value of the response ($\ln\left(y'/y\right)$).

$\mathbf{x}'$	z'	$\ln\left(y'/y\right)$
$x_{11}\ kx_{12}\ x_{13}$ $x_{21}\ kx_{22}\ x_{23}$	$z_2^{c\prime} = \ln k \frac{(x_{12}x_{22})^{1/2}}{(x_{11}x_{21})^{1/2}}$	$\beta_2^c \ln k - \beta_1^c\sqrt{\frac{4}{3}}\ln\sqrt{k}$
	$z_1^{c\prime} = \sqrt{\frac{4}{3}}\ln\frac{1}{\sqrt{k}}\frac{(x_{13}x_{23})^{1/2}}{(x_{11}x_{12}x_{21}x_{22})^{1/4}}$	
$\frac{1}{k}x_{11}\ x_{12}\ x_{13}$ $\frac{1}{k}x_{21}\ x_{22}\ x_{23}$	$z_2^{c\prime} = \ln k \frac{(x_{12}x_{22})^{1/2}}{(x_{11}x_{21})^{1/2}}$	$\beta_2^c \ln k + \beta_1^c\sqrt{\frac{4}{3}}\ln\sqrt{k}$
	$z_1^{c\prime} = \sqrt{\frac{4}{3}}\ln\sqrt{k}\frac{(x_{13}x_{23})^{1/2}}{(x_{11}x_{12}x_{21}x_{22})^{1/4}}$	
$x_{11}\quad x_{12}\quad x_{13}$ $\frac{1}{\sqrt{k}}x_{21}\ \frac{1}{\sqrt{k}}x_{22}\ \sqrt{k}x_{23}$	$z_{11}^{OR\prime} = \sqrt{\frac{1}{3}}\ln k\frac{(x_{11}x_{12})^{1/2}x_{23}}{(x_{21}x_{22})^{1/2}x_{13}}$	$\beta_{11}^{OR}\sqrt{\frac{1}{3}}\ln k -$
	$z_1^{r\prime} = \sqrt{\frac{3}{2}}\ln\frac{1}{\sqrt[6]{k}}\frac{(x_{21}x_{22}x_{23})^{1/3}}{(x_{11}x_{12}x_{13})^{1/3}}$	$\beta_1^r\sqrt{\frac{3}{2}}\ln\sqrt[6]{k} + \beta_1^c\sqrt{\frac{4}{3}}\ln\sqrt{k}$
	$z_1^{c\prime} = \sqrt{\frac{4}{3}}\ln\sqrt{k}\frac{(x_{13}x_{23})^{1/2}}{(x_{11}x_{12}x_{21}x_{22})^{1/4}}$	
$\sqrt{k}x_{11}\ \sqrt{k}x_{12}\ \frac{1}{\sqrt{k}}x_{13}$ $x_{21}\quad x_{22}\quad x_{23}$	$z_{11}^{OR\prime} = \sqrt{\frac{1}{3}}\ln k\frac{(x_{11}x_{12})^{1/2}x_{23}}{(x_{21}x_{22})^{1/2}x_{13}}$	$\beta_{11}^{OR}\sqrt{\frac{1}{3}}\ln k -$
	$z_1^{r\prime} = \sqrt{\frac{3}{2}}\ln\frac{1}{\sqrt[6]{k}}\frac{(x_{21}x_{22}x_{23})^{1/3}}{(x_{11}x_{12}x_{13})^{1/3}}$	$\beta_1^r\sqrt{\frac{3}{2}}\ln\sqrt[6]{k} - \beta_1^c\sqrt{\frac{4}{3}}\ln\sqrt{k}$
	$z_1^{c\prime} = \sqrt{\frac{4}{3}}\ln\frac{1}{\sqrt{k}}\frac{(x_{13}x_{23})^{1/2}}{(x_{11}x_{12}x_{21}x_{22})^{1/4}}$	
$\sqrt{k}x_{11}\ \frac{1}{\sqrt{k}}x_{12}\ x_{13}$ $x_{21}\quad x_{22}\quad x_{23}$	$z_{12}^{OR\prime} = \frac{1}{2}\ln k\frac{x_{11}x_{22}}{x_{12}x_{21}}$	$\beta_{12}^{OR}\frac{1}{2}\ln k - \beta_2^c\ln\sqrt{k}$
	$z_2^{c\prime} = \ln\frac{1}{\sqrt{k}}\frac{(x_{12}x_{22})^{1/2}}{(x_{11}x_{21})^{1/2}}$	
$x_{11}\quad x_{12}\quad x_{13}$ $\frac{1}{\sqrt{k}}x_{21}\ \sqrt{k}x_{22}\ x_{23}$	$z_{12}^{OR\prime} = \frac{1}{2}\ln k\frac{x_{11}x_{22}}{x_{12}x_{21}}$	$\beta_{12}^{OR}\frac{1}{2}\ln k + \beta_2^c\ln\sqrt{k}$
	$z_2^{c\prime} = \ln\sqrt{k}\frac{(x_{12}x_{22})^{1/2}}{(x_{11}x_{21})^{1/2}}$	

tion i has a greater influence on the model estimation at the location i. The regression parameters are therefore estimated using a weighted least squares formula

$$\hat{\boldsymbol{\beta}}_i = \left(V'W_iV\right)^{-1}V'W_i\mathbf{y}, \tag{16}$$

where V is an n times $m+1$ data matrix, $\mathbf{y} = (Y_1, \ldots, Y_n)'$ is vector of values of the response variable and W_i is the diagonal matrix

$$W_i = \begin{pmatrix} w_1^i & 0 & \cdots & 0 \\ 0 & w_2^i & \cdots & 0 \\ \cdots & & \ddots & 0 \\ 0 & 0 & \cdots & w_n^i \end{pmatrix}. \tag{17}$$

Table 5 List of kernel functions frequently used for computation of weights in GWR (Gollini et al. 2015). w_j^i denotes the weight of the jth observation in the ith regression model, d_{ij} is the geographical distance between the ith and jth location and b is the bandwidth

Global model	$w_j^i = 1$
Gaussian	$w_j^i = \exp\left(-\frac{1}{2}\left(\frac{d_{ij}}{b}\right)^2\right)$
Exponential	$w_j^i = \exp\left(-\frac{\lvert d_{ij}\rvert}{b}\right)$
Bi-square	$w_j^i = \begin{cases}\left(1-\left(\frac{d_{ij}}{b}\right)^2\right)^2 & \text{if } \lvert d_{ij}\rvert < b,\\ 0 & \text{otherwise}\end{cases}$
Tri-cube	$w_j^i = \begin{cases}\left(1-\left(\frac{\lvert d_{ij}\rvert}{b}\right)^3\right)^3 & \text{if } \lvert d_{ij}\rvert < b,\\ 0 & \text{otherwise}\end{cases}$
Box-car	$w_j^i = \begin{cases}1 & \text{if} \lvert d_{ij}\rvert < b,\\ 0 & \text{otherwise}\end{cases}$

The diagonal elements w_j^i correspond to the weights of each data point j and are assigned according to its geographical (Euclidean) distance from the actual location i. The geographical weighting is based on the value of a kernel function. The most frequently used ones are summarised in Table 5 according to Gollini et al. (2015). When the *global model* incorporates all observations with the same weight, it corresponds to the classical ordinary least squares (OLS) approach. If the Gaussian or Exponential kernel is used, observations which are far from the ith location contribute to the model estimation with a smaller weight. The bi-square, tri-cube or box-car kernels use only those observations which are within some bandwidth b around the location i. The bandwidth b is one of the key parameters of the local model estimation. It can be defined as the distance from the centre point i and is then fixed for all local models. Alternatively, if the adaptive version of the algorithm is used, b can vary according to a given number of local data. The optimal bandwidth choice is usually found based on a cross-validation method or a model performance criterion like Akaike Information Criterion (AIC) or its corrected version (AICc) (see Brunsdon et al. (1998); Gollini et al. (2015) for details). The bandwidth can also be determined by an expert choice based on the knowledge of real spatial relationships given by the behaviour of the observed phenomenon (e.g. the influence of commuting between municipalities).

Local models are characterised with local estimates of the regression coefficients $\boldsymbol{\beta}_i$ supplemented by the respective t-statistics and corresponding local R_i^2 values for each model. These characteristics can be compared with those of the global model, and their proper geographical visualisation can facilitate the exploration of spatial non-stationarity of the modelled phenomenon (Matthews and Yang 2012). Moreover, the overall performance of local models can be expressed by the local R^2 values (or alternatively by AIC_{local}) which are based on the residuals in the centre location of each partial model.

4 Geographically Weighted Regression for Compositional Tables

Similar to other conventional statistical methods, the GWR approach to modelling spatial non-stationary relations is primarily designed for real-valued regressors. When a variable of interest is modelled according to a relative structure of some regressor variable, the use of an ilr coordinate representation of the structure is needed. Therefore, the global model for one- or two-dimensional compositional regressors can be described by Eqs. (8) and (13), respectively. Assuming the case of spatial dependent two-factorial compositional data, the global model can be further extended with a set of local models as follows:

$$\begin{aligned} \mathrm{E}(Y_i|\mathbf{z}_i) = \beta_{i,0} + \beta^r_{i,1} z^r_{i,1} + \cdots + \beta^r_{i,I-1} z^r_{i,I-1} + \beta^c_{i,1} z^c_{i,1} + \cdots \beta^c_{i,J-1} z^c_{i,J-1} \\ + \beta^{OR}_{i,11} z^{OR}_{i,11} + \cdots + \beta^{OR}_{i,I-1,J-1} z^{OR}_{i,I-1,J-1} \end{aligned} \tag{18}$$

or alternatively, for the logarithmic response,

$$\begin{aligned} \mathrm{E}\left[\ln(Y_i)|\mathbf{z}_i\right] = \beta_{i,0} + \beta^r_{i,1} z^r_{i,1} + \cdots + \beta^r_{i,I-1} z^r_{i,I-1} + \beta^c_{i,1} z^c_{i,1} + \cdots \beta^c_{i,J-1} z^c_{i,J-1} \\ + \beta^{OR}_{i,11} z^{OR}_{i,11} + \cdots + \beta^{OR}_{i,I-1,J-1} z^{OR}_{i,I-1,J-1}. \end{aligned} \tag{19}$$

The model coefficients are estimated using the weighted least squares formula (16), where the data matrix V is now an n times IJ matrix of coordinates defined by Eqs. (9), (10) and (11). Similar to the case of the global model, the performance of its local alternatives does not depend on the specific choice of the coordinate system. The interpretation of the regression coefficients should be carried out with respect to the complexity of the modelled structure and taking into account findings from Sect. 2.

5 Application

The practical part of this chapter focuses on a country example of Germany. The aim of the study is to explore and analyse the relationship between the risk of falling into poverty and the educational attainment structure of the 30–34-aged population and its variation over space.

Poverty and inequality are not only an issue of developing countries. The vulnerability towards falling into poverty is driven by many factors determined by national, sector-specific, community, household or individual characteristics. The level and depth of poverty also increase as a result of any crisis with strong economic impacts such as an increase in unemployment inducing a reduction of available income of households (Duiella and Turrini 2014).

From a global perspective, the number of people living in extreme poverty in Germany is negligible compared to developing countries. However, poverty in Europe is not typically measured in absolute terms. People are considered poor if they have lower income and fewer opportunities than other individuals living in the same society. In fact, the percentage of people living at risk of poverty and social exclusion in Germany in 2017 was 19.0%, almost one-fifth of the population.

Many research findings suggest that factors that often condition the number of people living at risk of poverty or social exclusion in Germany are being of young age, single, a lone parent. An important role play also regional aspects such as unemployment, low education and the region of East Germany (Brülle 2018). People with a lower level of educational background and limited access to the labour market are highly vulnerable to easily fall into this group. Moreover, high rental costs, especially in urban areas, pose an additional burden to many households (Pieper et al. 2019).

5.1 Data and Methods

The data were processed in the environment of the R Software (R Core Team 2020) with packages `robCompositions` for the computation of coordinates and `GWmodel` and `spdep` for the estimation of the global and local models (Gollini et al. 2015; Bivand et al. 2013). The maps were produced using the ArcGIS® software by Esri (2011).

Data used for the analysis are sourced from the regional database of EUROSTAT accessed and processed through the R package `eurostat` (Lahti et al. 2017). The data were collected by the Federal Statistical Office Germany through Labour Force Survey (EU-LFS) and Statistics on Income and Living Conditions (EU-SILC). Both types of information are available by nomenclature of territorial units for statistics (NUTS2) regions.

The indicator chosen to represent the poverty dimension is the share of people at risk of poverty or social exclusion, i.e. the at-risk-of-poverty rate (RoP). At risk-of-poverty are people with an equivalised disposable income below the risk-of-poverty threshold, which is set at 60% of the national median equivalised disposable income (after social transfers).

The compositional tables show the structure of the population aged 30–34 by sex and the highest level of education successfully completed. The level of education attained is classified by International Standard Classification of Education (ISCED). In this chapter, the classes chosen for further analysis are

- Less than primary, primary and lower secondary education (ED0-2),
- Upper secondary and post-secondary non-tertiary education (ED3-4) and
- Tertiary education (ED5-8).

Educational attainment data series are available from 2000 to 2017. The information on the risk of poverty or social exclusion by NUTS2 regions is available only for

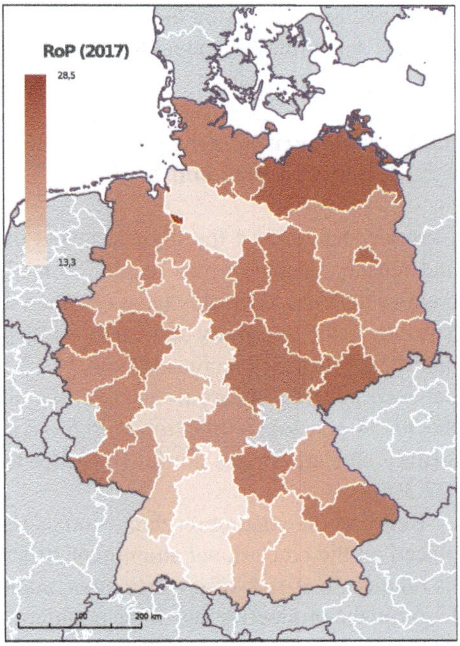

Fig. 2 People at risk of poverty or social exclusion in Germany by NUTS2 regions in 2017 (in percentages)

2016 and 2017. For the assessment of the regression models, the relationship between the dependent and independent variables is modelled for 36 German regions in the latest available year 2017.

5.2 *Poverty Analysis*

At-risk-of-poverty rates vary substantially across German regions. Values reported show the minimum of 13.3% in Lüneburg and the maximum of 28.5% in Bremen in 2017. Figure 2 demonstrates the distribution of people at risk of poverty or social exclusion in Germany by NUTS2 regions in 2017.

The highest at-risk-of-poverty rates are experienced by urban areas such as Bremen, Berlin and Hamburg, together with regions of the former East Germany, in particular Mecklenburg-Vorpommern, Chemnitz, Thuringia and Saxony-Anhalt. In contrast, Lüneburg followed by all regions of Baden-Württemberg have the lowest shares of people exposed to the risk of poverty or social exclusion.

Education can serve as an elevator for upward economic mobility. Particularly higher levels of education increase chances of participation in the labour market. The level of attained education influences income distribution and thus has a substantial effect on at-risk-of-poverty rates. As education plays an important role, the application of the methodology is illustrated on modelling people at risk of poverty or social exclusion by using the information on the structure of educational attainment level by sex within the population aged 30–34.

5.2.1 Comparison of Educational Attainment Structure in Germany Between 2000 and 2017

Germany has one of the highest levels of upper secondary education attainment, and tertiary education has expanded in the past decades. In 2019, 86.7% of the country's 25–64-year-olds have obtained at least an upper secondary qualification. The expansion of tertiary education is reflected by the number of people with tertiary qualifications which increased from 24.9% in 2004 to 29.9% in 2019. The differences between generations are noticeable: the share of people with tertiary education in the 30–34-year-old population is 35.5% compared to 26.9% of 55–64-year-olds. The tertiary education rates are fairly higher for men (32.4% in 2019) than for women (27.3%).

Building on existing research studies (Hanesch 2019), the focus of this article is placed on the educational attainment structure of population aged 30–34 to explore its relation to at-risk-of-poverty rates. Such constructed compositional tables which focus on the productive age of the population (30–34-year-olds) reveal patterns of the attained level of education with respect to gender characteristics. An example of such a compositional table is illustrated in Table 1 reflecting the educational attainment structure in Leipzig.

The coordinate representation of compositional tables, which is defined in Table 2 and whose schematic structure is given in Fig. 1, also enables us to compare the structure over time. The coordinates calculated for years 2000 and 2017 reveal a clear structural change in the German educational system.

Despite the lack of data for several regions, particularly interesting results are captured by coordinates z_1^c and z_{11}^{OR}, which quantify the relative prevalence of university-educated citizens and further compare their proportion within men and women.

The positive values of z_1^c indicate that the component of the population with tertiary education on average dominates the population with secondary or primary education across all women and men aged 30–34 in Germany. Figure 3 suggests that the importance of tertiary education has grown significantly in Germany in the past decades. While most of the German regions showed negative values in 2000 representing a relative dominance of secondary and primary education, this trend reversed in 2017. The majority of former Western German regions show a dominance of tertiary education in 2017 with the exception of Lower Saxony and Saarland. Both regions have undergone a structural change moving from mining-oriented economies towards high-technology industrial production. Negative values can also be observed in regions of former East Germany indicating a relative predominance of vocational qualifications across the population aged 30–34 in this area in 2017. Moreover, the relative dominance of the upper secondary attainment over the primary one in these regions is supported by positive values of the coordinate z_2^c.

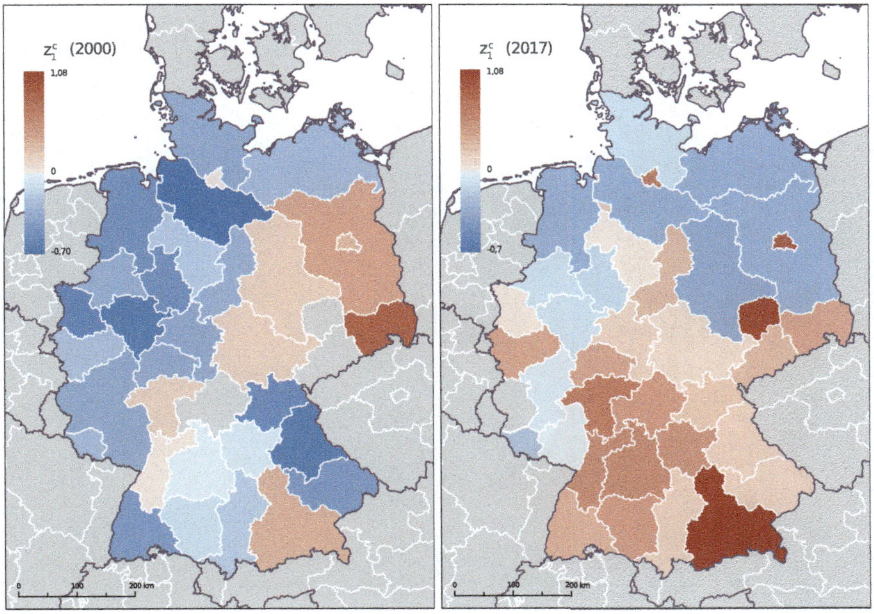

Fig. 3 Comparison of z_1^c in 2000 (left) and 2017 (right)

The coordinate z_{11}^{OR} examines the proportion of men with tertiary education with the proportion of women with tertiary education. The comparison of z_{11}^{OR} in 2000 and 2017 shows a much wider range of the coordinate values in 2000. This indicates greater disparities between men and women in the distribution of attained education in the population aged 30–34 in Germany in 2000. Positive values correspond to the situation when the proportion of men with tertiary education is higher in the 30–34-aged population than the proportion of women with tertiary education, which is a pattern observed in all regions of former West Germany. In past decades, the disparity has been significantly reduced as noticed in Fig. 4. No substantial change has been noticed for the territory of former East Germany where negative values of z_{11}^{OR} still prevail in most of the regions. In contrast, the dominance of tertiary education over the other education levels among 30–34-aged men has become similar to the situation of women in the west of Germany. The most balanced regions are, for instance, southern Bavaria, Hessen or the eastern part of Lower Saxony, which show z_{11}^{OR} values around 0.

5.2.2 The GWR Model

The introduced methodology is applied to explore whether the relationship of people at risk of poverty or social exclusion and the educational attainment structure by sex varies across German regions in the population of 30–34-years-olds. As a first

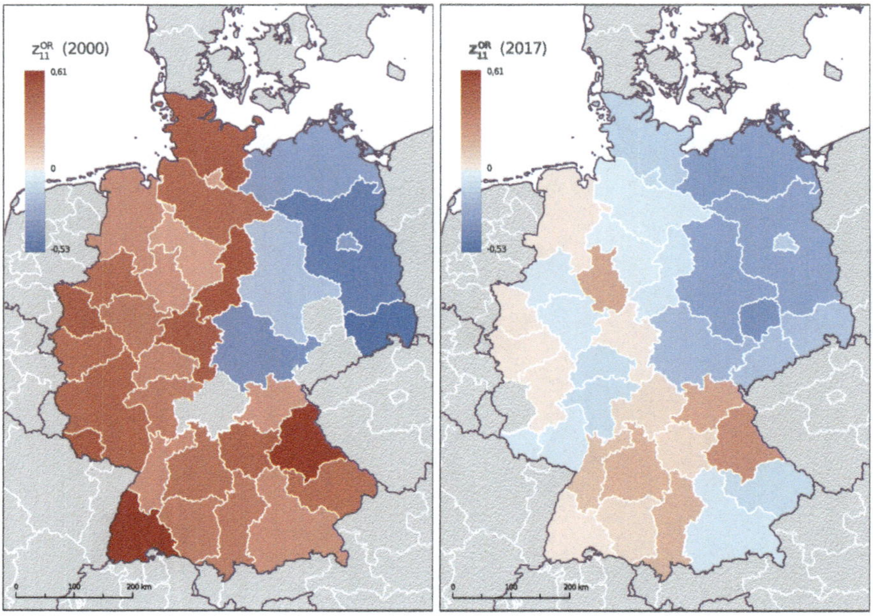

Fig. 4 Comparison of z_{11}^{OR} in 2000 (left) and 2017 (right)

Table 6 Results from the global OLS model. The R^2 and R^2_{adj} value are 0.370 and 0.265, residual sum of squares is 0.748 on 30 degrees of freedom and value of the F statistic is 3.527 on 5 and 30 degrees of freedom, AIC equals −23.292 and AICc −19.292

	β_0	β_1^r	β_1^c	β_2^c	β_{11}^{OR}	β_{12}^{OR}
Estimates	3.027	1.337	−0.040	−0.119	−0.913	1.250
Significance	* * *	*			* * *	**
Std. deviations	0.139	0.692	0.070	0.115	0.249	0.457
Coordinates		z_1^r	z_1^c	z_2^c	z_{11}^{OR}	z_{12}^{OR}

step, the linear regression model (14) with a compositional table as the explanatory variable was fitted to the examined data set. The global model diagnostics is shown in Table 6. According to the estimate of parameter β_0 of the global model and the interpretation of parameters defined in Sect. 2, the basic level of the RoP when all groups are equally frequent is approximately 20% (exp (3.027)).

The level of RoP is significantly affected by the OR coordinates z_{11}^{OR} and z_{12}^{OR} and therefore the comparison of the educational structure within the male and female population seems to be important. According to the parameter $\beta_{11}^{OR} = -0.91$ and findings listed in Table 3, a doubling of the proportion of men with tertiary education (related to the average proportion of those with the secondary or lower degree) in comparison to the same proportion for women is related to the decrease of the

RoP by approximately one-third ($\exp(-0.91\sqrt{1/3}\ln 2) = 0.69$). In order to keep the remaining coordinates fixed, this change can be caused by a simultaneous increase in groups of Male ED5-8 and Female ED0-2 and ED3-4 together with a proportional decrease in the remaining ones (see Table 3).

The second significant factor, which affects the overall value of at-risk-of-poverty rate, is the proportion of men with secondary education (within the group of all men who attained at least the upper secondary education) in comparison to the same proportion for women. According to the value of the parameter $\beta_{12}^{OR} = 1.25$ and Table 3, a doubling of the respective log-odds ratio is related to an increase of at-risk-of-poverty rate by 50% ($\exp(0.5 \cdot 1.25 \ln 2) = 1.54$). This growth can be caused by a simultaneous increase of ED0-2 women and ED3-4 men and a proportional decrease by ED3-4 women and ED0-2 men. In a less-complex situation, when the difference is caused by a change in just one category (an increase of the proportion of women with the lowest level of attained education or conversely in an increase of men with at least the secondary qualification), the change affects also the value of the coordinate z_2^c. According to Table 4, the overall change of at-risk-of-poverty rate is given as $0.5\beta_{12}^{OR} \ln k - \beta_2^c \ln \sqrt{k}$. Considering negative values of the parameter β_2^c, an increase of the proportion of ED0-2 women seems to be more critical as it would result in a growth of the at-risk-of-poverty rate by 60% ($\exp(0.5 \cdot 1.25 \ln 2 + 0.119 \ln \sqrt{2}) = 1.61$).

Moran's I test is employed to test the presence of spatial autocorrelation among residuals in our data set. We assume that the disturbances are normally distributed and therefore Moran's I test under normality is conducted (Leung et al. 2000). The resulting Moran's I statistic is 0.021 with the standard deviation of 0.448. The p-value of 0.327 suggests that the statistic does not significantly differ from zero and there is thus no spatial autocorrelation among the residuals. Nevertheless, we assume that the relationship between at-risk-of-poverty rates and the educational attainment structure might vary geographically. The GWR model for compositional tables is thus used to explore variations of this relationship across German regions.

It should be highlighted that the choice of the coordinate system does not affect the resulting outcome from either the global and local regression models. The coordinate system is chosen with respect to the convenience of results' interpretation. Our application builds on the coordinates constructed with the intention of exploring the relative dominance of attained tertiary education in the overall population aged 30–34 as an apparent determinant of at-risk-of-poverty rates.

The local regression model has been selected with the following settings of input parameters. The bi-square kernel type function was set to `ADAPTIVE` as it fits our data the best using the corrected AIC values (AICc). All 36 regions were considered in the estimation of each set of coefficients. The summary diagnostics of our local GWR model is provided in Table 7.

The comparison of both the global and the local regression models can be carried out through the AIC or AICc values (Fotheringham et al. 2002, p. 96). The AIC value of the local model $AIC_{local} = -36.756$ is compared to the global one $AIC_{global} = -23.292$, moreover the local and global corrected criterions $AICc_{local} = -14.210$

Table 7 Results from the local GWR model with the adaptive bandwidth of 36 (all regions), the given p-values are related to the Monte Carlo test of spatial variability of the coefficients. The R^2 and R^2_{adj} values are 0.514 and 0.254, residual sum of squares 0.577 with effective degrees of freedom 23.812. The effective number of parameters is 12.19, AIC equals −36.756 and AICc −14.210

	β_0	β_1^r	β_1^c	β_2^c	β_{11}^{OR}	β_{12}^{OR}
Min	2.778	−0.230	−0.233	−0.189	−0.946	0.283
1st Qu.	2.813	0.285	−0.146	−0.112	−0.869	0.658
Median	2.923	0.973	−0.060	−0.015	−0.764	1.248
3rd Qu.	3.025	1.419	0.012	0.083	−0.728	1.330
Max	3.122	1.479	0.019	0.148	−0.692	1.462
MCtest p-value	0.071	0.217	0.010	0.059	0.750	0.287
Coordinates		z_1^r	z_1^c	z_2^c	z_{11}^{OR}	z_{12}^{OR}

and $AICc_{global} = -19.292$ were compared. The resulting statistics indicate a lower value of AIC for the local model, however, the AICc shows slightly better performance for the global one. Alternatively, the coefficient of determination R^2 and its adjusted version R^2_{adj} are examined. The global R^2 is 0.370 and the local R^2 0.514, the global R^2_{adj} is 0.265 and the local R^2_{adj} 0.254, respectively. Although the coefficient R^2 has increased for the GWR model, the gain is probably driven by the difference in the number of parameters and degrees of freedom as the adjusted R^2_{adj} values remain almost unchanged.

Table 7 summarises the main statistics of estimated parameters generated using the local model. The median values of parameters preserve the same signs as the estimates resulting from the global model (Table 6). The local measure of goodness-of-fit R^2, whose geographical distribution is visualised in Fig. 5, varies between 0.48 and 0.58, which is satisfactory but with limitations. The R^2 values show a better fit for the model in the part of former East Germany and in the south of the country. In addition, each regression coefficient is accompanied by the p-value of the Monte Carlo test of its spatial variability (Brunsdon et al. 1998). The test results complement the ambiguous comparison of the local and the global model provided by AICs and R^2s. The majority of coefficients does not significantly vary across locations. However, the hypothesis that the parameter is globally fixed was rejected for β_1^c and values for β_0 and β_2^c are quite marginal. Hence, local differences are apparently driven by the effect of changes in the structure of the population with the highest level of attained education. Values of β_1^c indicate that the growth of the mean proportion of the population with tertiary education has a positive effect on at-risk-of-poverty rates in the north of Germany, while it leads to an at-risk-of-poverty rate decrease in the southern regions. Moreover, the parameter β_2^c suggests that the growth of the (mean) relative proportion of people with only primary qualification (in comparison to those with secondary education) contributes to higher at-risk-of-poverty rates in Northern Germany but it results in lower rates in the south.

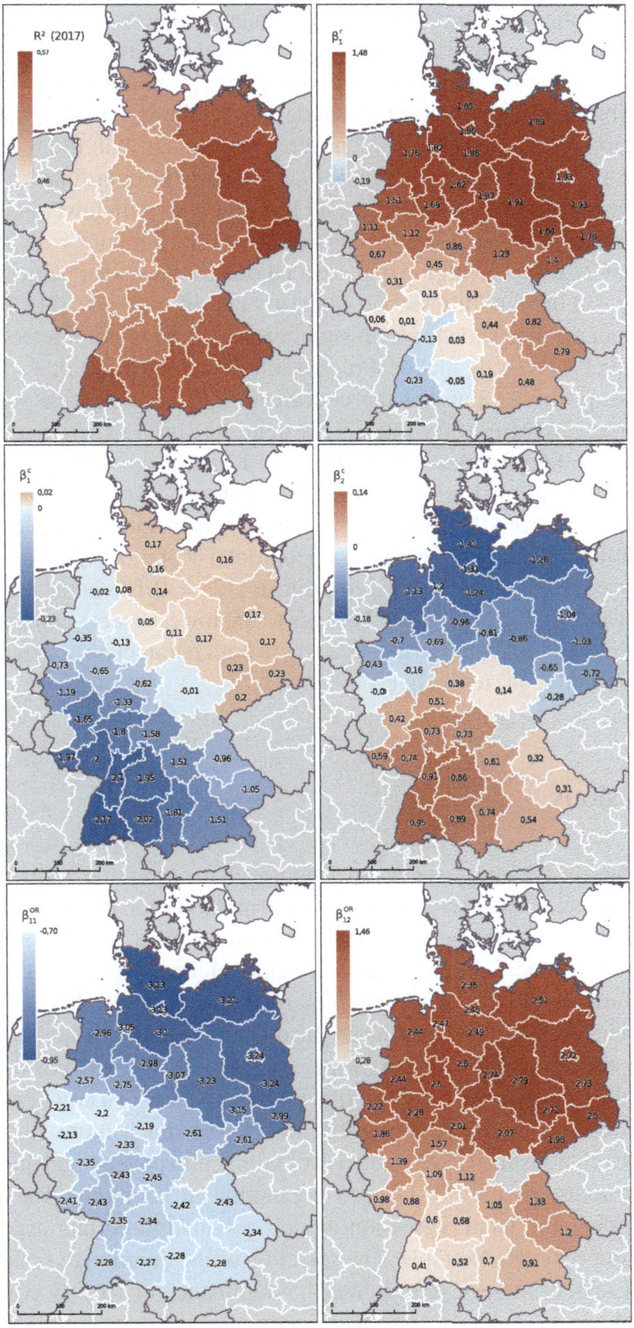

Fig. 5 Maps demonstrating estimates of regression parameters of the local regression models and values of the respective R^2 characteristic. The local t-value of the respective estimate is displayed in the centre of each region

According to the local regression model, apparent interactions between coordinates and at-risk-of-poverty rates are observed. Higher at-risk-of-poverty rates are associated with negative values of the coordinate z_{11}^{OR} and positive values of z_{12}^{OR}. Regions exhibiting this particular behaviour are Hamburg, Arnsberg and Saarland. All these regions reveal the at-risk-of-poverty rate higher than the overall national rate of 19.0% reported by Germany in 2017. However, the coordinates z_{11}^{OR} and z_{12}^{OR} attain values very close to zero, and they all lie in the area with lower goodness-of-fit measures R^2. The strongest association is observed in Hamburg with an at-risk-of-poverty rate of 21.6%. The region experienced one of the lowest values of the coordinate $z_{11}^{OR} = -0.085$ and the second highest value of $z_{12}^{OR} = 0.180$. Furthermore, Hamburg lies in the area showing high significant t-values of parameters (−2.96 and 2.44, respectively) given by the local regression models; see Fig. 5.

In contrast, positive values of the coordinate z_{11}^{OR} and negative values of z_{12}^{OR} lead to lower at-risk-of-poverty rates. This trend has been experienced by regions of Weser-Ems, Detmold, Kassel, Unterfranken, Karlsruhe and Freiburg. Except for Weser-Ems, all mentioned regions experienced at-risk-of-poverty rates below the German national rate of 19.0% in 2017. Examining a specific example of Detmold, the trend is further illustrated. The share of people at risk of poverty or social exclusion in Detmold was 16.9% in 2017. The coordinate $z_{11}^{OR} = 0.186$ is the second highest one, and the value of $z_{12}^{OR} = -0.045$ belongs to the lowest scores among all regions. Moreover, Detmold is situated in the area where both parameters are significant with t-values −2.75 and 2.44, respectively.

6 Conclusions

In this contribution, we introduced a novel approach to regression with spatial dependent compositional tables. Even though some methods for the analysis of spatial dependent compositional data have already been developed, all of them are primarily designed for its one-factorial (vector) version. Hence, this chapter serves as the first step towards modelling more complex relative structures. The main contribution of this chapter is that it serves as an introduction of a geographically weighted regression model designed in order to inspect the relationship between the dependent variable and the explanatory variable expressed in the form of a compositional table.

It turned out that the local regression model was much more suitable for the analysis of the relationship between the risk of falling into poverty and the educational attainment structure of the 30–34-aged population in Germany. The comparison of both the global and local regression models revealed important aspects of local variations in our real data example. The selection of the coordinate system based on the desired research interest, in our case the dominance of tertiary education over the entire population, enabled a convenient interpretation of the regression outcomes with respect to the variable of interest, the at-risk-of-poverty rates.

The future research initiatives might be devoted to the accommodation of spatial or spatiotemporal autoregressive models for the case of more-factorial compositional data as well as more complex or robust alternatives of the GWR model.

Acknowledgements This work was supported by The Czech Science Foundation under grant No.: 18-05432S.

References

M.S. Andreano, R. Benedetti, F. Piersimoni, A distance correlation index of spatial dependence for compositional data. Pap. Reg. Sci. **98**(6), 2371–2389 (2019)

L. Anselin, Spatial econometrics, in *A Companion to Theoretical Econometrics*, ed. by B.H. Baltagi (Blackwell Publishing Ltd., Hoboken, 2001)

L. Anselin, *Spatial Econometrics: Methods and Models*, vol. 4 (Springer Science & Business Media, Berlin, 2013)

R.S. Bivand, E. Pebesma, V. Gomez-Rubio, *Applied Spatial Data Analysis with R*, 2nd edn. (Springer, New York, 2013). http://www.asdar-book.org/

J. Brülle, *Poverty Trends in Germany and Great Britain* (Springer VS, Wiesbaden, 2018)

C. Brunsdon, S. Fotheringham, M. Charlton, Geographically weighted regression-modelling spatial non-stationarity. J. Royal Stat. Soc. Ser. D Stat. **47**(3), 431–443 (1998)

A.D. Cliff, K. Ord, Spatial autocorrelation: a review of existing and new measures with applications. Econ. Geogr. **46**(sup1), 269–292 (1970)

J. de Sousa, K. Hron, K. Fačevicová, P. Filzmoser, Robust principal component analysis for compositional tables. J. Appl. Stat. 1–20 (2020). https://doi.org/10.1080/02664763.2020.1722078

M. Duiella, A. Turrini, *Poverty developments in the EU after the crisis: a look at main drivers* (ECFIN Econ, Briefs (Directorate-General for Economic and Financial Affairs, European Commission, 2014)

J.J. Egozcue, V. Pawlowsky-Glahn, Groups of parts and their balances in compositional data analysis. Math. Geol. **37**, 795–828 (2005)

J.J. Egozcue, J.L. Díaz-Barrero, V. Pawlowsky-Glahn, Compositional analysis of bivariate discrete probabilities, in *Proceedings of CODAWORK'08, The 3rd Compositional Data Analysis Workshop*, ed. by J. Daunis i Estadella, J.A. Martín-Fernández (University of Girona, Spain, 2008)

J.J. Egozcue, V. Pawlowsky-Glahn, M. Templ, K. Hron, Independence in contingency tables using the simplicial geometry. Commun. Stat. Theory Methods **44**(18), 3978–3996 (2015)

ESRI, ArcGIS desktop: release 10 (Environmental Systems Research Institute, Redlands, CA, 2011)

K. Fačevicová, K. Hron, V. Todorov, M. Templ, Compositional tables analysis in coordinates. Scand. J. Stat. **43**(4), 962–977 (2016)

K. Fačevicová, K. Hron, V. Todorov, M. Templ, General approach to coordinate representation of compositional tables. Scand. J. Stat. **45**(4), 879–899 (2018)

P. Filzmoser, K. Hron, M. Templ, *Applied Compositional Data Analysis* (Springer, Cham, 2018)

A.S. Fotheringham, Trends in quantitative methods I: stressing the local. Prog. Hum. Geogr. **21**(1), 88–96 (1997)

A.S. Fotheringham, C. Brunsdon, M. Charlton, *Geographically Weighted Regression: The Analysis of Spatially Varying Relationships* (Wiley, Chichester, 2002)

I. Gollini, L. Binbin, M. Charlton, C. Brunsdon, P. Harris, GWmodel: an R package for exploring spatial heterogeneity using geographically weighted models. J. Stat. Softw. **63**(17), 1–50 (2015)

W. Hanesch, *ESPN thematic report on in-work poverty Germany (Technical report* (European Commission, European Social Policy Network (ESPN), Brussels, 2019)

P. Harris, C. Brunsdon, M. Charlton, Geographically weighted principal components analysis. Int. J. Geogr. Inf. Sci. **25**(10), 1717–1736 (2011)

K. Hron, P. Filzmoser, K. Thompson, Linear regression with compositional explanatory variables. J. Appl. Stat. **39**(5), 1115–1128 (2012)

S. Kalogirou, Testing local versions of correlation coefficients. Rev. Reg. Res.: Jahrb. Regionalwissenschaft **32**(1), 45–61 (2012)

L. Lahti, J. Huovari, M. Kainu, P. Biecek, eurostat R package (2017). https://journal.r-project.org/archive/2017/RJ-2017-019/index.html. Version 3.6.1

T.J. Leininger, A.E. Gelfand, J.M. Allen, J.A. Silander, Spatial regression modeling for compositional data with many zeros. J. Agric. Biol. Environ. Stat. **18**(3), 314–334 (2013)

Y. Leung, C.-L. Mei, W.-X. Zhang, Testing for spatial autocorrelation among the residuals of the geographically weighted regression. Environ. Plan. A: Econ. Space **32**(5), 871–890 (2000). https://doi.org/10.1068/a32117

S. Matthews, T.-C. Yang, Mapping the results of local statistics: using geographically weighted regression. Demogr. Res. **S13**(6), 151–166 (2012)

J. Morais, C. Thomas-Agnan, M. Simioni, Interpretation of explanatory variables impacts in compositional regression models. Austrian J. Stat. **47**(5), 1–25 (2018)

K. Nordhausen, H. Oja, P. Filzmoser, C. Reimann, Blind source separation for spatial compositional data. Math. Geosci. **47**(7), 753–770 (2015)

V. Pawlowsky-Glahn, A. Buccianti, *Compositional Data Analysis, Theory and Applications* (Wiley Ltd, Hoboken, 2011). ISBN 9781119976462

V. Pawlowsky-Glahn, J.J. Egozcue, R.A. Olea, E. Pardo-Igúzquiza, Cokriging of compositional balances including a dimension reduction and retrieval of original units. J. South. Afr. Inst. Min. Metall. **115**, 59–72 (2015). ISSN 2225-6253

J. Pieper, U. Schneider, W. Schröder, G. Stilling, 30 Jahre Mauerfall - Ein viergeteiltes Deutschland. Der Paritätische Armutsbericht 2019. Technical report (Der Paritätische Gesamtverband, Berlin, 2019)

B. Pirzamanbein, J. Lindström, A. Poska, M.-J. Gaillard, Modelling spatial compositional data: reconstructions of past land cover and uncertainties. Spat. Stat. **24**, 14–31 (2018)

R Core Team, R: a language and environment for statistical computing (R Foundation for Statistical Computing, Vienna, Austria, 2020). https://www.R-project.org/

C. Reimann, M. Ayras, V.A. Chekushin, I. Bogatyrev, R. Boyd, P. de Caritat, R. Dutter, T.E. Finne, J.H. Halleraker, Ø. Jæger, *Environmental Geochemical Atlas of the Central Barents Region*. Norges geologiske unders t'kelser (1997)

M. Templ, K. Hron, P. Filzmoser, robCompositions: an R-package for robust statistical analysis of compositional data, in *Compositional Data Analysis, Theory and Applications*, ed. by V. Pawlowsky-Glahn, A. Buccianti, Chapter 25 (Wiley, Ltd, Hoboken, 2011), pp. 341–355. ISBN 9781119976462

W.R. Tobler, A computer movie simulating urban growth in the Detroit region. Econ. Geogr. **46**(sup1), 234–240 (1970)

R. Tolosana-Delgado, V. Pawlowsky-Glahn, J.J. Egozcue, Indicator kriging without order relation violations. Math. Geosci. **40**, 327–347 (2008)

T. Yoshida, M. Tsutsumi, On the effects of spatial relationships in spatial compositional multivariate models. Lett. Spat. Resour. Sci. **11**(1), 57–70 (2018)

Factor Analysis of Compositional Data with a Total

Carles Barceló-Vidal and Josep Antoni Martín-Fernández

Abstract The sample space of a manifest random vector is of crucial importance for a latent variable model. Compositional data require an appropriate statistical analysis because they provide the relative importance of the parts of a whole. Any statistical model including variables created using the original parts should be formulated according to the geometry of the simplex. Methods based on log-ratio coordinates give a consistent framework for analyzing this type of data. Here, we introduce an approach that includes both the orthonormal log-ratio coordinates and an auxiliary variable carrying absolute information and illustrate it through the factor analysis of two real datasets.

1 Compositional Analysis with a Total

A real vector $\mathbf{w}$ of $\mathbb{R}_+^D$ (the strictly positive octant of $\mathbb{R}^D$) is a *D-part* composition (Aitchison 1986) when the components $w_1, \ldots, w_D$ of $\mathbf{w}$ describe the disjoint *parts of a whole*. A dataset $\mathbf{W} = \{\mathbf{w}_1, \ldots, \mathbf{w}_N\}$ consisting of N compositions is called *compositional data* (CoDa). Barceló-Vidal and Martín-Fernández (2016) coined the term *compositional analysis* (CoAn) to refer to the analysis of the relative information provided by compositional data, that is, the log-ratio analysis proposed in Aitchison (1986), also know as CoDA. In geochemistry, the original units in CoDa represent relative magnitudes (e.g., proportions, percentages, ppm, and mg/L), however, in a number of other fields they frequently represent absolute magnitudes (e.g., grams, euros and liters). In CoDa, one differentiates between *full composition* and *subcomposition*, respectively, when the parts $w_1, \ldots, w_D$ constitute (or not) an exhaustive partition of a whole. That is, a subcomposition is a composition not including all parts into which the whole has been divided. For full compositions, the absolute total amount of $\mathbf{w}$ can be a constant K, equal to $K = w_1 + \ldots + w_D$. In this case, the

C. Barceló-Vidal (✉) · J. A. Martín-Fernández
Department IMAE, Campus Montilivi, Edif. P4, University of Girona, Girona 17003, Spain
e-mail: carles.barcelo@udg.edu

J. A. Martín-Fernández
e-mail: josepantoni.martin@udg.edu

P. Filzmoser et al. (eds.), *Advances in Compositional Data Analysis*,
https://doi.org/10.1007/978-3-030-71175-7_7

composition $\mathbf{w}$ is only characterized by the *relative* information because the absolute values of the components are not informative. For example, in full compositions from daytime-use surveys, the constant absolute total is $K = 24\,\text{h}$ (or $1440\,\text{min}$). Another typical example are the data from experiments with mixtures in industry where $K = 1$ (or 100). However, often CoDa are formed by subcompositions. This case occurs frequently in geochemistry or air pollution data, among others, where the CoDa are concentrations of a few specific chemical elements. In other situations, such as in economics or waste management, among others, the whole simply does not take a constant value across the samples. In the scenarios where the total is of interest, a *D-part* composition provides two different types of information: *relative* and *absolute* (Martín-Fernández et al. 2020). For example, for a three-part subcomposition of three chemical elements (in mg/L), the relative information provided by the vectors $(0.13, 0.18, 0.51)^{\text{t}}$ and $(1.3, 1.8, 5.1)^{\text{t}}$ is the same, but the two subcompositions provide different absolute information because the concentrations in $(1.3, 1.8, 5.1)^{\text{t}}$ are ten times the concentration in $(0.13, 0.18, 0.51)^{\text{t}}$.

The fundamentals of CoAn were introduced in Aitchison (1986) and have been completed and justified mathematically in Pawlowsky-Glahn and Buccianti (2011), Barceló-Vidal and Martín-Fernández (2016), among many other papers. In this type of analysis, one is dealing with the *relative* information of compositions. Because the components of a composition represent relative magnitudes, any change in one of the parts necessarily causes the relative change in one or more of the other components. This lack of independence between the components is evident if $\mathbf{w}$ is a full composition with a fixed total. Importantly, it is also present even when one is dealing with subcompositions. The prevention of applying the standard procedures of statistical analysis for CoDa can be overcome if one analyzes the (log-)ratios w_i/w_j between the components of the compositions instead of their individual values (Aitchison 1986). Note that ratios and logarithms force us to avoid the case of zero values. The reason why a zero value is present in CoDa is informative and determines the approach to be applied (Palarea-Albaladejo and Martín-Fernández 2015). A CoAn based on the ratios of the components of the compositions ensures the *scale invariance* property (Barceló-Vidal and Martín-Fernández 2016). That is, the results of this type of analysis are the same if one replaces any D-part vector $\mathbf{w}_i$ of the dataset by the vector $k_i\mathbf{w}_i$, for any value $k_i > 0$, $i = 1, \ldots, N$. Consequently, the compositions $\mathbf{w}$ and $k\mathbf{w}$ are members of the same equivalence class providing the same relative information, that is, the information given by the ratios between the components. In mathematical terms, the sample space of CoDa is not $\mathbb{R}_+^D$, it is the quotient space $\mathbb{R}_+^D/\sim$, where $\sim$ symbolizes the compositional equivalence relation. Despite the D-part vector chosen to represent the equivalence class being irrelevant, it is usual to choose the representative whose components' sum is equal to 1. The *operation closure* $\mathcal{C}\mathbf{w} = \mathbf{w}/\sum_{j=1}^{D} w_j$ (Aitchison 1986) provides this representative. This criterion could be generalized to representatives with a sum equal to 100 or any other positive value. The set of vectors of $\mathbb{R}_+^D$, whose sum is equal to one, is the *unit simplex* (Aitchison 1986):

$$\mathcal{S}^D = \{\mathbf{x} = (x_1, \ldots, x_D)^{\text{t}} : x_1 > 0, \ldots, x_D > 0; x_1 + \ldots + x_D = 1\}.$$

Another useful procedure for selecting a representative is the *hyperbolic criterion* (Barceló-Vidal and Martín-Fernández 2016). This procedure assigns to a composition $\mathbf{w}$ the intersection of the ray going from the origin through $\mathbf{w}$ and the hyperbolic surface in $\mathbb{R}_+^D$ implicitly defined by the equation $\prod_{i=1}^D w_i = 1$. That is, the representative of the equivalence class is $\mathbf{w}/g(\mathbf{w})$, where $g(\mathbf{w}) = (\prod_{j=1}^D w_j)^{1/D}$ is the geometric mean of the components of vector $\mathbf{w}$. Importantly, the log-transformed vector of this representative, $\ln(\mathbf{w}/g(\mathbf{w}))$, is equivalent to the *centered log-ratio* transformation (clr) introduced by Aitchison (1986): $\operatorname{clr}\mathbf{w} = \ln(\mathbf{w}/g(\mathbf{w}))$.

2 Relative and Absolute Information in Compositional Data

Since CoAn ignores the information about the total abundance $w_1 + \cdots + w_D$ present in a composition $\mathbf{w}$, some authors proposed various alternatives to include this supplementary information in the analysis (see Pawlowsky-Glahn et al. 2015; Coenders et al. 2017; Simonacci and Gallo 2017; Martín-Fernández et al. 2020). Obviously, these alternatives only make sense if the compositions are not closed. We will agree to use the term *compositional analysis with a total* (CoAnTo) to refer to the analysis which, in addition to the analysis of the relative information provided by a compositional dataset, also includes the analysis of the total abundance into compositions. In this chapter, we propose a methodological approach to CoAnTo applied to a factorial analysis (FA).

Let $\{\mathbf{v}_1, \ldots, \mathbf{v}_{D-1}\}$ be an orthonormal basis of the subspace $\mathcal{V} = \{\mathbf{y} \in \mathbb{R}^D : \sum_{i=1}^D y_i = 0\}$ of dimension $D-1$. The *orthonormal* log-ratio coordinates (olr) of a composition $\mathbf{w}$ in $\mathcal{S}^D$ are the coordinates of $\operatorname{clr}\mathbf{w}$ relative to $\{\mathbf{v}_1, \ldots, \mathbf{v}_{D-1}\}$ (Martín-Fernández 2019). The matrix expression of the corresponding coordinate vector $\operatorname{olr}\mathbf{w}$ is equal to

$$\operatorname{olr}\mathbf{w} = \mathbf{V}^{\mathrm{t}} \cdot \operatorname{clr}\mathbf{w} = \mathbf{V}^{\mathrm{t}} \cdot \mathbf{G}_D \cdot \ln\mathbf{w}\ , \tag{1}$$

where the $D \times (D-1)$ matrix $\mathbf{V}$ is equal to $\mathbf{V} = (\mathbf{y}_1 : \cdots : \mathbf{y}_{D-1})$, in that $\mathbf{V}^{\mathrm{t}} \cdot \mathbf{V} = \mathbf{I}_{D-1}$ and $\mathbf{V} \cdot \mathbf{V}^{\mathrm{t}} = \mathbf{G}_D$, with $\mathbf{G}_D$ being the $D \times D$ *centering* matrix. Remember that $\mathbf{G}_D = \mathbf{I}_D - \frac{1}{D}\mathbf{J}_D$, with $\mathbf{J}_D$ being the $D \times D$ matrix of units. The olr- and clr-scores of a composition $\mathbf{w}$ are related since it holds that $\operatorname{clr}\mathbf{w} = \mathbf{V} \cdot \operatorname{olr}\mathbf{w}$. Because the sum of the elements of any row of the matrix $\mathbf{V}^{\mathrm{t}} \cdot \mathbf{G}_D$ in (1) is equal to zero, any component of $\operatorname{olr}\mathbf{w}$ can be written as a log-contrast, i.e., as a linear combination $\sum_j a_j \ln w_j$, with $\sum_j a_j = 0$.

Since the dimension of $\mathcal{V}$ is $D-1$, the dimension of $\mathcal{V}^\perp$ is equal to 1, with the basis being the unit-norm vector $\mathbf{t}_D = \frac{1}{\sqrt{D}}\mathbf{1}_D$, where $\mathbf{1}_D$ is the D-vector of units. Then, any vector $\mathbf{y}$ in $\mathbb{R}^D$ can be uniquely decomposed as the sum $\mathbf{y}_{\mathcal{V}} + \mathbf{y}_{\mathcal{V}^\perp}$ of two orthogonal vectors, one belonging to $\mathcal{V}$ and the other to $\mathcal{V}^\perp$:

$$\mathbf{y}_{\mathcal{V}} = \mathbf{G}_D \cdot \mathbf{y} \quad \text{and} \quad \mathbf{y}_{\mathcal{V}^\perp} = \frac{1}{\sqrt{D}}\mathbf{J}_D \cdot \mathbf{y} = \frac{\sum_{j=1}^{D} y_j}{\sqrt{D}}\mathbf{1}_D\,.$$

Given an orthonormal basis $\{\mathbf{v}_1, \ldots, \mathbf{v}_{D-1}\}$ of the subspace $\mathcal{V}$, one can complete the set of $D-1$ vectors to an orthonormal basis of $\mathbb{R}^D$ by adding the vector $\mathbf{t}_D = \frac{1}{\sqrt{D}}\mathbf{1}_D$. Let $\mathbf{Y} = (\mathbf{v}_1 : \ldots : \mathbf{v}_{D-1} : \mathbf{t}_D) = (\mathbf{V} : \mathbf{t}_D)$ be the $D \times D$ matrix whose columns are the vectors of the basis of $\mathbb{R}^D$. Since it is an orthonormal basis, it holds that $\mathbf{Y}^{\mathrm{t}} \cdot \mathbf{Y} = \mathbf{Y} \cdot \mathbf{Y}^{\mathrm{t}} = \mathbf{I}_D$. The vector of the coordinates of $\ln \mathbf{w}$ relative to basis $\mathbf{Y}$ can be decomposed into two parts in the following manner:

$$\mathbf{Y}^{\mathrm{t}} \cdot \ln \mathbf{w} = (\operatorname{olr} \mathbf{w} : \operatorname{tlog} \mathbf{w})\,, \quad \text{where} \quad \operatorname{tlog}\, \mathbf{w} = \frac{\sum_{j=1}^{D} \ln w_j}{\sqrt{D}}\,. \tag{2}$$

The scale invariance property holds that

$$\mathbf{Y}^{\mathrm{t}} \cdot \ln(k\mathbf{w}) = (\operatorname{olr}(k\mathbf{w}) : \operatorname{tlog}(k\mathbf{w})) = \left(\operatorname{olr} \mathbf{w} : (\operatorname{tlog} \mathbf{w} + \sqrt{D}\ln k)\right),$$

for any $k > 0$. The $D-1$ components of $\operatorname{olr} \mathbf{w}$ in (2) provide the *relative* information of the composition $\mathbf{w}$ while the last term $\operatorname{tlog} \mathbf{w}$ provides the *absolute* information. We will refer to this term as *total of logarithms* of $\mathbf{w}$. Observe that it is equal to the logarithm of the *multiplicative-total* $\mathrm{t}(\mathbf{w}) = \prod_{j=1}^{D} w_j^{1/\sqrt{D}}$ introduced in Pawlowsky-Glahn et al. (2015). Figure 1 shows how the information provided by a D-part composition $\mathbf{w}$ is decomposed. The log-transformed vector $\ln \mathbf{w}$ decomposes as a sum of two vectors which are in orthogonal spaces:

$$\ln \mathbf{w} = \ln(\mathbf{w}/g(\mathbf{w})) + \ln g(\mathbf{w}) \cdot \mathbf{1}_D = \operatorname{clr} \mathbf{w} + \ln g(\mathbf{w}) \cdot \mathbf{1}_D.$$

For the vector $\operatorname{clr} \mathbf{w}$, the relative information can be expressed in terms of an olr vector of coordinates. The coordinate of the vector $\ln g(\mathbf{w}) \cdot \mathbf{1}_D$ in the direction of the vector $\mathbf{t}$ gives the absolute information associated with the value $\operatorname{tlog} \mathbf{w}$. All the points on the vertical pink dashed line share the same relative information (clr scores) because they belong to the same equivalence class. For example, the clr-scores of compositions $(0.13, 0.18, 0.51)^{\mathrm{t}}$ and $(1.3, 1.8, 5.1)^{\mathrm{t}}$ are both equal to $(-0.56, -0.24, 0.80)^{\mathrm{t}}$, whereas the absolute information provided by them is different because $\ln g(0.13, 0.18, 0.51)^{\mathrm{t}} = -1.48$ and $\ln g(1.3, 1.8, 5.1)^{\mathrm{t}} = 0.83$.

3 Factorial Compositional Analysis

In statistics, the sample space of a manifest random vector $\mathbf{x}$ is of crucial importance for a latent variable model. For example, consider a typical factor-analysis (FA) model (Krzanowski 2000; Kline 1994) defined as

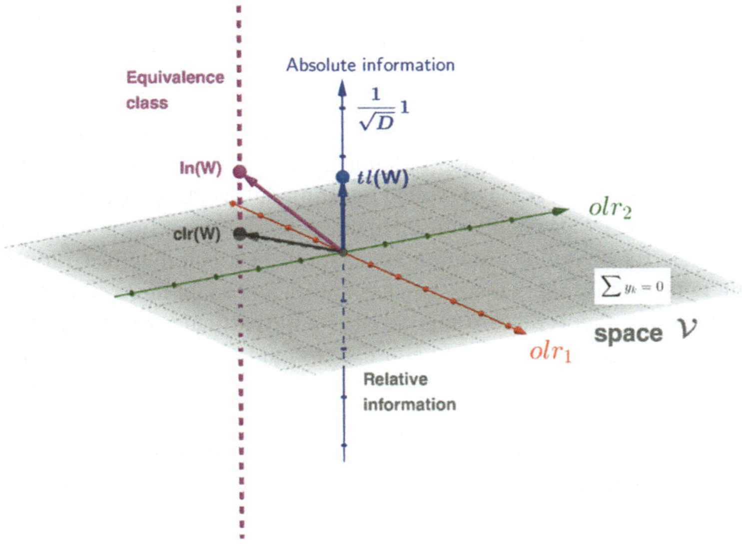

Fig. 1 Decomposition into relative and absolute information of a composition. All the compositions in the same equivalence class (*pink dashed line*) have the same relative information (clr(**w**)) but different absolute information (tl(**w**)). The gray plane represents the space $\mathcal{V}$ where an olr basis can be defined

$$\mathbf{x} - \boldsymbol{\mu} = \boldsymbol{\Gamma} \cdot \mathbf{z} + \mathbf{e}, \tag{3}$$

where $\boldsymbol{\mu}$ is the expectation of $\mathbf{x}$, $\boldsymbol{\Gamma} = (\gamma_{ij})$ is a $D \times Q$ matrix ($Q < D$) of constants (*loadings*), $\mathbf{z} = (z_1, \ldots, z_Q)^t$ is a latent random vector (*common factors* or simply *factors*) with mean zero and a covariance matrix $\mathbf{I}_Q$ (*identity* $Q \times Q$ matrix), and $\mathbf{e}$ is a random vector (*residual* or *specific factors*) with mean zero and a diagonal covariance matrix $\boldsymbol{\Psi} = \text{diag}(\psi_1^2, \ldots, \psi_D^2)$. Note that the structure of this matrix indicates that the variables e_j, $j = 1, \ldots, D$, are uncorrelated with each other. Another crucial assumption about these random variables is that the e_j, $j = 1, \ldots, D$, are uncorrelated with the z_i, $i = 1, \ldots, Q$.

When the manifest vector $\mathbf{x}$ is compositional, the model in (3) should provide the same information regardless of the representative selected for the composition. Consequently, when one uses the representative in the unit simplex, it holds that $\sum_{j=1}^{D} x_j = 1$ and $\sum_{j=1}^{D} \mu_j = 1$. Therefore, (3) it holds that

$$\sum_{j=1}^{D} (\gamma_{j1} z_1 + \gamma_{j2} z_2 + \ldots + \gamma_{jQ} z_Q + e_j) = 0\,,$$

and therefore,

$$e_j = \sum_{j=1}^{D} (\gamma_{j1} z_1 + \gamma_{j2} z_2 + \ldots + \gamma_{jq} z_q) - \sum_{k \neq j} e_k .$$

The equality above suggests that the typical FA model is not appropriate for CoDa because the assumptions about uncorrelation are not fulfilled. Many authors (Mooijaart et al. 1999) suited this kind of typical linear model, as PCA, factor-analysis, and least squares, to the CoDa peculiarities. However, these typical linear techniques are not appropriate for modeling common associations between manifest variables. At this point, two possible alternatives appear: modify and adapt more complex typical models, and apply models that are appropriate for the particular geometry of the simplex. Although we admit that the first alternative could provide reasonable results in some scenarios, we prefer the second because we believe that the more coherent the method, the more reasonable the results.

The mimetic translation of (3) to simplex leads us to define a latent variable model for a D-part composition as

$$\mathbf{w} \ominus \mathbf{g} = \mathbf{\Lambda} \boxdot \mathbf{u} \otimes \boldsymbol{\varepsilon} , \tag{4}$$

where $\mathbf{w}$ and $\mathbf{u}$ are random vectors in $\mathcal{S}^D$ and $\mathcal{S}^Q$ $(Q < D)$, respectively. Moreover, $\mathbf{g} = (g_1, \ldots, g_D) \in \mathcal{S}^D$ is the expectation of $\mathbf{w}$, $\mathbf{\Lambda} = (\lambda_{ij})$ is a $D \times Q$ matrix of constants (*loadings*), and $\boldsymbol{\varepsilon} \in \mathcal{S}^D$ is the *residual* random composition. As introduced by Aitchison (1986), and recently analyzed by Martín-Fernández et al. (2020), each expected value g_i, $i = 1, \ldots, D$ is estimated by the geometric mean of the corresponding part. The symbol $\otimes$ corresponds to the *perturbation* operation defined in the simplex, and $\ominus$ is its inverse operation. The matrix product in the simplex is $\mathbf{\Lambda} \boxdot \mathbf{u} = C(\prod_{j=1}^{Q} u_j^{\lambda_{1j}}, \ldots, \prod_{j=1}^{Q} u_j^{\lambda_{Dj}})^{\mathrm{t}} = C(\exp(\sum_{j=1}^{Q} \lambda_{1j} \ln u_j, \ldots, \sum_{j=1}^{Q} \lambda_{Dj} \ln u_j)^{\mathrm{t}})$. Note that the scale invariance principle requires that $\sum_{j=1}^{Q} \lambda_{kj} = 0$, for $k = 1, \ldots, D$. In consequence, the core of the model is formed by log-contrasts of a latent random composition $\mathbf{u}$.

Since the covariance structure of a compositional random vector is defined from the covariance matrix of some of its log-ratio transformations (either the clr-, alr- or olr-transformation), the difficulties appear when we try to define directly on the simplex the conditions that random compositions $\mathbf{u}$ and $\mathbf{e}$ in (4) must meet so that the model can be considered an FA model. A possible solution would be to define an FA model for a random D-part composition $\mathbf{w}$ from a standard latent variable model for clr $\mathbf{w}$, i.e., as

$$\operatorname{clr} \mathbf{w} - \operatorname{clr} \mathbf{g} = \Gamma \cdot \mathbf{z} + \mathbf{e} . \tag{5}$$

The fact that the covariance matrix of clr $\mathbf{w}$ is singular prevents fulfilling the theoretical requirements of latent variable models, since the covariance of the right side of (5) is equal to $\mathbf{\Gamma} \cdot \mathbf{\Gamma}^{\mathrm{t}} + \mathbf{\Psi}$: a matrix which is always non-singular. Filzmoser and Hron (2009) try to overcome this theoretical obstacle by allowing the covariance matrix of residuals to be non-diagonal as long as the off-diagonal elements of the matrix become small.

The non-isometric character of the alr transformations advises us not to define the latent variable model for a random D-part composition $\mathbf{w}$ from alr $\mathbf{w}$. Thus, it seems that we are forced to define the FA model using olr $\mathbf{w}$, i.e., the orthogonal log-ratio transformation of $\mathbf{w}$.

A latent variable model for a random D-part composition $\mathbf{w}$ will be defined as

$$\text{olr}\,\mathbf{w} - \text{olr}\,\mathbf{g} = \widetilde{\boldsymbol{\Gamma}} \cdot \widetilde{\mathbf{z}} + \widetilde{\mathbf{e}}\,, \tag{6}$$

where $\mathbf{g}$ is the expectation of $\mathbf{w}$, $\widetilde{\boldsymbol{\Gamma}} = (\widetilde{\gamma}_{ij})$ is a $(D-1) \times Q$ matrix $(Q < D-1)$ of constants (*loadings*), $\widetilde{\mathbf{z}} = (\widetilde{z}_1, \ldots, \widetilde{z}_Q)^{\mathrm{t}}$ is a latent random vector (*factors*) with mean zero and a covariance matrix $\mathbf{I}_Q$ (*identity* $Q \times Q$ matrix), and $\widetilde{\mathbf{e}}$ is a random vector (*residual*) with mean zero and a covariance matrix $\widetilde{\boldsymbol{\Psi}} = \text{diag}(\widetilde{\psi}_1^2, \ldots, \widetilde{\psi}_{D-1}^2)$. Moreover, the unobservable random vectors $\widetilde{\mathbf{z}}$ and $\widetilde{\mathbf{e}}$ must be uncorrelated.

The estimated FA model (6) depends on the orthonormal basis of the subspace $\mathcal{V}$ to which the olr-coordinates of compositions are referred. This effect is not an exclusive difficulty for CoDa because it is also present in the real space with the Euclidean geometry. Specifically, the application of the maximum likelihood method to estimate the FA model gives very different results depending on the orthonormal basis used, even when the desired number of factors is fixed in advance. The differences not only affect the factors, but also the percentage of total variance explained by the estimated loading matrix. The same applies if the principal factor method is applied to perform the estimation. Only the principal component (PC) method is independent of the orthonormal basis, although if orthogonal rotation of loadings is allowed the distribution of the total explained variance between the factors changes depending on the basis. This behavior of the FA model (6) indicates that selecting the olr-basis plays a crucial role to facilitate the interpretation of the FA results. In this context, if one knows a priori the existence of a certain relationship between some parts of $\mathbf{w}$, the corresponding olr-score that involves these parts should be defined. On the other hand, in the absence of expert information, it is recommended to use the PC of the clr-scores of CoDa to create an olr-basis based on a sequential binary partition (Pawlowsky-Glahn et al. 2015). Optionally, one can apply an automatic algorithm to create an olr-basis based on the maximization of the variance accounted (Martín-Fernández et al. 2018).

Let $\mathbf{W} = \{\mathbf{w}_1, \ldots \mathbf{w}_n\}$ be a compositional dataset. Assume that an FA model with Q factors is found to be reasonable, i.e., most of the (co)variations of the olr-coordinates of the compositions of $\mathbf{W}$ are explained by the Q fixed latent factors $\widetilde{z}_1, \ldots, \widetilde{z}_Q$ of $\widetilde{\mathbf{z}}$. For $f = 1, \ldots, Q$, the $\widetilde{z}_f$-scores of the compositions of $\mathbf{W}$ could be calculated using the regression method described in Johnson and Wichern (1998). Since the scores $\widetilde{z}_f$ are linear combinations of the standardized olr-coordinates of the compositions, according to (1) they can be expressed using a log-contrast $\sum_j a_{fj} \ln w_j$, with $\sum_j a_{fj} = 0$, for $f = 1, \ldots, Q$.

The next natural step is to try to understand what these factors represent. It makes sense to compute the correlations of $\widetilde{z}_f$ with olr$_j\mathbf{w}$ first. This is done for $f = 1, \ldots, Q$ and for $j = 1, \ldots, D-1$. From these correlations it will be possible to establish which of the olr-scores, $\mathrm{olr}_1\mathbf{w}, \ldots, \mathrm{olr}_{D-1}\mathbf{w}$, play a role in the unobserved common factors $\widetilde{z}_1, \ldots, \widetilde{z}_Q$. Obviously, to interpret the manifest relations it is necessary to know the expressions of the log-contrasts associated with the olr-scores. The interpretation might be difficult when the expressions of the log-contrasts involve a lot of parts of $\mathbf{w}$. Another way to interpret a factor $\widetilde{z}_f$ is from the sign and the magnitude of the correlations of the $\widetilde{z}_f$-scores with $\ln\left(w_1/\sum_i w_i\right), \ldots, \ln\left(w_D/\sum_i w_i\right)$.

To illustrate the procedure of factor analysis in CoAn, we use the results from the December 2017 elections to the Parliament of Catalonia, an autonomous community of Spain. The votes have been recorded for the 42 regions into which Catalonia is divided. The dataset (Cat2017) contains the number of valid votes for the following parties and coalitions: *Ciutadans-Partido de la Ciutadania* (Cs), *Junts per Catalunya* (JC), *Esquerra Republicana de Catalunya* (ERC), *Partit dels Socialistes de Catalunya* (PSC), *Catalunya en Comú-Podem* (CeCP), *Candidatura d'Unitat Popular* (CUP), and *Partit Popular* (PP). Several minor parties and coalitions without representatives in the parliament have been amalgamated into a single category called *Other* (Oth). Some parties (JC, ERC, and CUP) favor the complete independence of Catalonia from Spain, others (PSC and CeCP) want to negotiate a new relationship with the central Spanish government including a possible federal solution, while a third group of parties (Cs and PP) favors maintaining the status quo or even a reduction in Catalonia's autonomous powers. The results of the December 2017 elections led to a clear break between the three *separatist* parties (JC, ERC, and CUP) and the rest of the parties (Cs, PSC, CeCP, and PP), known as *unionist*. Some authors claim that this political fracture in parliament is also reflected in the population (Oller et al. 2019).

The eight components of the composition $\mathbf{w}$ correspond to the votes obtained by the parties and coalitions. According to the ideology of the parties, we define the olr-coordinates as follows:

$$
\begin{aligned}
\mathrm{olr}_1\mathbf{w} &= \sqrt{\frac{15}{8}}\ln\frac{(\mathrm{JC}\cdot\mathrm{ERC}\cdot\mathrm{CUP})^{1/3}}{(\mathrm{PP}\cdot\mathrm{Cs}\cdot\mathrm{PSC}\cdot\mathrm{CeCP}\cdot\mathrm{Oth})^{1/5}},\\
\mathrm{olr}_2\mathbf{w} &= \sqrt{\frac{2}{3}}\ln\frac{(\mathrm{JC}\cdot\mathrm{ERC})^{1/2}}{\mathrm{CUP}},\\
\mathrm{olr}_3\mathbf{w} &= \sqrt{\frac{1}{2}}\ln\frac{\mathrm{ERC}}{\mathrm{JC}},\\
\mathrm{olr}_4\mathbf{w} &= \sqrt{\frac{4}{5}}\ln\frac{\mathrm{Oth}}{(\mathrm{PP}\cdot\mathrm{Cs}\cdot\mathrm{PSC}\cdot\mathrm{CeCP})^{1/4}},\\
\mathrm{olr}_5\mathbf{w} &= \ln\frac{(\mathrm{PP}\cdot\mathrm{Cs})^{1/2}}{(\mathrm{PSC}\cdot\mathrm{CeCP})^{1/2}},\\
\mathrm{olr}_6\mathbf{w} &= \sqrt{\frac{1}{2}}\ln\frac{\mathrm{Cs}}{\mathrm{PP}},\\
\mathrm{olr}_7\mathbf{w} &= \sqrt{\frac{1}{2}}\ln\frac{\mathrm{PSC}}{\mathrm{CeCP}},
\end{aligned}
\qquad (7)
$$

where $\text{olr}_1\mathbf{w}$ compares the geometric mean of votes going to the independentist parties with those for the non-independentist parties. The coordinates $\text{olr}_2\mathbf{w}$ and $\text{olr}_3\mathbf{w}$ compare the votes between the three separatist parties, while the remaining olr-coordinates compare the votes between the non-independentist parties.

We applied the method of *principal factor analysis* (PFA) without rotation of the factors to the olr-coordinates of the compositions of the dataset Cat2017 using the routines provided in Revelle (2019). Figure 2 simultaneously represents the loadings of the variables and the scores of the samples. We refer to this figure as a biplot, where the "bi" refers to the two kinds of information (samples and variables) shown in the graphical representation. The number of factors was 2, which resulted in a good percentage (73.8%) of explained variability, most of which (64.8%) is concentrated in the first factor $\tilde{z}_1$. The analysis of communalities indicates that the variables best represented by the two factors are olr_1 ($h^2 = 0.83$) and olr_6 ($h^2 = 0.79$). The worst represented are olr_7 ($h^2 = 0.19$) and olr_5 ($h^2 = 0.25$). The factor $\tilde{z}_1$ is highly positively correlated with the olr_1-coordinate ($r = 0.91$) and negatively correlated with both the olr_3-coordinate ($r = -0.67$) and the olr_6-coordinate ($r = -0.65$). Similarly, it is positively correlated with the percentages of valid votes obtained by the independentist parties and negatively correlated with all the other parties. Thus, the first factor mainly measures the percentages of votes obtained by the parties in favor of the independence of Catalonia compared to those obtained by the unionist parties. The second factor $\tilde{z}_2$, which only explains 9.0% of the variability of the dataset, is positively correlated with the olr_4-coordinate ($r = 0.72$) and negatively correlated with the olr_2-coordinate ($r = -0.59$). From the perspective of the electoral results, the interpretation of $\tilde{z}_2$ is difficult.

4 Factor Analysis of Compositional Data with a Total

Let $\mathbf{w}$ be a random composition in $\mathbb{R}^D_+$, where the components of $\mathbf{w}$ represent absolute magnitudes. According to Mateu-Figueras et al. (2013), one can perform a standard factor analysis directly on the log-transformed vector $\ln \mathbf{w}$. According to (3), the expression of a latent model for the random vector $\ln \mathbf{w}$ will be

$$\ln \mathbf{w} - \ln \mathbf{g} = \mathbf{\Gamma} \cdot \mathbf{z} + \mathbf{e} \,, \tag{8}$$

where $\mathbf{g}$ is the expectation of $\mathbf{w}$, $\mathbf{\Gamma} = (\gamma_{ij})$ is a $D \times Q$ matrix of constants (*loadings*), and $\mathbf{e} \in \mathbb{R}^D$ is the *residual* random composition. This model does not allow us to know to what extent the compositional nature of $\mathbf{w}$ is present in the resulting factors.

We can see this in the results of the standard FA performed on the log-transformed dataset Cat2017 (Fig. 3), that is, the log-transformed values of the absolute number of votes obtained by the parties in each of the 42 regions of Catalonia. The first factor z_1 captures almost all the variability of the dataset (96.8%). It is highly positively correlated ($0.945 \leq r \leq 0.994$) with all the log-transformed variables. This fact links z_1 to the total number of voters in each region, regardless of the distribution of

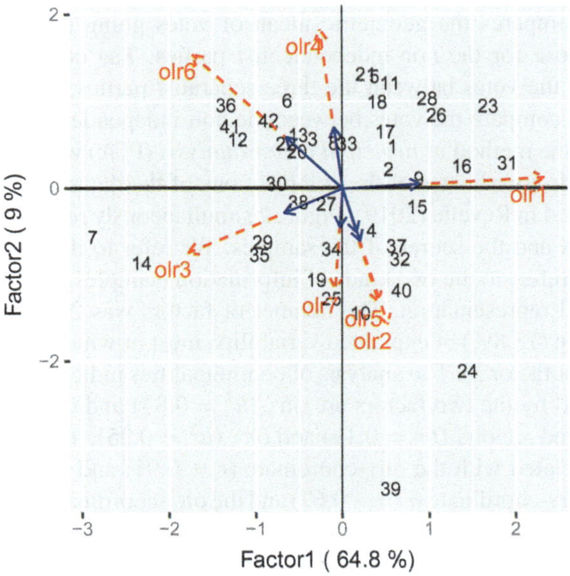

Fig. 2 Biplot for the two factors of FA for olr-coordinates of the 42 compositions of the dataset Cat2017. The method applied was PFA without rotation of the factors. The coordinates of each blue vector correspond to the factor loadings of each variable. These vectors are scaled to the dashed red vectors to assist the reader with the interpretation of the plot. The center of the plot corresponds to 63.9% of the votes to independentist parties. This percentage decreases if we move to the left along the Factor 1 axis. Label 7 (at the left end of this axis) corresponds to *Aran*, the region with the lowest percentage of votes to independentist parties (30.9%). Label 31 (in the right part of the first axis) corresponds to *Priorat*, the second region with the highest percentage of votes to independentist parties (81.3%)

votes among the different parties. To assist the reader with the interpretation of the axis, in the following lines we describe the extreme samples. So, for example, label 41 (to the right of the first FA axis in Fig. 3) corresponds to *Vallès Occidental*, one of the most populated regions of Catalonia, with over a half million voters in the 2017 elections. In contrast, label 31 (left side of the first axis) corresponds to *Priorat*, one of the Catalan regions with the fewest voters (6202) in the 2017 elections. The small variability (1.8%) associated with the second factor z_2 and the low correlation ($0.05 \leq |r| \leq 0.30$) of z_2 with all log-transformed variables prevents an interpretation of it being given. However, the log-variables associated with the independentist parties {JC, ERC, CUP} are on the positive part of the second FA axis, whereas the unionist parties are on the negative side. That is, the FA scores of the second axis are an ordination of the regions as regards these two groups of parties. Indeed, label 30 (top of the second axis) corresponds to the region *Pla de l'Estany* which has the maximum percentage of votes for independentist parties (82.6%). On the other hand, the region *Aran* with the minimum percentage of votes to these parties (30.9%) is at the bottom of the second axis. However, the relative distribution of votes

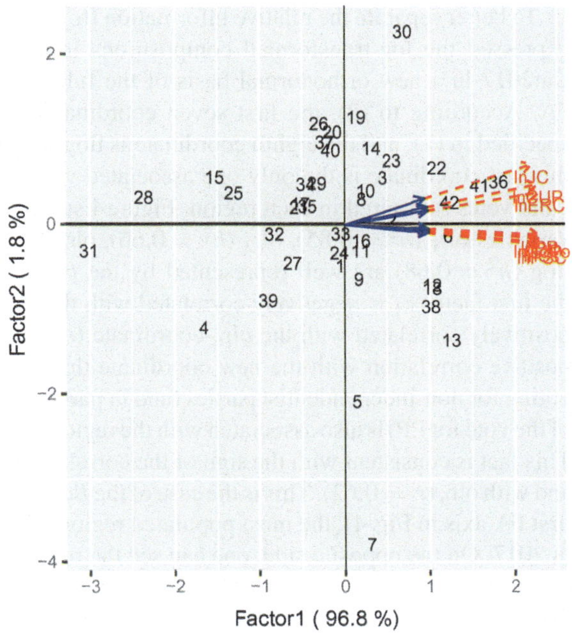

Fig. 3 Biplot for the two factors of FA for the log-transformed dataset Cat2017, that is, the variables are the logarithm of absolute valid votes obtained by the party. The method applied was PFA without rotation of the factors. All variables are well represented by the two factors ($h^2 > 0.960$, for all variables). The coordinates of each blue vector correspond to the factor loadings of each variable. These vectors are scaled to the dashed red vectors to assist the reader with the interpretation of the plot

among the different parties is hidden by the large effect of the absolute information. For this reason, in this section we propose a procedure to separate the relative and absolute information contained in the latent variables from an FA.

The FA model of (8) can be decomposed into two parts:

$$\begin{aligned} \mathbf{Y}^{\mathrm{t}} \cdot \ln \mathbf{w} - \mathbf{Y}^{\mathrm{t}} \cdot \ln \mathbf{g} &= \mathbf{Y}^{\mathrm{t}} \cdot \boldsymbol{\Gamma} \cdot \mathbf{z} + \mathbf{Y}^{\mathrm{t}} \cdot \ln \mathbf{e} \,, \\ \begin{pmatrix} \operatorname{olr} \mathbf{w} \\ \operatorname{tlog} \mathbf{w} \end{pmatrix} - \begin{pmatrix} \operatorname{olr} \mathbf{g} \\ \operatorname{tlog} \mathbf{g} \end{pmatrix} &= \begin{pmatrix} \mathbf{V}^{\mathrm{t}} \\ \mathbf{t}_D^{\mathrm{t}} \end{pmatrix} \cdot \boldsymbol{\Gamma} \cdot \mathbf{z} + \begin{pmatrix} \operatorname{olr} \mathbf{e} \\ \operatorname{tlog} \mathbf{e} \end{pmatrix} \,, \end{aligned} \tag{9}$$

being split into the *relative* model

$$\operatorname{olr} \mathbf{w} - \operatorname{olr} \mathbf{g} = \mathbf{V}^{\mathrm{t}} \cdot \boldsymbol{\Gamma} \cdot \mathbf{z} + \operatorname{olr} \mathbf{e} \,, \tag{10}$$

and the *absolute* model

$$\operatorname{tlog} \mathbf{w} - \operatorname{tlog} \mathbf{g} = \mathbf{t}_D^{\mathrm{t}} \cdot \boldsymbol{\Gamma} \cdot \mathbf{z} + \operatorname{tlog} \mathbf{e} \,. \tag{11}$$

Note that the assumptions for the residuals required by the latent models are not fulfilled in the relative part (10) of the model because the covariance matrix $\operatorname{cov}(\operatorname{olr} \mathbf{e}) = \mathbf{V}^{\mathrm{t}} \cdot \operatorname{cov}(\ln \mathbf{e}) \cdot \mathbf{V}$ is not a diagonal matrix, that is, the residuals are correlated.

To better separate the relative information from the absolute information, we have expressed the log-transformed compositions $\ln \mathbf{w}_i$ $(i = 1, \ldots, 42)$ of the dataset Cat2017 in a new orthonormal basis of the full space $\mathbb{R}^D_+$, before performing the FA. According to (2), the first seven coordinates of $\ln \mathbf{w}$ are the olr-coordinates specified in (7), and the eighth coordinate is $\text{tlog}\,\mathbf{w} = (\sum_{j=1}^{8} \ln w_j)/\sqrt{8}$. Therefore, this last coordinate is the only one associated with the absolute information of the valid votes registered in each region. Figure 4 summarizes the result of the FA. The variables olr_1 $(h^2 = 0.85)$, olr_3 $(h^2 = 0.66)$, olr_6 $(h^2 = 0.61)$, olr_7 $(h^2 = 0.63)$ and tlog $(h^2 = 0.68)$ are well represented by the two factors. As in the FA in Fig. 2, the first factor z_1 is negatively correlated with the olr_1-coordinate $(r = -0.86)$ and positively correlated with the olr_6-coordinate $(r = 0.74)$. But z_1 also shows a high positive correlation with the new coordinate $\text{tlog}\,\mathbf{w}$ $(r = 0.82)$. This indicates that voting for non-independentist parties (and in particular voting for Cs to the detriment of the vote for PP) is also associated with the regions with the largest number of voters. This fact is consistent with the sign of the correlations of tlog with olr_1 $(r = -0.68)$ and with olr_6 $(r = 0.62)$. This is the case of the *Barcelonès* (label 14, to the right of the first FA axis in Fig. 4), the most populated region of Catalonia, with 1252139 voters in 2017. On the opposite side, one can see the region *Alta Ribagorça* (label 5), with only 2295 voters and a high percentage of votes for independentist parties. Although the correlations of the second factor z_2 with olr_3 and olr_7 are significant $(r = 0.65$ and $r = 0.75$, respectively), the interpretation of z_2 is difficult from a perspective of the electoral results. However, the fact that both olr_3 and olr_7 have the same direction as regards the second axis (negative side) suggests that they have the same behavior. The variable olr_3 measures the ratio JC/ERC, both independentist parties, whereas the variable olr_7 is associated with the ratio PSC/CeCP. Consequently, the regions in the negative part of the second axis record more votes for PSC and ERC as regards CeCP and JC, respectively. On the other hand, regions that are in the positive part of the second axis record more votes for CeCP and JC. The region *Aran* (label 7) is a good example for the first case, whereas the region *Osona* (label 26) is a representative of the second case.

5 Example

Filzmoser et al. (2018) analyze the dataset "expendituresEU", which is available in the R package (Templ et al. 2011), by carrying out a PC analysis of the relative information. According to Templ et al. (2011), the source of the dataset is Eurostat and includes the "Mean consumption expenditure of households at EU level. The final consumption expenditure of households encompasses all domestic costs (by residents and non-residents) for individual needs". The dataset consists of 27 observations (in Euros), from the European countries concerning the following 12 parts: Food, Alcohol, Clothing, Housing, Furnishings, Health, Transport, Communications, Recreation, Education, Restaurants, and Other. Because the original units of the raw data are Euros, the sum is not constant across the countries. In fact, the range

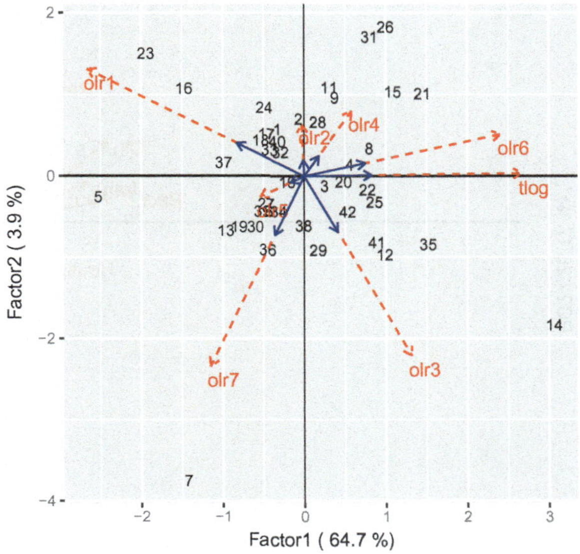

Fig. 4 Biplot for the two factors of FA for olr and tlog coordinates of the 42 compositions of the dataset Cat2017. The method applied was PFA without rotation of the factors. The coordinates of each blue vector correspond to the factor loadings of each variable. These vectors are scaled to the dashed red vectors to assist the reader with the interpretation of the plot

is quite wide because the minimum is Romania with €5,325 and the maximum is Luxemburg where the total annual average expenditure is €51,933. The question that remains to analyze is if the absolute information provided by the CoDa is relevant in the analysis.

Because we have no previous expert knowledge about the relative information, we select the olr-basis using the biplot (Fig. 5) and the automatic procedure of principal balances (Martín-Fernández et al. 2018). Note that the biplot in Fig. 5 was obtained using PC of the clr-scores of the "expendituresEU" dataset. Figure 5 suggests that one can split the 12 parts into four groups: {Education}, {Housing},{Food, Alcohol, Clothing, Health, Communication}, and {Furnishings, Transport, Recreation, Restaurants, Other}. An olr-basis should contain these groups of parts. However, the basis must be formed by 11 elements, and the information provided by the clr-biplot (Fig. 5) is not sufficient to easily define the complete basis. Table 1 shows the matrix of signs used to form the olr-basis provided by the principal balances algorithm (Martín-Fernández et al. 2018). The value 1 indicates that the part is on the numerator of the log-ratio, whereas the parts with the value -1 are in the denominator. A 0 corresponds to parts not involved in the log-ratio. In Table 1, the elements olr_k, $k = 1, \ldots, 11$ are listed in a hierarchical order, following a sequential binary partition. However, the automatic algorithm creates the olr-basis according to the criteria of a maximum variance accounted for by the direction of the vector of the basis. This variance is shown in the "%Var" column. The maximum variance is 37.16%, which is accounted for by the direction olr_7 associated with the ratio $\frac{Housing \cdot Furnishings \cdot Transport \cdot Recreation \cdot Other}{Education}$. The amount of variance retained by olr_7 is similar to the variance accounted by the axis PC1 (42.41%, Fig. 5). The difference between the variance accounted for by olr_6 (13.52%) and by the axis PC2 (28.17%)

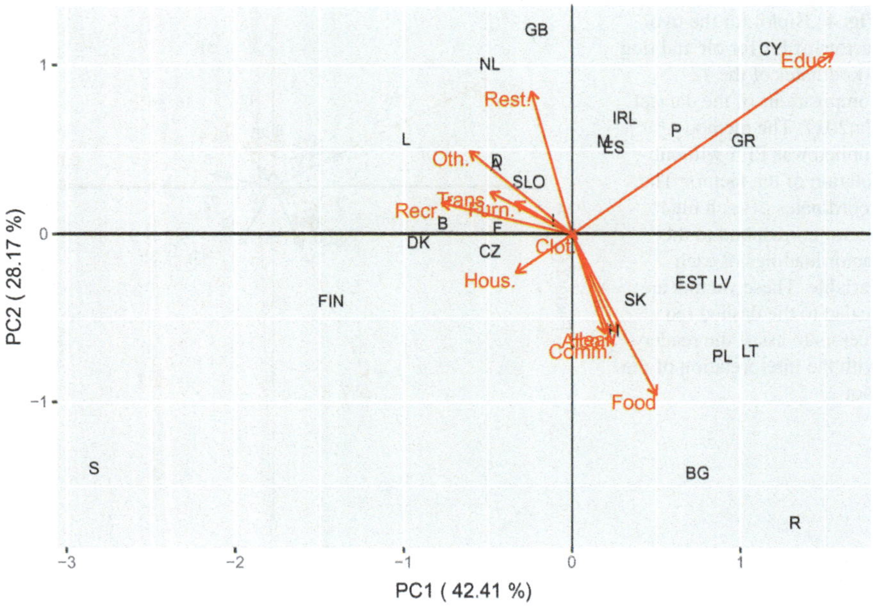

Fig. 5 PCA-biplot for clr-scores of dataset "ExpendituresEU"

Table 1 Matrix of signs to form the olr-basis provided by the principal balances algorithm for the dataset "ExpendituresEU". The column "%Var" shows the variance accounted for by the olr-vector

	Food	Alco.	Clot.	Hous.	Furn.	Heal.	Trans.	Comm.	Recr.	Educ.	Rest.	Oth.	%Var.
olr_1	−1	−1	−1	−1	−1	−1	−1	1	−1	−1	−1	−1	5.13
olr_2	−1	1	−1	−1	−1	−1	−1	0	−1	−1	−1	−1	6.08
olr_3	−1	0	1	−1	−1	1	−1	0	−1	−1	1	−1	10.65
olr_4	0	0	1	0	0	−1	0	0	0	0	−1	0	4.19
olr_5	0	0	0	0	0	1	0	0	0	0	−1	0	10.91
olr_6	1	0	0	−1	−1	0	−1	0	−1	−1	0	−1	13.52
olr_7	0	0	0	1	1	0	1	0	1	−1	0	1	37.16
olr_8	0	0	0	1	−1	0	−1	0	−1	0	0	−1	6.48
olr_9	0	0	0	0	−1	0	−1	0	1	0	0	−1	2.17
olr_{10}	0	0	0	0	1	0	1	0	0	0	0	−1	2.82
olr_{11}	0	0	0	0	1	0	−1	0	0	0	0	0	0.89

is larger than that for the first axis. In this case, olr_6 compares elements of the group {Furnishings, Transport, Recreation, Restaurants, Other} and the part Food.

Using (2) with the olr-basis of Table 1, the FA model in (9) can be created for the "ExpendituresEU" dataset with a total. The PFA method without rotation was applied when testing the hypothesis that two factors are sufficient to provide the significance of the model (p-value = 0.02) (Revelle 2019). Figure 6 shows the structure of the associations between factors and variables using a graph. The variance

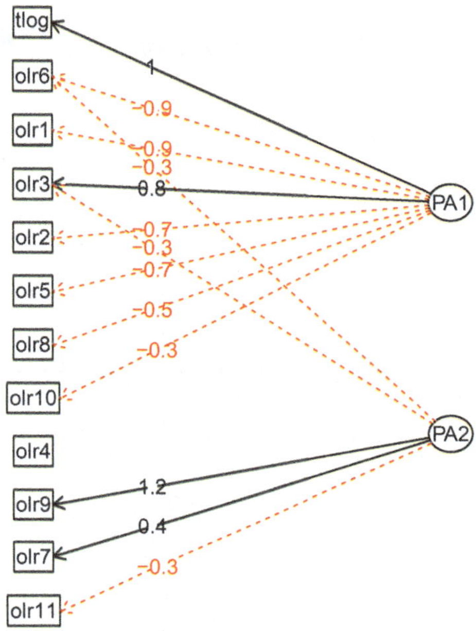

Fig. 6 A graphic representation of the two factors from the dataset "ExpendituresEU" showing the factor structure provided by the PFA method without rotation. The numerical values are the corresponding loadings. Loadings with absolute values below 0.3 are not represented. The variances explained by the factors are 78.62% (PA1) and 3.80% (PA2)

explained by the first factor is 78.62%, whereas the second factor explains only 3.80%. The numbers over the links are the corresponding loadings of each factor for each variable. The graph includes only links where the absolute values of the loading are over 0.3. The variables are shown in decreasing order of the loadings. The variable olr_4 is not linked with the factors suggesting that the ratio $\frac{Clothing}{Health \cdot Restaurants}$ (Table 1) has no contribution to the structure. The largest positive loadings for factor PA1 correspond to variables tlog ($h^2 = 1.00$) and olr_3 ($h^2 = 0.66$), whereas the largest negative loadings are, respectively, for olr_6 ($h^2 = 0.93$), olr_1 ($h^2 = 0.78$), olr_2 ($h^2 = 0.5$), and olr_5 ($h^2 = 0.45$), all of them involving ratios to compare parts of the group {Food, Alcohol, Clothing, Health, Communication} to parts in the group {Furnishings, Transport, Recreation, Restaurants, Other} (Table 1; Fig. 5). This interpretation suggests the label *essential daily items* for the first factor. The label *enough money for recreation* might be adequate for the factor PA2 because it is mainly represented by the variable olr_9 ($h^2 = 1.58$) that compares the expenditure on Recreation to the average expenditure on Furnishings, Transport, and Other (Table 1).

Figure 7(left) shows the position of each country as regards its FA scores provided by the PFA method without rotation (Revelle 2019). This figure includes the relative information associated with the olr-coordinates and the absolute information measured by the variable tlog. In this biplot, the variables are represented by rays according to their loadings. The bar plot in Fig. 7(right) represents the expenditure of the countries in decreasing order. The bar plot is useful for interpreting the biplot because the order of the countries is the same as in the variable tlog. The first axis (PA1) is opposing large/small values in variables tlog and olr_3, respectively, to small/large

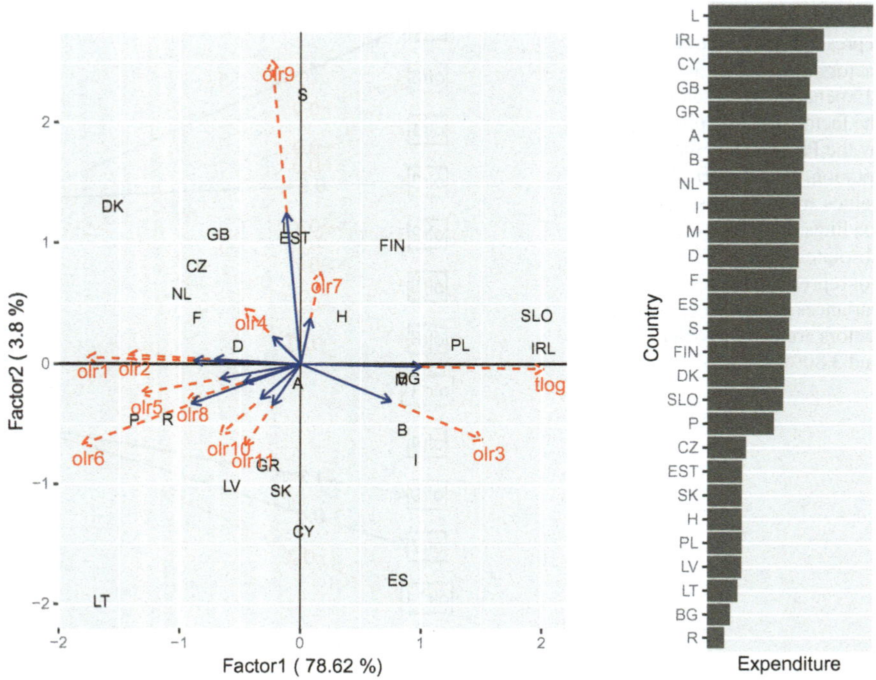

Fig. 7 (Left) Biplot for the two factors of FA for olr-coordinates of the 27 compositions of the dataset "ExpendituresEU". The method applied was PFA without rotation of the factors. The coordinates of each blue vector correspond to the factor loadings of each variable. These vectors are scaled to the dashed red vectors to assist the reader with the interpretation of the plot. (Right) Expenditure by country in decreasing order

values for variables olr_6, olr_1, olr_2, and olr_5. For example, the FA unstandardized scores of Ireland (IRL), the second country as regards to Expenditure (Fig. 7(right)), are (25.28, −0.93). This position is explained by the absolute information of "IRL" and because this country takes some small value for the variables olr_6, olr_1, olr_2, and olr_5. In particular, the smallest percentile for Ireland is 18.5% for the variable olr_6. This variable is associated with the ratio Food over the average of the parts {Housing, Furnishing, Transport, Recreation, Education, Other}. On the other hand, the country Denmark (DK), with FA scores (−20.30, 1.45), takes the percentile 48.1% for the variable tlog (Fig. 7(right)) and 40.7% for the variable olr_3. Therefore, the position of "DK" is explained by its large values in some of the variables olr_6, olr_1, olr_2, and olr_5. Indeed, "DK" takes the percentile 81.5% in the variable olr_2, that is, Danish people take large values for the ratio Alcohol over the rest of the parts except Communication (Table 1). In addition, "DK" takes the percentile 88.9% for the variable olr_9 associated with the second FA axis. This variable measures the values of the ratio Recreation over the average of {Furnishing, Transport, Other}.

6 Final Concluding Remarks

It is well known that CoAn is the analysis recommended when CoDa have a constant-sum constraint or the analyst is interested only in analyzing the relative information. For the other scenarios, when CoDa do not have a constant-sum constraint and the total is of interest, the absolute information associated with the multiplicative total can be added to the analysis in accordance with the geometry of the simplex. The fundamentals of the Aitchison geometry allow the factor analysis model to be decomposed into the relative and absolute information parts. This decomposition is based on the decomposition of the log-transformed compositions into the olr-coordinates and the log-score of the multiplicative total. As a result, the analyst can evaluate the association of each type of information to the compositional parts. In this chapter, we have illustrated this new approach by performing a classical factor analysis of two real datasets. The approach should be extended to more sophisticated models in structural equation modeling.

Acknowledgements This work has been partially financed by the CODAMET project (Ministerio de Ciencia, Innovación y Universidades; Ref: RTI2018-095518-B-C21).

References

J. Aitchison, *The Statistical Analysis of Compositional Data*. Monographs on Statistics and Applied Probability (Chapman & Hall Ltd., London, UK, 1986), p. 416. (Reprinted in 2003 with additional material by The Blackburn Press)

C. Barceló-Vidal, J.A. Martín-Fernández, The mathematics of compositional analysis. Austrian J. Stat. **45**, 57–71 (2016)

G. Coenders, J.A. Martín-Fernández, B. Ferrer-Rosell, When relative and absolute information matter: compositional predictor with a total in generalized linear models. Stat. Model. **17**(6), 494–512 (2017)

P. Filzmoser, K. Hron, Robust factor analysis for compositional data. Comput. Geosci. **35**(9), 1854–1861 (2009)

P. Filzmoser, K. Hron, M. Templ, *Applied Compositional Data Analysis: With Worked Examples in R* (Springer, Switzerland, 2018), p. 278

R.A. Johnson, D.W. Wichern *Applied Multivariate Statistical Analysis*, 4th edn. (Prentice-Hall, New York, 1998), p. 816

P. Kline, *An Easy Guide to Factor Analysis* (Routledge, London, UK, 1994), p. 194. (Reprinted 2005)

W. Krzanowski, *Principles of Multivariate Analysis*, 2nd edn. Oxford Statistical Science Series 23, (Clarendon Press, Oxford, UK, 2000), p. 563

J.A. Martín-Fernández, Comments on: compositional data: the sample space and its structure, TEST **28**(3), 653–657

J.A. Martín-Fernández, J.J. Egozcue, R.A. Olea, V. Pawlowsky-Glahn, Units recovery methods in compositional data analysis. Nat. Resour. Res. https://doi.org/10.1007/s11053-020-09659-7

J.A. Martín-Fernández, V. Pawlowsky-Glahn, J.J. Egozcue, R. Tolosana-Delgado, Advances in principal balances for compositional data. Math. Geosci. **50**(3), 273–298

G. Mateu-Figueras, V. Pawlowsky-Glahn, J.J. Egozcue, The principle of working on coordinates, in ed. by V. Pawlowsky-Glahn, A. Buccianti (2011), pp. 31–42; 378

G. Mateu-Figueras, V. Pawlowsky-Glahn, J.J. Egozcue, The normal distribution in some constrained sample spaces. SORT **37**(1), 29–56 (2013)

A. Mooijaart, P.G.M. Van der Heijden, L.A. Van der Ark, A least squares algorithm for a mixture model for compositional data. Comput. Stat. Data Anal. **30**, 359–379 (1999)

J.M. Oller, A. Satorra, A. Tobeña, Pathways and legacies of the secessionist push in Catalonia: linguistic frontiers, economic segments and media roles within a divided society. A Policy Network Paper (2019), p. 27. https://policynetwork.org/publications/papers/pathways-and-legacies-of-the-secessionist-push-in-catalonia/

J. Palarea-Albaladejo, J.A. Martín-Fernández, Compositions—R package for multivariate imputation of nondetects and zeros in compositional data sets. Chemom. Intell. Lab. Syst. **143**, 85–96 (2015)

V. Pawlowsky-Glahn, A. Buccianti (eds.), *Compositional Data Analysis: Theory and Applications* (Wiley, Hoboken, 2011), p 378

V. Pawlowsky-Glahn, J.J. Egozcue, D. Lovell, Tools for compositional data with a total. Stat Model **15**, 175–190 (2015)

W. Revelle, *psych: Procedures for Personality and Psychological Research* (Northwestern University, Evanston, Illinois, USA, 2019). https://CRAN.R-project.org/package=psychVersion=1.9.12. Accessed 30 April 2020

V. Simonacci, M. Gallo, Statistical tools for student evaluation of academic educational quality. Qual. Quant. **51**(2), 565–579 (2017)

M. Temp, K. Hron, P. Filzmoser, robCompositions: an R-package for robust statistical analysis of compositional data, in V. Pawlowsky-Glahn, A. Buccianti (2011), pp. 341–355; 378

A Compositional Three-Way Approach for Student Satisfaction Analysis

Michele Gallo, Violetta Simonacci, and Valentin Todorov

Abstract Three-way rating data on student satisfaction contain the scores assigned by students to a set of items measuring different aspects of educational quality at different time points. Such data provide information on the magnitude of satisfaction as well as information on how aspects vary with respect to each other and how they contribute to the total satisfaction of each student. Data magnitude is predominant in determining variability patterns, thus, any standard tool applied to these arrays only yields a one-dimensional solution measuring scale differences. The relative changes among items go completely undetected unless a compositional approach is used in combination with a multilinear tool. A case study on student satisfaction is presented to demonstrate that this method provides an insightful analysis of the role played by each aspect in generating satisfaction throughout faculties and years.

1 Introduction

One of the open challenges for Compositional Data (CoDa) experts is to convey the full potential of this approach to different fields of research. For this reason, the definition of CoDa has undergone a significant evolution throughout the years, progressively encompassing a broader category of data structures.

Starting from Aitchison's early descriptions (Aitchison 1982, 1986), in which CoDa are presented as vectors of non-negative values with a sum constraint, an extended definition has emerged. Now, CoDa are simply described as positive observations carrying relative information. For more recent definitions, we can refer to Pawlowsky-Glahn et al. (2015a) and Pawlowsky-Glahn et al. (2015b). In the latter

M. Gallo (✉) · V. Simonacci
University of Naples "L'Orientale" - DISUS, P.zza S.Giovanni 30, Naples 80134, Italy
e-mail: mgallo@unior.it

V. Simonacci
e-mail: vsimonacci@unior.it

V. Todorov
United Nations Industrial Development Organization (UNIDO), VIC, Vienna, Austria
e-mail: v.todorov@unido.org

P. Filzmoser et al. (eds.), *Advances in Compositional Data Analysis*,
https://doi.org/10.1007/978-3-030-71175-7_8

contribution, a further step is taken to show that the presence of relative information in the data does not necessarily mean that scale is irrelevant, i.e., abundances can still be of interest.

This concept can be extended even further as to say that any phenomenon described by non-negative values can be considered compositional if one is interested in detecting relative variation patterns. In this sense, the application of a CoDa approach hinges also on the research query and not only on the inherent structure of the data at hand. This step is fundamental especially for social science applications, in which datasets are multifold and noisy.

This wide-ranging definition is at the basis of several studies in which Compositional Data Analysis (CoDA) is proposed as a strategy for uncovering trade-off patterns (Bergeron-Boucher et al. 2018; Di Palma and Gallo 2019; Simonacci and Gallo 2019). Similarly, in Simonacci and Gallo (2017), a compositional approach was suggested along with PCA (CoDa-PCA) for studying rating scales as a way to understand the composition of student satisfaction.

Following this contribution, this work further demonstrates the usefulness of CoDA outside its ordinary bounds by proposing its application to a complex dataset with a tridimensional structure. In particular, this work proposes another study on student satisfaction recorded throughout time. A compositional CANDECOMP/PARAFAC (CP) model is performed to analyze student satisfaction data aggregated by faculty at different time points. This decomposition tool is chosen because it is capable of unveiling common latent constructs in trilinear data.

A CoDA approach answers different questions with respect to the standard CP model. By removing the bias of total satisfaction for each faculty, the focus is moved to faculty and time differences in terms of relative item variation.

In Sect. 2, a brief overview of the background for service quality studies and three-way rating data is provided. In Sect. 3, the compositional methodology for three-way rating scale data is presented; in Sect. 4, the case study and the principal results are illustrated; and Sect. 5 includes the final discussion.

2 Background

Three-way survey data collect scores on a rating scale recorded for I subjects (First mode) on a set of J items (Second mode) at different occasions (Third mode). Such data can be arranged in an array with three dimensions $\mathscr{X} \in \mathbb{R}^{I \times J \times K}$ and generic element x_{ijk}.

Depending on the role of subjects in the research query, rating data can be analyzed differently. If individual differences are deemed uninteresting, $\mathscr{X}$ can be simply converted into covariance or correlation matrices. Alternatively, if subjects are important to the analysis, survey data can be modeled with multilinear methods such as the Tucker3 and the CP models (Tucker 1966; Carroll and Chang 1970; Harshman 1970) to preserve separate mode variability.

The CP model is preferable if the latent constructs are proportional throughout occasions. Kroonenberg (2008) shows how both techniques can be successfully used to study three-way rating data depending on their characteristics.

Item mode normalization is not necessary if rating scales have the same range. Centering by item should be performed only for Likert scales, i.e., non-invertible monopolar or unidirectional scales characterized by only one adjective.

Data on service quality, such as student satisfaction studies, are presented as rating scale scores because satisfaction is considered a non-directly observable construct which can only be measured through the assessment of its many observable outcomes.

Despite the differences between approaches (Marsh 1982; Parasuraman et al. 1988; Cronin and Taylor 1992; Abdullah 2006), satisfaction/quality studies all aim to identify a hidden construct through multiple item questionnaires. Data collected in this fashion present three important features that make their statistical analysis problematic: (i) a multivariate structure; (ii) ordinal nature of rating scales (Stevens 1946); and (iii) subjective measurement scales. If data are also observed under different conditions or time points, their analysis becomes even more challenging because of the tridimensional structure of the variability.

In behavioral sciences, the most widely used methodology for dealing with two-way rating scale data is the Item Response Theory (IRT) approach (Andrich 1978; Rasch 1980; Wright and Masters 1982). IRT models were specifically developed in the framework of psychometrics with the purpose of building an interval scale for a non-directly observable one-dimensional variable.

At the basis of this approach, there is the idea that questionnaire/test scores are the results of both subject and item parameters. Therefore, to make inferences on the response structure of similar populations, these two sets of parameters should be modeled separately through probabilistic functions expressed as success to failure ratios on a logit scale.

This methodology allows testing for the goodness of the questionnaire and provides a ranking of abilities and difficulties on an interval scale. However, it presents some restrictions. It assumes that the latent construct is one-dimensional and focuses on the identification of a generalized pathway rather than on describing the structure of preferences emerging from the data.

Generalized versions of IRT models have been proposed in the literature for satisfaction data collected over a third dimension (von Davier et al. 2007; Carstensen and Rost 2007). However, their application never took off due to their complexity.

A novel approach based on CoDa-PCA for dealing with two-way rating scale preferences in the context of student satisfaction studies is suggested in Simonacci and Gallo (2017).

CoDA was originally devised to deal with the problematic structure of bounded data, which present a negative bias in their covariance structure and are forced in a subspace of the real space, called simplex (Egozcue et al. 2011). To understand why this approach can prove useful for rating scale data, which are not naturally constrained, some important considerations should be made.

Let us consider the responses of students or groups of students to a quality of teaching assessment questionnaire. This dataset contains two types of information:

(i) how much each student is satisfied in general; and (ii) how different aspects of the service contribute to forming the satisfaction of students. The first aspect can be captured by standard analysis, while the second aspect necessitates a scale-free approach.

By considering total satisfaction as a negligible aspect which can be studied separately with descriptive tools, students can be compared with respect to their preference structure in terms of ratios. The compositional methodology is, thus, an acceptable substitute to IRT for assessing item structure independently of students' levels of satisfaction.

A generalization of CoDa-PCA to higher orders is a natural extension of this methodology. In this work, three-way rating data are treated as compositions are decomposed with the CP model to study the complex structures within the multivariate phenomenon of student satisfaction.

3 Methods

3.1 Three-Way Rating Data as Compositions

Let us consider the responses to a survey on academic educational quality. Each student is required to assign a score to $1, \ldots, j, \ldots, J$ items that describe the service by selecting between a given number of $1, \ldots, l, \ldots, L$ ordinal categories, expressed as positive integers. The survey is administered at $1, \ldots, k, \ldots, K$ time points, thus, two-way numeric matrices can be built for each time instance, $\mathbf{X}_k \in \mathbb{R}^{(n_k \times J)}$ where n_k is the number of students. This number changes throughout time ($n_k \neq n_{k+1}$), thus the data present a multi-sample design and cannot be arranged in a tridimensional array.

Homogenous groups of students (in terms of service catering) can be identified by averaging the scores of students within the same faculty. In this manner, the same $1, \ldots, i, \ldots, I$ subjects, hereafter faculties, are considered at each time point so that each $\mathbf{X}_k \in \mathbb{R}^{(n_k \times J)}$ is transformed in $\mathbf{V}_k \in \mathbb{R}^{(I \times J)}$.

Such K matrices are part of a fully-crossed design and can be considered frontal slices of a three-way array $\mathcal{V} \in \mathbb{R}^{(I \times J \times K)}$. Here the generic row vector is denoted by $\mathbf{v}_{ik}(1 \times J)$ and contains the average scores for the i-*th* faculty at time k to the J items; its total sum κ_{ik} can range between J and $J * L$ and represents the total level of satisfaction.

The single J scores are highly correlated with each other because they all express an aspect of satisfaction, and in this sense, they are biased by κ_{ik}. An exploratory analysis on such data would only show these high correlations.

The total value κ_{ik} masks the relations among various aspects of satisfaction, which are only retrievable by focusing on ratios between parts with CoDA.

A relative variation approach complies with the scale invariance property. With the closure operator defined as $C(\mathbf{v}_{ik}) = \mathbf{v}_{ijk} / \sum_{j=1}^{J} v_{ijk}$, a scale invariant method considers two vectors $\mathbf{v}_{ik}$, $\mathbf{v}_{i'k}$ equivalent if $C(\mathbf{v}_{ik}) = C(\mathbf{v}_{i'k}) = \dot{\mathbf{v}}_{ik}$, i.e., there is a positive scalar λ such that $\dot{\mathbf{v}}_{ik} = \lambda \dot{\mathbf{v}}_{i'k}$.

To exemplify, let us consider two row vectors with $J = 3$ items expressed on a Likert scale with four possible outputs $l = 1, 2, 3, 4$, for example, $\mathbf{v}_{11} = [1, 1, 2]$ and $\mathbf{v}_{21} = [2, 2, 4]$. The information contained in each vector can be decomposed into two parts: the total course satisfaction given by $\kappa_{11} = 4$ and $\kappa_{21} = 8$, respectively, and the preference structure described by the closed vectors: $\dot{\mathbf{v}}_{11} = \dot{\mathbf{v}}_{21} = [0.25, 0.25, 0.5]$.

We see that if the research query aims to investigate the differences among levels of satisfaction, and satisfaction is considered one-dimensional, the focus is on the first element. If one wants to carry out a deeper analysis and study the multivariate structure of satisfaction, the focus is on the second part of information. In terms of preference patterns and possible trade-off among different aspects of teaching, the second element becomes relevant and a compositional analysis is required.

Consequently, the array of scores can be treated as an array of compositions, conceived as observations carrying relative information with each rating scale describing a part of total course satisfaction. Their sample space, called simplex, is formed by equivalence classes of proportional vectors (Pawlowsky-Glahn et al. 2015b) and can be defined as

$$\mathbb{S}^J = \{[\dot{v}_{i1k}, \ldots, \dot{v}_{ijk}, \ldots, \dot{v}_{iJk}] : \dot{v}_{i1k} \geq 0, \ldots, \dot{v}_{ijk} \geq 0, \ldots, \dot{v}_{iJk} \geq 0; \\ \dot{v}_{i1k} + \ldots + \dot{v}_{ijk} + \ldots + \dot{v}_{iJk} = 1\} \quad (1)$$

This vector space is characterized by its own geometric rules called Aitchison geometry (Egozcue et al. 2011).

A common practice to model compositions is to first project them onto the unconstrained real space by expressing them in logarithms of ratios between parts. Several types of such transformations have been proposed in the literature. The choice of the most appropriate one is connected to the data and to the statistical analysis one wishes to perform.

The centered log-ratio (*clr*) transformation is the most straight-forward in terms of both computational simplicity and ease of interpretation. It establishes an isometric mapping between $\mathbb{S}^J$ and a hyperplane of $\mathbb{R}^J$ by considering the ratios between the elements of a compositional vector and its geometric mean.

An array of composition can be easily transformed in an array of *clr* coefficients $\mathscr{Z} \in \mathbb{R}^{I \times J \times K}$ with frontal slices $\mathbf{Z}_k \in \mathbb{R}^{(I \times J)}$ and generic element $z_{ijk} = log(\frac{v_{ijk}}{g(\mathbf{v}_{ik})})$, with $g(\mathbf{v}_{ik}) = \sqrt[J]{\prod_{j=1}^{J} v_{ijk}}$.

This simplicity comes at the price of two major issues: (i) the new coefficients yield a singular covariance matrix; (ii) the transformation does not provide an orthonormal basis in the simplex.

Despite these shortcomings *clr* coefficients are used in this work, following Gallo (2013), Gallo and Simonacci (2013), Engle et al. (2014), Gallo (2015). This choice is dictated by the nature of the model of interest, i.e., the CP model.

In this model, results are not interpreted in terms of variability structures but in terms of real underlying factors describing the phenomenon. In this perspective referring to the original parts for understanding the meaning of the latent construct is the best option. In addition, the theoretical framework of this decomposition, discussed in detail in the following section, is unaffected by singularity and yields non-orthogonal components, thus *clr* limitations are not an issue.

A possible alternative is represented by the isometric log-ratio (*ilr*) coordinates introduced by Egozcue et al. (2003), which provide an isometric isomorphism, i.e., a mapping between $\mathbb{S}^J$ and a hyperplane of $\mathbb{R}^{J-1}$ resulting in a set of coordinates with respect to an orthonormal basis on the simplex. Differently from *clr*, this kind of transformation has well-defined properties and can be used even when singularity becomes problematic as for robust procedures (Di Palma et al. 2018).

A way to obtain orthonormal coordinates is to get a set of balances with Sequential Binary Partition (SBP) (Egozcue and Pawlowsky-Glahn 2005) which subdivides compositions into non-overlapping groups. This approach is based on the capability of creating a meaningful SBP, generally with the help of expert knowledge, and does not adapt well for large data (Quinn 2018). Considering that the focus here is on interpreting latent constructs and that a three-way framework could complicate the discussion on choosing SBP, this approach is not pursued but would prove useful for further analysis (see Sect. 5).

It is also important to note that when three-way data represent samples of compositional tables, for which the goal is to study the association between two sets of factors, the best approach is presented in Fačevicová et al. (2016) and (2018).

3.2 CP Model for CoDa

A three-way array of *clr* coefficients $\mathscr{Z} \in \mathbb{R}^{I \times J \times K}$ can be decomposed by means of the CP model. This multilinear tool aims to provide a low rank approximation of the array in order to synthesize its diagonal information by decomposing it into the sum of a given number of triads $1, \ldots, f, \ldots, F$ and an array of residuals $\mathscr{E} \in \mathbb{R}^{I \times J \times K}$. Each triad is an outer product of the loading vectors, $\mathbf{a}_f$, $\mathbf{b}_f$ and $\mathbf{c}_f$, one for each mode. Ideally F is minimal, namely, it corresponds to the tensorial rank r of the array, i.e., the real number of directions that fully describe the trilinear phenomenon. In this case, the CP model is a rank decomposition (Kolda and Bader 2009). Using this definition, the model can be written as

$$\mathscr{Z} = \hat{\mathscr{Z}} + \mathscr{E} = \sum_{f=1}^{F} \mathbf{a}_f \circ \mathbf{b}_f \circ \mathbf{c}_f + \mathscr{E} \tag{2}$$

where $\hat{\mathscr{Z}}$ is the structural image of the original array $\mathscr{Z}$ and the symbol $\circ$ represents the vector outer product. Alternatively, using a more intuitive matrix-wise notation, the decomposition can also be written as

$$\mathbf{Z}_k = \hat{\mathbf{Z}}_k + \mathbf{E}_k = \mathbf{A}\mathbf{D}_k\mathbf{B}^t + \mathbf{E}_k \quad k = 1, \ldots, K \tag{3}$$

Here, the loading vectors composing the triads are arranged as columns of the three loading matrices $\mathbf{A} \in \mathbb{R}^{I\times F}$, $\mathbf{B} \in \mathbb{R}^{J\times F}$ and $\mathbf{C} \in \mathbb{R}^{K\times F}$, respectively, while $\mathbf{D}_k$ is a diagonal matrix holding the k-th row vector of $\mathbf{C}$ and $\mathbf{E}_k \in \mathbb{R}^{I\times J}$ is the k-th frontal slice of the array of residuals.

The main theoretical difference between the CP model and the Tucker3 is that its focus is not on explaining variability structures but rather on identifying the real underlying constructs that influence the phenomenon. Each triad refers to a construct common to all modes, and it is studied on its own because components are not orthogonal.

The solution is usually estimated by means of the Alternating Least Squares algorithm PARAFAC-ALS (Harshman 1970), a three-step iterative procedure that minimizes the following least squares objective function:

$$\min_{\mathbf{A},\mathbf{B},\mathbf{C}} = \sum_{k=1}^{K} \|\mathbf{Z}_k - \mathbf{A}\mathbf{D}_k\mathbf{B}^t\|^2 \tag{4}$$

where $\|\cdot\|$ is the Frobenius norm.

Due to the row-centering of the *clr* transformation, the restriction $\mathbf{1B} = \mathbf{0}$ is imposed, where $\mathbf{1}$ is a J vector of ones and $\mathbf{0}$ is an F vector of zeros. This condition does not cause problems to the model because from a mathematical stand point is automatically ensured. However, the perfect collinearity of the data may still create some estimations inefficiencies (Kiers 1998) especially for large data-sets and if the model is not a rank decomposition (i.e., $F \neq r$). In these cases, it may be better to use alternative estimating procedures (for more details see Gallo et al. 2018).

For a correct interpretation of CP loadings, the rules presented in Aitchison and Greenacre (2002) for PCA results can be easily extended. *Clr* coefficients represent the deviations from the mean on a logarithmic scale, thus, also in a CP model each column of $\mathbf{B}$ is a linear combination of compositional parts and can be formulated as a logcontrast:

$$\sum_{j=1}^{J} b_{jf} \log\left(\frac{v_{ijk}}{g(\mathbf{v}_{ik})}\right) = \sum_{j=1}^{J} b_{jf}\, z_{ijk} \text{ where } b_{1f} + \ldots + b_{jf} + \ldots + b_{Jf} = 0 \tag{5}$$

To study the relation among two parts v_{ijk} and $v_{ij\prime k}$ of a compositional row vector $\mathbf{v}_{ik}$, the following correspondence relation stands

$$\log\left(\frac{v_{ij\prime k}}{g(\mathbf{v}_{ik})}\frac{g(\mathbf{v}_{ik})}{v_{ijk}}\right) = z_{ij\prime k} - z_{ijk} \tag{6}$$

To clarify, let us consider an illustrative three-part composition $[\alpha, \beta, \gamma]$:

$$b_{\alpha_f}\log(\alpha) + b_{\gamma_f}\log(\gamma) - (b_{\alpha_f} + b_{\gamma_f})\log(\beta) = constant \tag{7}$$

which, using exponential notation becomes

$$\left(\frac{\alpha}{\beta}\right)^{b_{\alpha f}} \propto \left(\frac{\beta}{\gamma}\right)^{b_{\gamma f}} \tag{8}$$

or for multiple part compositions, we can write

$$\frac{(\alpha)^{b_{\alpha f}}(\gamma)^{b_{\gamma f}}}{(\beta)^{b_{\beta f}}} = constant \tag{9}$$

Aitchison and Greenacre (2002) also provide specific interpretational rules for *clr* biplots based on the link between arrows' end point representing compositional parts. These links are an approximation of the relative variation of the corresponding pairwise log-ratio, while the cosine of the angle between them is an approximation of their correlation.

The fundamental difference in interpretation between PCA and CP loadings of compositional parts is that for a standard CP model the extracted component are not orthogonal and fit partitioning between components cannot be determined. This means that a bi-dimensional representation with two loadings plotted against each other is not recommended.

The preferred plotting tool for the CP model is the per-component plot. This representation shows only one of the extracted components at a time, which is illustrated as a common direction along which the loadings of the three modes of the analysis are plotted together. Conventionally loadings are scaled to unit mean-square to account for a difference in the number of levels. The first mode coefficients are plotted on the right, the second mode on the left, and the third mode in the middle. The latent construct must be interpreted taking into account all modes. In this plot, the distances among parts, or one-dimensional links, are the distances between the logcontrast coefficients for that component and can be interpreted as in *clr* biplots. The only difference is that instead of being an approximation of the total variability, they are an approximation of the variability connected to the phenomenon captured by that axis. For more details on per-component plots and their interpretation see Kroonenberg (2008).

4 Case Study

4.1 Dataset

In Italy, Law No. 370/99 requires universities to compulsorily and independently develop and implement a system of students' opinion polling on the quality of teaching. In this perspective, the National Agency for the Evaluation of Universities and Research (ANVUR) was founded in 2006. This agency is responsible for carrying out a periodic assessment of the Italian higher education system, and its primary purpose is to systematically monitor academic educational quality through the principles of accreditation, evaluation, and self-assessment (ANVUR 2013).

With the Presidential Decree 76/2010, the ANVUR was also assigned the task of defining the general outlines and standardized methodologies for the evaluation of academic institutions and degree programs by also promoting the active involvement of students. The primary tool employed for implementing quality assessment is represented by yearly student questionnaires as the opinions of students on courses and degree programs represent essential information to identify critical aspects and margins for improvement.

Such surveys are structured to provide information on student satisfaction to get a useful insight into their perceptions and preferences with the ultimate goal of better tailoring the teaching supply to the population of students and their needs. In this sense finding appropriate tools for studying this information is of paramount importance for academic institutions. However, universities generally use only univariate statistical tools to analyze satisfaction data.

The current standardized questionnaire is submitted for every course, and it is organized into five sections to evaluate satisfaction within different macro-categories: degree program, course characteristics, teacher, equipment, and overall satisfaction. The survey consists of 18 standardized questions (the same for all universities) reported in Table 1.

Each item is assessed by the students on a ten-level Likert scale in which 1 = "Strongly Disagree" and 10 = "Strongly Agree".

Following ANVUR guidelines, the University of Florence has been monitoring the students' opinions on teaching quality through these standardized questionnaires and has made its data available for assessment (VALMON 2018).

The original dataset with micro-data on student responses has been aggregated reporting the mean values of 10 faculties: Agriculture, Architecture, Economy and Management, Law, Engineering, Psychology, Health, Mathematics, Political Sciences, Humanities and Education. The responses of students interviewed for 6 consecutive academic years (2012–2017) are considered. Consequently, the data can be arranged in a three-way array of faculties by items by years obtaining $\mathcal{V} \in \mathbb{R}^{10\times 18\times 6}$.

This dataset is analyzed with both a standard CP and a compositional CP on *clr* coefficients. The computations in this paper were performed using the R package *rrcov3way* (Todorov et al. 2020), which is freely available from the *Comprehensive R Archive Network* (CRAN) at https://cran.r-project.org/. This package provides Tucker3 and CP functions with a robust option for standard and compositional three-way data as well as specific plotting functions for resulting objects.

4.2 CP Results: Rank Selection

The latent structure of the three-way rating dataset is explored using the standard and the compositional CP model. To this end, the first step is to identify the correct

Table 1 ANVUR survey questions to investigate student satisfaction

Tag	Questions
Q-1 General Workload or G_W	Is the overall workload of the courses officially scheduled in the reference period acceptable?
Q-2 General Organization of G_O	Is the overall organization of the courses officially scheduled in the reference period acceptable?
Q-3 Individual Time or I_T	Is the timetable for the lessons designed to allow time for individual studying?
Q-4 Starting knowledge or S_K	Is your preliminary knowledge sufficient for a general understanding of the topics of the exams?
Q-5 Topic or TOP	Are the covered topics new or supplementary with respect to the knowledge already acquired?
Q-6 Work load or W_L	Is the work load appropriate to the number of credits assigned to the course?
Q-7 Teaching material or T_M	Is the teaching material (provided or recommended) suitable for the studied subject?
Q-8 Other activities or O_A	Are supplementary activities (exercises, tutorials, workshops, etc. ...) useful to the learning experience?
Q-9 Exam procedures or E_P	Are exam procedures clearly defined?
Q-10 Respected hours or R_H	Are teaching hours respected?
Q-11 Motivation or MOT	Is the teacher able to spark your interest in the subject?
Q-12 Clarity or CLA	Is the teacher able to explain studied topics clearly?
Q-13 Availability or AVA	Is the teacher actually available for further explanations?
Q-14 Helpfulness or HEL	Is the teacher available and exhaustive on requests for clarification?
Q-15 Classroom or CLS	Are the classrooms in which the lessons are held adequate (can you see, hear, find a place to sit)?
Q-16 Equipment or EQU	Are the premises and equipment for supplementary teaching activities (exercises, workshops, seminars, etc.) adequate?
Q-17 Interest or INT	Are you interested in the topics covered?
Q-18 Satisfaction or SAT	Overall, are you satisfied with this course?

rank of the model for both procedures, i.e., the dimensionality that fully explains the trilinear information in the data without modeling excess noise. Here, the convex hull procedure (Ceulemans and Kiers 2006) is used to evaluate different dimensionalities and results are reported in Fig. 1 for the standard CP (left) and for the compositional approach (right).

If we look at the convex hull for raw data, we notice right away that an $F = 1$ dimensionality is more than sufficient to explain total variability, ensuring a fit of 99.93%. Additional components add redundant information: if an $F = 2$ component model is considered, for example, a mere increase of 0.046% is recorded and its first component corresponds exactly to the component of the $F = 1$ model, with a Tucker's congruence coefficient of 0.99. This shows that even in the two-component model all variability is explained by the same one-dimensional construct. Thus, raw data appear to be characterized by only one direction capable of describing the phenomenon.

Tucker's congruence coefficient is an index of factor similarity (Lorenzo-Seva and Ten Berge 2006), which can be interpreted as the cosine of the angle between two axes. Here it is computed for each of the three modes and then the three scalars obtained are multiplied. It can be exemplified for the first mode in the following manner:

$$\phi\left(\mathbf{a}_1^{(F=1)}, \mathbf{a}_1^{(F=2)}\right) = \frac{\sum_{i=1}^{I} a_{i1}^{(F=1)} a_{i1}^{(F=2)}}{\sqrt{\sum_{i=1}^{I} a_{i1}^{2(F=1)} \sum_{i=1}^{I} a_{i1}^{2(F=2)}}} \tag{10}$$

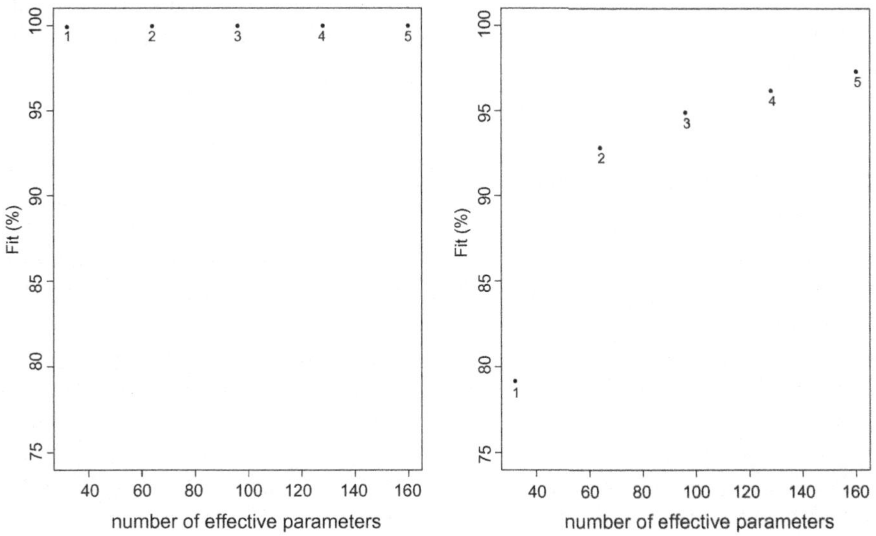

Fig. 1 Convex hull curve for dimensionality selection: the total fit for the five model dimensionalities $F = 1, \ldots, 5$ is plotted against the associated number of effective parameters to assess the increments in fit provided by an increase in model complexity. The plot is displayed for raw data (left) and *clr* coefficients (right)

where $a_{i1}^{(F=1)}$ and $a_{i1}^{(F=2)}$ are the generic elements of the first-mode components $\mathbf{a}_1^{(F=1)}$ and $\mathbf{a}_1^{(F=2)}$ of the $F = 1$ and the $F = 2$ models, respectively.

If the focus is shifted to the CP on *clr* coefficients results appear significantly different. In this instance, the convex hull procedure suggests a two-component complexity. Between an $F = 1$ and $F = 2$ dimensionalities a relevant jump in fit is recorded (13.7%), and it is reasonable to consider the structure not one-dimensional.

The total percentage of fit explained by the $F = 2$ model equals 92.85%, which refers to the solution as a whole because, due to the non-orthonormal axes of the CP model, fit percentage cannot be interpreted in an additive fashion. However, it is interesting to note that if Tucker's congruence coefficient is calculated between the first and second component of the $F = 2$ compositional model following Eq. 10, they are nearly independent (0.037), thus the two dimensions represent uncorrelated constructs and the structure is two-dimensional.

To check how much the compositional results differ from the raw solution, Tucker's coefficient can also be calculated among the construct of the raw solution and the two dimensions of the *clr* case. It was found that both components of the compositional solution are nearly independent from the raw data dimension with values of 0.038 and 0.001, respectively. The structure unveiled by the compositional CP technique is completely different from the standard approach, and it can be interesting to compare their findings.

The drop in total fit recorded in the CoDA approach is due to the *clr* preprocessing, very much as it happens after other types of preprocessing that remove uninteresting information from the data such as mean and variability differences. In this sense, the drop in fit should not to be considered pejorative. On the contrary, the extremely high fit in the standard model is an indication that information on size is disguising other important patterns.

4.3 CP Results: Comparative Solutions

The most useful graphical tool for interpreting CP solutions is the per-component plot. Since each component of the model represents the same latent construct, the scores for the first, second, and third modes can be studied simultaneously and plotted together. In this manner intermodal relationships can be assessed and the latent entities, which the loadings represent, can be studied in detail. The scores of the three modes should be read only in reference to the direction of the component (vertical axis).

Per-component plots are thus built for the one component of the raw CP model and for the two components of the compositional one, to compare their results.

Let us focus on the per-component plot on raw scores presented in Fig. 2. The interpretation of the first component is of immediate reading as it provides an approximation of the ranking in terms of total satisfaction for all three modes, i.e., the construct can be label "Student satisfaction". This can be easily deduced from the fact

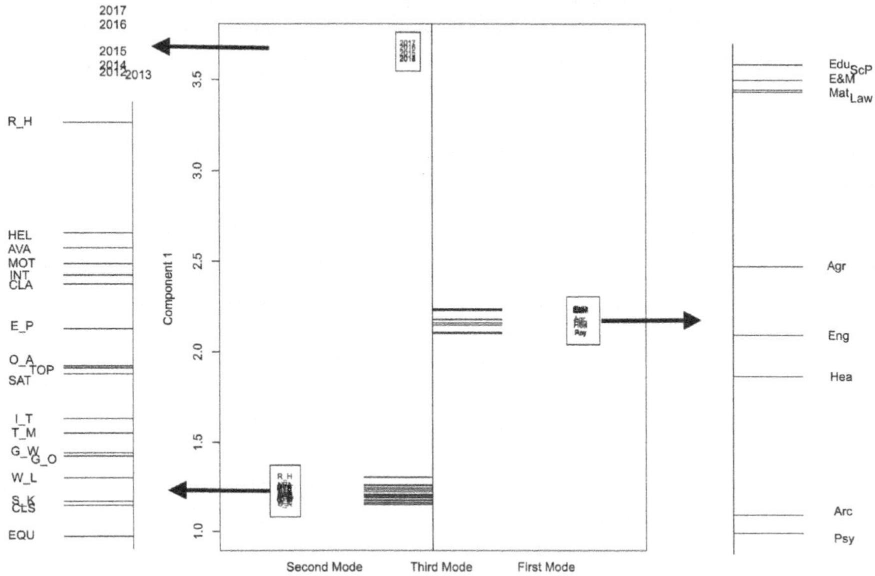

Fig. 2 Per-component plot of raw data: the coefficients of the first CP triad estimated on raw data, $\mathbf{a}_1$ (on the right—first mode), $\mathbf{b}_1$ (on the left—second mode) and $\mathbf{c}_1$ (in the middle—third mode) are scaled to unit mean-squared lengths and plotted together along the same direction to study interactions

that all three modes present concordant coefficients and their direction is consistent with element dimensionality.

In particular, the following information can be extrapolated for each mode. If the different aspects that characterize the service (items) are considered, it is possible to see how throughout faculties and years R_H records the highest score followed by HEL, AVA, MOT, INT, and CLA even though for the remaining item differences are not as marked. The items with which students appear less satisfied are EQU, CLS, and S_K.

In terms of faculties, three groups can be detected: the "most satisfied students" cluster belonging to the faculties "Edu", "ScP", "E&M", "Mat", and "Law", which present very similar profiles, the "least satisfied students" cluster attending the faculties "Psy" and "Arc" and a group with intermediate scores consisting of "Hea", "Eng" and "Agr".

If time scores are considered, we find that the most relevant difference is recorded between 2012 and 2017. Since, in general, year points record coefficients with scores moving along the component direction, the plot shows the existence of a trend in favor of increased satisfaction.

The interpretation of the axis as a measure of satisfaction can be demonstrated by looking at total scores. If item ranking is calculated by simply summing all the scores by item throughout the years, these results yield the same ranking provided in Fig. 2. Similarly, if total scores are aggregated by faculty, the ranking also corresponds.

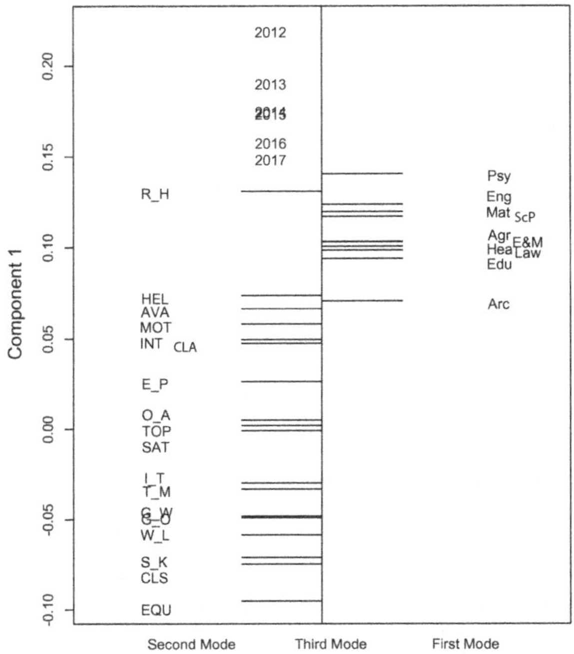

Fig. 3 Per-component plot of *clr* data, first axis: the coefficients of the first CP triad estimated on *clr* transformed data, $\mathbf{a}_1$ (on the right—first mode), $\mathbf{b}_1$ (on the left—second mode) and $\mathbf{c}_1$ (in the middle—third mode) are scaled to unit mean-squared lengths and plotted together along the same direction to study interactions

Only the positions of "ScP" and "Edu" are reversed, but their scores are basically overlapping.

The pre-component plot on raw data summarizes all the information extracted in a simple and effective manner. However, the information it conveys, as much as it appears to be nearly 100% of the total inertia, concerns only one predominant aspect, i.e., the total level of satisfaction. Thus, it only allows answering questions such as (i) which aspects of the service generate the most satisfaction? (ii) In which years the level of satisfaction is higher? (iii) The most satisfied students belong to which faculty?

The visualization of this information, albeit very important, hardly justifies the use of a complex tool such as the CP model because it can easily be obtained through simple univariate statistics, even if in a less precise and organized fashion.

In the compositional CP model, a two-dimensional structure is uncovered instead. Like in the previous analysis the first step is to analyze the per-component plots. This plot for the first construct is reported in Fig. 3. Results can be read in a similar fashion, however now the focus is shifted from the level of satisfaction to the composition of satisfaction: how the different aspects of the service, represented by each item, are combined together to make up the latent trait.

Let us look at item scores. In this plot, due to the *clr* transformation, the distances of item points from zero yield a ranking of item proportions in terms of variability explained, that is the absolute loadings provide the compositional weight of each aspect of satisfaction. In addition, since the sum of item coefficients is equal to zero, the scores for R_H, HEL, AVA, and MOT present values that counterbalance the scores registered for EQU, CLS, and S_K. Other aspects have small coefficients, especially TOP, O_A and SAT, which present close-to-zero values.

This structure highlights how student satisfaction stems mainly from the evaluation of the professional characteristics of the teacher (R_H, HEL, AVA, and MOT), which are balanced by attributes related to facilities (EQU, CLS) and personal readiness (S_K and items connected to the university organization). Other aspects have a reduced impact on students' assessments.

To fully describe the construct in compositional terms it is important to interpret scores also based on links, i.e., distances among item points that represent approximations of their pairwise ratio. Close items vary similarly throughout years and faculties as their ratio remains nearly constant, see $log(G\_O/G\_W)$, $log(INT/CLA)$ and $log(S\_K/CLS)$.

It is possible to even detect groups with small within relative variations: (i) HEL, AVA, MOT, INT, and CLA; (ii) G_W, G_O, W_L, S_K and CLS which record high relative variation between them.

R_H and EQU are the most variable items with respect to all other items and each other, and their relative variability strongly characterizes the axis.

To numerically assess the counterbalancing relations described above, it is possible to look directly at loadings on the first component in Table 2 (first column). These values can be interpreted as in Eq. 9 as logcontrast weights.

This linear combination synthesizes a general pattern in which items with positive scores tend to vary in the same direction and opposite to the ones with negative scores. It can be easily seen how this structure simply quantifies the contrasting relationships described above and this first component can be read as the latent construct "teacher professionalism" against other aspects connected to university/course fittingness.

Differences in faculties can also be interpreted in terms of the latent aspect captured by the axis. There is a clear difference between what is recorded for the average student of "Psy" and that of "Arc", with all other faculties positioned in intermediate situations. Therefore, for "Psy", a more marked polarization toward teacher professionalism attributes is detected in the average student preference structure. This is not the case for the average "Arc" students whose preference structure, in relative terms, is less polarized.

In terms of time changes, the greatest difference is recorded between 2012 and 2017, thus there is a trend towards an increment in the relative weight of university fittingness aspects with respect to the professional characteristics of the teacher. In other words, the polarization toward teaching professionalism items appears to be flattening.

Each mode has been analyzed with respect to the axis, however, it is also possible to read intermodal relations. For example, the average "Psy" student has a level of satisfaction strongly characterized by the preference toward teachers' aspects with the maximum difference recorded for the pairwise log-ratio (R_H / EQU). These differences register higher values in 2012 and see a slow reduction year by year. If the preference structure of the average "Arc" student is considered, this appears

Table 2 Logcontrast coefficients

ITEM	Logcontrast1	Logcontrast2
G_W	−0.20414	−0.03997
G_O	−0.20799	−0.02341
I_T	−0.12617	−0.03768
S_K	−0.30106	−0.05569
TOP	0.008751	−0.05635
W_L	−0.24786	−0.06463
T_M	−0.14067	−0.1123
O_A	0.021102	−0.10114
E_P	0.111528	−0.04122
R_H	0.55561	−0.10362
MOT	0.247273	−0.12325
CLA	0.201552	−0.13635
AVA	0.282383	−0.05689
HEL	0.313496	−0.07938
CLS	−0.31682	0.600801
EQU	−0.40389	0.607711
INT	0.210658	−0.09362
SAT	−0.00375	−0.08303

much less polarized between teacher and non-teacher related items. In addition, these differences also record a reduction throughout the years.

After examining this first axis in detail, the second one can also be discussed. Figure 4 shows the latent construct emerging from the second *clr* component. Here the variability structure in terms of items is mainly given by the contraposition of two items related to university facilities EQU and CLS with all other aspects of service.

To give an interpretation coherent with a compositional approach, we can say that a second relative variation pattern distinguishes between facility-related items and everything else. This means that this secondary pattern allows to differentiate, or better to zoom in on the variability within the items at the denominator in the first logcontrast.

For this second component, the variability structure is much more polarized and a logcontrast for this axis has a simple structure, with only two aspects with high positive scores. For the original coefficients, which help quantify the proportional relations emerging from this pattern, see the second column in Table 2.

Looking at the pattern emerging from the first mode, it is possible to see how the differences among faculties appear much more marked and assume a prominent role in explaining axis variability. Therefore, the composition of satisfaction in terms of student perception of facility and non-facility-related items is highly variable among faculties.

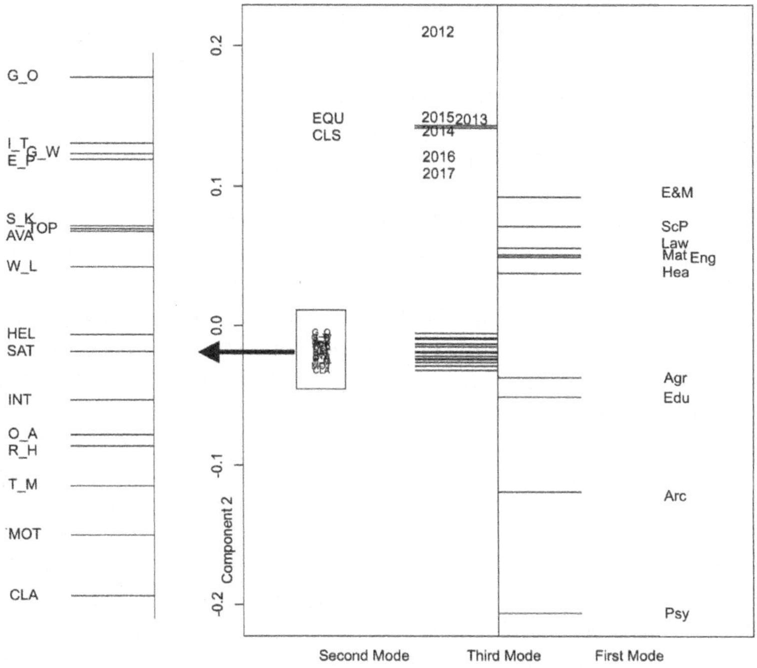

Fig. 4 Per-component plot of *clr* data, second axis: the coefficients of the second CP triad estimated on *clr* transformed data, $\mathbf{a}_2$ (on the right—first mode), $\mathbf{b}_2$ (on the left—second mode) and $\mathbf{c}_2$ (in the middle—third mode) are scaled to unit mean-squared lengths and plotted together along the same direction to study interactions

This noticeable divergence in faculty profiles can be explained by the fact that at the University of Florence faculties are assigned different lecture halls. Specifically, faculties such as "E&M", "ScP" and "Law" are based in buildings of newer constructions, fully equipped and characterized by modern structures, while other faculties such as "Arc" and "Psy" are based in remodeled historical buildings which lack the same efficient design.

These differences can be better described in terms of intermodal relationship. Some faculties such as "Psy" and "Arc" are characterized by a strong negative polarization for facilities-specific items which, however, decreases throughout the years. Other faculties such as "E_M "and "ScP" display a different preference structure in which facilities are able to generate more satisfaction in relative terms. Thus, it is the variability among these sub-groups of faculties with respect to facility features which characterizes this second pattern.

5 Conclusive Remarks and Discussion

Positive vectors describing a single construct can be decomposed into vectors of proportions multiplied by a scale factor. Depending on the research query, one chooses to focus on scale differences using standard statistical tools or on the proportional aspect by means of CoDA.

To exemplify the usefulness of a compositional approach for rating scale data, it is proposed to analyze the responses to a student satisfaction survey throughout years with both a standard and a compositional CP model.

The CP model is preferred here because it is the best tool to investigate trilinear data for common preference structures. These common directions represent real existing patterns in student responses. Their analysis allows assessing the usefulness of questionnaire items and understands the challenges faced by the University of Florence with respect to teaching quality.

It is shown that the standard CP only extracts trivial information contained in scale factors, thus, nothing can be said about the relative association among the multifold aspects of satisfaction.

The compositional CP is capable of showing which aspects have constant or highly changeable relative variation, unveiling the items that follow a similar behavior throughout the samples. In this manner, the role played by each question or group of questions in defining student satisfaction can be better understood. In addition, time and faculty changes can be interpreted with respect to these satisfaction patterns and proportionality.

If both logcontrasts are considered this analysis uncovers the presence of three subgroups of items: "teacher-related", "facility related", and "organizational aspects". This suggests that the satisfaction of the students is at least three-fold. Some items appear to be redundant, as they may be too generic such as SAT, TOP, and O_A. Given these considerations, surveys can be improved in light of these specifications in the future.

The "sub-groups" identified in this analysis, whose between variability is captured by the CP axes, can be of further use for identifying interpretable orthonormal coordinates. In particular, for cases in which expert knowledge and a priori assumptions cannot be made, exploratory tools may be of great help to find meaningful SBP for constructing balances. In Pawlowsky-Glahn and Egozcue (2011) an exploratory technique based on the CoDa Dendrogram is proposed, while in Pawlowsky-Glahn et al. (2015a) several approaches to obtain balances are suggested, one of which is based on the use of the *clr* biplot.

In a similar fashion, the results obtained in this work and presented through the per-component plot on *clr* transformed data can become a good starting point for constructing balances. The advantages of such an approach become even more relevant if a Tucker3 analysis is performed instead. In this case, the focus would no longer be on identifying common latent directions as much as on finding subspaces of approximation in which, through rotation, it is possible to maximize variability in one mode.

References

F. Abdullah, The development of HEdPERF: a new measuring instrument of service quality for the higher education sector. Int. J. Consum. Stud. **30**(6), 569–581 (2006)

J. Aitchison, The statistical analysis of compositional data. J. Roy. Stat. Soc.: Ser. B (Methodol.) **44**(2), 139–160 (1982)

J. Aitchison, The Statistical Analysis of Compositional Data. Monographs on Statistics and Applied Probability (Chapman & Hall Ltd., London, 1986) (Reprinted in 2003 with additional material by The Blackburn Press)

J. Aitchison, M. Greenacre, Biplots of compositional data. J. Roy. Stat. Soc.: Ser. C (Appl. Stat.) **51**(4), 375–92 (2002)

D. Andrich, A rating scale formulation for ordered response categories. Psychometrika **43**, 561–573 (1978)

ANVUR: Proposta operativa per l'avvio delle procedure di rilevamento dell'opinione degli studenti per l'A.A. 2013-2014. ANVUR, Roma (2013). Available at www.anvur.it

M.P. Bergeron-Boucher, V. Simonacci, J. Oeppen, M. Gallo, Coherent modeling and forecasting of mortality patterns for subpopulations using multiway analysis of compositions: an application to canadian provinces and territories. N. Am. Actuar. J. **22**(1), 92–118 (2018)

J.D. Carroll, J.J. Chang, Analysis of individual differences in multidimensional scaling via an N-way generalization of Eckart-Young decomposition. Psychometrika **35**(3), 283–319 (1970)

C.H. Carstensen, J. Rost, Multidimensional three-mode Rasch models, in *Multivariate and Mixture Distribution Rasch Models*, ed. by M. von Davier, C.H. Carstensen (Springer, New York, 2007), pp. 157–175

E. Ceulemans, H.A.L. Kiers, Selecting among three-mode principal component models of different types and complexities: a numerical convex hull based method. Br. J. Math. Stat. Psychol. **59**(1), 133–150 (2006)

J.J. Cronin Jr., S.A. Taylor, Measuring service quality: a reexamination and extension. J. Mark. **563**, 55–68 (1992)

M.A. Di Palma, P. Filzmoser, M. Gallo, K. Hron, A robust Parafac model for compositional data. J. Appl. Stat. **45**(8), 1347–1369 (2018)

M.A. Di Palma, M. Gallo, External information model in a compositional perspective: evaluation of campania adolescents' preferences in the allocation of leisure-time. Soc. Indic. Res. **146**(1–2), 117–133 (2019)

J.J. Egozcue, V. Pawlowsky-Glahn, G. Mateu-Figueras, C. Barceló-Vidal, Isometric logratio transformations for compositional data analysis. Math. Geol. **35**, 279–300 (2003)

J.J. Egozcue, V. Pawlowsky-Glahn, Groups of parts and their balances in compositional data analysis. Math. Geol. **37**(7), 795–828 (2005)

J.J. Egozcue, C. Barceló-Vidal, J.A. Martín-Fernéndez, E. Jarauta-Bragulat, J.L. Díaz-Barrero, G. Mateu-Figueras, A. Buccianti, Elements of simplicial linear algebra and geometry, in *Compositional Data Analysis: Theory and Applications*, eds. by V. Pawlowsky-Glahn, V. Buccianti (Wiley Online Library, 2011), pp. 141–157

M.A. Engle, M. Gallo, K.T. Schroeder, N.J. Geboy, J.W. Zupancic, Three-way compositional analysis of water quality monitoring data. Environ. Ecol. Stat. **21**(3), 565–581 (2014)

K. Fačevicová, K. Hron, V. Todorov, M. Templ, Compositional tables analysis in coordinates. Scand. J. Stat. **43**(4), 962–977 (2016)

K. Fačevicová, K. Hron, V. Todorov, M. Templ, General approach to coordinate representation of compositional tables. Scand. J. Stat. **45**(4), 879–899 (2018)

M. Gallo, Log-ratio and parallel factor analysis: an approach to analyze three-way compositional data, in *Advanced Dynamic Modeling of Economic and Social Systems*, ed. by A.N. Proto, M. Squillante, J. Kacprzyk (Springer, Heidelberg, 2013), pp. 209–221

M. Gallo, V. Simonacci, A procedure for the three-mode analysis of compositions. Electron. J. Appl. Stat. Anal. **6**(2), 202–210 (2013)

M. Gallo, Tucker3 model for compositional data. Commun. Stat.-Theory Methods **44**(21), 4441–4453 (2015)

M. Gallo, V. Simonacci, M.A. Di Palma, An integrated algorithm for three-way compositional data. Qual. Quant., 1–8 (2018)

R.A. Harshman, Foundations of the PARAFAC procedure: models and conditions for an 'explanatory' multi-modal factor analysis. UCLA working papers in phonetics **16**, 1–84 (1970)

H.A. Kiers, A three-step algorithm for CANDECOMP/PARAFAC analysis of large data sets with multicollinearity. J. Chemom.: J. Chemom. Soc. **12**(3), 155–71 (1998)

T.G. Kolda, B.W. Bader, Tensor decompositions and applications. SIAM Rev. **51**(3), 455–500 (2009)

P.M. Kroonenberg, *Applied Multiway Data Analysis*, vol. 702 (Wiley, Hoboken, 2008)

U. Lorenzo-Seva, J.M.F. Ten Berge, Tucker's congruence coefficient as a meaningful index of factor similarity. Methodology **2**(2), 57–64 (2006)

H.W. Marsh, SEEQ: a reliable, valid, and useful instrument for collecting students' evaluations of university teaching. Br. J. Educ. Psychol. **52**(1), 77–95 (1982)

A. Parasuraman, V.A. Zeithaml, L.L. Berry, Servqual: a multiple item scale for measuring customer perceptions of service quality. J. Retail. **64**(1), 12–40 (1988)

V. Pawlowsky-Glahn, J.J. Egozcue, Exploring compositional data with the CoDa-dendrogram. Austrian J. Stat. **40**(1&2), 103–113 (2011)

V. Pawlowsky-Glahn, J.J. Egozcue, D. Lovell, Tools for compositional data with a total. Stat. Model. **15**(2), 175–190 (2015)

V. Pawlowsky-Glahn, J.J. Egozcue, R. Tolosana-Delgado, *Modeling and Analysis of Compositional Data* (Wiley, Hoboken, 2015)

Quinn, T.P.: Visualizing balances of compositional data: a new alternative to balance dendrograms. F1000Research **7** (2018)

G. Rasch, *Probabilistic Models for Some Intelligence and Attainment Tests* (Danish Institute for Educational Research, Copenhagen, 1980)

V. Simonacci, M. Gallo, Statistical tools for student evaluation of academic educational quality. Qual. Quant. **51**(2), 565–579 (2017)

V. Simonacci, M. Gallo, Detecting public social spending patterns in Italy using a three-way relative variation approach. Soc. Indic. Res. **146**(1–2), 205–219 (2019)

S.S. Stevens, On the theory of scales of measurement. Science **103**, 677–680 (1946)

V. Todorov, V. Simonacci, M.A. Di Palma, M. Gallo, rrcov3way: robust methods for multiway data analysis, applicable also for compositional data. R package version 0.2-0 (2020). Available at http://CRAN.R-project.org/package=rrcov3way

L.R. Tucker, Some mathematical notes on three-mode factor analysis. Psychometrika **31**(3), 279–311 (1966)

VALMON: Gruppo di Ricerca sulla Valutazione ed il Monitoraggio delle Politiche e dei Servizi dell'Università degli Studi di Firenze. Copyright 2006 - VALMON s.r.l, Firenze (2018). Available at https://valmon.disia.unifi.it/sisvaldidat/unifi

M. von Davier, J. Rost, C.H. Carstensen, Introduction: extending the Rasch model, in *Multivariate and Mixture Distribution Rasch Models*, ed. by M. von Davier, C.H. Carstensen (Springer, New York, 2007), pp. 1–12

B.D. Wright, G.N. Masters, *Rating Scale Analysis* (MESA Press, Chicago, 1982)

Artificial Neural Networks to Impute Rounded Zeros in Compositional Data

Matthias Templ

Abstract Methods of deep learning have become increasingly popular in recent years, but they have not arrived in compositional data analysis. Imputation methods for compositional data are typically applied on additive, centered, or isometric log-ratio representations of the data. Generally, methods for compositional data analysis can only be applied to observed positive entries in a data matrix. Therefore, one tries to impute missing values or measurements that were below a detection limit. In this paper, a new method for imputing rounded zeros based on artificial neural networks is shown and compared with conventional methods. We are also interested in the question whether for ANNs, a representation of the data in log-ratios for imputation purposes is relevant. It can be shown that ANNs are competitive or even performing better when imputing rounded zeros of data sets with moderate size. They deliver better results when data sets are big. Also, we can see that log-ratio transformations within the artificial neural network imputation procedure nevertheless help to improve the results. This proves that the theory of compositional data analysis and the fulfillment of all properties of compositional data analysis is still very important in the age of deep learning.

1 Introduction

Compositional data and rounded zeros: Compositional data are quantitative descriptions of the parts of some whole, whereby the important information is not the absolute values but relative information. Measurements involving proportions, percentages, probabilities, concentrations are compositional data. An example of compositional data is the 24 h day categorized into sleeping time, working time, lunch break, and free time. There exists inner dependencies: if one sleeps longer at a day, automatically less time for other things is left. Also not all values are possible (e.g., a day is restricted to 24 h), and the compositions are represented in the simplex instead of

M. Templ (✉)
Institute of Data Analysis and Process Design, Zurich University of Applied Sciences, Rosenstrasse 3, CH-8401 Winterthur, Switzerland
e-mail: matthias.templ@zhaw.ch

P. Filzmoser et al. (eds.), *Advances in Compositional Data Analysis*,
https://doi.org/10.1007/978-3-030-71175-7_9

an Euclidean geometry. As for the previous example, the sleeping time plus working time cannot be larger than 24 h, for example. Statistical methods are mostly designed to be applied in the Euclidean geometry. A log-ratio presentation of the data is a way out to shift the original measurements from the simplex to the Euclidean geometry. However, zeros or rounded zeros are problematic when computing log-ratios. Either in the denominator, the ratio is not defined or in the numerator the log is not defined. Zeros are true values where the measurement reports zero. A rounded zero occurs whenever a measurement unit cannot measure a too small amount in a sample. For example, a measurement unit can only detect ppm's of a certain chemical element higher than a certain threshold—the detection limit—and reports zeros whenever there are fewer ppm's in the sample. The value is thus rounded to zero, but only because the measurement unit could not detect anything even though we know that there must be a certain contribution of this chemical element in the sample. Values below the detection limit are thus known under the name rounded zeros (Aitchison 1986; Martín-Fernández et al. 2003) and represent left censored data.

It is fully meaningful to impute rounded zeros with a reasonable small value and continue with the processing of a complete data set (Filzmoser et al. 2018). To guarantee imputation by a considerable small value, an upper bound should be considered—typically the detection limit of this variable. Any imputation method should then impute with values below this upper bound, but larger than zero.

To give an example as outlined also similarly in Filzmoser et al. (2018), a rounded zero value present for a variable, say *Arsenic* and the detection limit for *Aresenic* is 0.1 mg/kg. An imputed value should be thus in the interval $(0, 0.1)$. The imputation should consider the information about all other compositional parts, i.e., a multivariate approach is preferable. Furthermore, the log-ratios between other parts should remain unchanged after the imputation of the rounded zeros.

The imputation of rounded zeros can be considered as a very difficult problem with different solutions proposed in the literature (Martín-Fernández et al. 2003; Palarea-Albaladejo et al. 2007; Palarea-Albaladejo and Martín-Fernández 2008; Martín-Fernández et al. 2012; Palarea-Albaladejo and Martín-Fernández 2013; Palarea-Albaladejo et al. 2014; Martín-Fernández et al. 2015; van-den Boogaart et al. 2015; Templ et al. 2016; Chen et al. 2017). More on this in Sect. 2.

Artificial deep neural networks: A neural network is just a non-linear statistical model (Hastie et al. 2009), which is based on weighted linear combinations of the sample values. Within an ANN, the aim is to find millions of weights to get the best possible output from input data and multiple layers. The weights are updated iteratively. In the first iteration, the weights are random and the result is accordingly bad. For each iteration, we go in the "optimum direction", whereby a gradient method is used for this. It requires backpropagation with an optimizer (e.g., Adam) and a loss function (e.g., mean squared error). The quality of the predictions is evaluated based on a selected metric (e.g., mean absolute error) on validation and training data.

Often missing values are just initialized before the neural network is fitted (Smieja et al. 2018). Also, missing values may be incorperated in a classification problem without imputing them. The trick is to add an indicator matrix determining the position of missing values to the data matrix and then using a generative additive network (GAN). For more details, see Li et al. (2019); Yoon et al. (2018); Mattei and Frellsen (2018). Note that this is applied within several image inpainting methods (see, e.g., Xie et al. 2012). However, our aim is different. Rounded zeros are imputed with fixed values so that the imputed data set can be used by any researcher or practitioner for any analysis that needs complete data.

ANNs for missing values: Artificial neural networks have been rarely used for the imputation of missing values purposes. Maiti et al. (2008) used neural networks for an agriculture survey (and compared it with hot-deck imputation). Various other contributions took place around 2007/2008. Afterwards, Jerez et al. (2010) used a multi-layer perceptron method for the imputation of breast cancer data. McCoy et al. (2018) used variational autoencoders for imputation. Choudhury and Pal (2019) and Silva-Ramírez et al. (2015) used artificial neural networks. Another approach is to not explicitly impute missing values, but to add the missingness matrix. Hereby, a position of the original data matrix with missing value is denoted with a 1, and non-missing values are denoted with 0. This matrix is then integrated into the whole model. Li et al. (2019), Yoon et al. (2018) and Mattei and Frellsen (2018) used generative artificial networks for classification purposes under missing values. For special applications, these were also used by Lim (2019) and Arisdakessian et al. (2019) (RNA data). However, no ANN was used in the context of compositional data analysis and in the context of the imputation of rounded zeros in compositional data.

Outline: In Sect. 2, methods for the imputation of rounded zeros are listed, and Table 1 summarizes them. Section 3 introduces the specific type of artificial neural networks that is relevant in the context of imputation for compositional data. The parameters of our imputation procedure that have been made available in software are described in Sect. 4 for the imputation of rounded zeros in one variable. The focus of Sect. 5 is the adaptation to impute rounded zeros as an Expectation-Maximum (EM) method. This allows also to impute not only one variable at a time but also sequentially all compositional parts with rounded zeros in a data set. Two different approaches are presented to replace rounded zeros in compositional data, one using a log-ratio transformation and the other without using a log-ratio representation. The outcome and comparison of the methods on real-world data sets are shown and described in Sect. 6. Section 7 concludes and summarizes the major findings.

2 General Imputation Methods and Imputation Methods for Rounded Zeros

Note that the simplest method is univariate and imputes zeros with 65% of the detection limit (DL). This minimizes the distortion of the covariance structure (Martín-Fernández et al. 2003; Martín-Fernández et al. 2011). However, this kind of mean

Table 1 Methods compared in this contribution. Univariate methods were not considered. We distinguish between methods for the imputation of missing values (not designed for detection limit problems) and rounded zeros (considering detection limites) and methods that do or do not take the compositional nature of compositional data into account

Method	References	Description
Non-compositional approaches not designed for detection limit problems (non-CoDa, non-DL)		
deepImp	Templ (2020)	ANNs in an EM-based implementation (serves as a benchmark for deepImp-dl)
kNN	Kowarik and Templ (2016)	k-nearest neighbor imputation
mice	van Buuren and Groothuis-Oudshoorn (2011)	Predictive mean matching in an EM-based implementation
missForest (rf)	Stekhoven and Bühlmann (2011)	Imputation using random forests in an EM-based implementation
Non-compositional approaches considering detection limits (non-CoDa, DL)		
deepImp-dl	Templ (2020)	ANNs in an EM-based implementation considering detection limits (serves as a benchmark for deepImpCoDa-dl)
Compositional methods ignoring not designed for detection limit problems (CoDa, non-DL)		
deepImpCoDa	this article	ANNs in an EM-based implementation considering compositinal data (serves as a benchmark for deepImpCoDa-dl)
aknn	Hron et al. (2010)	kNN imputatation for compositional data
impCoDa	Hron et al. (2010)	EM-based regression imputation using pivot coordinates
Compositional methods considering detection limits (CoDa, DL)		
deepImpCoDa-dl	This article	ANNs in an EM-based implementation considering detection limits
imputeBDL-lm	Templ et al. (2016)	EM-based regression imputation using pivot coordinates and considering detection limits
imputeBDL-pls	Templ et al. (2016)	EM-based partial least squares regression imputation using pivot coordinates and considering detection limits
imputeBDL-rob	Templ et al. (2016)	EM-based robust (MM) regression imputation using pivot coordinates and considering detection limits
impRZilr	Martín-Fernández et al. (2012)	EM-based partial least squares regression imputation using pivot coordinates and considering detection limits

imputation leads to an underestimation of the compositional variability. The variability can be increased by drawing random numbers from a uniform distribution in $(0, DL)$, with DL, the detection limit of a variable, but then the precision of the imputations lowers. Alternatively, Palarea-Albaladejo and Martín-Fernández (2013) and Palarea-Albaladejo et al. (2014) used a lognormal distribution, truncated by the threshold. In any case, all these methods represent univariate approaches to a multivariate compositional problem. Note that the methods from Palarea-Albaladejo and Martín-Fernández (2013) and Palarea-Albaladejo et al. (2014) need at least one fully observed variable. In our application, we compare only multiple and multivariate methods and leave out univariate methods.

EM-based model-based methods offer multivariate approaches for the imputation of rounded zeros and by using censored regression it can be guaranteed that imputed values do not exceed a defined threshold. A modified EM algorithm (Palarea-Albaladejo et al. 2007; Palarea-Albaladejo and Martín-Fernández 2008) in additive log-ratio coordinates (Aitchison 1986) was reformulated by using pivot log-ratio coordinates (Filzmoser et al. 2018) in Martín-Fernández et al. (2012). Both algorithms use censored multiple regression for each variable in an EM-based manner until the imputations stabilize. The algorithm is complex because detection limits must also be represented in isometric log-ratio coordinates and inverse isometric log-ratio transformation must happen in each step of the algorithm together with an adjustment of values to not lose the absolute information of the original data. This is necessary when we still want to know, for example, the exact household expenditures in expenditure data and not only the ratios between them. A detailed description of this complex algorithm can be found in Martín-Fernández et al. (2012); Filzmoser et al. (2018). The method itself needs also an initialization of rounded zeros (for example, using the 65% method) because otherwise log-ratios cannot be calculated from zero values. Instead of ordinary least squares regression, a robust estimator can be plugged in.

For high-dimensional data, two other methods to replace rounded zeros should be mentioned. Templ et al. (2016) uses partial least squares regression for imputing all variables sequentially and repeatedly in an EM-based algorithm. To decrease computation time, Chen et al. (2017) use a clustering approach to select variables within the algorithm.

In addition to methods suitable to handle rounded zeros by imputing below a given detection limit, we also want to compare popular imputation methods that do not take these limits into account. A major point of criticism of this contribution could be that methods for rounded zeros are compared with methods that do not take into account the special problem of rounded zeros. The motivation for this comparison comes from practice and is not scientifically based, but is based on the experience as package maintainer of VIM (Templ et al. 2012; Kowarik and Templ 2016), and robCompositions (Templ et al. 2011; Filzmoser et al. 2018)—both containing imputation methods. This experience says that readers need to see a comparison to understand that only (compositional) methods that take the problem of rounded zeros into account successfully impute rounded zeros. Up to our knowledge, articles on rounded zero imputation have only compared specific methods for rounded zeros, with the result that users without the obvious comparison still resort to unsuitable methods, because they are just trendy to use for more general imputation problems. Many users are also confused about whether they should use imputation methods for missing values in compositional data or methods for replacing rounded zeros. We show what happens when the wrong methods are selected and what the consequences are. Therefore, methods for rounded zeros of compositional data are also compared to very popular imputation methods: imputation using random forests implemented in R package `missForest` (Stekhoven and Bühlmann 2011) or the faster implementation in `missRanger` (Mayer 2019), predictive mean matching using R package `mice` (van Buuren and Groothuis-Oudshoorn 2011), k-nearest neighbor imputation of R package `VIM` (Kowarik and Templ 2016), missing value imputation using pivot coordinates as implemented in R package `robCompositions` in function `impCoda` (Hron et al. 2010) and from the same package a k-nearest neighbor method for missing values using Aitchison distances in function `kNNa` (Hron et al. 2010).

3 Artificial Neural Networks

In this section, the general construction of a neural network is explained, details about the neural network used for the imputation are given in the next section.

Neurons hold some information on each variable and a set of observations. Each neuron has an activation, depending on how the input information looks like. A layer in a deep neural network is a collection of neurons in one step. There are three types of layers: the input layer, the hidden layers, and the output layer. A neuron ν_1^l in the input layer, for example, takes the observations of a data set as activations and the hidden layers represent non-linear weighted observations.

The activations **a** should take values between zero and one. The (default) activation function reLU (rectified Linear Unit) (He et al. 2015) takes the value zero if the activation is negative. Most activation functions use a non-linear function mapping between the inputs and the response.

In the first hidden layer ι_l with $l = 2, ..., \Phi - 1$ with each neuron ν_υ with $\upsilon = 1, .., \phi_l$ represents very detailed information, while passed through the layers, the information gets generalized and in the end summed up to either one or multiple neurons depending on the type of variable. Φ represents the number of layers, including the input and the output layer, and ϕ the number of neurons within the layer l. Figure 1 shows a simplified illustration of a network with two hidden layers and few neurons.

The goal of training a network is to find the optimal weights $\omega_{\nu_\upsilon^l, \nu_\upsilon^{l+1}}$ for each connection between the neurons of two layers (see also Fig. 2). Then the activations of the neurons from the first layer are taken and the weighted sum according to these weights is computed.

Let $\mathbf{\Omega}$ be the matrix of weights in each layer. The weights of the lth layer are defined as $\boldsymbol{\omega}^l$. The activation values of the neurons in one layer are $\boldsymbol{\nu}^l = \nu_1^l, ..., \nu_{\phi_l}^l$

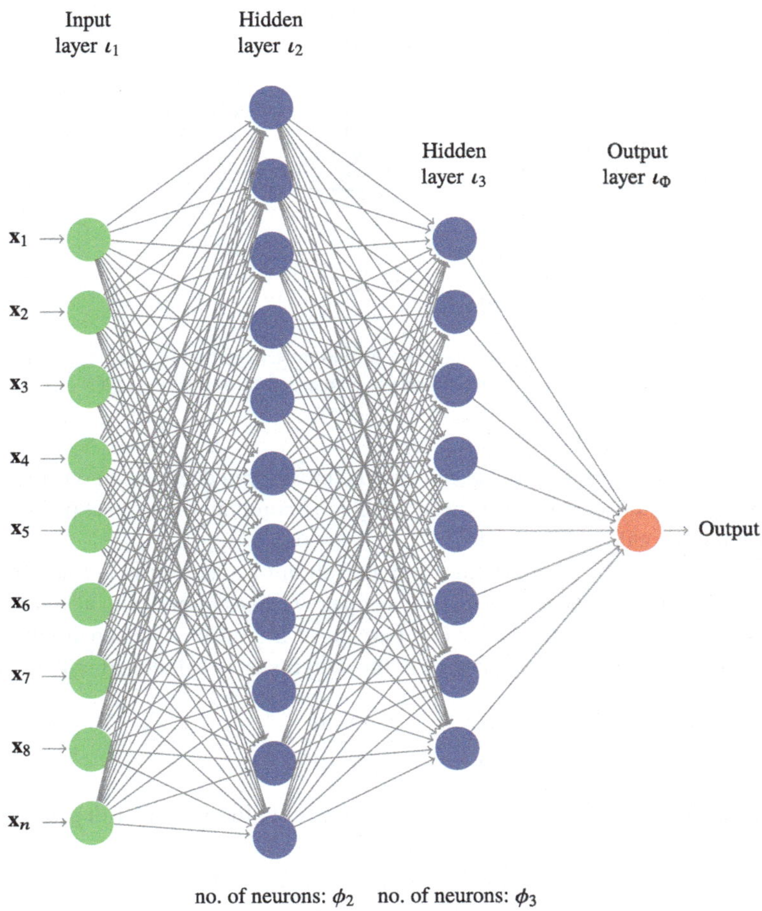

Fig. 1 Simplified schematic overview of a neural network with a few observations, two hidden layers with 13 and 9 neurons. Note that the default deep artificial neural network consists of n observations, 10 layers, 1000 neurons in the first hidden layer, 900 in the second, ..., 100 in the last hidden layer and the output depends on the scaling of the target variable. In total 3325601 parameters (mostly weights) are trained in the default network

while the biases within one layer are $\boldsymbol{\beta}^l = \beta_1^l, \ldots, \beta_{\phi_l}^l$. The formula for the transition of one layer to the other is then defined by $relu(\boldsymbol{\Omega} v^l + \boldsymbol{\beta})$.

After setting initially all weights randomly, a loss function of the network is defined. The output of the loss function is a single number judging the quality of the neural network. To lower the value of the loss function, an adaptive moment estimation called Adam (Kingma and Ba 2014; Ruder 2016) is used (other methods can be selected), which is a stochastic gradient descend method that uses adaptive learning rates for each parameter of the algorithm. With this gradient descend optimization all the weights $\boldsymbol{\Omega}$ are optimized to reach the next local minimum $-\nabla C(\boldsymbol{\Omega})$ resulting

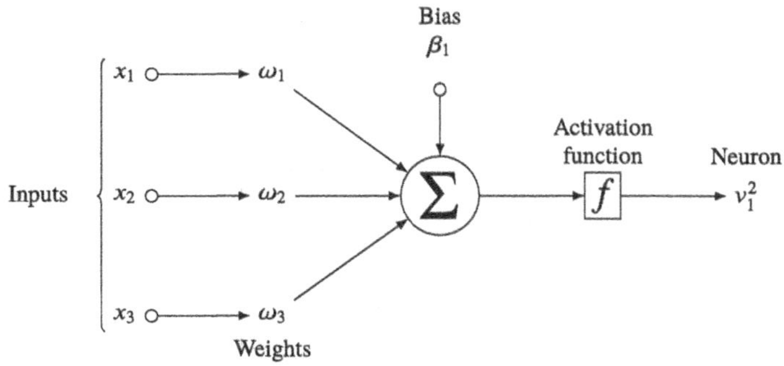

Fig. 2 This schematic illustration tells the story behind the calculation of one single neuron with respect to the first hidden layer. This neuron serves then as input for the next layer

in the most rapid decrease. This is causing the most rapid decrease in our loss function. The gradient used to adjust these weights is computed with backpropagation, whereby the weights get updated.

The weighted sum, which defines the activation of a neuron is defined by $\boldsymbol{z}^l = \boldsymbol{\omega}^l \boldsymbol{v}^{l-1} + \boldsymbol{\beta}^l$. The activation of a neuron $\boldsymbol{v}^l$ in layer l is then defined with $\boldsymbol{v}^l = relu(\boldsymbol{z}^l)$. The loss is therefore computed by comparison of $\boldsymbol{z}^l$ and $\boldsymbol{v}^l$. The sensitivity of the loss function C with regard to changes in the weights $\boldsymbol{\omega}^l$ is calculated with the chain rule

$$\frac{\partial C}{\partial \boldsymbol{\omega}^l} = \frac{\partial \boldsymbol{z}^l}{\partial \boldsymbol{\omega}^l} \frac{\partial \boldsymbol{v}^l}{\partial \boldsymbol{z}^l} \frac{\partial C}{\partial \boldsymbol{v}^l} \quad .$$

This means that the amount of this nudge influencing the last layer depends on the strength of the previous neuron, $\frac{\partial \boldsymbol{z}^l}{\partial \boldsymbol{\omega}^l} = \boldsymbol{v}^{l-1}$.

This is done over all training observations, $\frac{\partial C}{\partial \boldsymbol{\omega}^l} = \frac{1}{n} \sum_{k=0}^{n-1} (\frac{\partial C_k}{\partial \boldsymbol{\omega}^l})$. The gradient vector ∇C contains all the partial derivatives from the loss function with respect to all weights and biases. The sensitivity of the loss function to changes in the bias is computed equivalent with

$$\frac{\partial C}{\partial \boldsymbol{\beta}^l} = \frac{\partial \boldsymbol{v}^l}{\partial \boldsymbol{z}^l} \frac{\partial C}{\partial \boldsymbol{v}^l} \quad .$$

The above-described approach is then applied and computed through the complete neural network. The weights will be adjusted and the computation is iterated and repeated many times (also called the number of epochs, see below). More details about how neural networks are computed can be found, for example, in Nielsen (2015).

4 Artificial Neural Networks to Impute Rounded Zeros—Parameter Settings

In the following, an artificial neural network is described and adapted for imputation purposes, implemented in function *impNNetCoDa* in R package *deepImp*. This deep learning method works with a neural network based on *keras* (Allaire and Chollet 2019; Chollet et al. 2015) and *tensorflow* (Abadi et al. 2018, 2015).

Before describing the whole EM-based algorithm that automatically imputes all rounded zeros in all variables of a data set, the parameter choices available in our implementation are outlined.

Initialization of rounded zeros: In principle, any other method from before can be used to initialize the rounded zeros and several choices are available in our implementation. The default approach is to use a k-nearest neighbor approach that uses Aitchison distances and robust means (Hron et al. 2010).

Iterations: Initialized missing values are updated step by step with imputations. Whenever all variables with rounded zeros are updated once, one iteration has passed. This represents an outer loop of the algorithm. If the EM-based sequential algorithm does not converge after a specified number of `iterations`, then the algorithm is stopped and a warning is supplied.

Threshold for convergence: Given the rth iteration, a change in the imputed values compared to the previous iteration is measured. More precisely, the algorithm stops as soon as the sum of squared relative distances between the imputed values in the rth step and the $r-1$th step is below the specified threshold.

Number of epochs: The data set is not only sent once through the deep neural network but many times (number of epochs). This is needed to optimize the learning, which is an iterative process. Updating the weights in one pass (i.e., one epoch) is too little. Generally, it applies that too few epochs often lead to an underfit while too many epochs can lead to an overfit. The optimum number of epochs is data set dependent. In all our experiments, a choice of about 200–300 epochs gave the most promising results, when either looking at and comparing validation and training errors or when evaluating it directly on the validation criteria defined in Sect. 6.2. If the validation error is higher than the training error, this is an indication of overfitting. Figure 3 corresponds to the imputation of rounded zeros in the variable Ag in the Moss data set (see Sect. 6.1) using our artificial deep learning network with 10 layers and different number of epochs. No overfit is visible. Note that in our implementation 20% of the observations is used as the validation set by default (for each epoch).

The number of epochs is one parameter of our algorithm with a sensible default. However, it is better to set the number of epochs to a high value instead of too small value, because in any case, the next parameter provides a stopping rule to stop the training of the deep neural network earlier.

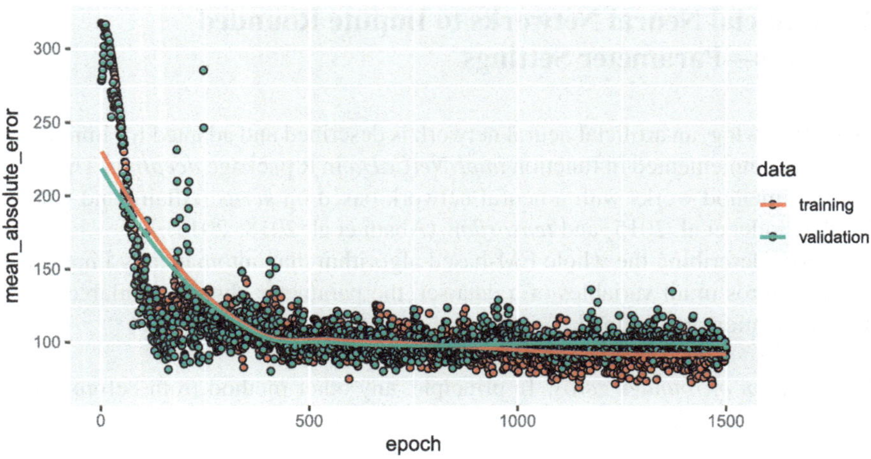

Fig. 3 Training and validation error for the imputation of variable Ag in the Moss data

For any given data set, the training and validation errors can be used to evaluate the correct number of epochs, but also to evaluate the number of layers and the number of neurons in the layers. If training and validation errors are quite different, too few layers, neurons, and/or epochs were selected. In Fig. 3, we see a perfect picture where both errors in the deep neural network, using the evaluation criterion *mean_absolute_error*, are about the same size. The optimal number of epochs is about 350, after that there is no real improvement (Fig. 4).

Regarding imputation, it is also interesting to look at an imputation error, as long as you know the true values (as in the numerical examples in Sect. 6). The imputation error over the validation measure CED (see Sect. 6.2) is lowest at about 200–300 epochs for all later used data.

Patient value: The aim is to stop at those number of epochs without any significant change of weights in the last epochs (patient value). Setting this value appropriately overcomes overfitting when specifying the number of epochs too high. For example, when the weights do not change for about 25 epochs (default), the training of the weights in the deep neural network would stop.

Dropout: This is the rate of input units in each layer, which are excluded in the neural network. The main purpose of using dropout is to protect against over-matching, but also to achieve different results when imputing again if the random number generator is not set to the same value. This is also the only way to go in the direction of multiple imputation. Proper multiple imputation (with correct coverage rates) is not possible to achieve in general using dropout.

Number and density of layers in the deep network: The optimal number of layers is data-dependent. In general, too many layers can cause an overfit, while too few layers

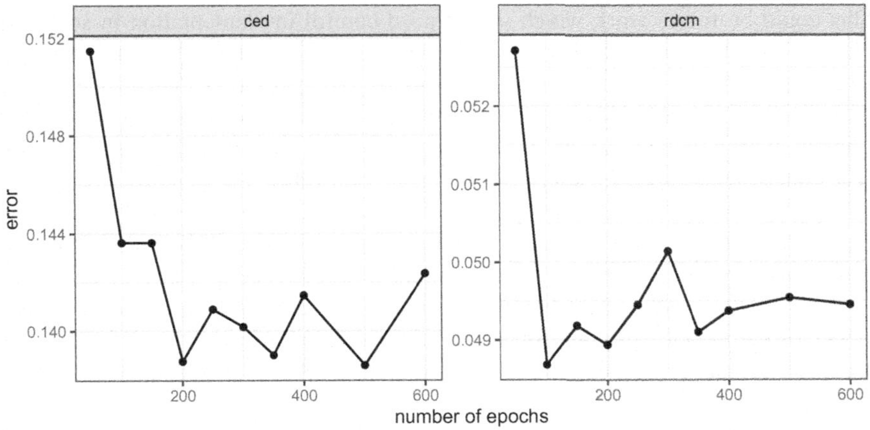

Fig. 4 Imputation error for different numbers of epochs for the imputation of variable Ag in the Moss data set. For our choice of the number of epochs for the numerical results section is based on the CED criteria (see Sect. 6.2)

can cause an underfit. Generally, it is better to add too many layers than too few. This is the reason why by default the number of layers in the deep neural network is set to 10.

The user can evaluate the optimal number of layers using cross-validation. The validation error is not very different from training terror, both errors are obtained from cross-validation.

Number of neurons in each layer: The optimal number of neurons in each layer depends on your data particularities. By default, the first hidden layers consist of 1000 neurons and then decrease by 100 in each following layer so that the 10th layer has 100 neurons. Note that the user can overwrite this default choice. Again cross-validation using different choices of neurons can give some hints about the optimal choice of the number of neurons.

Optimizer: The choice of the optimizer is crucial. It defines the gradient method used to iteratively update the weights in the neural network. Our default choice is Adam (Kingma and Ba 2014), a first-order stochastic gradient-based optimization based on adaptive estimates of lower order moments.

Other optimizers selectable are Stochastic gradient descent optimizer (SGD), RMSprop, Adagrad, Adadelta, Adamax, and Nadam (Nesterov Adam optimizer). Note that all methods have sensible defaults, e.g., on the learning rate.

Loss function: The choice of the loss function depends on the scale type of the response variable (the variable to be imputed). For nominal variables to be imputed, typically a cross-entropy is used, while for continuous variables a mean squared error is used. Note that we did not replace this with a compositional counterpart.

This could be future work which would need careful implementation in software. Please also note that no robust loss function can be integrated into Keras, because no quantiles or any robust measure can be used to construct a robust loss measure. Also, the deep neural networks are computationally expensive, thus replacing a mean squared error with robust measures, the computation time would increase heavily. As a consequence, all deep neural networks are non-robust and sensitive to outliers.

Evaluation metrics: An evaluation metric is used to evaluate the performance of your model. A metric function is thus similar to a loss function. The latter is used for training the model while the evaluation metrics are used to judge the performance of the model. One can use any of the loss functions as a metric function. In our software implementation, the default evaluation metric is the mean absolute error.

Activation: Especially for compositional data, a good choice of an activation function is important, because the weights for neurons get transformed by the activation function. For example, when using the activation *linear*, the outcome of the neural network for a 2-dimensional continuous data set in a regression context is approximately the same as for least squares regression. To allow for non-linear combination of neurons, we choose by default the activation *reLU*, see Vedaldi and Lenc (2015). One positive aspect of the reLU function is that it accelerates the convergence of stochastic gradient descent compared to other activation functions and it is computationally cheap (Krizhevsky et al. 2012). Other choices are, e.g., softmax, elu, selu, or sigmoid, to just name a few. For the last layer (that actually do the imputation), a different choice of activation must be taken, namely linear.

5 Imputation of Rounded Zeros Using Artificial Neural Networks in an Expectation-Maximum Style

The proposed approach uses one artificial network per variable to be imputed. Note that the structure of the output layer, the activation function, the loss, and evaluation function for the output layer depend on the distribution of a variable to be imputed. Naturally, this is a log-ratio, but in principle also external variables that are non-compositional could be integrated. Generally, for continuous measurements and log-ratios, the activation function relu (He et al. 2015) is used concerning all layers (default), the mean absolute error is used as a loss function (default) and the output layer consists of a single neuron and linear activation.

As already mentioned, the initialization of rounded zeros is necessary whenever artificial neural networks are applied to log-ratio representations instead of raw original values.

The algorithm works in a chain, i.e., it iterates sequentially applied to all variables until convergence (EM-based method).

All networks have $\Phi = 10$ layers (by default), which in experimental pre-studies showed to be sufficient to reach a convergent result. It uses fully connected layers with a dropout rate of 10% after each layer to overcome overfitting and to allow for a kind of multiple imputation. In the first hidden layer $l = 2$ there are $\phi_2 = 1000$ neurons, as well as in the others except for the output layer. Note that this default choice can be overwritten by the user.

The layers are applied one after another, while the activations $\mathbf{a}$ of neurons in a first layer determine the activations of the second layer. In a final output layer, the desired imputation of missing values takes place.

Let $\mathbf{X} = (\mathbf{x}_1, \mathbf{x}_2, \ldots, \mathbf{x}_D)$ be a $n \times D$-dimensional data matrix. $\mathbf{x}_k^{(obs)}$ denotes the observed values of the kth variable, while $\mathbf{x}_k^{(mis)}$ represents the missing values. We define two algorithms, while the first one serves as a benchmark when the compositional nature of compositional data is ignored.

Algorithm 1: Imputation of rounded zeros with ANNs (without using log-ratios)

Result: The imputed data matrix $\mathbf{X}_{(imp)}$ in original scale
$\mathbf{d} \leftarrow$ vector of detection limits (one entry for each variable);
$\mathbf{k} \leftarrow$ vector of indices with colums in $\mathbf{X}$ that contains missing values;
$\epsilon \leftarrow$ threshold for convergence with sufficient small value, e.g., set to 1;
$m \leftarrow$ number of rounded zeros in $\mathbf{X}$;
$r \leftarrow 0$ index for iterations and $maxiter$ the maximum number of iterations;
Initialization using aKNN (default) or similar methods ;
while $\Delta > \epsilon$ *and* $r < maxiter$ **do**
  $r = r + 1$;
  $\mathbf{X}^{(imp,old)} \leftarrow$ store previously imputed data matrix;
  for j *in* $\mathbf{k}$ **do**
    Fit the deep neural net: $\mathbf{x}_s \sim (\mathbf{x}_1, \mathbf{x}_2, ..., \mathbf{x}_D) \setminus \mathbf{x}_s$ and predict $\mathbf{x}_s^{(mis)}$
    $\mathbf{X}^{(imp,new)} \leftarrow$ update imputed data matrix with $\mathbf{x}_s^{(mis)}$;
    Imputed values above d_s are set to d_s ;
  end
  update $\Delta = \frac{1}{m}|\mathbf{X}^{(imp,new)} - \mathbf{X}^{(imp,old)}|$
end

The second algorithm makes use of pivot log-ratio transformations. One particular choice of an orthogonal basis using one original composition $(x_{i1}, ..., x_{iD})'$ is

$$z_{ij} = \sqrt{\frac{D-j}{D-j+1}} \ln \frac{x_{ij}}{\sqrt[D-j]{\prod_{k=j+1}^{D} x_{ik}}} \quad \text{for} \quad j = 1, \ldots, D-1. \tag{1}$$

These coordinates will be referred to as *pivot* (log-ratio) *coordinates* (Filzmoser et al. 2018).

As mentioned above, the pivot coordinates have the feature that part $\mathbf{x}_1$ only appears in coordinate $\mathbf{z}_1$. This is not the case for other parts; $\mathbf{x}_2$, for example, appears in $\mathbf{z}_1$ and in $\mathbf{z}_2$. Isolating one part into one coordinate is appealing, since $\mathbf{z}_1$ summarizes now all relative information (log-ratios) about $\mathbf{x}_1$.

It is possible to come back to the original parts (up to a scaling factor) by the inverse pivot (log-ratio) transformation (see also Filzmoser et al. 2018)

$$
\begin{aligned}
x_{i1} &= \exp\left(\frac{\sqrt{D-1}}{\sqrt{D}} z_{i1}\right),\\
x_{ij} &= \exp\left(-\sum_{k=1}^{j-1} \frac{1}{\sqrt{(D-k+1)(D-k)}} z_{ik} + \frac{\sqrt{D-j}}{\sqrt{D-j+1}} z_{ij}\right),\ j = 2, \ldots, D-1,\\
x_{iD} &= \exp\left(-\sum_{k=1}^{D-1} \frac{1}{\sqrt{(D-k+1)(D-k)}} z_{ik}\right),
\end{aligned}
\tag{2}
$$

The scaling factor is different for each composition in the imputation context. For more details on this factor, we refer to Templ et al. (2016).

Algorithm 2: Imputation of rounded zeros with ANNs (with log-ratios)

Result: The imputed data matrix $\mathbf{X}_{(imp)}$
Use the same parameters as described in Algorithm 1;
Initialization using aKNN (default) or similar methods;
while $\Delta > \epsilon$ *and* $r < maxiter$ **do**
  for j *in* **k** **do**
    Sort the data matrix so that $\mathbf{x}_j$ is the first column;
    $\mathbf{Z} \leftarrow$ apply a pivot log-ratio transformation on the sorted $\mathbf{X}$;
    $\mathbf{Z}^{(imp,old)} \leftarrow$ store previously imputed data matrix;
    $\phi \leftarrow$ represent the detection limit regarding $\mathbf{z}_1$ in pivot log-ratio coordinates ;
    Fit the deep neural net: $\mathbf{z}_1 \sim (\mathbf{z}_2, \mathbf{z}_3, ..., \mathbf{z}_{D-1})$ and predict $\mathbf{z}_1^{(mis)}$
    $\mathbf{Z}^{(imp,new)} \leftarrow$ update imputed data matrix with $\mathbf{z}_1^{(mis)}$;
    Imputed values above ϕ are set to ϕ ;
    Apply the inverse pivot log-ratio transformation ;
    Backsort to the original order of variables ;
    Adjust values so that absolute values are preserved ;
  end
  update ϵ
end

6 Numerical Results

The numerical results are all based on real data. To evaluate the imputation of rounded zeros on real data, detection limits for compositional parts are artificially chosen. This way the imputations can be compared with the real observed values. Note that in principle simulation studies could also be done with artificially generated data. However, it is more realistic to work with real data—where detection limit problems occur in principle—and to set detection limits there. Furthermore, deep learning methods require a lot of computing time (see Fig. 10), and therefore, it is unrealistic to perform large simulation studies with many replications or varying an extensive grid of parameters for one real data set.

6.1 Data

Kola C-horizon and Moss data: The *Kola Ecogeochemistry Project* was a geochemical survey of the Barents region whose aim was to reveal the environmental conditions in the European arctic. Soil samples were taken at different levels and are linked to spatial coordinates. In this paper, the C-horizon and the Moss data (Reimann et al. 2008) are used. The C-horizon data represent soil samples at approximately 1 1/2 m depth under the surface, the Moss data represent soil samples from the mosses on the surface. In the Moss data, one can see the environmental influence (e.g., pollution by nickel smelters) while in the C-horizon the type of rock has a decisive influence on the soil sample. The raw data sets, in which zeros are the result of element concentrations below the detection limit, are included in the R package `VIM` (Templ et al. 2012; Kowarik and Templ 2016). For this study the main elements (Al, Ca, Fe, K, Mg, Mn, Na, P, Si, Ti) are selected for the C-horizon data and for the Moss data we selected the elements Ag, Al, Ca, Cu, Fe, Mg, Na, Ni, Pb, and Si. These elements did not include any values below the detection limit, thus by artificially setting a detection limit, we can compare the true value with the imputed ones. The number of observations is 606 (C-horizon) and 594 (Moss data).

Gemas data: The Geochemical Mapping of Agricultural and Grazing land Soil (GEMAS) project provides geochemical data on agricultural and grazing land soil. It documents the concentration of almost 60 chemical elements at the scale of Europe.

Element concentrations of 2108 samples of agricultural soil (in 0–20 cm depth) and grazing land soil (0–10 cm) are available in package robCompositions (Templ et al. 2011; Filzmoser et al. 2018). The average sample density was 1 site per 2500 km^2. For our study we used all chemical elements of the data set: Al, Ba, Ca, Cr, Fe, K, Mg, Mn, Na, P, Si, Sr, Ti, V, Y, Zn, Zr.

TrondelagO data: A regional-scale geochemical survey of O-horizon samples in Nord-Trondelag including the concentrations of nine different plant types. The data set includes 754 observations, and we used the chemical concentrations on Ag, Al, As, Ca, Cr, Cu, Fe, Mg, Na, Ni, Pb, and S that represent those chemical elements that are often used for analysis (see, e.g., Templ et al. 2008; Filzmoser et al. 2018).

For all data sets, an artificial detection limit was introduced. For each variable in the data set, the 0.05 quantile represents the detection limit. This means that for each variable 5% of the smallest values are set to zero and are subsequently imputed with different methods.

6.2 Validation Criteria

The validation criteria defined in the following (see also Martín-Fernández et al. 2012; Templ et al. 2016) assume that the complete data information is available. The symbol * is used for imputed data or estimators thereof.

Relative difference between covariance matrices (RDCM): Let $\mathbf{S} = [s_{ij}]$ be the sample covariance matrix of the original observations in pivot log-ratio coordinates z_{ij} from Eq. (1) (or any other isometric log-ratio coordinates), and let $\mathbf{S}^* = [s^*_{ij}]$ denote the sample covariance matrix computed with the same isometric log-ratio coordinates where all the rounded zeros have been imputed. A measure of the relative difference between both covariance matrices (Hron et al. 2010), based on the Frobenius matrix norm $\|\cdot\|_F$, is

$$RDCM = \sqrt{\frac{1}{(D-1)^2}\sum_{i=1}^{D-1}\sum_{j=1}^{D-1}\left(s_{ij} - s^*_{ij}\right)^2} = \frac{1}{D-1}\frac{\|\mathbf{S} - \mathbf{S}^*\|_F}{\|\mathbf{S}\|_F} \quad . \tag{3}$$

Compositional error deviation (CED): Let n_M denotes the number of samples $\mathbf{x}_k$ containing at least one rounded zero, M the index set referring to such samples, and $\mathbf{x}^*_k$ the corresponding observation with imputed values. Then, the criterion

$$\frac{\frac{1}{n_M}\sum\limits_{k\in M} d_a(\mathbf{x}_k, \mathbf{x}^*_k)}{\max\limits_{\{\mathbf{x}_i, \mathbf{x}_j \in \mathbf{X}\}}\{d_a(\mathbf{x}_i, \mathbf{x}_j)\}} \tag{4}$$

measures the differences by using Aitchison distances (defined below). d_a stands for the Aitchison distance, which is defined for two compositions $\mathbf{x}$ and $\tilde{\mathbf{x}}$ as $d_A(\mathbf{x}, \tilde{\mathbf{x}}) = \left[\frac{1}{D}\sum_{i=1}^{D-1}\sum_{j=i+1}^{D}\left(\log\frac{x_i}{x_j} - \log\frac{\tilde{x}_i}{\tilde{x}_j}\right)^2\right]^{1/2}$ (Aitchison et al. 2000). This criterion represents a generalization of the measure defined in Hron et al. (2010) and it was used in Martín-Fernández et al. (2012) and Templ et al. (2016). Note that the denominator is the maximum distance in the original data set.

Curious Imputations: The number of imputations below 0 and above the detection limit represents the imputation of rounded zeros that are not valid.

6.3 Results

Looking at the result for the imputation of the C-horizon data in Fig. 5, it can be seen that, except for `deepImp-dl`, the best-judged methods are those which take the compositional nature of the data and the special problems for rounded zeros into account. The very best methods for the imputation of rounded zeros for the C-horizon data are `impRZilr`, `imputeBDL_pls` and `deepImpCoDa-dl`. Since `impRZilr` and `imputeBDL_pls` differ algorithmically only slightly and both methods represent an EM algorithm with partial least squares regression, it is not surprising that almost identical results are obtained. Slightly better than `deepImp-dl` is `deepImpCoDa-dl`. This shows that a pivot log-ratio transformation does bring a slight improvement even when using ANNs. The two methods are also best for the evaluation criterion CED, which expresses normalized Aitchison distances between observed and imputed values. Comparatively bad are methods that represent pure imputation methods without taking the problem of rounded zeros into account. The benchmarks `deepImpCoDa` and `deepImp` are comparably worse than their rounded zeros counterparts. Surprisingly, `knn` scores slightly better than `aknn`.

The results of the Moss data are shown in Fig. 6. The two new methods, `deepImpCoDa-dl` (ced: 0.1539, RDCM: 0.0516) and `deepImp-dl` (ced: 0.1584, RDCM: 0.0479) are by far the best, followed by `aknn`. (ced: 0.4053, RDCM: 0.0856). So there is no difference whether algorithm 1 or algorithm 2 is used with respect to

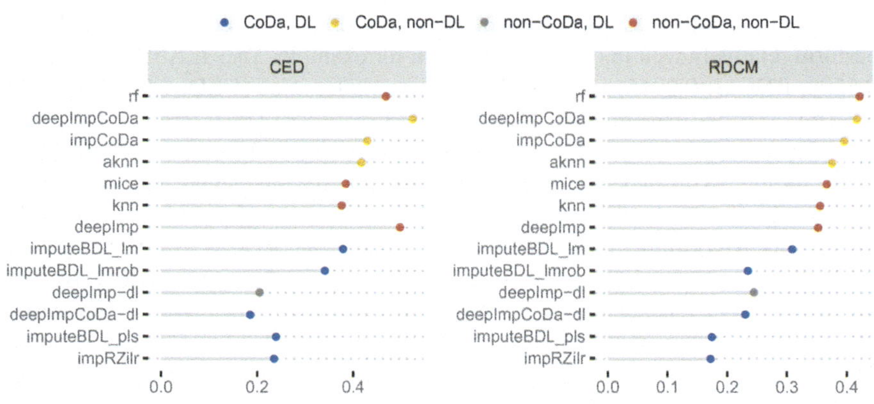

Fig. 5 Comparison of methods for the C-horizon data. Methods that take the compositional nature of the data and detection limits into account are shown in blue. The legend expresses the nature of the method. For example, *CoDa, DL* means the use of a compositional method that is designed to deal with detection limits. See Table 1 for details

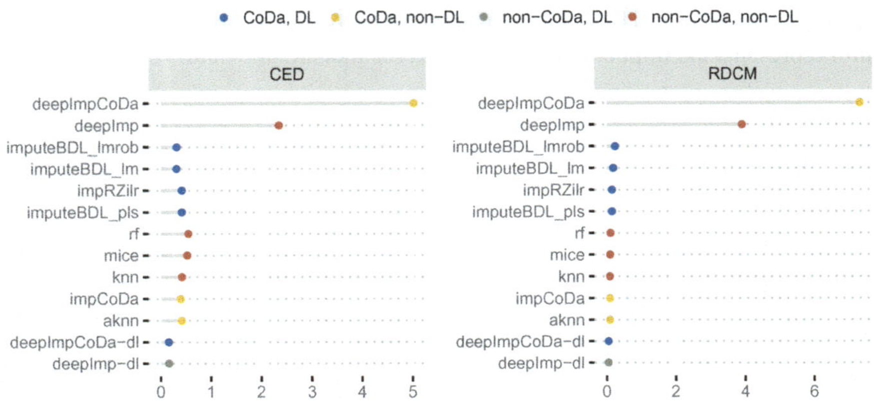

Fig. 6 Comparison of methods for the Moss data. Methods that take the compositional nature of the data and detection limits into account are shown in blue

ANNs, i.e., the deep neural network learns the restrictions of the simplex indirectly and also maps those possibly non-linear relationships which may become linear or better represented by a log-ratio transformation. The other methods for rounded zeros also perform relatively poorly, for example, `imputeBDL_pls` (ced: 0.4136, RDCM: 0.1545). The benchmark methods `deepImp` and `deepImpCoDa` work very poorly, neither of which addresses the problem of detection limits.

A very similar picture can be seen in the results for the TrondelagO data in Fig. 7. Again, the two algorithms 1 (Benchmark: `deepImp-dl`, CED: 0.2608, RDCM: 0.1645) and algorithm 2 (CoDa-Adaption: `deepImpCoDa-dl`, CED: 0.2539, RDCM: 0.1642) performed best by far. Again, a representation in pivot log-ratio coordinates is not important, and the deep neural network learns the relationships even in simplex. Compared to other methods, the rounded zeros methods also perform relatively well in the CED criterion, for example, `imputeBDL_pls` (ced: 0.6508, RDCM: 0.2915) is better than, for example, mice (CED: 0.9552, RDCM: 0.2708). The two benchmarks `deepImp` and `deepImpCoDa` are comparatively poor. The reason is that a few imputed values were imputed very poorly (Fig. 7).

By far the best rounded zero imputation method for the Gemas data (see Fig. 8) is Algorithm 2 (`deepImpCoDa-dl`, CED: 0.1721, RDCM: 0.0971), followed by Benchmark Algorithm 1 (`deepImp-dl`, CED: 0.2538, RDCM: 0.1195). All other methods are comparatively poor, for example, `imputeBDL_pls` (CED: 0.5201, RDCM: 0.5172). It is interesting that for the ANNs this time the pivot log-ratio transformations contribute to the improvement of the results. This means in this case that the ANNs did not learn the compositional structure of the data, but an isometric log-ratio transformation is needed to improve the results (compare `deepImpCoDa-dl` with `deepImp-dl` and `deepImpCoDa` with `deepImp`).

In Fig. 9, one can see which imputation methods do not take detection limits into account and very often they impute values outside the interval (0, detection limit). While this is not such dramatic with the Moss and TrondelagO data, it is evident

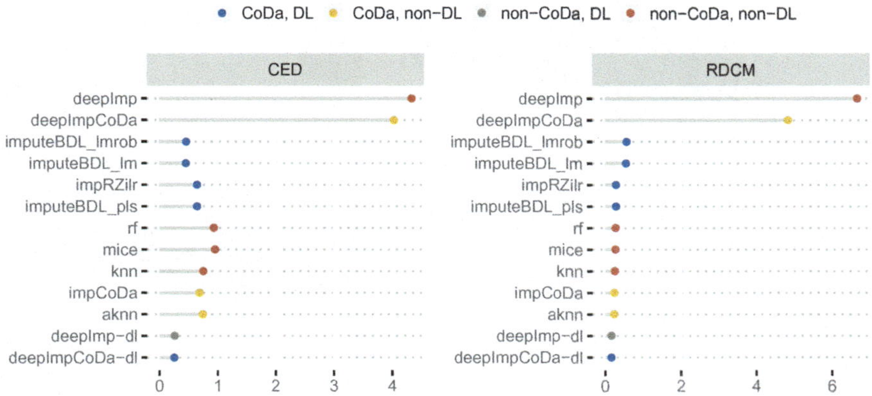

Fig. 7 Comparison of methods for the TrondelagO data. Methods that take the compositional nature of the data and detection limits into account are shown in blue

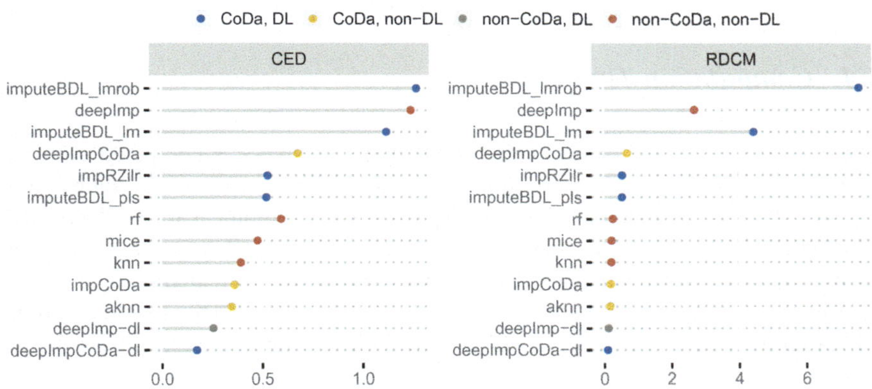

Fig. 8 Comparison of methods for the Gemas data. Methods that take the compositional nature of the data and detection limits into account are shown in blue

with the C-horizon and Gemas data. It is interesting to note that `deepImpCoD` imputes comparatively fewer values outside the intervals than `deepImp`, since no values below 0 can be imputed with this method. The use of methods that do not take detection limits into account is generally critical, since many values outside the definition range, censored with 0 and detection limit, are imputed. Note that values outside the interval (0, DL] can be shrinked to the limits. This means that imputed values above the detection limit are winsorized to the detection limit, and values less than or equal to zero are set to zero + constant, where typically the constant is often chosen very small. The latter implies that outliers may be generated by a log-ratio transformation.

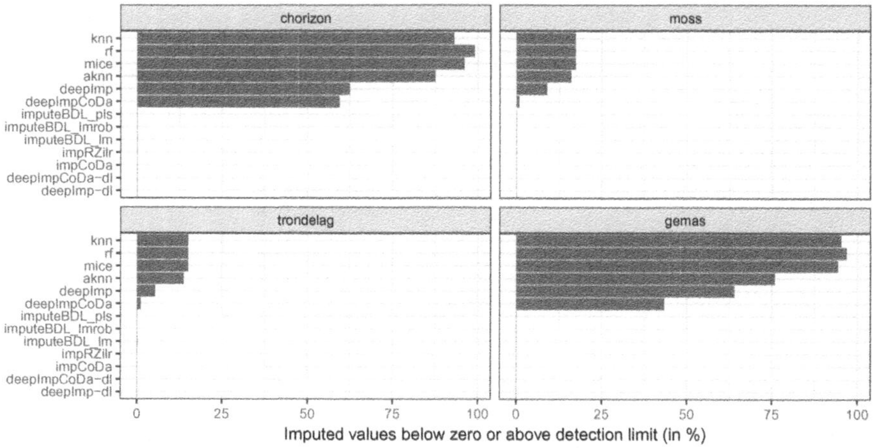

Fig. 9 Percent of imputed values above the detection limit or below 0

6.4 Computation Time

The computational cost of our ANN imputation framework is compared with the runtimes of other imputation procedures using a 2.6 GHz Intel Core i7 single cpu in Fig. 10. The kNN method implemented in package VIM (Templ et al. 2012; Kowarik and Templ 2016) is the fastest in general, however, this method does not consider the rounded zero problem adequately. The runtimes of the EM-based methods using pivot coordinates in Martín-Fernández et al. (2012); Templ et al. (2016) are computationally faster than the artificial neural network methods by approximately a factor of 10.

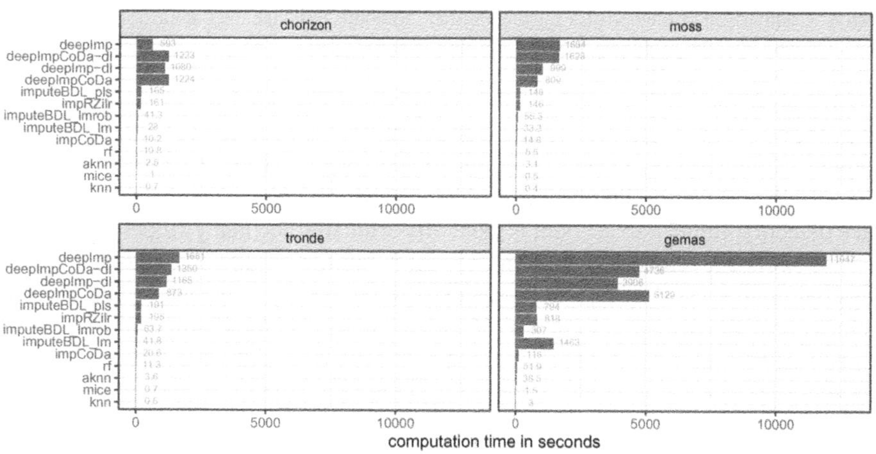

Fig. 10 Computation times on all data sets

Note that `imputeBDLs_rob` should be generally faster than `imputeBDLs_lm`, but in the EM-based approach, the linear model approach takes longer to converge as with the robust method, i.e., the imputations need longer to stabilize and more iterations are necessary. This is also the case with `impNNet`.

The increase of the quality of the imputation using our ANN framework for the imputation of rounded zeros has a price regarding the computation time.

However, if needed, there are several ways to speed up the computation of our ANN framework, namely to use fewer layers, fewer neurons per layer, fewer epochs, and/or fewer iterations.

7 Conclusion

Up to our knowledge and literature research on the date of the submission of this paper, this is the first peer-reviewed contribution in compositional data analysis with ANN's and, more general, using deep learning approaches. The proposed new methods for the imputation of rounded zeros were compared not only with other methods for the imputation of rounded zeros but also with popular conventional imputation methods. As already mentioned, this sounds a bit strange at first glance, since rounded zeros are exclusively censored data. The motivation for this comes from practical and subjective experience as package maintainer of imputation packages and author of many imputation functions. This subjective experience shows that many users and researchers haven't understood that the problem of rounded zeros requires specially designed methods and that users and researchers mostly make the wrong choice of imputation methods, although the articles and function descriptions are clear on the choice of methods. Also, with such a comparison, we highlight the compositional data problem to research areas where compositional data are not well known. For this reason, we will compare exemplary imputation methods with general imputation methods to make this more obvious and evident. Note that a practitioner who neglects compositional data analysis methods could criticize that the compositional data methods are probably only better because of the suggested compositinal evaluation measures. However, compositional data also require compositional validation criteria. Furthermore, the results shown in Fig. 9 are a strong argument against the use of non-compositional "classical" imputation methods because these methods are often used to impute above the detection limit.

Also, benchmark methods were determined. For example, Algorithm 1 is not adequate from a theoretical point of view, but it is interesting to compare it as a benchmark for Algorithm 2 to answer the question of how far results improve on real data when using log-ratio transformations.

The newly introduced method using ANNs (`deepImpCoDa-dl`) is not only competitive but consistently gave the best results for all three data sets. The Moss data has better correlations between the variables than the C-horizon data—this gives a boost to the ANNs. The TrondelagO data have a good group structure concerning different plant species, which the ANNs learn better than other methods. The Gemas data are significantly larger than the other data sets, which has an impact on the quality of the ANNs, which generally improve as more training, testing, and validation data are available. The presented imputation methodology with ANNs beats all other methods by far. Note that for other data sets it is likely to reveal different measures of performance caused by distinct multivariate correlation and dependency structures.

It is interesting to see that a representation in (pivot) log-ratio coordinates is not necessarily needed anymore, but the results hardly deteriorate if the data are not represented in log-ratio coordinates, because the limitations of the simplex and the coherences in the data are well learned in the deep neural network. Theoretically, however, the correct representation of the data also plays a role in deep neural networks, since the distances between observed values should be considered in a compositional way. We also see differences in the C-horizon results and especially in the results for the Gemas data set, where Algorithm 2 (with log-ratio transformations) performs better than the benchmark Algorithm 1 (without log-ratio transformations).

A drop of bitterness is the long computing times because an ANN with enough layers and neurons takes a long time to train. For typical data sets in geochemistry, for example, this is not a problem as long as no parameter optimization on the number of layers, neurons, loss function, evaluation function, activation function, epochs, etc., is made. In practice, the imputation is often performed exactly once, and therefore it doesn't matter if an algorithm is ready in a second or if it takes 4 h. In any case, we provide sensible defaults and stop whenever the evaluation criteria will not improve after a given number of epochs. Multiple imputation is possible in principle and can be controlled by the parameter `dropout`, which also can avoid overfitting.

The presented algorithms using ANNs are part of a larger R package for deep learning in imputation and compositional data analysis that is work in progress, and it will be published independently of this contribution.

We propose three topics for future research. Firstly, it would be interesting to compare the methods from the R package *zCompositions*. Although these methods are mostly of univariate nature or include a multiple EM-based method based on the alr transformation, in many cases rounded zeros are very well imputed. Secondly, large simulation studies may, under certain circumstances, lead to further knowledge about the quality of the methods. Thirdly, in this contribution, we only looked at some well-known evaluation measures such as the CED and RDCM but further analysis is possible. For example, the evaluation of classifiers on the imputed data can show which methods are more advantageous and lead to fewer misclassifications, and log-ratio biplots or comparison of results from a regression model can show what the effects of imputing rounded zeros are.

Acknowledgements I would like to thank Peter Filzmoser and Karel Hron for the many collaborations and the fruitful discussion on the topic of compositional data analysis, imputation, and rounded zeros. Furthermore, my thanks go to Eric Grunsky and Peter Filzmoser as well as to one unknown reviewer for their constructive and helpful comments on the initial submission.

References

M. Abadi, A. Agarwal, P. Barham, E. Brevdo, Z. Chen, C. Citro, G.S. Corrado, A. Davis, J. Dean, M. Devin, S. Ghemawat, I. Goodfellow, A. Harp, G. Irving, M. Isard, Y. Jia, R. Jozefowicz, L. Kaiser, M. Kudlur, J. Levenberg, D. Mané, R. Monga, S. Moore, D. Murray, C. Olah, M. Schuster, J. Shlens, B. Steiner, I. Sutskever, K. Talwar, P. Tucker, V. Vanhoucke, V. Vasudevan, F. Viégas, O. Vinyals, P. Warden, M. Wattenberg, M. Wicke, Y. Yu, and X. Zheng. TensorFlow: Large-scale machine learning on heterogeneous systems, 2015. URL https://www.tensorflow.org/. Software available from tensorflow.org

M. Abadi, A. Agarwal, P. Barham, E. Brevdo, Z. Chen, and C. Citro. TensorFlow: large-scale machine learning on heterogeneous systems, 2018. URL https://www.tensorflow.org/. Version: 1.10.0, Software available from tensorflow.org

J. Aitchison, *The Statistical Analysis of Compositional Data* (Chapman & Hall, London, 1986)

J. Aitchison, C. Barceló-Vidal, J.A. Martín-Fernández, V. Pawlowsky-Glahn, Logratio analysis and compositional distance. Math. Geol. **32**(3), 271–275 (2000)

J.J. Allaire and F. Chollet. *keras: R Interface to 'Keras'*, 2019. URL https://keras.rstudio.com. R package version 2.2.4.1.9001

C. Arisdakessian, O. Poirion, B. Yunits, X. Zhu, L.X. Garmire, Deepimpute: an accurate, fast, and scalable deep neural network method to impute single-cell rna-seq data. Genome Biol. **20**(1), 211 (2019). https://doi.org/10.1186/s13059-019-1837-6

J. Chen, X. Zhang, K. Hron, M. Templ, S. Li, Regression imputation with q-mode clustering for rounded zero replacement in high-dimensional compositional data. J. Appl. Stat. **45**(11), 2067–2080 (2017). https://doi.org/10.1080/02664763.2017.1410524

F. Chollet et al., Keras (2015). https://keras.io

S.J. Choudhury, N.R. Pal, Imputation of missing data with neural networks for classification. Knowl.-Based Syst. **182**, 104838 (2019.) ISSN 0950-7051. https://doi.org/10.1016/j.knosys.2019.07.009

P. Filzmoser, K. Hron, M. Templ, *Applied Compositional Data Analysis* (Springer International Publishing, 2018). ISBN 9783319964225. https://doi.org/10.1007/978-3-319-96422-5

T. Hastie, R. Tibshirani, J. Friedman, *The Elements of Statistical Learning*, 2nd edn. (Springer, New York, 2009). ISBN 978-0-387-84857-0

K. He, X. Zhang, S. Ren, J. Sun, Delving deep into rectifiers: surpassing human-level performance on ImageNet classification (2015)

K. Hron, M. Templ, P. Filzmoser, Imputation of missing values for compositional data using classical and robust methods. Comput. Stat. Data Anal. **54**(12), 3095–3107 (2010). ISSN 0167-9473. https://doi.org/10.1016/j.csda.2009.11.023

J.M. Jerez, I. Molina, P.J. García-Laencina, E. Alba, N. Ribelles, M. Martń, L. Franco, Missing data imputation using statistical & machine learning methods in a real breast cancer problem. Artif. Intell. Med. **50**(2), 105–115 (2010). ISSN 0933-3657. https://doi.org/10.1016/j.artmed.2010.05.002

D.P. Kingma, J. Ba. Adam: a method for stochastic optimization (2014). arXiv:abs/1412.6980

A. Kowarik, M. Templ, Imputation with the R package VIM. J. Stat. Softw. **74**(7), 1–16 (2016). https://doi.org/10.18637/jss.v074.i07

A. Krizhevsky, I. Sutskever, G.E. Hinton. Imagenet classification with deep convolutional neural networks (2012). Internet Resource, accessed on 14 Jan 2019

S.C-X. Li, B. Jiang, B.M. Marlin, Misgan: learning from incomplete data with generative adversarial networks (2019). http://arxiv.org/abs/1902.09599

Y.C. Lim, Learning generative models from incomplete data. Technical report CMU-CS-19-120, School of Computer Science, Computer Science Department, Carnegie Mellon University, Pittsburgh, PA 15213 (2019)

T. Maiti, C.P. Miller, P.K. Mukhopadhyay, Neural network imputation: an experience with the national resources inventory survey. J. Agric., Biol., Environ. Stat. **13**(3), 255–269 (2008). ISSN 10857117

J. Martín-Fernández, K. Hron, P. Templ, M. Filzmoser, J. Palarea-Albaladejo, Model-based replacement of rounded zeros in compositional data: classical and robust approaches. Comput. Stat. Data Anal. **56**(9), 2688–2704 (2012). https://doi.org/10.1016/j.csda.2012.02.012

J.A. Martín-Fernández, C. Barceló-Vidal, V. Pawlowsky-Glahn, Dealing with zeros and missing values in compositional data sets using nonparametric imputation. Math. Geol. **35**(3), 253–278 (2003)

J.A. Martín-Fernández, J. Palarea-Albaladejo, R.A. Olea, Dealing with zeros, in *Compositional Data Analysis: Theory and Applications*, ed. by V. Pawlowsky-Glahn, A. Buccianti (Wiley, Chichester, 2011), pp. 43–58

J.A. Martín-Fernández, K. Hron, M. Templ, P. Filzmoser, J. Palarea-Albaladejo, Bayesian-multiplicative treatment of count zeros in compositional data sets. Stat. Model. **15**(2), 134–158 (2015)

P-A. Mattei, J. Frellsen, missiwae: deep generative modelling and imputation of incomplete data (2018). ArXiv:abs/1812.02633

M. Mayer, missRanger: fast imputation of missing values (2019). https://CRAN.R-project.org/package=missRanger. R package version 2.1.0

J.T. McCoy, S. Kroon, L. Auret, Variational autoencoders for missing data imputation with application to a simulated milling circuit. IFAC-PapersOnLine **51**(21), 141 – 146 (2018). ISSN 2405-8963. https://doi.org/10.1016/j.ifacol.2018.09.406; in *5th IFAC Workshop on Mining, Mineral and Metal Processing MMM 2018*

M.A. Nielsen, *Neural Networks & Deep Learning*, vol. 25 (Determination Press, USA, 2015)

J. Palarea-Albaladejo, J.A. Martín-Fernández, A modified em alr-algorithm for replacing rounded zeros in compositional data sets. Comput. Geosci. **34**(8), 902–917 (2008)

J. Palarea-Albaladejo, J.A. Martín-Fernández, Values below detection limit in compositional chemical data. Anal. Chim. Acta **764**, 32–43 (2013)

J. Palarea-Albaladejo, J.A. Martín-Fernández, J. Gómez-García, A parametric approach for dealing with compositional rounded zeros. Math. Geol. **39**(7), 625–645 (2007)

J. Palarea-Albaladejo, J.A. Martín-Fernández, R.A. Olea, A bootstrap estimation scheme for chemical compositional data with nondetects. J. Chemom. **28**(7), 585–599 (2014)

C. Reimann, P. Filzmoser, R.G. Garrett, R. Dutter, *Statistical Data Analysis Explained: Applied Environmental Statistics with R* (Wiley, Chichester, 2008)

S. Ruder, An overview of gradient descent optimization algorithms (2016). http://arxiv.org/abs/1609.04747

E-L. Silva-Ramírez, R. Pino-Mejías, M. López-Coello, Single imputation with multilayer perceptron and multiple imputation combining multilayer perceptron and k-nearest neighbours for monotone patterns. Appl. Soft Comput. **29**, 65–74 (2015.) ISSN 1568-4946. https://doi.org/10.1016/j.asoc.2014.09.052

M. Smieja, U. Struski, J. Tabor, B. Zieliski, P. Spurek, Processing of missing data by neural networks, in *Proceedings of the 32nd International Conference on Neural Information Processing Systems, NIPS'18* (Curran Associates Inc, Red Hook, NY, USA, 2018), pp. 2724–2734

D.J. Stekhoven, P. Bühlmann, Missforest—non-parametric missing value imputation for mixed-type data. Bioinformatics **28**(1), 112–118, 10 (2011). ISSN 1367-4803. https://doi.org/10.1093/bioinformatics/btr597

M. Templ, deepImp: imputation with deep learning methods (2020). https://bitbucket.org/matthias-da/deepimp/. R package version 1.0.0

M. Templ, P. Filzmoser, C. Reimann, Cluster analysis applied to regional geochemical data: problems and possibilities. Appl. Geochem. **23**(8), 2198 – 2213 (2008). ISSN 0883-2927. https://doi.org/10.1016/j.apgeochem.2008.03.004. http://www.sciencedirect.com/science/article/pii/S088329270800125X

M. Templ, K. Hron, P. Filzmoser, *robCompositions: An R-package for Robust Statistical Analysis of Compositional Data* (Wiley, Hoboken, 2011), pp. 341–355. ISBN 9781119976462. http://dx.doi.org/10.1002/9781119976462.ch25

M. Templ, A. Alfons, P. Filzmoser, Exploring incomplete data using visualization techniques. Adv. Data Anal. Classif. **6**(1), 29–47 (2012). https://doi.org/10.1007/s11634-011-0102-y

M. Templ, K. Hron, P. Filzmoser, A. Gardlo, Imputation of rounded zeros for high-dimensional compositional data. Chemom. Intell. Lab. Syst. **155**, 183–190 (2016). https://doi.org/10.1016/j.chemolab.2016.04.011. http://www.sciencedirect.com/science/article/pii/S0169743916300958

S. van Buuren, K. Groothuis-Oudshoorn, mice: multivariate imputation by chained equations in R. J. Stat. Softw. **45**(3), 1–67 (2011). http://www.jstatsoft.org/v45/i03/

K.G. van-den Boogaart, R. Tolosana-Delgado, M. Templ, Regression with compositional response having unobserved components or below detection limit values. Stat. Model. **15**(2), 191–213 (2015)

A. Vedaldi, K. Lenc, Matconvnet: convolutional neural networks for MATLAB, in *Proceedings of the 23rd ACM International Conference on Multimedia* (ACM, 2015), pp. 689–692

J. Xie, L. Xu, E. Chen, Image denoising and inpainting with deep neural networks, in *Proceedings of the 25th International Conference on Neural Information Processing Systems - Volume 1, NIPS'12* (Curran Associates Inc, Red Hook, NY, USA, 2012), pp. 341–349

J. Yoon, J. Jordon, M. van der Schaar, GAIN: missing data imputation using generative adversarial nets (2018). http://arxiv.org/abs/1806.02920

Compositional DuPont Analysis. A Visual Tool for Strategic Financial Performance Assessment

Elisabet Saus–Sala, Àngels Farreras–Noguer, Núria Arimany–Serrat, and Germà Coenders

Abstract *DuPont analysis* is a classical tool for assessing the determinants of financial performance of firms. It is based on financial ratios comparing revenues with costs (the so-called *margin ratio*), revenues with assets (*turnover ratio*), and debt with assets (*leverage ratio*). DuPont analysis thus focuses on comparing accounting values in relative terms and lends itself naturally to compositional analysis. In this chapter, we show how to graphically display firms according to margin, turnover and leverage by means of a standard compositional biplot, and how to cluster firms into strategic groups by means of k-means compositional cluster analysis. Practitioners who prefer to stick to the classic definitions of industry or cluster-level financial ratios can compute them with the usual formulae from the centre of the composition, i.e. from the industry or cluster geometric averages rather than the totals or arithmetic averages commonly used. An illustration is presented with farm-tourism firms.

E. Saus–Sala · À. Farreras–Noguer
Department of Business Studies, University of Girona, c./ Universitat de Girona 10, 17003 Girona, Spain
e-mail: elisabet.saus@gmail.com

À. Farreras–Noguer
e-mail: angels.farreras@udg.edu

N. Arimany–Serrat
Department of Economics and Business, University of Vic-UCC, c./ Sagrada Familia 7, 08500 Vic, Spain
e-mail: nuria.arimany@uvic.cat

G. Coenders (✉)
Department of Economics, University of Girona, c./ Universitat de Girona 10, 17003 Girona, Spain
e-mail: germa.coenders@udg.edu

P. Filzmoser et al. (eds.), *Advances in Compositional Data Analysis*,
https://doi.org/10.1007/978-3-030-71175-7_10

1 Financial Ratios as Carriers of Relative Information About Accounting Values

Financial ratios (i.e. ratios comparing the relative values of accounts or account categories in financial statement analysis) enable researchers and professionals to evaluate a firm's position compared to its counterparts in the industry or to itself along time, taking into account differences or evolution in firm size. Since the beginning of the last century, standard financial ratios have been used both in management performance and strategic assessment, and in research relating them to other financial or non-financial variables (Barnes 1987; Faello 2015; Horrigan 1968; Tascón et al. 2018; Willer do Prado et al. 2016). In particular, the well-known method of *DuPont analysis* decomposes the most common measure of financial performance, *Return on Equity* (ROE), into three very relevant standard financial ratios: *turnover*, *margin* and *leverage*, thus making for a simple financial performance assessment tool. The aim of this chapter is to put forward some simple and visual tools for DuPont analysis from a compositional perspective.

When treated as variables in statistical analyses, standard financial ratios have long been reported to have a number of serious statistical and practical problems, including, among others, asymmetry, outliers, non-linearity, redundancy and even dependence of the results on the arbitrary decision regarding which account category appears in the numerator of the ratio and which in the denominator (Faello 2015; Linares-Mustarós et al. 2018; Lev and Sunder 1979; Cowen and Hoffer 1982; So 1987; Carreras-Simó and Coenders 2020). These problems are particularly serious for the multivariate statistical methods we use in this chapter, namely, principal component analysis and cluster analysis (Linares-Mustarós et al. 2018; Cowen and Hoffer 1982; Carreras-Simó and Coenders 2020).

Financial ratios constitute a genuine case of researchers' and professionals' interest in relative rather than absolute values and thus a natural field of application of *Compositional Data* (CoDa). Moving away from the classical and restrictive definitions of CoDa as parts of a whole or as fixed-sum data, CoDa are contemporarily defined in more practical and general terms as arrays of strictly positive numbers for which ratios between them are considered to be relevant (Egozcue and Pawlowsky-Glahn 2019). DuPont analysis fulfils this definition to the letter.

The CoDa methodology offers a number of advantages in statistical analysis of financial statements, as compared to standard financial ratios. Among other features, CoDa treats accounts in a symmetric fashion in such a way that results do not depend on numerator and denominator permutation. CoDa also tends to reduce non-linearity and outliers and treats redundancy by acknowledging the fact that no analysis will require more ratios than account categories (Linares-Mustarós et al. 2018; Carreras-Simó and Coenders 2020).

Starting with a widespread acceptance of CoDa in geology (Martín-Fernández et al. 2001; Pawlowsky-Glahn and Buccianti 2002; Buccianti et al. 2006), accessible handbooks (Pawlowsky-Glahn and Buccianti 2011; Pawlowsky-Glahn et al. 2015a) have contributed to extending its use to many other scientific fields, including the social sciences (Coenders and Ferrer-Rosell 2020). Recently, the CoDa methodology has often been applied in finance to answer research questions concerning the relative importance of magnitudes. Examples include crowdfunding (Davis et al. 2017), financial markets (Ortells et al. 2016; Wang et al. 2019), municipal budgeting (Voltes-Dorta et al. 2014), insurance (Belles-Sampera et al. 2016; Boonen et al. 2019; Verbelen et al. 2018), stock market return forecasting (Kokoszka et al. 2019), exchange-rate arbitrage (Maldonado et al. 2021) and product portfolios (Joueid and Coenders 2018). As of now there are few applications in accounting and financial statement analysis (Linares-Mustarós et al. 2018; Carreras-Simó and Coenders 2020; Creixans-Tenas et al. 2019). To the best of our knowledge, CoDa methods have never been applied in DuPont analysis.

This chapter is organized as follows. First, we provide a non-technical introduction to DuPont analysis. Next, we consider the necessary accounts in DuPont analysis as a composition, including the needed log-ratio transformations. Then we present the application data we use, drawn from financial statements in the farm-tourism industry. Next, we present the CoDa biplot as a visualization tool for firms' turnover, margin and leverage. Then we present CoDa cluster analysis as a tool for drawing strategic groups with similar turnover, margin and leverage profiles. Clusters may be characterized both in terms of standard financial ratios and compositional financial ratios. Finally, we conclude and comment on possible extensions.

2 DuPont Analysis

DuPont analysis was developed in 1914 by Donaldson Brown (Dale et al. 1980), and owes its name to the company where he was working at that time. It has continued to be in use ever since as a popular method for decomposing the firm's performance into a small number of standard financial ratios.

Performance can be defined as a firm's ability to generate profit from its investments through an economic activity involving transformation, production or trade of goods or services. Performance is usually measured as a standard financial ratio of profit over investment. Depending on the way we conceptualize investment there are several measures of performance. One possible way is to use all firm's assets. This measure is independent on the so-called capital structure, which indicates how assets are financed with a combination of shareholders' investment (equity) and debt (liabilities). The measure is referred to as *Return on Assets* (ROA):

$$\text{ROA} = \frac{\text{profit}}{\text{assets}} = \frac{\text{revenues} - \text{costs}}{\text{assets}}. \tag{1}$$

Another way to conceptualize investment is to use shareholders' equity. The resulting measure is referred to as *Return on Equity* (ROE):

$$\text{ROE} = \frac{\text{profit}}{\text{equity}} = \frac{\text{revenues} - \text{costs}}{\text{assets} - \text{liabilities}}. \tag{2}$$

There are other possibilities to conceptualize investment in the denominator of the ratios, and yet several possibilities to conceptualize profit, revenues and costs in the numerator of the ratios. For a discussion of their pros and cons, which are out of the scope of this chapter, see Chen et al. (2014). Irrespective of the manner in which profit, revenues and costs are defined, ROA can be decomposed into the product of turnover and margin:

$$\text{ROA} = \text{turnover} \times \text{margin}. \tag{3}$$

Turnover indicates the effectiveness in the use of assets, as the revenues obtained per each asset unit:

$$\text{turnover} = \frac{\text{revenues}}{\text{assets}}. \tag{4}$$

Margin is a function of price, sales volume and costs, including material costs, labour costs, overheads, selling, general and administrative expenses, a.s.o:

$$\text{margin} = \frac{\text{profit}}{\text{revenues}} = \frac{\text{revenues} - \text{costs}}{\text{revenues}}. \tag{5}$$

If we consider the shareholders' perspective rather than total assets, we must resort to ROE in Eq. (2). This indicator shows the yearly rate of return of shareholders' investment; in other words, profit generated per invested monetary unit. This should be at least equal to the return of a low-risk asset in financial markets plus a risk premium.

The DuPont analysis method explains the origin of firm performance by decomposing ROE into ROA, and the multiplier effect of indebtedness, which is commonly called leverage or gearing. It is thus computed as the product of the three standard financial ratios in Eqs. (4), (5) and (7):

$$\text{ROE} = \text{turnover} \times \text{margin} \times \text{leverage} = \text{ROA} \times \text{leverage}. \tag{6}$$

$$\text{leverage} = \frac{\text{assets}}{\text{equity}} = \frac{\text{assets}}{\text{assets} - \text{liabilities}}. \tag{7}$$

Financial leverage indicates the factor by which ROE varies as a function of indebtedness. By means of debt, more assets can be acquired per each unit of equity. Unlike ROA, debt interest rates tend to be rather stable. If the returns of the additional assets are higher than the borrowing cost, then leverage increases ROE. Conversely, when return is low or negative, leverage makes things worse, because interest due has to be paid anyway. The word 'leverage' has been widely adopted because it acts like a lever in physics, amplifying the output force for a given input force.

The DuPont method is popular among researchers and professionals to explain the factors driving firms' financial performance (Chen et al. 2014; Parulekar 2015; Soliman 2008; Baležentis et al. 2019; Doorasamy 2016), both in small and medium enterprises and in large corporations traded in the main stock markets (Soliman

2008). The DuPont method makes it possible to uncover strategies to achieve higher performance based on higher turnover, higher margin or higher leverage. The values of the ratios involved in DuPont analysis are usually not comparable across industries, but they do make it possible to compare large and small firms in the same industry (Barnes 1987). A good example is the agriculture sector, in which very small family-owned farms coexist with large industrial agriculture farms (Baležentis et al. 2019).

3 Financial Statement Accounts as Compositional Data

The CoDa methodology provides a standard toolbox for statistical analyses concerning the relative importance of a set of magnitudes (Egozcue and Pawlowsky-Glahn 2019; Pawlowsky-Glahn and Buccianti 2011; Pawlowsky-Glahn et al. 2015a). All common statistical methods can be used on CoDa after an appropriate transformation, and taking care that interpretation always concerns relative rather than absolute values. This is the approach we follow in this chapter throughout. This section deals mainly with the transformations needed for two well-known multivariate methods (biplot and cluster analysis), whose interpretation will later be shown in the light of DuPont analysis.

A D-part composition is defined as a positive vector, the relative magnitude of whose parts is of interest to the researcher:

$$\mathbf{x} = (x_1, x_2, ..., x_D) \text{ with } x_j > 0, \ \ j = 1, 2, ..., D \ . \qquad (8)$$

In compositional financial statement analysis parts normally constitute values of accounts or, more commonly, broader account categories (Linares-Mustarós et al. 2018; Carreras-Simó and Coenders 2020; Creixans-Tenas et al. 2019) and have to be constructed in such a way that negative values are not possible (Creixans-Tenas et al. 2019). Even if sometimes standard financial ratios involve account categories which may be negative, their use is advised against in the accounting literature, because they can cause a discontinuity, outliers or even a reversal of interpretation when the account which may be negative is in the denominator (Lev and Sunder 1979). In general, account categories are negative because they imply some form of subtraction of other positive account categories, which are the ones to be used. This means that when building the margin ratio one should directly use revenues and costs rather than profit. This limitation implies no loss of information whatsoever. The always positive revenues over costs ratio in Eq. (9) can easily be shown to be just a transformation of the problematic profit over revenues ratio in Eq. (5) (Creixans-Tenas et al. 2019):

$$\frac{\text{revenues}}{\text{costs}} = \frac{\text{revenues}}{\text{revenues} - \text{profit}} = \frac{1}{\frac{\text{revenues}-\text{profit}}{\text{revenues}}} = \frac{1}{1 - \frac{\text{profit}}{\text{revenues}}} \,. \qquad (9)$$

Along similar lines, in order to compute a measure of leverage, one should use assets and liabilities rather than assets and equity, with no loss of information. Having taken this into account, only the following $D = 4$ positive account categories are needed for the purposes of DuPont analysis:

- x_1: revenues,
- x_2: costs,
- x_3: liabilities,
- x_4: assets.

3.1 Transformations

As stated above, the usual approach to CoDa is to use existent standard statistical methods on transformed data (Pawlowsky-Glahn and Buccianti 2011; Pawlowsky-Glahn et al. 2015a). Logarithms of ratios are the commonest transformation in CoDa and they solve the main problems of standard financial ratio usage in statistical analysis reported in the literature (Faello 2015; Linares-Mustarós et al. 2018; Lev and Sunder 1979; Cowen and Hoffer 1982; So 1987; Carreras-Simó and Coenders 2020). Log-ratios involving only pairs of account categories are the simplest. Assuming that only positive values are permitted, a standard ratio is bounded between zero and infinity. Conversely, a log-ratio is symmetric in the sense that its range is from minus infinity to plus infinity. Besides, permuting the numerator and the denominator affects no other property of the log-ratio than the sign:

$$\log\left(\frac{x_i}{x_j}\right) = -\log\left(\frac{x_j}{x_i}\right). \tag{10}$$

Furthermore, if one of the account categories being compared is close to zero, it may lead to an outlying standard ratio when placed in the denominator and to a typical ratio when placed in the numerator. For log-ratios, placement makes no difference (Linares-Mustarós et al. 2018; Carreras-Simó and Coenders 2020; Creixans-Tenas et al. 2019).

Some log-ratios between pairs of account categories are especially interesting in DuPont analysis. By definition, turnover compares revenues (x_1) and assets (x_4):

$$y_1 = \log\left(\frac{x_1}{x_4}\right). \tag{11}$$

As argued when constructing Eq. (9), comparing revenues (x_1) and costs (x_2) provides a notion of margin:

$$y_2 = \log\left(\frac{x_1}{x_2}\right). \tag{12}$$

Finally, comparing liabilities (x_3) and assets (x_4) provides a notion of leverage:

$$y_3 = \log\left(\frac{x_3}{x_4}\right). \tag{13}$$

In contrast with standard financial ratios, we term y_1 to y_3 compositional financial ratios.

Even if there are potentially $D(D-1)/2$ distinct log-ratios among pairs of D account categories, they necessarily contain redundant information (Creixans-Tenas et al. 2019; Greenacre 2019). For instance, once the researcher has the log-ratios y_1 to y_3, a log-ratio comparing costs (x_2) and assets (x_4) is no longer needed, as it results from the previous three:

$$y_1 - y_2 = \log\left(\frac{x_1}{x_4}\right) + \log\left(\frac{x_2}{x_1}\right) = \log\left(\frac{x_2}{x_4}\right). \tag{14}$$

The redundancy problems encountered in the financial literature when using a very large number of ratios arise from the fact that some are unavoidably exact functions of other ratios (Chen and Shimerda 1981). According to Barnes (1987), p. 456:

> To identify those ratios which contain complete information about a firm while minimising duplication cannot be achieved purely by logic; in fact it is largely an empirical matter.

Log-ratios in CoDa do not necessarily need to be computed only from selected pairs of account categories (Egozcue et al. 2003). Actually, D so-called *centred log-ratios* contain all information about the relative importance of D account categories (Aitchison 1983), thus minimizing redundancy. Even if they do not have any particular interpretation in the accounting field by themselves, centred log-ratios can be readily submitted to the multivariate descriptive statistical methods used in this chapter (Carreras-Simó and Coenders 2020) whose results are the ones to interpret with an accounting focus. Centred log-ratios compare each account category with the geometric average of all for each individual firm. For the four account categories that we have are computed as follows:

$$\ln\left(\frac{x_1}{\sqrt[4]{x_1x_2x_3x_4}}\right), \ln\left(\frac{x_2}{\sqrt[4]{x_1x_2x_3x_4}}\right), \ln\left(\frac{x_3}{\sqrt[4]{x_1x_2x_3x_4}}\right), \ln\left(\frac{x_4}{\sqrt[4]{x_1x_2x_3x_4}}\right). \tag{15}$$

In contrast to standard multivariate statistical analysis, it must be noted that in compositional multivariate analyses raw log-ratios are always used. In other words, centred log-ratios must not be standardized to unit variance, because doing so results in losing their geometric properties.

3.2 *Centre*

The average value of a composition is referred to as *centre* of the composition (Pawlowsky-Glahn and Buccianti 2011). For the purpose of highlighting relative rather than absolute differences, it is computed from the geometric averages of all individual firms for each account category (not to be mistaken with the geometric averages of all account categories for each individual firm in Eq. (15)). In order to provide a comparable scale, geometric averages are then normalized to a unit sum, which is referred to as *closure* in CoDa.

The centre computed as a geometric average under the CoDa approach makes it possible to compute standard financial ratios at industry level in a straightforward manner. The geometric average has the attractive property that the ratio of two geometric averages equals the geometric average of the ratios of the two involved account categories. Let $g_a\left(x_i\right)$ be the geometric average of account category i over a sample of n firms:

$$g_a\left(x_i\right)/g_a\left(x_j\right) = g_a\left(x_i/x_j\right). \tag{16}$$

The arithmetic average does not have this property. Computing first arithmetic averages at industry level and then standard financial ratios between those averages may stand in contradiction with the results of computing first standard ratios at firm level and then the ratio arithmetic averages.

4 Data

ROE is the variable around which all efforts must be pivoted in order to make a business viable. For this purpose, within farm-tourism firms, differentiation strategies, such as rural quality labels, are being sought in order to provide an increase in value, and the consequent increase in income (Bilbao and Valdés 2016).

Regarding the costs that this type of business must incur, we must differentiate those that come directly from the initial investment (e.g. acquiring or restoring the farm building), including sometimes very high finance costs, from those that will be generated continuously in the operation of the business itself. The exhaustive control of both variables can make a difference in the survivability of the tourist farm (Castro 2019), especially if family owned. Recovery of family heritage (manors and rustic houses sometimes centuries old) constitutes both one of the values of farm tourism and one of the entrepreneurial goals and motivations, which have to be well understood for the sake of appealing investment and making it profitable (Castro 2019).

Therefore, an analysis of the factors that have an impact on profitability in farm tourism is essential. Most studies on profitability in farm tourism carried out to date are based on survey questionnaires (Ye et al. 2019; Kallmuenzer and Peters 2018; Costa 2018; Polo Peña et al. 2016) in which owners are asked to provide at best a qualitative or subjective appraisal of ROE. These studies need to be expanded on through using 'hard' accounting data. This is what the application presented in this chapter is about.

The accounting data were obtained from the SABI (Iberian Balance sheet Analysis System, accessible at https://sabi.bvdinfo.com/) database, developed by INFORMA D&B in collaboration with Bureau Van Dijk. Search criteria were active firms on 19

Table 1 Sample statistics for standard financial ratios (left) and centred log-ratios (right) with x_1 to x_4 in the numerator

	St. turnover	St. margin	St. leverage	St. ROE	x_1	x_2	x_3	x_4
Average	1.65	−0.64	4.05	0.21	−0.29	−0.19	−0.11	0.59
SD	2.94	3.70	11.73	2.57	0.97	0.84	1.23	1.08
Skewness	3.07	−6.62	1.74	3.95	−0.43	−0.10	−0.36	0.19
Kurtosis	10.97	44.48	8.08	26.70	−0.35	−0.90	−0.43	0.32

December 2019 in the Catalonia autonomous region, Spain, belonging to the NACE Rev. 2 classification code 5520 'holiday and other short-stay accommodation' and dedicated to farm tourism (Eurostat 2008). All observations fulfilling these criteria were selected ($n = 46$ firms). The last available yearly data were selected for each firm, which, in most cases, were those in 2018.

Table 1 presents descriptive statistics for the application data and shows the centred log-ratios in Eq. (15) to be much better suited for multivariate statistical analysis than the standard financial ratios used in DuPont analysis in Eqs. (4), (5), (7) and (2), having, as expected, much less asymmetry (i.e. skewness) and outliers at the tails of the distribution (i.e. kurtosis) (Linares-Mustarós et al. 2018; Cowen and Hoffer 1982; Carreras-Simó and Coenders 2020).

All subsequent analyses were carried out with CoDaPack, an intuitive menu-driven freeware for CoDa developed by the *Research Group in Statistics and Compositional Data Analysis* at the University of Girona. It can be downloaded at http://www.compositionaldata.com/codapack.php.

5 CoDa Biplot as a Visual Representation of DuPont Analysis

The so-called *CoDa biplot* is a standard *covariance biplot* based on a principal component analysis using the centred log-ratios in Eq. (15) as data (Aitchison 1983; Aitchison and Greenacre 2002; Egozcue and Pawlowsky-Glahn 2011, 2016). The centred log-ratios corresponding to the account categories appear as rays in the biplot, and the firms appear as points. More interestingly for the purpose of financial statement analysis (Carreras-Simó and Coenders 2020), the direction defined by the link between the vertices of a pair of rays can be interpreted as the logarithm of the financial ratio between the two corresponding account categories. Thus, the orthogonal projection of all firms along the direction defined by the link between the vertices of a pair of account categories shows an approximate ordering of firms according to the financial ratio between them. In this way, the CoDa biplot is a visual representation of firms according to any of the $D(D-1)/2$ possible financial ratios computed from any two account categories, of which researchers can choose the most informative ones according to their objectives. In the case of DuPont analysis, turnover, margin and leverage as defined in Eqs. (11)–(13) can be interpreted in a straightforward manner.

The ability to visually interpret ratios between any two account categories is of great interest in financial statement analysis. The y_1 (turnover), y_2 (margin) and y_3 (leverage) directions have been drawn above the biplot provided by CoDaPack (Fig. 1). Given the positions of the revenue and asset vertices, if we follow the y_1-direction upwards and to the left we encounter firms with ever-increasing turnover. If we follow the y_2-direction downwards and to the left we encounter firms with ever-increasing margin. If we follow the y_3-direction upwards we encounter firms with ever-increasing leverage.

For instance, firm 33 has a very high margin, a very low turnover and a very low leverage; firm 34 has a very low margin and a very high leverage; firm 19 has a very high turnover and a very high leverage; firm 42 a very low turnover and firm 15 a very high turnover. The compositional biplot thus becomes an intuitive and useful tool

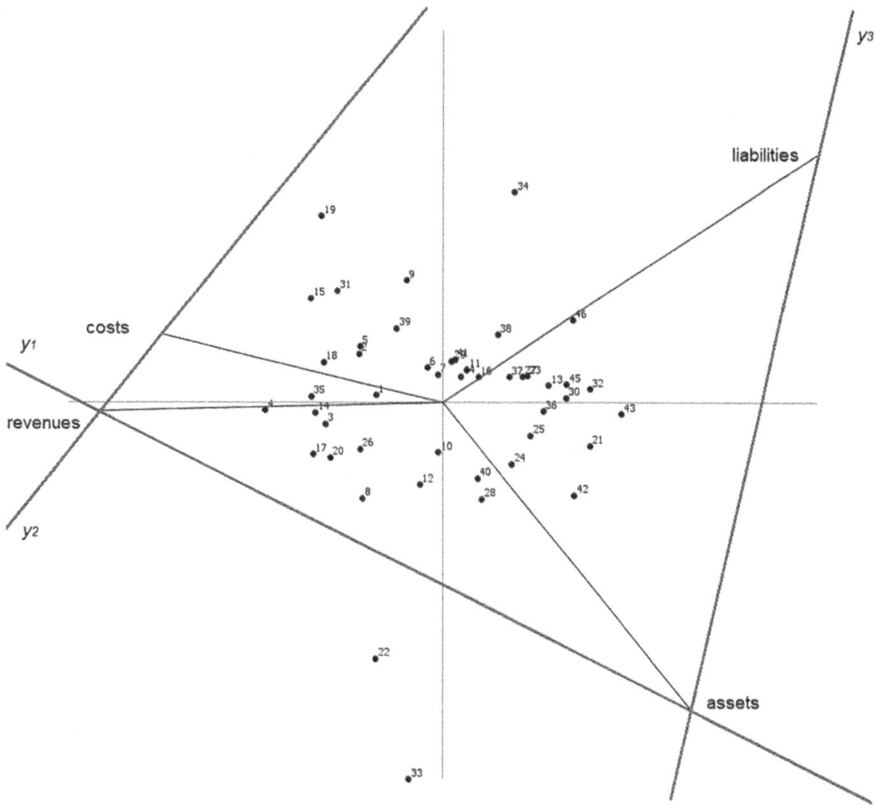

Fig. 1 Compositional biplot with relevant ratios between pairs of account categories

Table 2 Cluster and overall geometric averages normalized to unit sum

Cluster	n	Revenues	Costs	Liabilities	Assets
1	2	0.064	0.029	0.003	0.904
2	25	0.062	0.079	0.366	0.493
3	19	0.385	0.378	0.079	0.158
Overall	46	0.176	0.193	0.210	0.422

for strategic analysis (Carreras-Simó and Coenders 2020) as it allows researchers to quickly and visually identify the strategic choices of each individual firm competing on the basis of turnover, on the basis of margin or on the basis of leverage. Firms situated close together are similar with respect to those choices. Finally, firms situated close to the origin of coordinates behave like the 'average firm' in the industry, in terms of turnover, margin and leverage. In more precise terms, this industry average corresponds to firms whose relative importance of x_1 to x_4 is equal to the centre of the composition, which is in the last row of Table 2. Examples of such are firms 4, 6 and 7.

A usual measure of biplot accuracy is the percentage of total centred log-ratio variance explained by the two biplot dimensions, which in our application is very high at 96.5%.

6 CoDa Clustering as a Tool to Find Strategic Groups in DuPont Analysis

Compositional cluster analysis can be used to identify groups of firms with similar financial performance or distress profiles (Linares-Mustarós et al. 2018), in the DuPont analysis case, strategic groups competing with similar mixes of turnover, margin and leverage. The centred log-ratios in Eq. (15) ensure the meaningfulness of Euclidean distances among firms. Thus, standard clustering methods based on Euclidean distances such as the k-means algorithm can be routinely used after transforming the data as centred log-ratios (Pawlowsky-Glahn and Buccianti 2002; Aitchison et al. 2000; Martín-Fernández et al. 1998). The usual criteria to determine the k number of clusters may be used. For illustrative purposes, we select $k = 3$ clusters, which maximize the average silhouette width statistic at 0.487 (Kaufman and Rousseeuw 1990).

Table 2 shows the cluster and overall geometric averages of the account categories normalized (closed) to unit sum. From this table, both the standard and compositional financial ratios can be computed. Researchers who are less familiar with the CoDa methodology may prefer to interpret standard ratios and may do so in the usual manner. For instance, for cluster 1 the computations are as follows:

Table 3 Standard and compositional financial ratios computed from cluster and overall geometric averages in Eqs. (17)–(24)

Cluster	St. turnover	St. margin	St. leverage	St. ROE	y_1	y_2	y_3
1	0.071	0.547	1.003	0.039	−2.648	0.792	−5.708
2	0.126	−0.274	3.882	−0.134	−2.073	−0.242	−0.298
3	2.437	0.018	2.000	0.089	0.891	0.018	−0.693
Overall	0.417	−0.097	1.991	−0.080	−0.875	−0.092	−0.698

$$\text{Standard turnover ratio} = 0.064/0.904 = 0.071, \quad (17)$$
$$y_1 = \log\left(0.064/0.904\right) = -2.648, \quad (18)$$
$$\text{Standard margin ratio} = \left(0.064 - 0.029\right)/0.064 = 0.547, \quad (19)$$
$$y_2 = \log\left(0.064/0.029\right) = 0.792, \quad (20)$$
$$\text{Standard leverage ratio} = 0.904/\left(0.904 - 0.003\right) = 1.003, \quad (21)$$
$$y_3 = \log\left(0.003/0.904\right) = -5.708, \quad (22)$$
$$\text{ROE} = \left(0.064 - 0.029\right)/\left(0.904 - 0.003\right) = 0.039, \quad (23)$$
$$\text{ROE} = 0.071 \times 0.547 \times 1.003 = 0.039. \quad (24)$$

Results are in Table 3. The industry is composed of two major strategic groups and a smaller one. Cluster 2 (25 firms, 54%) has the highest average leverage and the lowest average margin, which is actually negative. Cluster 3 (19 firms, 41%) has the highest average turnover. The very small Cluster 1 (2 firms, 4%) has the lowest average turnover, the highest average margin and the lowest average leverage. Identical conclusions are drawn from the standard and the compositional financial ratios.

Cluster 3 has the highest ROE (8.9%) and finds itself in a very favourable financial scenario. The standard leverage ratio shows that firms in this cluster have adequate indebtedness: on average the weights of debt and equity are equal. The favourable ROE results from this cluster having the highest turnover: the standard turnover ratio shows that the average firm in this cluster is able to generate 2.437 monetary units of revenue for each asset monetary unit. This notwithstanding, the low standard margin ratio ultimately penalizes the attained ROE.

Cluster 2 contains the worst performing firms, with negative average ROE resulting from their negative average margin. Besides, it has the highest leverage, resulting from very high liabilities as compared to assets; in other words, very high indebtedness. This has two unfavourable consequences: the high resulting finance costs further reduce profit, and, being a multiplier, high leverage is detrimental to ROE when margin is negative. The relatively low turnover also reveals a low capacity of assets to generate revenue. However, the most serious problem in this cluster is the negative standard margin ratio. This cluster requires a more profound study of the cost structure, in order to determine at which level costs contribute to the negative margin. It is not the same to encounter firms whose operating costs alone already exceed revenues, as firms with operating profits burdened with finance costs. All things considered, cluster 2 represents a group of farm-tourism firms under serious financial distress.

Cluster 1 features intermediate ROE values. Firms in this cluster have virtually no debt, a very low turnover and attain their performance mainly on the basis of high margin. A close study of this cluster could provide hints and future directions for the industry, in terms of which efforts are successful in improving margin; in other words, in increasing sales while keeping costs down.

The last row of Table 3 contains the representative farm-tourism ratios at the whole industry level. Generally speaking, the farm-tourism industry has negative average margin and ROE, which is consistent with previous research in farm tourism. This notwithstanding, cluster 1 and cluster 3, which amount to nearly half of the firms, are relatively well off. The ultimate objective of cluster analysis is precisely to uncover such heterogeneous patterns within an industry. Financial diagnosis thus does not need to limit itself to the industry as a whole.

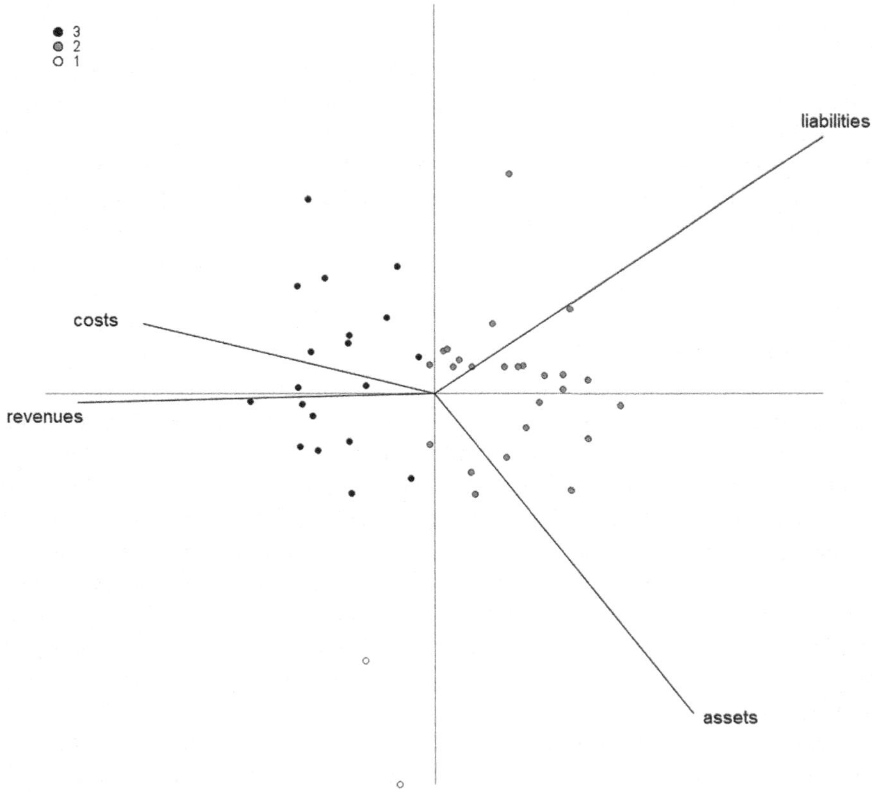

Fig. 2 Compositional biplot by cluster

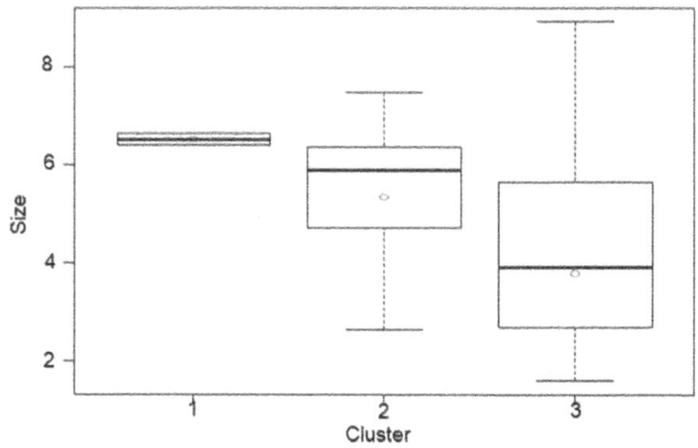

Fig. 3 Box plot of firm size (log assets) by cluster

Figure 2 shows the compositional biplot redrawn with firms labelled by cluster, which confirms the above conclusions. For instance, Cluster 1 is in the lower part of the biplot, and can be interpreted in reference to the directions defined by the log-ratios y_1, y_2 and y_3 as that with the highest margin and the lowest leverage; cluster 3 belongs to the left side of the biplot and thus corresponds to firms with the highest turnover.

Finally, clusters based on CoDa can be related to external variables in the usual manner. If we consider firm size measured, for instance, as log assets, size box plots can be drawn for each cluster separately. Figure 3 shows that the worst performing cluster 2 predominantly contains medium-sized firms. Cluster 1 contains mostly large firms and cluster 3 mostly small firms. Being a medium-sized farm thus seems to be detrimental to performance.

7 Conclusions

Compositional Data (CoDa) methods can be used to advantage in order to distil the relative information in accounts which lies at the ground of researchers' and professionals' interests when performing financial statement analysis by means of financial ratios. In this manner, researchers solve the problems reported in the literature when submitting standard financial ratios to a multivariate statistical analysis (Linares-Mustarós et al. 2018; Cowen and Hoffer 1982; Carreras-Simó and Coenders 2020; Creixans-Tenas et al. 2019). The common notion of industry or group centre has to be expressed as account geometric averages, which have the attractive property in industry and strategic group analysis that the geometric average of ratios equals the ratio between geometric averages.

Classical descriptive multivariate analysis tools such as principal component analysis, biplots and cluster analysis are meaningful and can be applied on CoDa in a completely standard manner after transforming the accounting data by means of centred log-ratios. In particular, the principal component analysis biplot visualizes all firms with respect to the ratio of any pair of account categories, and cluster analysis makes it possible to draw groups with similar financial performance profiles. Once geometric averages of accounting values have been computed for each cluster, interpretation may revert back to standard financial ratios, if the researcher so wishes, for an ease of communication with the accounting and management community. As a matter of fact, in classical financial statement analysis, the raw composition in Eq. (8) and Table 2 is hardly ever interpreted, so that interpreting the results back in standard financial ratios makes more sense than the usual CoDa approach of interpreting them back in the simplex (Mateu-Figueras et al. 2011).

In particular, and in spite of its age, DuPont analysis continues to be a well-liked, simple and powerful tool. Based on only four broad categories of accounts and three ratios, it decomposes the standard ratio of profitability on equity into its three key underlying concepts: turnover, margin and leverage. The biplot provides a visual representation of individual firms along directions representing each of the said key concepts while cluster analysis draws strategic groups competing on the basis of different combinations of turnover, margin and leverage. This combination of tools has enabled us to perform a fruitful and much-needed diagnosis of financial performance in the farm-tourism industry as a whole and group-wise. As a whole, the industry faces a challenging situation, having negative ROE on average. Group-wise, we encounter a large cluster amounting to more than half of the firms under severe financial distress, with negative margin and very high indebtedness.

Several extensions are possible. In the first place, both clusters and principal component dimensions can be related to external non-financial variables in the usual way (Pawlowsky-Glahn and Buccianti 2002; Egozcue and Pawlowsky-Glahn 2011) for the purpose of statistical inference, for instance, by means of multiple regression models or analysis of variance models, complementing the box plots used in this chapter. In this manner, researchers can find, for instance, if younger and older firms have similar or different turnover, margin and leverage, or if multinational firms tend to belong to a different cluster from domestic firms. Secondly, any positive financial statement accounts and any ratios can be submitted to a biplot or to a cluster analysis. Analyses involving a larger number of more detailed account categories than those involved in DuPont analysis are indeed possible. Thirdly, if data are available for several years, firm trajectories can be depicted in the biplot (Carreras-Simó and Coenders 2020) and group transitions can be studied in the cluster analysis. Finally, if the researcher is not only interested in the relative information contained in financial ratios, firm size, measured for instance as total assets or as total revenues, can be combined with compositional financial ratios as an additional variable in the analysis rather than as an external variable. This is known as *compositional data with a total* (Pawlowsky-Glahn et al. 2015b; Ferrer-Rosell et al. 2016). Along similar lines, non-financial non-negative variables like the number of employees can be included as an additional part in the composition, and thus participate in the log-ratios of interest, indicating, for instance, revenue per employee, assets per employee or labour cost per employee (Linares-Mustarós et al. 2018).

Acknowledgements This work was supported by the Spanish Ministry of Science, Innovation and Universities/FEDER (grant RTI2018–095518–B–C21), the Spanish Ministry of Health (grant CIBERCB06/02/1002) and the Catalan Government (grants 2017SGR656, 2017SGR386 and 2017SGR155).

References

J. Aitchison, Principal component analysis of compositional data. Biometrika **70**(1), 57–65 (1983)

J. Aitchison, M. Greenacre, Biplots of compositional data. J. Roy. Stat. Soc. C App. **51**(4), 375–392 (2002)

J. Aitchison, C. Barceló-Vidal, J.A. Martín-Fernández, V. Pawlowsky-Glahn, Logratio analysis and compositional distances. Math. Geol. **32**(3), 271–275 (2000)

T. Baležentis, A. Galnaitytė, I. Kriščiukaitienė, V. Namiotko, L. Novickytė, D. Streimikiene, R. Melnikiene, Decomposing dynamics in the farm profitability: An application of index decomposition analysis to Lithuanian FADN sample. Sustainability **11**(10), 2861 (2019)

P. Barnes, The analysis and use of financial ratios: a review article. J. Bus. Finan. Account. **14**(4), 449–461 (1987)

J. Belles-Sampera, M. Guillén, M. Santolino, Compositional methods applied to capital allocation problems. J. Risk **19**(2), 15–30 (2016)

C. Bilbao, L. Valdés, Evaluation of the profitability of quality labels in rural tourism accommodation: a hedonic approach using propensity score matching. Appl. Econ. **48**(34), 3253–3263 (2016)

T. Boonen, M. Guillén, M. Santolino, Forecasting compositional risk allocations. Insur. Math. Econ. **84**, 79–86 (2019)

A. Buccianti, G. Mateu-Figueras, V. Pawlowsky-Glahn, *Compositional Data Analysis in the Geosciences: From Theory to Practice* (Geological Society, London, 2006)

M. Carreras-Simó, G. Coenders, Principal component analysis of financial statements. A compositional approach. Rev. Métodos Cuant. Econ. Empresa **29**, 18–37 (2020)

C. Castro, Rural tourism in Northern Portugal-Motivations and barriers. RITUR **9**(2), 217–240 (2019)

K.H. Chen, T.A. Shimerda, An empirical analysis of useful financial ratios. Financ. Manag. **10**(1), 51–60 (1981)

L. Chen, S. Wang, Z. Qiao, DuPont model and product profitability analysis based on activity-based costing and economic value added. Eur. J. Bus. Manag. **6**(30), 25–35 (2014)

G. Coenders, B. Ferrer-Rosell, Compositional data analysis in tourism. Review and future directions. Tour. Anal. **25**(1), 153–168 (2020)

A.C. Costa, Influence of sociodemographic characteristics on the financial performance of micro and small businesses in rural tourism. Espacios **39**(14), 12 (2018)

S.S. Cowen, J.A. Hoffer, Usefulness of financial ratios in a single industry. J. Bus. Res. **10**, 103–118 (1982)

J. Creixans-Tenas, G. Coenders, N. Arimany-Serrat, Corporate social responsibility and financial profile of Spanish private hospitals. Heliyon **5**(10), e02623 (2019)

E. Dale, R.S. Greenwood, R.G. Greenwood, Donaldson brown: GM's pioneer management theorist and practioner. Acad. Manag. Proc. **1980**(1), 119–123 (1980)

B.C. Davis, K.M. Hmieleski, J.W. Webb, J.E. Coombs, Funders' positive affective reactions to entrepreneurs' crowdfunding pitches: the influence of perceived product creativity and entrepreneurial passion. J. Bus. Ventur. **32**(1), 90–106 (2017)

M. Doorasamy, Using DuPont analysis to assess the financial performance of the top 3 JSE listed companies in the food industry. Invest. Manag. Financ. Innov. **13**(2), 29–44 (2016)

J.J. Egozcue, V. Pawlowsky-Glahn, Basic concepts and procedures, in *Compositional Data Analysis. Theory and Applications*, eds. by V. Pawlowsky-Glahn, A. Buccianti (Wiley, New York, 2011)

J.J. Egozcue, V. Pawlowsky-Glahn, What are compositional data and how should they be analyzed? BEIO **32**(1), 5–29 (2016)

J.J. Egozcue, V. Pawlowsky-Glahn, Compositional data: the sample space and its structure. TEST **28**(3), 599–638 (2019)

J.J. Egozcue, V. Pawlowsky-Glahn, G. Mateu-Figueras, C. Barceló-Vidal, Isometric logratio transformations for compositional data analysis. Math. Geol. **35**(3), 279–300 (2003)

Eurostat, NACE Rev. 2 – Statistical classification of economic activities in the European Community. Eurostat methodologies and working papers. European Communities, Luxembourg (2008)

J. Faello, Understanding the limitations of financial ratios. Acad. Account. Financ. Stud.J. **19**(3), 75–85 (2015)

B. Ferrer-Rosell, G. Coenders, G. Mateu-Figueras, V. Pawlowsky-Glahn, Understanding low cost airline users' expenditure patterns and volume. Tour. Econ. **22**(2), 269–291 (2016)

M. Greenacre, Variable selection in compositional data analysis using pairwise logratios. Math. Geosci. **51**(5), 649–682 (2019)

J.O. Horrigan, A short history of financial ratio analysis. Account. Rev. **43**, 284–294 (1968)

A. Joueid, G. Coenders, Marketing innovation and new product portfolios. A compositional approach. J. Open Innov. Technol. Mark. Complex. **4**, 19 (2018)

A. Kallmuenzer, M. Peters, Entrepreneurial behaviour, firm size and financial performance: the case of rural tourism family firms. Tour. Recreat. Res. **43**(1), 2–14 (2018)

L. Kaufman, P.J. Rousseeuw, *Finding Groups in Data: An Introduction to Cluster Analysis* (Wiley, New York, 1990)

P. Kokoszka, H. Miao, A. Petersen, H.L. Shang, Forecasting of density functions with an application to cross-sectional and intraday returns. Int. J. Forecast. **35**(4), 1304–1317 (2019)

B. Lev, S. Sunder, Methodological issues in the use of financial ratios. J. Account. Econ. **1**(3), 187–210 (1979)

S. Linares-Mustarós, G. Coenders, M. Vives-Mestres, Financial performance and distress profiles. From classification according to financial ratios to compositional classification. Adv. Account. **40**, 1–10 (2018)

W.L. Maldonado, J.J. Egozcue, V. Pawlowsky-Glahn, No-arbitrage matrices of exchange rates: some characterizations. Int. J. Econ. Theory (2021). https://doi.org/10.1111/ijet.12249

J.A. Martín-Fernández, C. Barceló-Vidal, V. Pawlowsky-Glahn, A critical approach to non-parametric classification of compositional data, in *Advances in Data Science and Classification*, eds. by A. Rizzi, M. Vichi, H.H. Bock (Springer, Berlin, 1998)

J.A. Martín-Fernández, R.A. Olea-Meneses, V. Pawlowsky-Glahn, Criteria to compare estimation methods of regionalized compositions. Math. Geol. **33**(8), 889–909 (2001)

G. Mateu-Figueras, V. Pawlowsky-Glahn, J.J. Egozcue, The principle of working on coordinates, in *Compositional Data Analysis. Theory and Applications*, eds. by V. Pawlowsky-Glahn, A. Buccianti (Wiley, New York, 2011)

R. Ortells, J.J. Egozcue, M.I. Ortego, A. Garola, Relationship between popularity of key words in the Google browser and the evolution of worldwide financial indices, in *Compositional Data Analysis*. Springer Proceedings in Mathematics & Statistics, vol. 187, eds. by J.A. Martín–Fernández, S. Thió–Henestrosa (Springer, Cham, 2016)

P.D. Parulekar, Dupont analysis for selected five nifty fifty companies. SFIMAR Res. Rev. **10**(2), 8–16 (2015)

V. Pawlowsky-Glahn, A. Buccianti, Visualization and modeling of sub-populations of compositional data: statistical methods illustrated by means of geochemical data from fumarolic fluids. Int. J. Earth Sci. **91**(2), 357–68 (2002)

V. Pawlowsky-Glahn, A. Buccianti, *Compositional Data Analysis. Theory and Applications* (Wiley, New York, 2011)

V. Pawlowsky-Glahn, J.J. Egozcue, R. Tolosana-Delgado, *Modelling and Analysis of Compositional Data* (Wiley, Chichester, 2015a)

V. Pawlowsky-Glahn, J.J. Egozcue, D. Lovell, Tools for compositional data with a total. Stat. Model **15**(2), 175–190 (2015b)

A.I. Polo Peña, D.M. Frías Jamilena, M.Á. Rodríguez Molina, The effect of customer orientation on smaller sized service firms and on the market: a multilevel application embracing firms and customers. J. Small Bus. Manag. **54**(2), 566–581 (2016)

J.C. So, Some empirical evidence on the outliers and the non-normal distribution of financial ratios. J. Bus. Finan. Account. **14**(4), 483–496 (1987)

M.T. Soliman, The use of DuPont analysis by market participants. Account Rev. **83**(3), 823–853 (2008)

M.T. Tascón, F.J. Castaño, P. Castro, A new tool for failure analysis in small firms: frontiers of financial ratios based on percentile differences (PDFR). Span. J. Financ. Account. **47**(4), 433–463 (2018)

R. Verbelen, K. Antonio, G. Claeskens, Unravelling the predictive power of telematics data in car insurance pricing. J. Roy. Stat. Soc. C App. **67**(5), 1275–1304 (2018)

A. Voltes-Dorta, J.L. Jiménez, A. Suárez-Alemán, An initial investigation into the impact of tourism on local budgets: a comparative analysis of Spanish municipalities. Tour. Manag. **45**, 124–133 (2014)

H. Wang, S. Lu, J. Zhao, Aggregating multiple types of complex data in stock market prediction: a model-independent framework. Knowl. Based Syst. **164**(15), 193–204 (2019)

J. Willer do Prado, V. de Castro Alcântara, F. de Melo Carvalho, K. Carvalho Vieira, L.K. Cruz Machado, D. Flávio Tonelli, Multivariate analysis of credit risk and bankruptcy research data: a bibliometric study involving different knowledge fields (1968–2014). Scientometrics **106**(3), 1007–1029 (2016)

S. Ye, H. Xiao, L. Zhou, Small accommodation business growth in rural areas: effects on guest experience and financial performance. Int. J. Hosp. Manag. **76**, 29–38 (2019)

Spatial Statistics for Distributional Data in Bayes Spaces: From Object-Oriented Kriging to the Analysis of Warping Functions

Alessandra Menafoglio

This work is dedicated to Vera Pawlowsky-Glahn whose work is far reaching, and inspiring for an entire generation of researchers.

Abstract In the presence of increasingly massive and heterogeneous spatial data, geostatistical modeling of distributional observations plays a key role. Choosing the "right" embedding space for these data is of paramount importance for their statistical processing, to account for their nature and inherent constraints. The Bayes space theory is a natural embedding space for (spatial) distributional data and was successfully applied in varied settings. The aim of this work is to review the state-of-the-art methods for spatial dependence modeling and prediction of distributional data, while shedding light on the strong links between Compositional Data Analysis, Functional Data Analysis, and, more generally, Object-Oriented Data Analysis, in the context of spatial statistics. We propose extensions of these methods to the multivariate setting, and discuss the applicability of the Bayes space approach to the spatial modeling of phase variability in Functional Data Analysis.

1 Introduction

Modern geoscience applications rely on increasingly massive and heterogeneous spatial data. Even though several field studies are still characterized by a rather sparse spatial sampling—constrained by geographical and economic resources—the *nature* of available records at the sampling sites is nowadays increasingly complex and often high dimensional.

A. Menafoglio (✉)
MOX, Department of Mathematics, Politecnico di Milano, Milano, Italy
e-mail: alessandra.menafoglio@polimi.it

P. Filzmoser et al. (eds.), *Advances in Compositional Data Analysis*,
https://doi.org/10.1007/978-3-030-71175-7_11

In this broad framework, when massive or heterogeneous records are available at the sites, raw data may be represented more effectively and efficiently through *summaries* than by the complete collection of single data values. This is the case of water-resources monitoring, where long time series of records of geochemical concentrations often need to be effectively and exhaustively summarized to allow for decision-making, protection, and planning by the governments and local authorities. Another instance is the case of soil particle-size data, where the interest is usually posed on the overall density of the grain sizes, rather than on the dimension of the single grains constituting the sample. These data are then used, e.g., for soil-type classification or to infer permeability or porosity of the medium under study. As a matter of fact, in a world where information is becoming complex, abundant, but spatially sparse, geostatistical modeling of distributional data has the clear potential to play a key role.

Distributional data are compositional in their nature. This is widely recognized in the multivariate context, where discrete probability densities are typically described as data conveying only relative information (Aitchison 1982; Pawlowsky-Glahn and Buccianti 2011)—i.e., compositional vectors—and thus embedded in the simplex endowed with the Aitchison geometry (Pawlowsky-Glahn and Egozcue 2001). Performing the statistical analysis in this space allows attaining the typical goals of classical multivariate (geo)statistics, while respecting the compositional nature of the data and the so-called *principles* of compositional data analysis (scale invariance, relative scale, permutation invariance, and sub-compositional coherence, see, e.g., Pawlowsky-Glahn et al. (2015)). In this setting, notable works presenting geostatistical methods for multivariate compositions are found, e.g., in Tolosana-Delgado et al. (2008, 2011), moving from the pioneering work of Pawlowsky (1984, 1989) (see (Tolosana-Delgado et al. 2019) for a recent review).

The literature focusing on methods for the analysis of datasets of continuous distributions is more recent, and stands at the very interface between compositional data analysis (CoDa), Functional Data Analysis (FDA, Ramsay and Silverman (2005)), and Object-Oriented Data Analysis (OODA, Marron and Alonso (2014)). FDA is the area of statistics which studies methods for infinite-dimensional data (i.e., *functional data*) such as curves, surfaces, or images. It is now recognized as part of OODA, a more recent area of research focused on the study of more complex data objects, such as constrained data, tensors, trees, or networks. Early works in FDA were mostly focused on developing methods for square-integrable functional data (i.e., data in the space L^2), allowing, e.g., for dimensionality reduction, inference, and prediction (Ramsay and Silverman 2005; Horváth and Kokoszka 2012), in some cases in a geostatistical framework (e.g., Delicado (2011)). These methods inevitably have strong limitations when applied to constrained data such as distributions (see, e.g., Delicado (2007, 2011); Menafoglio et al. (2014, 2018b)), as they do not account for the *nature* of the data, i.e., for their characterizing features.

In the last years, the statistical community has experienced an increased awareness of the importance of the choice of the *feature space* for the data, that is, the mathematical space where the statistical analysis is embedded. This is particularly true in the context of distributional data, where the feature space should properly represent the nature of the data, and their inherent constraints. This renovated awareness has been strongly resonating with the approach of CoDa. In fact, the strong geometrical perspective upon which these research fields are grounded has paved the way for the construction of a fruitful ground where to build sound statistical frameworks for the analysis of (spatial) distributional data.

The aim of this contribution is to shed further light on the strong links between CoDa, FDA, and, more generally, OODA, in the context of spatial statistics. The focus shall be posed on methods for dependence modeling and spatial prediction, in the presence of univariate and multivariate random fields of distributional data. To this purpose, we shall first introduce the principle and basic notions of the theory of Bayes spaces, which represent the generalization of the Aitchison geometry to the infinite-dimensional setting (Sect. 2). We shall illustrate the fundamentals of the kriging methodology to univariate distributional data in Sect. 3, and discuss its extension to the multivariate setting, with a co-kriging approach, in Sect. 4. In Sect. 5, we shall draw a connection between these geostatistical settings and a topic of key importance in the literature on FDA, which is the modeling of phase and amplitude variability in (spatial) functional data. Here, we shall show how the theory of Bayes spaces may be helpful in unexpected areas, well beyond the distributional setting.

2 Density Data as Elements of Bayes Spaces

Bayes spaces represent the generalization of Aitchison's simplex to the infinite-dimensional setting. Formally, they are spaces whose elements are σ-finite positive measures, which are assumed to convey exclusively relative information. In most applications, Bayes spaces are understood as functional spaces, whose elements are (identifiable with) probability density functions (PDFs) over a compact support I. For the sake of simplicity, in the following, we shall always assume that $I = [a, b]$ is an interval, and denote by f, g positive functions defined over I, representing the densities of absolutely continuous measures, resp., μ_f, μ_g, with respect to the Lebesgue measure. One should, however, note that the theory of Bayes spaces is much more general than that here recalled, see, e.g., van den Boogaart et al. (2010, 2014).

Bayes spaces were originally introduced in Egozcue et al. (2006) and later extended in van den Boogaart et al. (2010, 2014); they provide a mathematical and geometrical framework where to embed *sound* analyses of continuous and discrete distributional data. Their construction relies on the key observation that PDFs can be interpreted as *functional compositional data*, i.e., vectors with infinitely many infinitesimal parts. In this sense, the analysis of PDF data should follow the same key principles as CoDa. For instance, they should be *scale invariant*, i.e., their results should be invariant to a rescaling of the data. This principle yields the definition of the equivalence relation underlying Bayes spaces, according to which two elements f and g are equivalent (denoted by $f =_{\mathcal{B}} g$) if they are proportional (almost everywhere) through a positive constant c. The remaining principles of CoDa are satisfied thanks to the geometrical structure of the Bayes space, particularly the operations and the metric (Pawlowsky-Glahn et al. 2015).

In this work, we shall focus on a particular Bayes space, which is the Bayes Hilbert space $\mathcal{B}^2$ defined by Egozcue et al. (2006). It is the space of (equivalence classes of) positive functions defined on I, with square-integrable logarithm, that is,

$$\mathcal{B}^2 = \{f > 0 \text{ s.t. } \int_I [\ln f(t)]^2 dt < \infty\}.$$

Here, we remark that more general definitions of $\mathcal{B}^2$ would be possible, by extending the domain of definition of the densities to infinite supports and by considering reference measures other than the Lebesgue one (see, e.g., van den Boogaart et al. (2014); Talská et al. (2020)). For the sake of simplicity in notation, given an equivalence class in $\mathcal{B}^2$, we shall always refer to the representative which integrates to unity (if well defined)—and take the representative for the equivalent class itself, unless the distinction is needed.

Given two elements f, g of $\mathcal{B}^2$, and a scalar $\alpha \in \mathbb{R}$, the operations in $\mathcal{B}^2$ are defined as perturbation and powering as

$$(f \oplus g)(t) = \frac{f(t)g(t)}{\int_I f(\tau)g(\tau)d\tau}, \quad (\alpha \odot f)(t) = \frac{(f(t))^\alpha}{\int_I (f(\tau))^\alpha d\tau}, \quad t \in I. \tag{1}$$

It is immediate to verify that the origin of the space—and neutral element of perturbation—is the uniform density $e(t) = 1/(b-a)$. More in general, the origin of a Bayes space is the reference measure set to define the space itself. It is worth noting that the operations defined in (1) allow for an interesting interpretation in terms of information theory. In the light of the Bayes theorem, perturbation is figured as a sum of information—where the information included in f is added to the information included in g, and vice versa. Similarly, for powering, where the information in f is *inflated* by a constant value α. More generally, interpretations in terms of information theory as those provided in Egozcue and Pawlowsky-Glahn (2018) for vector data are indeed possible in the Bayes space. These were partially explored in Egozcue et al. (2013), where the authors investigate with particular emphasis on the exponential families. Note that the latter families are conjugate priors in a Bayesian setting. Thus, it should not come as a surprise that exponential families were proved to represent affine subspaces of Bayes spaces. This property was further explored in Hron et al. (2016), in the context of principal component analysis of density data.

Egozcue et al. (2006) proved that operations $(\oplus, \odot)$ defined in (1) equip the space $\mathcal{B}^2$ with a vector structure. On $\mathcal{B}^2$ one may, however, define a Hilbert space structure by considering the inner product

$$\langle f, g \rangle_{\mathcal{B}} = \frac{1}{2(b-a)} \int_I \int_I \ln \frac{f(t)}{f(s)} \ln \frac{g(t)}{g(s)} dt ds, \quad f, g \in \mathcal{B}^2, \tag{2}$$

which induces the norm

$$\|f\|_{\mathcal{B}} = \left[\frac{1}{2(b-a)}\int_I\int_I\left(\ln\frac{f(t)}{f(s)}\right)^2 dtds\right]^{1/2}, \quad f, g \in \mathcal{B}^2. \qquad (3)$$

A closer inspection to the mathematical form of (2)–(3) shows that these definitions are strictly adherent to the *principles* of CoDa. Indeed, they only depend on the *log-ratios* among (infinitesimal) components $\ln(f(t)/f(s))$, $\ln(g(t)/g(s))$, which form the only relevant information when analyzing data in $\mathcal{B}^2$.

The space $\mathcal{B}^2$ being a Hilbert space, most methods of multivariate statistics based on a geometrical perspective can be directly extended. Relevant research efforts in this direction have been made in the recent years, particularly in terms of developing methods for dimensionality reduction (simplicial functional principal component analysis, Hron et al. (2016)), testing (Pini et al. 2018), regression (Talská et al. 2018), and quality control (Menafoglio et al. 2018b). All these methods can typically be formulated directly in the geometry of the Bayes Hilbert space. However, for convenience of computation, one may resort to *log-ratio transformations*, that allow mapping the space $\mathcal{B}^2$ in (a subspace of) the space of square-integrable functions L^2, where most computational methods of FDA have been developed. Among these transformations, the one by far most used is the *centered log-ratio transformation* (clr), which is defined, for f in $\mathcal{B}^2$, as

$$\text{clr}(f)(t) = f_c(t) = \ln f(t) - \frac{1}{b-a}\int_I \ln f(\tau)d\tau. \qquad (4)$$

The clr transformation defines an isometric isomorphism (i.e., a bijective map which preserves distances and angles) between the space $\mathcal{B}^2$ and the space L_0^2 of square-integrable functions with zero-integral. It is indeed immediate to verify that (a) $\int_I clr(f)(t)dt = 0$ for any $f \in \mathcal{B}^2$ (zero-integral), and that (b) operations and inner products in $\mathcal{B}^2$ are preserved under the clr transformation (isometric property), i.e.,

$$\text{clr}(f \oplus g)(t) = f_c(t) + g_c(t), \quad \text{clr}(\alpha \odot f)(t) = \alpha \cdot f_c(t), \quad t \in I$$
$$\langle f, g\rangle_{\mathcal{B}} = \langle f_c, g_c\rangle_{L^2} = \int_I f_c(t)g_c(t)dt.$$

Note that the zero-integral constraint of clr is similar as that characterizing the multivariate definition of centered log-ratio transformation (see, e.g., Pawlowsky-Glahn et al. (2015)). In particular, it may cause model singularities. One may otherwise use different transformations, such as an isometric log-ratio (ilr) transformation defined by the orthogonal projection of the element f over any orthonormal basis of $\mathcal{B}^2$ (e.g., defined by the dataset principal components). Although basis projections are well defined in $\mathcal{B}^2$ and suitable bases have been studied and successfully used in recent works (e.g., (Egozcue et al. 2006; Hron et al. 2016; Menafoglio et al. 2016b, 2018b)), we shall not discuss this aspect here, as the class of methods discussed in this work is only laterally affected by this issue.

3 Geostatistical Analysis of Densities in Bayes Spaces

Whenever data are spatially distributed, suitable paradigms of analysis need to be developed able to take advantage of the spatial dependence among data for modeling and prediction purposes. In this section, we recall some key ideas of spatial modeling in the presence of distributional data in Bayes space, which can be seen as generalization stemming from the pioneering works of Pawlowsky (1989).

We denote by $\mathcal{X}_s$ a distributional observation at location s in the study domain $D \subset \mathbb{R}^d$. For instance, the element $\mathcal{X}_s$ may represent the particle-size density at a location s in an aquifer domain D as in Menafoglio et al. (2014, 2016c) and Fig. 1a. In (Menafoglio et al. 2018a), the data object $\mathcal{X}_s$ refers to the PDF of the concentration of oxygen dissolved in water at a monitoring site of an estuarine system, as estimated from a 20-year time series (Fig. 1b). The observation $\mathcal{X}_s$ shall be here assumed to be a random element taking values in the space $\mathcal{B}^2$.

Given a distributional dataset $\mathcal{X}_{s_1}, \ldots, \mathcal{X}_{s_n}$ observed over n sampling locations $s_1, \ldots, s_n$, the key goal of geostatistics is typically twofold, namely, (a) estimate the dependence among observations at different locations and (b) perform prediction at unsampled locations in the system. In compositional geostatistics, these goals can be attained by first modeling a set of variograms and cross-variograms suitably defined from data log-ratio transformations, and then consistently defining linear unbiased predictors from the data (i.e., compositional (co)kriging, see Tolosana-Delgado et al. (2019)). In Bayes spaces, one can proceed according to two main approaches, which are described hereafter.

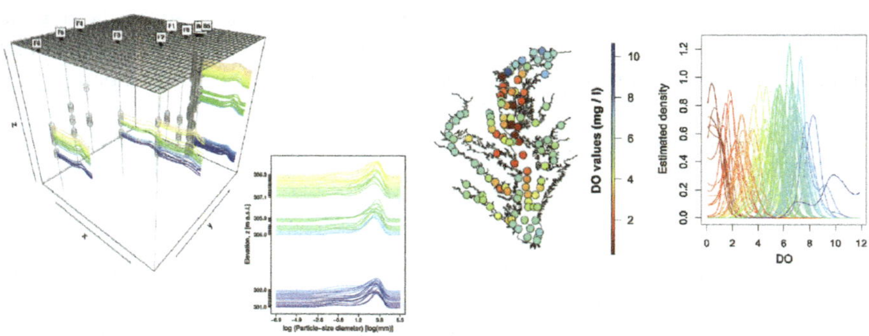

Fig. 1 Examples of distributional data in the geosciences. **a** Particle-size densities in a 3D aquifer domain, analyzed in Menafoglio et al. (2014, 2016b, c). Colors are given according to the elevation [m a.s.l.] of the sampling site. **b** Densities of dissolved oxygen in the Chesapeake Bay, as analyzed by Menafoglio et al. (2018a): sample sites (left) and densities (right); colors are given according to the mean value of dissolved oxygen at the site

3.1 First Project, Then Analyze

The first possible approach to the analysis of spatial distributional data grounds on a preliminary dimensionality reduction. Data are first projected over an orthonormal, finite-dimensional basis (e.g., through principal component analysis), yielding the representation

$$\mathcal{X}_{s_i} = \sum_{k=1}^{K} x_k(s_i)\phi_k, \quad i = 1, \ldots, n,$$

where $x_k(s_i)$ are coordinates over the basis and $\{\phi_k, k = 1, \ldots, K\}$ is an orthonormal basis of $\mathcal{B}^2$, truncated to a suitable order K. The data are then expressed through the coefficient vectors $\boldsymbol{x}(s_i) = (x_1(s_i), \ldots, x_K(s_i))'$ and analyzed through classical multivariate geostatistical methods (Chilès and Delfiner 1999). This approach has a flavor similar to the ilr transformation used in the multivariate context. However, unlike classical ilr, it suffers from one major drawback, that is, the need to choose a truncation order K for the basis representation. In principle, the data being infinite dimensional, any truncation order would imply a loss of information. In practice, data can never be represented as infinite dimensional within a finite memory, which implies that the truncation order can be set to a finite threshold without any loss of information. In fact, $K = n - 1$ principal components are sufficient to perfectly represent a dataset of n distributional data (see, e.g., Horváth and Kokoszka (2012)). Nonetheless, the choice of K is affected by a trade-off between the ability to well represent the data (attained for high values of K), and the computational resources available to perform the multivariate geostatistical analysis of vectors $\boldsymbol{x}(s_i)$. In fact, when n is moderately large, using $K = n - 1$ is impracticable from the modeling viewpoint, inevitably entailing a limitation in the information which can be accounted for in the analysis.

3.2 Stay in the (Infinite) Simplex

A second possible approach to the problem can be pursued by following the paradigm of Object-Oriented Spatial Statistics (O2S2, see Menafoglio and Secchi (2017) and Menafoglio and Secchi (2019) for recent reviews). This is a methodological area recently developed within OODA, which precisely aims to develop data analytics tools for complex data with a spatial dependence. Within this framework, a new class of geostatistical methods was developed, mostly of which do not require any dimensionality reduction to perform the analysis. We shall here briefly recall the main methodological concepts underpinning the theory of spatial modeling and kriging developed in O2S2, which are instrumental to the extensions presented in Sect. 4.

In the context of O2S2, the dataset $\mathcal{X}_{s_1}, \ldots, \mathcal{X}_{s_n}$ is modeled as a partial observation of a random field $\{\mathcal{X}_s, s \in D\}$ valued in $\mathcal{B}^2$. The first-order properties of the field are described by the *mean* of $\{\mathcal{X}_s, s \in D\}$, which is defined as, in the Fréchet sense,

$$m_s = \arg\min_{x \in \mathcal{B}^2} \|\mathcal{X}_s \ominus x\|^2_{\mathcal{B}}, \quad s \in D, \tag{5}$$

where the operation $\ominus$ is defined as $f \ominus g = f \oplus [(-1) \odot g]$ for f, g in $\mathcal{B}^2$. In a stationary random field, the mean m_s does not depend on the spatial location s. Note that the definition of mean given in (5) clearly recalls the concept of compositional center discussed in Pawlowsky-Glahn and Egozcue (2002), but now extended to a more general setting.

To define a measure of spatial dependence for the field $\{\mathcal{X}_s, s \in D\}$, one may resort to an extension of the classical variogram, which is called *trace-variogram* and is defined, under stationarity and isotropy conditions, as

$$2\gamma(\|s_1 - s_2\|_d) = \mathbb{E}[\|\mathcal{X}_{s_1} \ominus \mathcal{X}_{s_2}\|^2_{\mathcal{B}}], \tag{6}$$

for s_1, s_2 in $D \subset \mathbb{R}^d$ and $\|s_1 - s_2\|_d$ denoting the distance between s_1 and s_2. It is worth noting that definition (6) can be restated in terms of the *metric variance* of the increment $\mathcal{X}_{s_1} \ominus \mathcal{X}_{s_2}$ around the origin, in the same flavor as Aitchison (2001); Pawlowsky-Glahn and Egozcue (2002). Extending the definition given in these works to our setting, the metric variance of a random element $\mathcal{X}$ in $\mathcal{B}^2$ with respect to the center ξ is given by

$$\text{Mvar} = \mathbb{E}[\|\mathcal{X} \ominus \xi\|^2_{\mathcal{B}}],$$

and defines a *global* notion of spread for the element $\mathcal{X}$ with respect to the center ξ. The trace-variogram defined in (6) then defines a *global* notion of spatial dependence, which is characterized by similar properties as its scalar counterpart (e.g., positivity, conditional negative semi-definiteness, see (Menafoglio et al. 2013)). It is also similarly interpreted. For instance, a stationary field is characterized by a trace-variogram stabilizing around an asymptote for diverging distances, whereas the distance at which the trace-variogram reaches the sill defines the range of influence of the data. Alternatively, one may define a trace-covariogram as

$$C(\|s_1 - s_2\|_d) = \mathbb{E}[\langle \mathcal{X}_{s_1} - m_{s_1}, \mathcal{X}_{s_2} - m_{s_2} \rangle_{\mathcal{B}}], \tag{7}$$

which is consistent with trace-variogram (6), and fully analog to the classical covariogram (a.k.a. covariance function, Cressie (1993)).

In practice, the use of variogram is preferred over the covariogram, but it is rarely a priori known. More often, it needs to be estimated from the data. For this purpose, one may use an empirical estimator

$$2\widehat{\gamma}(h) = \frac{1}{|N(h)|} \sum_{(i,j) \in N(h)} \|\mathcal{X}_{s_i} \ominus \mathcal{X}_{s_j}\|^2_{\mathcal{B}},$$

where $N(h)$ stands for the set of data pairs separated approximately by a distance h, and $|N(h)|$ for the cardinality of this set. To the empirical estimate $2\widehat{\gamma}(h)$, one can then fit any of the widely used parametric variogram model (spherical, exponential, Matérn, see, e.g., Chilès and Delfiner (1999)).

Having estimated the trace-variogram, prediction of an unobserved element $\mathcal{X}_{s_0}$ at a target site s_0 can be performed by a kriging predictor in $\mathcal{B}^2$ (a.k.a. *object-oriented kriging*), as proposed by Menafoglio et al. (2013, 2014). This is found as the best linear combination of the observations $\bigoplus_{i=1}^{n} \lambda_i \odot \mathcal{X}_{s_i}$ subject to unbiasedness. That is, the optimal scalar weights $\lambda_1^*, \ldots, \lambda_n^*$ are found by solving the optimization problem

$$\min_{\lambda_1,\ldots,\lambda_n \in \mathbb{R}} \mathbb{E}\left[\left\| \mathcal{X}_{s_0} \ominus \bigoplus_{i=1}^{n} \lambda_i \odot \mathcal{X}_{s_i} \right\|_{\mathcal{B}}^2\right] \quad \text{subject to} \quad \mathbb{E}\left[\bigoplus_{i=1}^{n} \lambda_i \odot \mathcal{X}_{s_i}\right] = m_{s_0}. \tag{8}$$

Note that the kriging predictor $\mathcal{X}_{s_0}^* = \bigoplus_{i=1}^{n} \lambda_i^* \odot \mathcal{X}_{s_i}$ defined by (8) is optimal in the sense of minimization of the metric variance (in $\mathcal{B}^2$) of the prediction error, under unbiasedness. It is easy to prove that problem (8) admits a unique solution, which is found by solving a linear system analog to that of the scalar case, provided that the trace-variogram is used in place of the usual variogram; for further details, we refer the reader to Menafoglio et al. (2014).

For the scope of this work, it is worth noting that the framework discussed in this section is valid not only for data embedded in $\mathcal{B}^2$, but also for any feature space $\mathcal{H}$ equipped with a separable Hilbert space structure $(+, \cdot, \langle \cdot, \cdot \rangle_{\mathcal{H}})$, either finite or infinite dimensional. In particular, this framework may also be used for datasets of georeferenced p-parts compositions, with $p < \infty$. In this case, the feature space $\mathcal{H}$ would be defined as the p-dimensional simplex S^p, equipped with the Aitchison geometry, which is indeed a finite-dimensional Hilbert space. In this view, the two approaches discussed in Sects. 3.1 and 3.2 can also be seen as alternative viewpoints for the analysis of multivariate spatial compositional data.

A natural question which arises is whether any of the two approaches here detailed would be preferable in the finite-dimensional case, and which are the drawbacks of either approaches. As far as the dimensionality of the problem is concerned (see the discussion in Sect. 3.1), the case of S^p is clearly less critical, as typical applications would not require to perform a prior dimensionality reduction (i.e., to set $K < p$). Still, this would not be the case for very high-dimensional problems—which are encountered in geochemical applications—for which retaining the full information content in the data (i.e., setting $K = p$) may represent a computational and modeling challenge. On the other hand, it is possible to show that the form of the linear predictor obtained by co-kriging the ilr coordinates (i.e., the approach discussed in Sect. 3.1) is more general than the linear combination of the observations advocated by the O2S2 approach, entailing potential better results—in terms of prediction error—of the former method over the latter. This is only partially true in the infinite-dimensional context, where Menafoglio and Petris (2016) proved that, under some conditions, the two approaches are mathematically equivalent. In this context, a number of simulation studies showed the practically equivalent performances of the two methods (see, e.g., Menafoglio et al. (2016a)) in terms of prediction errors—with a sensible gain in terms of computational and modeling effort for the O2S2 approach.

4 Toward Multivariate Distributional Geostatistics

The aim of this section is to discuss on the multivariate extensions of the frameworks introduced in Sect. 3. Here, at least two notions of *multivariate* are possible: (i) the observed object $\mathcal{X}_s$ is the PDF of a p-dimensional random vector $\boldsymbol{X}$ valued in $\mathbb{R}^p$ and (ii) the object $\mathcal{X}_s = (\mathcal{X}_{s,1}, \ldots, \mathcal{X}_{s,p})$ is a collection of p univariate PDFs. In other words, case (i) consists of the analysis of joint PDFs, whereas case (ii) focuses on collections of p marginal PDFs. These cases deserve separate treatment, and will be discussed in greater details in the next subsections. Although the methodological contexts for these analyses are different, both cases are of interest from the application viewpoint. For instance, case (i) may be used to analyze the joint probabilistic structures of environmental variables. This would be of interest, e.g., in the study of the probabilistic distribution of dissolved oxygen in the Chesapeake Bay (see Fig. 1b) which is known to be closely linked to water temperature (see, e.g., Menafoglio et al. (2018c) for a spatial analysis of this association). On the other hand, case (ii) would be the reference framework for the analysis of population pyramids, which describe the age distribution of male and female populations in a given administrative unit; this kind of data was analyzed, e.g., in Delicado (2011), Hron et al. (2016).

4.1 Analyzing Joint PDFs

From the mathematical viewpoint, the object $\mathcal{X}_s$ of the analysis of type (i) is a probability density function defined over a p-dimensional domain. Thus, the element $\mathcal{X}_s$ can be embedded in a Bayes Hilbert space $\mathcal{B}^2(\mathcal{I})$, defined over a p-dimensional compact set $\mathcal{I} \subseteq \mathbb{R}^p$. The space $\mathcal{B}^2(\mathcal{I})$ can be built similarly as done in Sect. 2, provided that a multidimensional domain is considered in each definition. In particular, $\mathcal{B}^2(\mathcal{I})$ is defined as the space of (equivalence classes of) positive functions with logarithm in $L^2(\mathcal{I})$, equipped with operations

$$(f \oplus g)(\mathbf{t}) =_{\mathcal{B}} f(\mathbf{t})g(\mathbf{t}); \quad (\alpha \odot f)(\mathbf{t}) =_{\mathcal{B}} f(\mathbf{t})^{\alpha}, \quad \mathbf{t} = (t_1, \ldots, t_p) \in \mathcal{I},$$

for f, g in $\mathcal{B}^2(\mathcal{I})$, and α in $\mathbb{R}$, and inner product

$$\langle f, g \rangle_{\mathcal{B}} = \frac{1}{2|\mathcal{I}|} \int_{\mathcal{I}} \int_{\mathcal{I}} \ln \frac{f(\mathbf{t})}{f(\boldsymbol{\tau})} \ln \frac{g(\mathbf{t})}{g(\boldsymbol{\tau})} d\mathbf{t} d\boldsymbol{\tau},$$

$|\mathcal{I}|$ denoting the Lebesgue measure of $\mathcal{I}$. Analyses of datasets $\mathcal{X}_{s_1}, \ldots, \mathcal{X}_{s_n}$ in $\mathcal{B}^2(\mathcal{I})$ can thus be performed by simply using one of the frameworks detailed in Sect. 3, provided that the geometry of $\mathcal{B}^2(\mathcal{I})$ is used instead. Variogram modeling and kriging is possible in this context, following the same steps as the case of $\mathcal{B}^2$.

We remark that, although the spatial analysis may be framed within the same type of mathematical setting as the univariate case, joint PDFs are characterized by very peculiar features. For instance, the association between the underlying variables here plays a key role in terms of inferential results and interpretations, which is only partially highlighted by the O2S2 theory developed so far. A decomposition of bivariate PDFs in terms of interaction and independent parts has been recently proposed in Hron et al. (2020), extending the work of Egozcue et al. (2015) to the Bayes space setting. This latter framework allows for a powerful characterization and quantification of the association between the underlying variables through a concept of *simplicial deviance*, fully compatible with the Bayes theory. These recent developments, grown on the fruitful ground of CoDa, are seen by the author as an area of high potential for the further development of O2S2, as they would allow for enhanced modeling and results' interpretation, in a context of great data complexity.

4.2 Multivariate Marginal PDFs

When analyses of collections of p univariate PDFs are concerned, an extension of the framework introduced in Sect. 3 is required to account for the *multivariate* structure of the data. A similar extension was considered in FDA by Grujic et al. (2018), with application to multi-fidelity modeling in computer experiments with functional response (in L^2). We shall consider here a similar extension, but focused on multivariate vectors of density data. The object $\mathcal{X}_s = (\mathcal{X}_{s,1}, \ldots, \mathcal{X}_{s,p})$ is here interpreted as a point in a Cartesian space $(\mathcal{B}^2)^p = \mathcal{B}^2 \times \mathcal{B}^2 \times \cdots \times \mathcal{B}^2$, which in turn shall be equipped with vector/matrix operations and an inner product, based on the geometrical structure of $\mathcal{B}^2$. For the sake of simplicity, we focus hereafter on the case $p = 2$, the case of $p > 2$ being treatable in a similar manner. We shall also assume that both elements of $\boldsymbol{f} = (f_1, f_2)$ are defined over the same domain $I = [a, b]$; unless otherwise specified, this assumption is made without loss of generality.

On $(\mathcal{B}^2)^2$, we define the operations in an element-wise fashion. Let $\boldsymbol{f} = (f_1, f_2)$, $\boldsymbol{g} = (g_1, g_2)$ be elements of $(\mathcal{B}^2)^2$, and α a scalar. We define perturbation $\oplus$ and powering $\odot$ as

$$(\boldsymbol{f} \oplus \boldsymbol{g})_k(t) = (f_k \oplus g_k)(t), \quad (\alpha \odot \boldsymbol{f})_k(t) = (\alpha \odot f_k)(t), \quad k = 1, 2,\ t \in I. \tag{9}$$

Space $(\mathcal{B}^2)^2$ equipped with operations $(\oplus, \odot)$ defined in (9) is a vector space. It is a (separable) Hilbert space, when endowed with the inner product

$$\langle \boldsymbol{f}, \boldsymbol{g} \rangle_{(\mathcal{B}^2)^2} = \sum_{k=1}^{2} \langle f_k, g_k \rangle_{\mathcal{B}}. \tag{10}$$

Alternative forms of the inner product (10) are possible, while preserving the Hilbert property of the space. For instance, one may consider a weighted sum

$$\langle \boldsymbol{f}, \boldsymbol{g} \rangle_{(\mathcal{B}^2)^2, w} = \sum_{k=1}^{2} w_k \langle f_k, g_k \rangle_{\mathcal{B}}, \tag{11}$$

where the positive weights w_i control the *importance* of the i-th component on the inner product. As a matter of fact, this could be useful when the variability (in terms of metric variance, see Sect. 3) of one component prevails on the variability of the other. In this case, including a weight in the scalar product (11) may help reducing the masking effect between the different scales of variability of the data—similarly as the standardization does in multivariate statistics. For the sake of simplicity, the following developments will be referred to the case of unweighted inner product (10), the case of weighted inner product being developed similarly.

We consider a multivariate random field $\{\mathcal{X}_s, s \in D\}$ in $(\mathcal{B}^2)^2$, and describe its first-order properties element-wise, defining a mean vector

$$m_s = (m_{s,1}, m_{s,2}),$$

whose elements $m_{s,i}$ are the metric centers of the respective univariate fields $\{\mathcal{X}_{s,k}, s \in D\}$ (as defined by (5)). Note that vector m_s is itself the metric center of the random vector $\mathcal{X}_s$, i.e.,

$$m_s = \arg \min_{x \in (\mathcal{B}^2)^2} \|\mathcal{X}_s \ominus x\|^2_{(\mathcal{B}^2)^2}, \quad s \in D.$$

Under stationarity, the mean vector m_s does not depend on the spatial coordinate s.

Similarly as in the classical setting, describing the second-order properties of the field requires defining the marginal variability of each univariate field $\{\mathcal{X}_{s,k}, s \in D\}$, $k = 1, 2$, and the cross-variability of the two fields. These can be defined by analogy with the classical case (Chilès and Delfiner 1999), and again exploiting the geometrical structure of $\mathcal{B}^2$ and $(\mathcal{B}^2)^2$. Under stationarity and isotropy of the field, we thus define the marginal trace-covariograms C_{11}, C_{22} and the cross-trace-covariogram C_{12} as

$$C_{kl}(\|s_1 - s_2\|_d) = \mathbb{E}[\langle \mathcal{X}_{s_1,k} \ominus m_{s_1,k}, \mathcal{X}_{s_2,l} \ominus m_{s_2,l} \rangle_{\mathcal{B}}], \tag{12}$$

where s_1, s_2 is any pair of locations in D and $\|s_1 - s_2\|_d$ denotes their distance. Note that, for $k = l$, definition (12) coincides with the univariate definition (7).

Denote by $s_{1,k}, \dots, s_{n_k,k}$ the set of sites where the k-th element of the vector field is observed, for $k = 1, 2$; with a slight abuse of notation, we denote by $\mathcal{X}_{s_1,k}, \dots, \mathcal{X}_{s_{n_k},k}$

the actual observations at these sites, $k = 1, 2$. In the simplest case, at each location s_i, both elements $\mathcal{X}_{s_i,1}, \mathcal{X}_{s_i,2}$ are observed (a.k.a. isotopic data sampling); in this case, $n_1 = n_2 = n$ and the sampled sites $s_{i,1}, s_{i,2}$ coincide for $i = 1, \ldots, n$. However, the setting here described is also applicable to the heterotopic data sampling (i.e., if $s_{i,1} \neq s_{i,2}$, or $n_1 \neq n_2$).

To predict the unobserved element $\mathcal{X}_{s_0,k}$ at the target site s_0, we consider a co-kriging predictor in $\mathcal{B}^2$. It is the best linear unbiased predictor, among the linear predictors of the form $\bigoplus_{l=1}^{2} \bigoplus_{i=1}^{n_l} \lambda_{li} \odot \mathcal{X}_{s_i,l}$. The optimal weights λ_{li}^*, $l = 1, 2$, $i = 1, \ldots, n_l$ are found by minimizing, under unbiasedness, the metric variance of the prediction error around the origin, i.e., by solving

$$\min_{\lambda_{l,i} \in \mathbb{R}, l=1,2, i=1,\ldots,n_l} \quad \mathbb{E}\left[\left\| \mathcal{X}_{s_0} \ominus \bigoplus_{l=1}^{2} \bigoplus_{i=1}^{n_l} \lambda_{li} \odot \mathcal{X}_{s_i,l} \right\|_{\mathcal{B}}^2\right]$$

$$\text{subject to} \qquad \mathbb{E}\left[\bigoplus_{l=1}^{2} \bigoplus_{i=1}^{n_l} \lambda_{li} \odot \mathcal{X}_{s_i,l}\right] = m_{s_0,k}.$$

Following the same derivations as described by Grujic et al. (2018), it is easy to prove that the unique solution of problem (13) is found by solving the following linear system:

$$\begin{aligned}
&\sum_{l=1}^{2} \sum_{i=1}^{n_l} \lambda_{li} C_{ll'}(\|s_{i,l} - s_{i',l'}\|_d) + \mu_{l'} = C_{l'k}(\|s_{i,l'} - s_0\|_d), \quad l' = 1, 2;\ i' = 1, \ldots, n_l; \\
&\sum_{i=1}^{n_k} \lambda_{ki} = 1; \\
&\sum_{i=1}^{n_l} \lambda_{li} = 0, \quad l \neq k;
\end{aligned} \tag{13}$$

where μ_1, μ_2 are Lagrange multipliers. Note that system (13) is analogous to the finite-dimensional counterpart. Its solution yields the co-kriging prediction $\mathcal{X}_{s_0,k}^* = \bigoplus_{l=1}^{2} \bigoplus_{i=1}^{n_l} \lambda_{li}^* \odot \mathcal{X}_{s_i,l}$. Likewise, in the univariate case, the uncertainty associated with the optimal predictor $\mathcal{X}_{s_0,k}^*$ is found as the value of the objective functional (13) (i.e., the metric variance of the error) at the optimum.

It is worth noting that the framework discussed so far can be used as long as both the elements $\mathcal{X}_{s,1}, \mathcal{X}_{s,2}$ forming the vector $\mathcal{X}_s$ belong to the same space $\mathcal{B}^2$ (or more generally to the same Hilbert space). In particular, it requires the elements $\mathcal{X}_{s,1}, \mathcal{X}_{s,2}$ to be defined over the same domain I. Indeed, definition (12) is meaningful as far as the inner product between the (centered) elements $\mathcal{X}_{s,1}, \mathcal{X}_{s,2}$ is defined. Providing sound definitions of cross-variability between elements defined over different domains $I_1 \neq I_2$—or different Hilbert spaces—is still an open field of research. A solution to this issue may be to consider the same setting explored in Sect. 3.2, but embedded in the (multivariate) Hilbert space $(\mathcal{B}^2)^2$. A formal comparison between the latter setting and that detailed in this section is yet to be fully addressed in the literature. In the view of the author, this is one of the high-potential challenges at the cutting edge of CoDa and O2S2, which is still nearly unexplored.

5 Beyond Distributional Data: Modeling Phase and Amplitude in FDA

In this section, we discuss a key topic in FDA, which is the analysis of phase and amplitude variability in datasets of functional observations. Although at a first sight the problem may seem totally unrelated with the scope of this work, we shall show that the theory of Bayes spaces appears extremely useful in this context too, demonstrating a great potential impact in the geosciences.

Given a dataset of functional observations $\mathcal{Y}_1, \ldots, \mathcal{Y}_n$ defined on a domain $\mathcal{T} \subset \mathbb{R}$, $\mathcal{Y}_i : \mathcal{T} \to \mathbb{R}$, one may distinguish between two types of data variability: a *phase* variation and an *amplitude* variation. The former refers to the *horizontal* variability of the data, i.e., to the possible misalignment of the functional observations, as that shown in Fig. 2a. The latter refers to the *vertical* variability, which affects the range of values $\mathcal{Y}_i(t)$ taken by the data.

A large body of literature in FDA has been devoted to methods allowing one to decouple these two sources of variability. From a mathematical viewpoint, the problem of *alignment* (a.k.a. *registration*) of functional data consists of finding a set warping functions ω_i, $i = 1, \ldots, n$ (Fig. 2b), allowing one to transform the original misaligned data into a set of *aligned* functions $\mathcal{Z}_i = \mathcal{Y}_i(\omega_i(t))$, characterized by amplitude variability only (Fig. 2c). The problem of alignment is not easy, and affected by identifiability issues. For instance, if the functional form of the ω's is too flexible, alignment may lead to data distortions, thus influencing also their amplitude variability. These types of challenges have been thoroughly analyzed in the literature; the reader may refer, e.g., to Marron et al. (2015) and references therein.

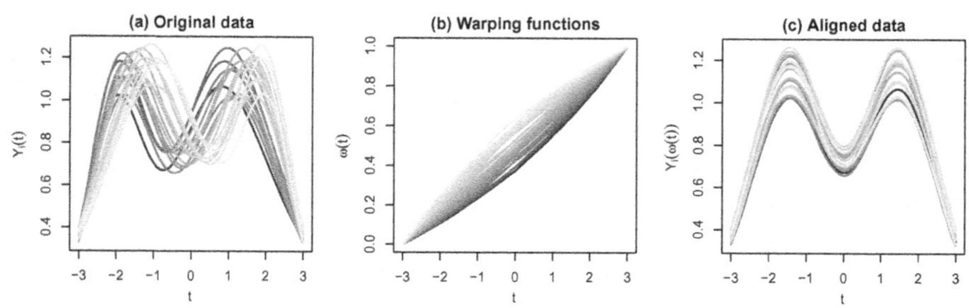

Fig. 2 Alignment in functional data analysis. **a** Functional data characterized by phase and amplitude variabilities; **b** warping functions; **c** aligned functional data. Adapted from Kurtek et al. (2011)

Early works in this field used to consider the alignment step as a preprocessing step, allowing one to get rid of the phase variability—described by the warping functions $\omega_1, \ldots, \omega_n$—and to focus on amplitude variability only. Nowadays, it is widely recognized that phase variability may be even *the* key feature of the process being studied. A notable example of this is the Berkeley growth study (Tuddenham and Snyder 1954), which concerns the analysis of height growth curves in boys and girls aged between 0 and 18 years. Here, the misalignment between curves describes a difference in the *biological clock* of the individuals being studied (see, e.g., Ramsay and Silverman 2005; Sangalli et al. 2010), which in turn is key to understand the variability of growth processes.

The increased awareness on the importance of modeling phase variability in functional data has recently motivated research efforts for the analysis of (i) datasets formed by warping functions $\omega_1, \ldots, \omega_n$ and (ii) multivariate datasets $(\omega_1, \mathcal{Z}_1), \ldots, (\omega_n, \mathcal{Z}_n)$ built upon considering phase and amplitude components jointly. Note that the analysis of datasets containing warping functions should take into account the *nature* of these data, as advocated by the geometrical view pursued along this work. Indeed, a warping function ω_i maps the domain $\mathcal{T}$ into a transformed domain $\mathcal{I}$; here, the transformed domain $\mathcal{I}$ has not a meaning in an absolute sense, but it is conventionally set to the interval $\mathcal{I} = [0, 1]$. In this case, the warping function ω_i should be a positive, smooth, and invertible function, constrained to be monotonically increasing—as to avoid artifacts or distortions on $\mathcal{Z}_i = \mathcal{Y}_i(\omega_i)$ merely induced by the alignment step. Among the broad range of literature works addressing this problem (see, e.g., (Marron et al. 2015; Srivastava et al. 2007; Srivastava and Klassen 2016) and references therein), we mention the proposals of Hadjipantelis et al. (2015) and Happ et al. (2019), as these shed further light on the potential powerful links between FDA and CoDa. These authors propose analyzing warping functions by considering their derivatives $\mathcal{X}_i = \frac{d\omega_i}{dt}$. Note that $\mathcal{X}_i$ is a positive function, constrained to integrate to unity—if the codomain of ω is $\mathcal{I} = [0, 1]$. However—borrowing a CoDa expression—any other *total* for $\mathcal{X}_i$ would provide the same type of *relative* information on the contribution of portions of the domain to the phase variability of the i-th objects. In other words, a rescaling of $\mathcal{X}_i$ by a constant c would just imply a rescaling of the interval $\mathcal{I}$ to $\mathcal{I}' = [0, c]$, without any conceptual impact on the interpretation of the warping itself. Hadjipantelis et al. (2015); Happ et al. (2019) thus proposed to analyze $\mathcal{X}_1, \ldots, \mathcal{X}_n$ as compositions, embedding them in a Bayes space. More precisely, Hadjipantelis et al. (2015) consider a discretized version of $\mathcal{X}_1, \ldots, \mathcal{X}_n$ and include them in a mixed-effect model for Mandarin Chinese phonemes, by using a discrete clr transformation. Happ et al. (2019) propose to model $\mathcal{X}_1, \ldots, \mathcal{X}_n$ as *generalized distributions* and embed them in a Bayes Hilbert space. These are then analyzed jointly with the aligned data, through a multivariate principal component analysis of the joint dataset $(clr(\mathcal{X}_1), \mathcal{Z}_1), \ldots, (clr(\mathcal{X}_n), \mathcal{Z}_n)$, developed within a Cartesian space $L^2 \times L^2$.

The bridge being built between the theory of Bayes spaces and that of functional alignment, it is natural to consider a broad spectra of applications where the analysis of phase variability could benefit from a spatial statistics viewpoint. For instance, in the Earth Sciences, Happ et al. (2019) consider a dataset of misaligned seismic records. Here, misalignment between measures provides indications on local effects of the propagation of seismic waves due to geomorphology of the study area, which are crucial to predict the reaction of buildings and infrastructures to shaking events. Embedding these types of analyses in the geostatistical framework described in Sect. 3 would allow for a spatial characterization of amplification/attenuation patterns from sampled seismic records, with great potential impact in seismic engineering. The range of applicability of this class of methods is far reaching, and not limited to seismic modeling. Other fields of potential application are found, e.g., in subsurface modeling, climatology, and oceanography, where the misalignment among functional observations indeed represent physic-driven delays in the observed phenomenon.

6 Conclusions

In this work, we discussed on spatial statistics approaches when data can be meaningfully embedded in a Bayes space. Generalizing Aitchison's simplex to the infinite-dimensional setting, spatial analysis of distributional data can borrow strength from the broad range of methodologies well established in CoDa. In this broad framework, we aimed to shed light on the points of contact between CoDa, FDA, and OODA, in the context of spatial statistics, through the foundational role of the *geometry* of the statistical analysis. The mutual relation between these methodological fields is far reaching, and may contribute to move forward the frontier of spatial statistics, not only for finite- or infinite-dimensional compositional data. Indeed, the powerful viewpoint of Bayes spaces represents a fruitful ground where to build new paradigms of analysis well beyond distributional data, as demonstrated by the analysis of warping functions described in Sect. 5. Ultimately, in the view of the author, the cross-fertilization between these communities has the clear potential to help advancing our knowledge, inspired by the sparks of original research and enlightening ideas of the pioneers of these areas.

References

J. Aitchison, The statistical analysis of compositional data. J. R. Stat. Soc. Ser. B (Methodol.) **44**(2), 139–177 (1982)

J. Aitchison, *Algebraic Methods in Statistics*, chapter Simplicial inference. M. Viana and D. Richards edition (American Mathematical Society, New York, 2001)

J.P. Chilès, P. Delfiner, *Geostatistics: Modeling Spatial Uncertainty* (Wiley, New York, 1999)

N. Cressie, *Statistics for Spatial Data* (Wiley, New York, 1993)

P. Delicado, Functional k-sample problem when data are density functions. Comput. Stat. **22**, 391–410 (2007)

P. Delicado, Dimensionality reduction when data are density functions. Comput. Stat. Data Anal. **55**(1), 401–420 (2011)

J.J. Egozcue, J.L. Díaz-Barrero, V. Pawlowsky-Glahn, Hilbert space of probability density functions based on Aitchison geometry. Acta Math. Sinica, Engl. Ser. **22**(4), 1175–1182 (2006)

J.J. Egozcue, V. Pawlowsky-Glahn, Evidence functions: a compositional approach to information. SORT: Stat. Oper. Res. Trans. **42**(2), 101–124 (2018)

J.J. Egozcue, V. Pawlowsky-Glahn, R. Tolosana-Delgado, M.I. Ortego, K.G. van den Boogaart, Bayes spaces: use of improper distributions and exponential families. Revista de la Real Academia de Ciencias Exactas, Fisicas y Naturales. Serie A. Matematicas **107**(2), 475–486 (2013)

J.J. Egozcue, V. Pawlowsky-Glahn, M. Templ, K. Hron, Independence in contingency tables using simplicial geometry. Commun. Stat. Theory Methods **44**, 3978–3996 (2015)

O. Grujic, A. Menafoglio, G. Yang, J. Caers, Cokriging for multivariate Hilbert space valued random fields: application to multi-fidelity computer code emulation. Stoch. Env. Res. Risk Assess. **32**, 1955–1971 (2018)

P.Z. Hadjipantelis, J.A.D. Aston, H.G. Müller, J.P. Evans, Unifying amplitude and phase analysis: a compositional data approach to functional multivariate mixed-effects modeling of Mandarin Chinese. J. Am. Stat. Assoc. **110**(510), 545–559 (2015)

C. Happ, F. Scheipl, A.A. Gabriel, S. Greven, A general framework for multivariate functional principal component analysis of amplitude and phase variation. Stat **8**(e220), 1–12 (2019)

L. Horváth, P. Kokoszka, *Inference for Functional Data with Applications*. Springer Series in Statistics (Springer, 2012)

K. Hron, A. Menafoglio, M. Templ, K. Hrůzová, P. Filzmoser, Simplicial principal component analysis for density functions in Bayes spaces. Comput. Stat. Data Anal. **94**, 330–350 (2016)

K. Hron, J. Machalová, A. Menafoglio, Bivariate densities in Bayes spaces: orthogonal decomposition and spline representation. Technical report, Politecnico di Milano (2020). MOX-report 83/2020

S. Kurtek, A. Srivastava, W. Wu, Signal estimation under random time-warpings and nonlinear signal alignment, in *NIPS* (2011)

J.S. Marron, A.M. Alonso, Overview of object oriented data analysis. Biom. J. **56**(5), 732–753 (2014)

J.S. Marron, J.O. Ramsay, L.M. Sangalli, A. Srivastava, Functional data analysis of amplitude and phase variation. Stat. Sci. **30**(4), 468–484 (2015)

A. Menafoglio, G. Petris, Kriging for Hilbert-space valued random fields: the operatorial point of view. J. Multivar. Anal. **146**, 84–94 (2016). https://doi.org/10.1016/j.jmva.2015.06.012

A. Menafoglio, P. Secchi, Statistical analysis of complex and spatially dependent data: a review of object oriented spatial statistics. Eur. J. Oper. Res. **258**(2), 401–410 (2017)

A. Menafoglio, P. Secchi, O2s2: a new venue for computational geostatistics. Appl. Comput. Geosci. **2**, 100007 (2019)

A. Menafoglio, P. Secchi, M. Dalla Rosa, A Universal Kriging predictor for spatially dependent functional data of a Hilbert Space. Electron. J. Stat. **7**, 2209–2240 (2013)

A. Menafoglio, A. Guadagnini, P. Secchi, A Kriging approach based on Aitchison geometry for the characterization of particle-size curves in heterogeneous aquifers. Stoch. Env. Res. Risk Assess. **28**(7), 1835–1851 (2014)

A. Menafoglio, O. Grujic, J. Caers, Universal kriging of functional data: trace-variography vs cross-variography? Application to forecasting in unconventional shales. Spat. Stat. **15**, 39–55 (2016a)

A. Menafoglio, A. Guadagnini, P. Secchi, Stochastic simulation of soil particle-size curves in heterogeneous aquifer systems through a Bayes space approach. Water Resour. Res. **52**(8), 5708–5726 (2016b)

A. Menafoglio, P. Secchi, A. Guadagnini, A class-kriging predictor for functional compositions with application to particle-size curves in heterogeneous aquifers. Math. Geosci. **48**, 463–485 (2016c)

A. Menafoglio, G. Gaetani, P. Secchi, Random domain decompositions for object-oriented kriging over complex domains. Stoch. Env. Res. Risk Assess. **32**, 3421–3437 (2018a)

A. Menafoglio, M. Grasso, P. Secchi, B.M. Colosimo, Profile monitoring of probability density functions via simplicial functional PCA with application to image data. Technometrics **60**(4), 497–510 (2018b)

A. Menafoglio, D. Pigoli, P. Secchi, Kriging riemannian data via random domain decompositions. Technical report, Politecnico di Milano (2018c). MOX-report 64/2018

V. Pawlowsky, On spurious spatial covariance between variables of constant sum. Sciences de la Terre série informatique **21**, 107–113 (1984)

V. Pawlowsky, Cokriging of regionalized compositions. Math. Geol. **21**(5), 513–521 (1989)

V. Pawlowsky-Glahn, A. Buccianti, *Compositional Data Analysis. Theory and Applications* (Wiley, 2011)

V. Pawlowsky-Glahn, J.J. Egozcue, Geometric approach to statistical analysis in the symplex. Stoch. Env. Res. Risk Assess. **15**, 384–398 (2001)

V. Pawlowsky-Glahn, J.J. Egozcue, BLU estimators and compositional data. Math. Geol. **34**(3), 259–274 (2002)

V. Pawlowsky-Glahn, J.J. Egozcue, R. Tolosana-Delgado, *Modelling and Analysis of Compositional Data* (Wiley, 2015)

A. Pini, A. Stamm, S. Vantini, Hotelling's T2 in separable Hilbert spaces. J. Multivar. Anal. **167**, 284–305 (2018)

J. Ramsay, B. Silverman, *Functional Data Analysis*, 2nd edn. (Springer, New York, 2005)

L.M. Sangalli, P. Secchi, S. Vantini, V. Vitelli, K-mean alignment for curve clustering. Comput. Stat. Data Anal. **54**(5), 1219–1233 (2010)

A. Srivastava, E.P. Klassen, *Functional and Shape Data Analysis* (Springer, New York, 2016)

A. Srivastava, I. Jermyn, S. Joshi, Riemannian analysis of probability density functions with applications in vision, in IEEE, editor, *2007 IEEE Conference on Computer Vision and Pattern Recognition*, vol. 44, pp. 1–8 (2007)

R. Talská, A. Menafoglio, J. Machalová, K. Hron, E. Fišerová, Compositional regression with functional response. Comput. Stat. Data Anal. **123**, 66–85 (2018)

R. Talská, A. Menafoglio, K. Hron, J.J. Egozcue, J. Palarea-Albaladejo, Weighting the domain of probability densities in functional data analysis. Stat (2020). https://doi.org/10.1002/sta4.283

R. Tolosana-Delgado, V. Pawlowsky-Glahn, J.J. Egozcue, Indicator kriging without order relation violations. Math. Geosci. **40**(3), 327–347 (2008)

R. Tolosana-Delgado, K.G. van den Boogaart, V. Pawlowsky-Glahn, *Geostatistics for Compositions*, pp. 73–86, Pawlowsky-Glahn & Buccianti edition (Wiley, 2011)

R. Tolosana-Delgado, U. Mueller, K.G. van den Boogaart, Geostatistics for compositional data: an overview. Math. Geosci. **51**, 485–526 (2019)

R.D. Tuddenham, M.M. Snyder, Physical growth of California boys and girls from birth to eighteen years. Univ. Calif. Publ. Child Dev. **1**, 183–364 (1954)

K.G. van den Boogaart, J.J. Egozcue, V. Pawlowsky-Glahn, Bayes Hilbert spaces. Aust. N. Z. J. Stat. **56**, 171–194 (2014)

K.G. van den Boogaart, J.J. Egozcue, V. Pawlowsky-Glahn, Bayes linear spaces. SORT **34**(2), 201–222 (2010)

Spatial Simultaneous Autoregressive Models for Compositional Data: Application to Land Use

Christine Thomas-Agnan, Thibault Laurent, Anne Ruiz-Gazen, Thi Huong An Nguyen, Raja Chakir, and Anna Lungarska

Abstract Econometric land use models study determinants of land use shares of different classes: "agriculture", "forest", "urban" and "other" for example. Land use shares have a compositional nature as well as an important spatial dimension. We compare two compositional regression models with a spatial autoregressive nature in the framework of land use. We study the impact of the choice of coordinate space and prove that a choice of coordinate representation does not have any impact on the parameters in the simplex as long as we do not impose further restrictions. We discuss parameters interpretation taking into account the non-linear structure as well as the spatial dimension. In order to assess the explanatory variables impact, we compute and interpret the semi-elasticities of the shares with respect to the explanatory variables and the spatial impact summary measures.

C. Thomas-Agnan (✉) · A. Ruiz-Gazen · T. H. A. Nguyen
Toulouse School of Economics, University of Toulouse Capitole, Toulouse, France
e-mail: christine.thomas@tse-fr.eu

A. Ruiz-Gazen
e-mail: Anne.ruiz-gazen@tse-fr.eu

T. H. A. Nguyen
e-mail: nguyen_huongan@yahoo.com

T. Laurent
Toulouse School of Economics, CNRS, University of Toulouse Capitole, Toulouse, France
e-mail: thibault.laurent@tse-fr.eu

T. H. A. Nguyen
Danang University of Architecture, Da Nang, Vietnam

R. Chakir · A. Lungarska
INRAE, AgroParisTech, Economie Publique, Université Paris-Saclay, 78850 Thiverval-Grignon, France
e-mail: raja.chakir@inra.fr

A. Lungarska
e-mail: lungarska@gmail.com

P. Filzmoser et al. (eds.), *Advances in Compositional Data Analysis*,
https://doi.org/10.1007/978-3-030-71175-7_12

1 Introduction

Land use and land use changes are among the main human pressures on the environment in terms of biodiversity loss, carbon cycle and water quality (Foley et al. 2005; Pielke 2005; Lal 2004; Verburg et al. 2013). Given the environmental impacts and societal issues associated to land use, numerous empirical land use models have been developed among different disciplines. The aim of these models is to support future land use planning and environmental impact assessments of land use change. All these models have mainly two different research focuses: explaining the underlying process behind land use or predicting spatial land use patterns or dynamics. The distinction between these two approaches is often not easy (Shmueli 2010). In summary, the purpose of econometric models, mostly developed by economists, is to test theoretical results and identify the economic processes behind observed land use changes and patterns. This leads to the search for parsimonious models, i.e., models with limited number of explanatory variables suggested by the theoretical model. Statisticians, geographers and other researchers outside of economics typically have focused on prediction models of land use (Munroe and Müller 2007; Veldkamp and Lambin 2001). These models are useful for prediction of land use patterns but they provide little insight into the underlying economic and other processes that generate these patterns. There is clearly no land use model which is preferable in absolute terms. The choice of the best model is largely dependent on the research question, data availability and calculation costs.

In terms of methodology, econometric land use studies can be classified into different groups depending on the type of data used (aggregated data vs. micro-geographic data), the type of model (spatially explicit vs. aspatial model) and the land use categories under consideration (rural vs. urban, agriculture vs. forest and developed vs. non-developed).

Among land use models, we restrict attention to econometric formulations based on simultaneous spatial autoregressive models (SAR) because they are common in the econometric land use literature (see for instance LeSage and Pace 2009) and easy to fit. There are alternative approaches based on conditional autoregressive models such as Pirzamanbein et al. (2018), Leininger et al. (2013) but they require MCMC techniques, see Nguyen et al. (2019) for further details.

Our starting point is the application to land use in Chakir and Lungarska (2017), Lungarska and Chakir (2018) which is based on a spatial error model (SEM) in alr coordinates. The objective, using Nguyen et al. (2019), is to propose and compare alternative models based on alternative coordinate systems (ilr instead of alr) and alternative model formulations in the same family (LAG instead of SEM as we will explain later on). Using the same dataset, we estimate both models and question model specification and interpretation. The two approaches differ in several important dimensions. First of all, the first approach specifies the model in an alr coordinate space whereas the second one uses an ilr coordinate space, see, e.g., Pawlowsky-Glahn et al. (2015) for the definition of alr and ilr transformations. Secondly the first approach uses a spatial error model formulation for each coordinate separately while the second approach uses a joint multivariate spatial lag model. Finally, the first approach performs a separate maximum likelihood for each coordinate whereas the second one uses the simultaneous spatial two-stage least squares method (acronym S2SLS) which will be presented in Sect. 2. This confrontation raises many questions: in particular is it possible to write a simplex formulation of the first model? and since parameters of the models are linked to a particular transformation, what are the links between parameters associated to different transformations? For models using ilr coordinates, the link between parameters in the simplex space and in coordinate space is well known (Chen et al. 2017). We extend these formulas to models using alr coordinates.

Section 3.2 is devoted to the interpretation of model parameters. The first approach uses plots of the fitted shares as a function of each explanatory variable. Morais and Thomas-Agnan (2020) prove that semi-elasticities are natural tools for the interpretation of model parameters in simplicial regression models with a compositional dependent variable because they are derived from simplicial derivatives and because they describe variations in the simplex rather than in coordinate space. However their framework does not encompass the case of spatial models. We thus derive an approximation formula for semi-elasticities in the present models which is illustrated later in Sect. 4.3. In Sect. 4, after presenting the data set and paying attention to the treatment of missing values according to compositional data analysis (CoDa) principles, we analyze it with the different models and propose a comparison.

2 Multivariate SAR Models

In this section, after recalling some classical facts about compositional data analysis, we recall the models from Chakir and Lungarska (2017), Nguyen et al. (2019) with a common notation and refer to the first one as to the UniSEM model and to the second one as to the MultiLAG model.

2.1 Notations

In the Compositional Data analysis (CoDa) literature, a D-composition $\mathbf{u}$ is a vector with positive components conveying relative information which can be represented by an element of the so-called simplex space $\mathbf{S}^D$ defined by Aitchison (1986):

$$\mathbf{S}^D = \left\{ \mathbf{u} = (u_1, \ldots, u_D)^T : u_m > 0, m = 1, ..., D; \sum_{m=1}^{D} u_m = 1 \right\},$$

where ^{T} is the transpose operator. $\mathbf{S}^D$ can be equipped with a vector structure and the compositional/Aitchison inner product.

The analysis of compositional data makes use of some classical log-ratio transformations which are the additive log-ratio (alr), the centered log-ratio (clr) and the isometric log-ratio (ilr) transformations. In this paper, we will use alr and ilr transformations.

For the reference level D, let us recall that the alr transformation $\mathbf{u}^*$ of a vector $\mathbf{u}$ in $\mathbf{S}^D$ is defined by

$$\mathbf{u}^* = \text{alr}_D(\mathbf{u}) = (\ln(u_1/u_D), \ldots, \ln(u_{D-1}/u_D))$$

and its inverse transformation alr_D^{-1} is given by

$$\mathbf{u} = \text{alr}_D^{-1}(\mathbf{u}^*) = C(\exp(u_1^*), \ldots, \exp(u_{D-1}^*), 1),$$

where $C(.)$ denotes the closure operation defined by

$$C(\mathbf{w}) = \left(\frac{w_1}{\sum_{m=1}^{D} w_m}, \cdots, \frac{w_D}{\sum_{m=1}^{D} w_m} \right),$$

for any vector $\mathbf{w} \in \mathbb{R}_+^D$. A similar definition can be given changing the reference level D into any other level $m = 1, \ldots, D-1$.

A given additive log-ratio transformation for reference level $m = 1, \ldots, D$ can also be expressed using a $(D-1) \times D$ matrix $\mathbf{F}_m$ by $\text{alr}_m(\mathbf{u}) = \mathbf{F}_m \ln(\mathbf{u})$ for a vector $\mathbf{u} \in \mathbf{S}^D$. For example, in the case $m = D$, $\mathbf{F}_D = \begin{bmatrix} \mathbf{I}_{D-1} & -\mathbf{j}_{D-1} \end{bmatrix}$, where $\mathbf{j}_{D-1}$ denotes the $(D-1)$-dimensional column vector of ones and $\mathbf{I}_{D-1}$ is the $(D-1) \times (D-1)$ identity matrix.

Similarly, in order to define an ilr transformation, let $\mathbf{V}$ be a $D \times (D-1)$ contrast matrix associated to a given orthonormal basis $(\mathbf{e}_1, \cdots, \mathbf{e}_{D-1})$ of $\mathbf{S}^D$ by $\mathbf{V} = \text{clr}(\mathbf{e}_1, \cdots, \mathbf{e}_{D-1})$, where clr is understood columnwise and where $g(\mathbf{u}) = (\prod_{i=1}^{D} u_i)^{1/D}$ is the geometric mean of the components of $\mathbf{u} \in \mathbf{S}^D$:

$$\text{clr}(\mathbf{u}) = \left(\ln \frac{u_m}{g(\mathbf{u})} \right)_{m=1,\ldots,D}.$$

For each such matrix $\mathbf{V}$, an isometric log-ratio transformation (ilr) is then defined by

$$\mathbf{u}^* = \text{ilr}_V(\mathbf{u}) = \mathbf{V}^T \ln(\mathbf{u}),$$

where the logarithm of $\mathbf{u} \in \mathbf{S}^D$ is understood componentwise. Note that we use the same star notation throughout the paper to indicate either an ilr or an alr transformed vector. It is to emphasize the similarity of treatment and the meaning will be clear from the context. The inverse transformation is given by

$$\mathbf{u} = \text{ilr}_V^{-1}(\mathbf{u}^*) = C(\exp(\mathbf{V}\mathbf{u}^*)).$$

Moreover, the compositional product of a matrix by a composition vector is denoted by $\boxdot$. It is defined for a $D \times D$ matrix $\mathbf{B} = (b_{lm}),\ l, m = 1, \ldots D$ such that $\mathbf{B}\mathbf{j}_D = \mathbf{0}_D$ and $\mathbf{B}^T\mathbf{j}_D = \mathbf{0}_D$ (with $\mathbf{0}_D$ a D-dimensional vector of zeros) and for a vector $\mathbf{u} \in \mathbf{S}^D$ by

$$\mathbf{B} \boxdot \mathbf{u} = C\left(\prod_{m=1}^{D} u_m^{b_{1m}}, \cdots, \prod_{m=1}^{D} u_m^{b_{Dm}}\right)^T.$$

As detailed in Pawlowsky-Glahn et al. (2015), the previous expression comes from the usual product matrix definition in the coordinate space for an ilr_V transformation: $\mathbf{B} \boxdot \mathbf{u} = \mathrm{ilr}_V^{-1}\left(\mathbf{B}^*\mathrm{ilr}_V(\mathbf{u})\right)$ where $\mathbf{B}^* = \mathbf{V}^T\mathbf{B}\mathbf{V}$. We also have

$$\mathbf{B} \boxdot \mathbf{u} = \mathrm{ilr}_V^{-1}\left(\mathbf{V}^T\mathbf{B}\ln(\mathbf{u})\right).$$

Note that this matrix product in the simplex does not depend on the particular choice of contrast matrix $\mathbf{V}$. This product also verifies the following property (see the proof in Appendix 6.1):

$$\mathbf{B} \boxdot \mathbf{u} = \mathrm{alr}_m^{-1}\left(\mathbf{F}_m\mathbf{B}\ln(\mathbf{u})\right).$$

In what follows, the data we consider is made of samples of composition vectors and is stored in a $n \times D$ matrix $\mathbf{Y} = (\mathbf{Y}_{il}), i = 1, \ldots, n,\ l = 1, \ldots, D$. We propose to generalize the matrix product to a $n \times D$ matrix $\mathbf{Y}$ of composition vectors multiplied by a $D \times D$ matrix $\mathbf{B}$ (such that $\mathbf{B}\mathbf{j}_D = \mathbf{0}_D$ and $\mathbf{B}^T\mathbf{j}_D = \mathbf{0}_D$) by applying the usual matrix product to each row of $\mathbf{Y}$. We have

$$\mathbf{Y} \boxdot \mathbf{B} = \mathrm{ilr}_V^{-1}\left(\ln(\mathbf{Y})\mathbf{B}\mathbf{V}\right) = \mathrm{alr}_m^{-1}\left(\ln(\mathbf{Y})\mathbf{B}\mathbf{F}_m^T\right). \tag{1}$$

For a $n \times D$ matrix of compositions $\mathbf{Y}$, and a $n \times n$ weight matrix $\mathbf{W}$, (Nguyen et al. 2019) define the following operation $\triangle$ as a map from the cartesian product of simplex spaces $(\mathbf{S}^D)^n$ to itself specified by

$$\mathbf{W}\triangle\mathbf{Y} = \mathrm{ilr}_V^{-1}(\mathbf{W}\mathrm{ilr}_V(\mathbf{Y})) = \mathrm{ilr}_V^{-1}(\mathbf{W}\ln(\mathbf{Y})\mathbf{V}). \tag{2}$$

The operation $\mathbf{W}\triangle\mathbf{Y}$ defines a $n \times D$ matrix of compositions whose components are weighted geometric means of the $\mathbf{Y}$ values weighted by the $\mathbf{W}$ weights.

2.2 *The MultiLAG Model*

In Nguyen et al. (2019), the authors propose to define the MultiLAG model for a simplex-valued dependent variable. In the present paper, we consider a simpler version of this model which involves in each ilr coordinate equation spatial lags of the other coordinates but not the other coordinates themselves.

Let us consider a sample of size n. Let $\mathbf{Y}$ be a $n \times D$ matrix of dependent compositional vectors and $\mathbf{X}$ be a $n \times K$ matrix of classical (non-compositional) explanatory variables. An extension to models including compositional explanatory variables would be straightforward with some adjustments for the interpretation section. Let $\mathbf{W}$ be a $n \times n$ spatial weight matrix specifying the neighborhood structure. For n spatial locations, the elements w_{ij} of the matrix $\mathbf{W}$ are measures of proximity between locations i and j. The model can be written in the simplex as follows:

$$\mathbf{Y} = (\mathbf{W} \triangle \mathbf{Y}) \boxdot \mathbf{R} \oplus \mathbf{X} \odot \boldsymbol{\beta} \oplus \boldsymbol{\epsilon}, \tag{3}$$

where $\mathbf{R}$ is a $D \times D$ matrix of parameters such that $\mathbf{R}\mathbf{j}_D = \mathbf{0}_D$ and $\mathbf{R}^T\mathbf{j}_D = \mathbf{0}_D$, $\boldsymbol{\beta}$ is a $K \times D$ matrix of parameters. $\boldsymbol{\epsilon}$ is a $n \times D$ matrix of compositional errors satisfying the following conditions. Denoting by $\boldsymbol{\epsilon}_{.l}$ the columns of $\boldsymbol{\epsilon}$ and by $\boldsymbol{\epsilon}_{i.}$ its rows, we assume that $\mathbb{E}(\boldsymbol{\epsilon}^*_{i.}\boldsymbol{\epsilon}^{*\,T}_{j.}) = \boldsymbol{\Sigma}^*$ if individuals i and j are equal and $\mathbf{0}$ if they are different, where $\boldsymbol{\Sigma}^*$ is a $(D-1) \times (D-1)$ covariance matrix. Note that the sum in the simplex sense ($\oplus$) over individuals of the rows of the matrix $\boldsymbol{\epsilon}$ is the neutral element of the simplex.

Following Nguyen et al. (2019), the model can be written in ilr coordinate space with a coordinate-wise formulation:

$$\mathbf{Y}^*_{.l} = \sum_{m \in S_l^{\mathbf{WY}^*}} \mathbf{R}^*_{ml}\mathbf{W}\mathbf{Y}^*_{.m} + \mathbf{X}_{S_l^{\mathbf{X}}}\boldsymbol{\beta}^*_{S_l^{\mathbf{X}}} + \boldsymbol{\epsilon}^*_{.l},\ l = 1, \ldots, D-1, \tag{4}$$

where $\mathbf{R}^*$ and $\boldsymbol{\beta}^*$ are the parameters of the model in coordinate space and where $\mathbf{Y}^*_{.l}$ and $\boldsymbol{\epsilon}^*_{.l}$ denote, respectively, $\mathrm{ilr}_V(\mathbf{Y}_{.l})$ and $\mathrm{ilr}_V(\boldsymbol{\epsilon}_{.l})$ for any given contrast matrix $\mathbf{V}$. We allow for using a different set of explanatory variables in each equation. For this reason, $S_l^{\mathbf{Y}^*}$, $S_l^{\mathbf{X}}$, $S_l^{\mathbf{WY}^*}$ denote the sets of indices of the variables which appear in the lth equation for $\mathbf{Y}^*$, $\mathbf{X}$, $\mathbf{WY}^*$, respectively.

It is important to note that $\mathbf{R}^*$ and $\boldsymbol{\beta}^*$ depend upon the specific choice of $\mathbf{V}$ despite the fact that we do not indicate it in the notation for simplification purposes. The relation between $\mathbf{R}^*$ and $\mathbf{R}$, and between $\boldsymbol{\beta}^*$ and $\boldsymbol{\beta}$ will be explored in Sect. 3. This model can be easily estimated via the S2SLS method in a multivariate fashion. Two-stage least squares also called instrumental variables regression is popular in econometrics when an explanatory variable of interest is correlated with the error term and it has been adapted to the spatial models by Kelejian and Prucha (1998). Ignoring the dependence between coordinates and performing a univariate estimation coordinate by coordinate would result in the same parameters estimates but incorrect standard errors estimates.

In view of Sect. 4, we need another formulation of the MultiLAG spatial model in ilr coordinate space with a matrix formulation as in Kelejian and Prucha (2004) since Nguyen et al. (2019) only give a coordinate-wise formulation. Let $\mathbf{Y}^*$ be the $n \times (D-1)$ matrix whose columns are the ilr coordinates $\mathbf{Y}^*_{.l}$. The matrix formulation of model (4) is given by

$$\mathbf{Y}^* = \mathbf{W}\mathbf{Y}^*\mathbf{R}^* + \left[\mathbf{X}_{S_1^X}\boldsymbol{\beta}^*_{S_1^X} \dots \mathbf{X}_{S_{D-1}^X}\boldsymbol{\beta}^*_{S_{D-1}^X}\right] + \boldsymbol{\epsilon}^*, \quad (5)$$

where $[A_1 \dots A_k]$ is a notation for a block matrix with blocks A_1 through A_k.

In order to be able to compute fitted values and marginal effects, we also need a formulation of the so-called reduced form of the model: it is an equivalent formulation of the model where the dependent variable appears only on the left-hand side of the equation. It is possible to do so using a vectorized formulation. For a matrix $\mathbf{A}$, let $\text{vec}_c\mathbf{A}$ be the column vectorization obtained by stacking the columns of $\mathbf{A}$.

Theorem 1 *The reduced form of model (4) expressed in vectorized formulation is given by*

$$vec_c\mathbf{Y}^* = \left(\mathbf{I}_{n(D-1)} - ((\mathbf{R}^*)^T \otimes \mathbf{W})\right)^{-1} \left[\mathbf{X}_{S_1^X}\boldsymbol{\beta}^*_{S_1^X} \dots \mathbf{X}_{S_{D-1}^X}\boldsymbol{\beta}^*_{S_{D-1}^X}\right]^T + vec_c\boldsymbol{\epsilon}^*, \quad (6)$$

where $\otimes$ denotes the Kronecker product of matrices.

The matrix $\mathbf{A}(\mathbf{W}) = (\mathbf{I}_{n(D-1)} - ((\mathbf{R}^*)^T \otimes \mathbf{W}))^{-1}$ is called the filter matrix. The proofs of Eq. (5) and Theorem 1 are in the appendix.

2.3 *The UniSEM Model*

Even though Chakir and Lungarska (2017) do not use the compositional data analysis formalism, their model can be formulated in a similar fashion as that of Nguyen et al. (2019). Indeed if we use the same star notation for denoting now an alr transformation instead of an ilr one, it is easy to see that the specification (3) in Chakir and Lungarska (2017) can be reformulated as

$$\begin{cases} \mathbf{Y}^*_{.l} &= \mathbf{X}_{S_l^X}\boldsymbol{\beta}^*_{S_l^X} + \boldsymbol{\epsilon}^*_{.l} \\ \boldsymbol{\epsilon}^*_{.l} &= \sum_{m \in S_l^{WY^*}} \mathbf{R}^*_{ml}\mathbf{W}\boldsymbol{\epsilon}^*_{.m} + \boldsymbol{\nu}^*_{.l} \end{cases} \quad (7)$$

with the restriction that the $\mathbf{R}^*$ matrix is diagonal and that the alr coordinates are independent (i.e., the errors $\boldsymbol{\nu}^*_{.l}$ are independent). Therefore the reduced form of the model can be derived as in Theorem 1. It is also easy to see that we can reformulate this model in the simplex as

$$\begin{cases} \mathbf{Y} &= \mathbf{X} \odot \boldsymbol{\beta} \oplus \boldsymbol{\epsilon} \\ \boldsymbol{\epsilon} &= (\mathbf{W} \triangle \boldsymbol{\epsilon}) \boxdot \mathbf{R} \oplus \boldsymbol{\nu} \end{cases}. \quad (8)$$

The motivation for returning to the simplex is twofold. In this work, we need to compare models which are using different alr and ilr transformations and correspond to different coordinate spaces: only the simplex parameters are comparable. Moreover Morais and Thomas-Agnan (2020) advocate the interpretation of parameters in the simplex as being more informative. Model (8) will be called the MultiSEM model when no restriction is made on the $\mathbf{R}^*$ matrix and UniSEM with the diagonality restriction as in Chakir and Lungarska (2017).

2.4 Summary of Models

We just recalled the definition of the MultiLAG and the UniSEM models which are solutions for defining a compositional dependent variable regression model with spatial dependence. The MultiLAG takes into account the possible dependence between coordinates. The submodel of MultiLAG obtained by constraining the coordinates to be independent as in the UniSEM will be called the UniLAG. For the same type of data, an ordinary regression model for compositional dependent variable without spatial dependence could also be considered and we will later refer to it as the OLM model for ordinary linear model (see Filzmoser et al. 2018). Fitting the UniLAG or the UniSEM just requires code for spatial univariate regression models as in the R package *spdep* (Bivand and Wong 2018). Fitting the MultiLAG with the S2SLS method just requires code for multivariate regression such as the R function *lm* but we keep the implementation of MultiSEM for future work because of the lack of specific code for this case.

3 Model Specification and Interpretation

3.1 Influence of the Transformation

The impact of the coordinate system on the model is crucial for compositional data: indeed a model formulation involving some restrictions on the parameters which are dependent upon a particular choice of coordinate system is inappropriate since it means that the model does not have an intrinsic simplex formulation. Chakir and Lungarska (2017) specify the model in coordinate space after choosing a particular alr transformation. We have shown in Sect. 2 that this model can also be formulated in the simplex. Nguyen et al. (2019) first specify the model in coordinate space with a particular ilr transformation, but then show that the model can also be specified in the simplex. If the formulation of a model is done in the simplex, it is independent of any choice of transformation. A natural question is then to establish the relationships between parameters in the simplex and parameters in coordinate space.

Nguyen et al. (2019) show that the relationship between the simplex parameters of (3) $\mathbf{R}$ and $\boldsymbol{\beta}$ and the coordinate parameters of (4) $\mathbf{R}^*$ and $\boldsymbol{\beta}^*$ is then

$$\begin{cases} \mathbf{R} = \mathbf{V}\mathbf{R}^*\mathbf{V}^T \\ \boldsymbol{\beta} = \mathrm{ilr}_V{}^{-1}(\boldsymbol{\beta}^*) \end{cases} \tag{9}$$

which can also be written as $\mathbf{V}^T\mathbf{R} = \mathbf{R}^*\mathbf{V}^T$.

Consider now the UniSEM model specified by (8). Let us derive the correspondence between the parameters in alr space and in the simplex. The following theorem establishes the equivalent of (9) for alr. Our statements and proofs below concern the alr_D transformations but are easily generalized to any alr.

Theorem 2 *If* $\mathbf{R}$ *and* $\boldsymbol{\beta}$ *are the simplex parameters of model (8) and if* $alr_D(\mathbf{R}) = \mathbf{R}^*$ *and* $alr_D(\boldsymbol{\beta}) = \boldsymbol{\beta}^*$*, then*

$$\begin{cases} \mathbf{R} = \mathbf{K}_D\mathbf{R}^*\mathbf{F}_D \\ \boldsymbol{\beta} = alr_D{}^{-1}(\boldsymbol{\beta}^*), \end{cases} \tag{10}$$

where $\mathbf{F}_D$ *is the matrix associated to the alr transformation* alr_D *and* $\mathbf{K}_D$ *is defined by*

$$\mathbf{K}_D = \begin{bmatrix} \mathbf{I}_{D-1} - \mathbf{j}_{D-1}\mathbf{j}_{D-1}^T/D \\ -\mathbf{j}_{D-1}^T/D \end{bmatrix}.$$

Proof Let us start with model (3) and apply the alr_D transformation to both sides of the equation. Using Eq. (1), it is easy to see that

$$\mathrm{alr}_D(\mathbf{R} \boxdot \mathbf{W} \triangle \mathbf{Y} \oplus \mathbf{X} \odot \boldsymbol{\beta} \oplus \boldsymbol{\epsilon}) = \mathbf{F}_D\mathbf{R}\ln(\mathbf{W}\triangle\mathbf{Y}) + \mathbf{X}\mathrm{alr}_D(\boldsymbol{\beta}) + \mathrm{alr}_D(\boldsymbol{\epsilon}). \tag{11}$$

Comparing to (11) the right-hand side of (4) where star would mean the alr_D transformation, we get that $\mathrm{alr}_D(\boldsymbol{\beta})$ corresponds to $\boldsymbol{\beta}^*$(which proves the second part of (10)) and that $\mathbf{F}_D\mathbf{R} = \mathbf{R}^*\mathbf{F}_D$. Then Lemma 1 in Appendix 6.3 allows to show that $\mathbf{F}_D\mathbf{R} = \mathbf{R}^*\mathbf{F}_D$ is equivalent to the first equation of (10).

Note that these relationships between simplex-space parameters and coordinate-space parameters induce the same relationships between their estimated counterparts for the MultiLAG model when estimated by S2SLS (see Nguyen et al. 2019). It also holds for the UniSEM estimated by maximum likelihood since maximum likelihood is preserved by a one to one transformation. Finally, we see that if we specify the model in the simplex, the choice of a particular transformation for estimation does not change the final result in the simplex. However a restriction like $\mathbf{R}^*$ is diagonal, used in Chakir and Lungarska (2017), in a given coordinate space (ilr or alr) does not imply that the same is true in another coordinate space. In this sense one can say that it is not a simplex assumption.

Moreover another issue with the choice of coordinate system is the relationships between the mean and variance parameters associated to each system. For the case of two ilr transformations, the result can be found in Pawlowsky-Glahn et al. (2015). Let us derive the same type of relationship for the case of two alr transformations and for the case of an alr and an ilr transformation.

Theorem 3 *If we assume that* $alr_D(\mathbf{X}) \sim \mathcal{N}_{S_D}(\boldsymbol{\mu}_{alr_D}, \boldsymbol{\Sigma}_{alr_D})$, *and if* $\mathbf{V}$ *is a contrast matrix, then the means and variances of* $ilr_V(\mathbf{X})$ *and* $alr_j(\mathbf{X})$ *are related as follows to those of* $alr_D(\mathbf{X})$

$$\boldsymbol{\mu}_{ilr_V} = \mathbf{V}^T \mathbf{K}_D \boldsymbol{\mu}_{alr_D} \tag{12}$$

$$\boldsymbol{\Sigma}_{ilr_V} = \mathbf{V}^T \mathbf{K}_D \boldsymbol{\Sigma}_{alr_D} (\mathbf{V}^T \mathbf{K}_D)^T \tag{13}$$

$$\boldsymbol{\mu}_{alr_j} = \mathbf{F}_j \mathbf{K}_D \boldsymbol{\mu}_{alr_D} \tag{14}$$

$$\boldsymbol{\Sigma}_{alr_j} = \mathbf{F}_j \mathbf{K}_D \boldsymbol{\Sigma}_{alr_D} (\mathbf{F}_j \mathbf{K}_D)^T. \tag{15}$$

Proof We recall that $\text{alr}_j(\mathbf{X}) = \mathbf{F}_j \ln(\mathbf{X})$ for $j = 1, \ldots, D$ and $\text{ilr}_V(\mathbf{X}) = \mathbf{V}^T \ln(\mathbf{X})$. Let us prove that

$$\text{ilr}_V(\mathbf{X}) = \mathbf{V}^T \mathbf{K}_D \text{alr}_D(\mathbf{X}) \tag{16}$$

$$\text{alr}_j(X) = \mathbf{F}_j \mathbf{K}_D \text{alr}_D(\mathbf{X}). \tag{17}$$

It is easy to prove that $\mathbf{F}_D \mathbf{K}_D = \mathbf{V}^T \mathbf{V} = \mathbf{I}_{D-1}$ and $\mathbf{K}_D \mathbf{F}_D = \mathbf{K}_j \mathbf{F}_j = \mathbf{V}\mathbf{V}^T$. Thus,

$$\begin{aligned} \text{ilr}_V(\mathbf{X}) &= \mathbf{V}^T \mathbf{V} \text{ilr}_V(\mathbf{X}) \\ &= \mathbf{V}^T \mathbf{V} \mathbf{V}^T \ln(\mathbf{X}) \\ &= \mathbf{V}^T \mathbf{K}_D \mathbf{F}_D \ln(\mathbf{X}) \\ &= \mathbf{V}^T \mathbf{K}_D \text{alr}_D(\mathbf{X}) \end{aligned}$$

and

$$\begin{aligned} \text{alr}_j(\mathbf{X}) &= \mathbf{F}_j \mathbf{K}_j \text{alr}_j(\mathbf{X}) \\ &= \mathbf{F}_j \mathbf{K}_j \mathbf{F}_j \ln(\mathbf{X}) \\ &= \mathbf{F}_j \mathbf{K}_D \mathbf{F}_D \ln(\mathbf{X}) \\ &= \mathbf{F}_j \mathbf{K}_D \text{alr}_D(\mathbf{X}) \end{aligned}$$

Equations (12) and (13) (resp. (14) and (15)) are then directly derived from (16) (resp. (17)).

This theorem implies that if Σ_{alr_D} is diagonal, then Σ_{ilr_V} is not necessarily diagonal and Σ_{alr_j} neither, implying that this restriction is not simplex compatible.

3.2 Model Interpretation

Concerning the impact of explanatory variables, we first focus on looking at how the fitted shares in the simplex depend upon the values of a particular explanatory variable. For the UniSEM model, the question is quite simple: the fitted values in coordinate space are straightforward to compute and their inverse alr transformation yields the fitted shares.

For the MultiLAG model, thanks to Theorem 1, it is first easy to derive the fitted values of $\mathbf{Y}^*$ in coordinate space by substituting the parameters by their estimates and we get

$$\mathrm{vec}_c \hat{\mathbf{Y}}^* = \left(\mathbf{I}_{n(D-1)} - ((\hat{\mathbf{R}}^*)^T \otimes \mathbf{W})\right)^{-1} \begin{bmatrix} \mathbf{X}_{S_1} \hat{\boldsymbol{\beta}}_1^* \\ \cdot \\ \mathbf{X}_{S_{D-1}} \hat{\boldsymbol{\beta}}_{D-1}^* . \end{bmatrix} \tag{18}$$

Applying the inverse ilr transformation to (18), we can compute the fitted shares in the simplex and use them to illustrate the impact of the covariates by a graph as follows. Since this investigation is relative to a single variable at a time, we can drop the variable index in this paragraph to simplify the derivation. With this notation, for a given covariate $\mathbf{X}$, x_j denotes the value of $\mathbf{X}$ at location j.

For a given spatial unit and a given covariate, we imagine changing the value of that particular covariate at that particular location holding everything else constant. We denote by $\mathbf{Y}_{im}(x_j + \delta)$ the value of the component m of the share vector $\mathbf{Y}$ for location i obtained by increasing by δ the value of $\mathbf{X}$ at location j (note that we only indicate the location that changes). We then compute the new fitted shares $\hat{\mathbf{Y}}_{im}(x_j + \delta)$ for a grid of points $x_j + \delta$ in the range of the covariate of interest. Finally we draw a scatterplot of the fitted shares as a function of the grid points. Some figures in Sect. 4 illustrate this approach.

A further step in this investigation implies computing the semi-elasticities in the simplex as in Morais and Thomas-Agnan (2020). This comprises two issues. The first issue is the computation of spatial impacts in coordinate space and the second one the evaluation of semi-elasticities in the simplex.

For the first issue, indeed parameters in a simultaneous spatial autoregressive model cannot be interpreted as in ordinary linear models due to the non-linear filter matrix in the expected value of the dependent variable (see LeSage and Pace 2009). We obtain the impacts in coordinate space thanks to the reduced form of the model (6).

For a given explanatory variable $\mathbf{X}$ and a given component m, the impact on $\mathbf{Y}^*_{im}$ of changing $\mathbf{X}$ at location j is defined by $\frac{\partial \mathbb{E}\mathbf{Y}^*_{im}}{\partial x_j}$. If $a_{im:jm}(\mathbf{W})$ denote the generic element of $A(\mathbf{W})$, for location i, j, component m, we have

$$\frac{\partial \mathbb{E}\mathbf{Y}^*_{im}}{\partial x_j} = \boldsymbol{\beta}^*_m a_{im:jm}. \tag{19}$$

If a variable $\mathbf{X}$ does not appear in the equation of coordinate m, then set the corresponding parameter to zero (this will not happen in our application). The estimated impacts are then obtained by substituting the parameters with their estimates.

For the second issue, the formulas in Morais and Thomas-Agnan (2020) cannot be applied directly due to the fact that the UniSEM and MultiLAG models involve this non-linear filter. Extensions of Morais and Thomas-Agnan (2020) adapted to the presence of a spatial filter can be found in Thomas-Agnan et al. (2020). In the present work, we propose an approximation of these semi-elasticities as follows. Morais and Thomas-Agnan (2020) show that the effect of increasing a classical explanatory variable $\mathbf{X}$ by an additive amount δ on unit j has a multiplicative impact on the components of the dependent vector $(\mathbf{Y}_{i1} \ldots \mathbf{Y}_{iD})$ (for statistical unit i):

$$\mathbb{E}\mathbf{Y}_{im}(x_j + \delta) \sim \mathbb{E}\mathbf{Y}_{im}(x_j)\left(1 + \text{se}_{ij;m}\right), \tag{20}$$

where

$$\text{se}_{ij;m} = \frac{\partial \ln \mathbb{E}\mathbf{Y}_{im}}{\partial x_j}(x_j)$$

is classically called semi-elasticity of the mth component $\mathbf{Y}_{im}$ ($m = 1, \ldots, D$ and $i, j = 1, \ldots, n$) with respect to $\mathbf{X}$. Morais and Thomas-Agnan (2020) show that in the non-spatial model these semi-elasticities depend on the share vector at the point of interest but not on the location at which we increase the explanatory variable. Formula (20) yields a way to estimate these approximate semi-elasticities using a finite difference approach for a small value of δ and using fitted shares $\hat{\mathbf{Y}}_{im}(x_j + \delta)$. For each data point we obtain a full structure of size $n^2(D - 1)$ of semi-elasticities. To summarize the spatial impacts, LeSage and Pace (2009) propose summary measures called the direct, indirect and total impacts. Here, one first needs to re-arrange the semi-elasticities vectors thus computed at the individual level into D matrices of size $n \times n$ (one matrix of semi-elasticities per component). For each component m, the average of the diagonal terms will represent the average direct impact (acronym ADI) $\text{ADI}_m = \sum_{i=1}^{n} \text{se}_{ii;m}/n$. The average indirect impact (acronym AII) should be the average of the sum of the extra-diagonal terms. In order to alleviate its evaluation, we propose to restrict the sum to neighboring points as follows:

$$\text{AII}_m = \frac{1}{n}\sum_{i=1}^{n}\sum_{j \neq i} \text{se}_{ij;m} \sim \frac{1}{n}\sum_{i=1}^{n}\sum_{j \neq i, j \in N(i)} \text{se}_{ij;m}, \tag{21}$$

where $N(i)$ is the set of first order neighbors of a given location i. Because this averaging will hide some variability, we propose to introduce the local contributions to the direct impact and indirect impact respectively as $\text{se}_{ii;m}$ and $\sum_{j \neq i, j \in N(i)} \text{se}_{ij;m}$ (see Sect. 4 for their illustration).

4 Data and Results

4.1 Land Use Data

The dataset and descriptive statistics of the variables used in this study are presented in Table 1. More detail can be found in the supplementary material available online

Table 1 Descriptive statistics for land use shares and for the explanatory variables

	Description	Mean	St.dev.	Min	q.25%	q.50%	q.75%	Max
Land use shares								
Agriculture	Share of crops and pastures	0.60	0.29	0	0.39	0.67	0.85	1.00
Forest	Share of forest	0.26	0.22	0	0.07	0.20	0.41	1.00
Other	Share of other uses	0.09	0.17	0	0.00	0.01	0.07	1.00
Urban	Share of urban	0.05	0.09	0	0.01	0.02	0.05	0.99
Explanatory variables								
Shadow price	Agricultural land shadow price (AROPAj, spatialized) (k€/ha)	0.55	0.22	0.00	0.40	0.48	0.68	1.11
Forest revenue	Forest revenues (FFSM++, regional) (€/ha)	137.68	66.51	28.93	85.72	127.75	187.99	308.04
Pop. Income	Population revenues (INSEE, commune) (k€/year/household)	12.31	3.24	0.00	10.38	11.97	13.98	41.80
Pop. density	Population density (INSEE, "carottage") (household/ha)	5.43	2.27	2.75	4.48	4.90	5.58	58.72
Slope	Terrain average slope (GTOPO30) (%)	4.33	6.15	0.00	1.14	2.10	4.23	47.72
Texture	Soil's texture classes				1	2	3	4
	Number of cells				1242	4820	3120	579

(see http://www.thibault.laurent.free.fr/code/land_use_spat_coda/). Land use data are collected from the Corine Land Cover (CLC) databases for France at the scale of $100\,\text{m} \times 100\,\text{m}$ (1 ha) for the year 2000 (except for the Paris area). Land uses are grouped into four categories: "agriculture", "forest", "urban" and "other" where "agriculture" is aggregated from crops and pastures. We calculate the land use shares for each grid cell of $8\,\text{km} \times 8\,\text{km}$. Land use shares are expressed as the sum of the land use classes in hectares divided by the surface of the grid cell. We are using the simple zero replacement strategy suggested by Aitchison (1986) and fully described in Martín-Fernández et al. (2003). We choose to use a weight matrix $\mathbf{W}$ based on first order queen contiguity, i.e., two polygons are neighbors if they share a common side or a common vertex.

According to the economic literature, standard drivers of land use allocation, consistent with the von Thünen and Ricardian conceptual models, are the measures of the net return to each use and biophysical suitability. However, net returns are not easily observable for all land uses and one solution is to use proxies. For example, due to data limitations, a measure of the average net return to urban land is difficult to construct. An alternative is to use a variable that proxies for these returns. One measure frequently used in land use share models is population density, which indicates the pressures for urbanization. Chakir and Lungarska (2017) compared three possible proxies of agricultural land rent: agricultural land price, farm income and the land shadow price. These comparisons are made based on various criteria including: consistency with the theoretical hypothesis, prediction quality and specification tests. They conclude that shadow price proxy provides stable and intuitive results. In this paper, we consider the same explanatory variables as in Lungarska and Chakir (2018) but we replace the spatial Durbin error model by the spatial error model because the lagged explanatory variables of the spatial Durbin error model resulted in collinearity issues with the S2SLS method.

Following Pawlowsky-Glahn and Egozcue (2011), to guide our choice of ilr transformation (contrast matrix), we look at the biplot associated with the first two principal components (see the supplementary material available online).

The first two principal axes explain 94% of the variability in the data. On the first one, the land use "other" is opposed to the rest of the land uses while on the second one, the land use "forest" is opposed to "urban". We thus propose the following contrast matrix: the first ilr coordinate opposes the land use "other" to the geometric mean of the other land uses while the second opposes "agricultural" to the geometric mean of "forest" and "urban" and the third "forest" to the "urban" land use.

4.2 Model Results

In this section, we fit the UniSEM model and the MultiLAG model, but also consider two other models based on the same variables introduced in Sect. 2.4: the Ordinary Linear Model (OLM) and the UniLAG model which can both be fit separately for each equation by maximum likelihood.

Table 2 shows the results of the OLM and the MultiLAG regression models, the most simple and the most complex of our models. Results for the other models can be found in the supplementary material. Most explanatory variables are significant for the three ilr coordinates. The Moran test (see for instance Cliff and Ord 1981) on the OLM residuals is highly significant motivating the need for a spatial model. Most of the spatial autocorrelation parameters of $\mathbf{R}^*$ are significant too. However it is not possible to directly compare the parameters of the OLM with those of the MultiLAG due to the spatial filter (see LeSage and Pace 2009) and the parameters of the MultiLAG with those of the UniSEM due to the fact that they use different coordinate spaces, hence the need to go back to the simplex.

Following Egozcue et al. (2012), we propose to compare the fit of these models using a mean square error measure built with the Aitchison geometry (norm denoted by $\| \cdot \|_A$):

$$\text{MSE} = \frac{1}{n} \sum_{i=1}^{n} \| \hat{\mathbf{Y}}_i \ominus \mathbf{Y}_i \|_A^2 = \frac{1}{n} \sum_{i=1}^{n} \| \text{ilr}_V(\hat{\mathbf{Y}}_i) - \text{ilr}_V(\mathbf{Y}_i) \|^2 \tag{22}$$

Table 2 Results of the OLM versus MultiLAG

	OLM in the ILR space			MultiLAG in the ILR space		
	ilr_1	ilr_2	ilr_3	ilr_1	ilr_2	ilr_3
Intercept	−0.369*	3.582***	−6.353***	1.498***	2.499***	−4.657***
	(0.168)	(0.089)	(0.13)	(0.318)	(0.18)	(0.271)
Shadow price	0.137	0.515***	0.755***	−0.117	0.09	0.214*
	(0.138)	(0.073)	(0.107)	(0.121)	(0.069)	(0.103)
Forest revenues	−0.008***	0	0.001***	−0.001**	0	0.001***
	(0)	(0)	(0)	(0)	(0)	(0)
Population density	0.121***	−0.138***	0.279***	0.018	−0.119***	0.223***
	(0.013)	(0.007)	(0.01)	(0.013)	(0.007)	(0.011)
Population income	−0.137***	−0.117***	0.191***	−0.096***	−0.074***	0.158***
	(0.009)	(0.005)	(0.007)	(0.013)	(0.007)	(0.011)
Slope	0.203***	−0.097***	−0.124***	0.072***	−0.029***	−0.081***
	(0.005)	(0.003)	(0.004)	(0.008)	(0.005)	(0.007)
Texture 2	−0.737***	0.815***	0.743***	−0.442***	0.242***	0.344***
	(0.086)	(0.045)	(0.066)	(0.085)	(0.048)	(0.073)
Texture 3	−1.562***	0.944***	0.691***	−0.71***	0.347***	0.369***
	(0.092)	(0.049)	(0.071)	(0.092)	(0.052)	(0.078)
Texture 4	−1.81***	1.324***	0.405***	−0.846***	0.666***	0.122
	(0.136)	(0.072)	(0.105)	(0.123)	(0.07)	(0.105)
$\mathbf{R}^*_{1.}$	–	–	–	0.784***	0.052*	0.14***
				(0.037)	(0.021)	(0.032)
$\mathbf{R}^*_{2.}$	–	–	–	−0.031	0.677***	0.275***
				(0.066)	(0.037)	(0.056)
$\mathbf{R}^*_{3.}$	–	–	–	0.229***	0.178***	0.44***
				(0.044)	(0.025)	(0.038)
$\mathbf{\Sigma}^*_{1.}$	7.216	−0.233	−0.544	4.461	0.158	−0.668
$\mathbf{\Sigma}^*_{2.}$	−0.233	2.002	−0.075	0.158	1.434	−0.481
$\mathbf{\Sigma}^*_{3.}$	−0.544	−0.075	4.312	−0.668	−0.481	3.238
Nb. Obs.	9760	9760	9760	9760	9760	9760
Moran's *I* test	76.02***	64.29***	70.14***	–	–	–
LMlag	5834***	4008***	4448***	–	–	–

Note $^{*}p < 0.1$; $^{**}p < 0.05$; $^{***}p < 0.01$

Table 3 MSE for all models

Coordinates	Univariate					Multivariate	
	OLM	SEM	LAG(TS)	LAG(TC)	LAG(BP)	LAG(TC)	LAG(TS)
ilr1	7.21	4.52	4.49	6.99	4.36	9.13	4.55
ilr2	2.00	1.38	1.43	1.99	1.43	2.58	1.53
ilr3	4.31	2.91	3.01	4.40	2.97	4.49	3.57
Sum	13.52	8.81	8.93	13.38	**8.76**	16.20	9.65

for any contrast matrix $\mathbf{V}$. Note that since the goal is to evaluate a prediction ability of the model, instead of using a fitted value understood as an estimate of $\mathbb{E}(\mathbf{Y}_i)$, we rather use a prediction of $\mathbf{Y}_i$. There are several ways to compute predicted values in spatial models, therefore we indicate which formula we use among best prediction (BP), trend signal (TS) and trend corrected (TC) (see Goulard et al. 2017 for detail). We use univariate predictors due to the lack of code to evaluate multivariate versions. For the MultiLAG model, because the ilr is isometric with respect to the Aitchison geometry, it is enough to compute the MSE in coordinate space. For the UniSEM model, this is not true anymore since the alr is not isometric. For this reason, we first compute the fitted response in the simplex and then compute the MSE in an ilr coordinate space (and it does not depend upon which particular one). Table 3 summarizes the MSE of all considered models. The MSE of UniSEM, UniLAG (using TS prediction) and UniLAG (using BP prediction) are very close to each other. The UniSEM model reaches the smallest MSE for the second and third ilr. Results also show that the total MSE of the UniLAG with the BP prediction method (see Goulard et al. 2017) is the smallest among all models.

4.3 *Interpretation*

In this section, we try to better understand the impact of population density changes on land use using the techniques proposed in Sect. 3.2. We first interpret the fitted shares and the semi-elasticities thanks to very simple but informative scatterplots. Then we illustrate the interpretation of the direct and indirect impacts in the MultiLAG model using maps.

4.3.1 Fitted Shares and Semi-elasticities

We first select four particular grid cells, shown on Fig. 1a, based on the fact that they are typical of a dominant share (more than 80%) of each particular land use (M_u for "urban", M_a for "agriculture", M_f for "forest" and M_o for "other"). We then draw the fitted shares scatterplot presented in Sect. 3.2 for a sequence of 100 equally

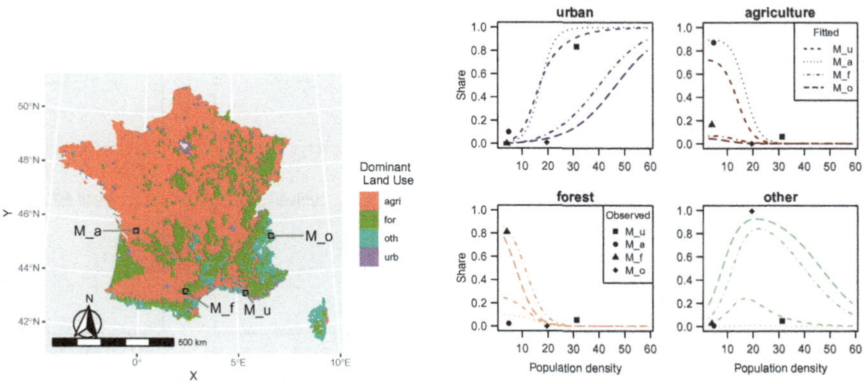

Four selected points with a respective dominance of "agriculture", "urban", "forest" and "other".

Fitted shares curves according to population density for the four selected points.

Fig. 1 Selected points with a dominant land use text and their fitted shares curves according to population density for the OLM model

spaced values from 2.75 (minimum) to 58.72 (maximum) of the variable population density while the other variables are fixed at the observed values for the four selected points. Figure 1b shows the fitted values obtained with the OLM model.

Each selected location is represented by a different dash type for the fitted shares and a different symbol for the observed shares. Population density shows a positive impact on "urban" use and a negative impact on both "agriculture" and "forest" uses. The effect on "other" use is first increasing and then decreasing. One can emphasize the variability of these curves according to the selected locations: "urban" use is gaining ground rapidly at the expense of "agriculture" for M_u and M_a. For M_f and M_o, it seems that the decline of the "forest" use is first beneficial to the "other" use before switching to the "urban" use.

Figure 2a shows the average fitted shares according to population density for the OLM, UniSEM, UniLAG and MultiLAG models. At the exception of the UniLAG model, we see that the three other models, OLM, UniSEM and MultiLAG, give relatively comparable results for the fitted "urban" and "agricultural" uses: in particular these two curves intersect around the point where density is equal to 20 household per hectare whereas the intersection occurs at 10 for the UniLAG. Only the MultiLAG model shows a less flat shape of "other" uses. Generally, an increase in population density results in an increase of "urban" areas at the expense of "agricultural" land. This phenomenon has been observed in France since many years in rural peri-urban areas with the sprawl of cities and road infrastructure (Chakir and Madignier 2006).

Figure 2b present the average semi-elasticities with respect to population density for the four models. They are computed with a value of $\delta = 10^{-10}$ for the approximation (20). The average semi-elasticities have the same shape for the UniSEM and MultiLAG models. The semi-elasticities are always positive for "urban" land

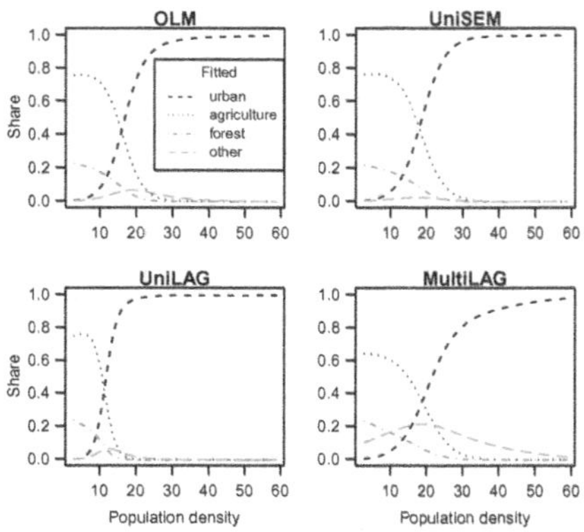

Fitted shares according to population density for OLM, SEM, UniLAG and MultiLAG models.

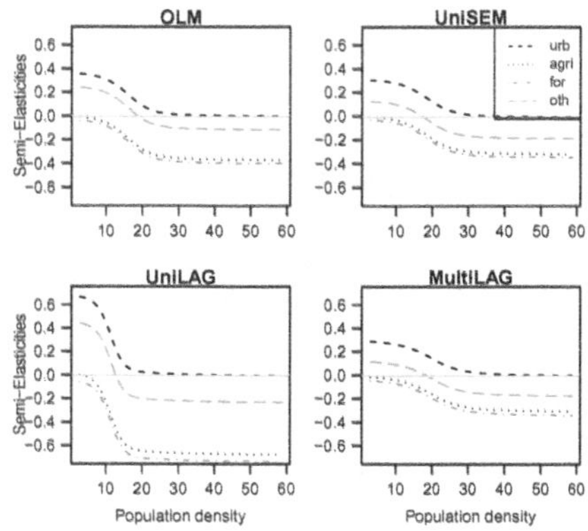

Average semi-elasticities with respect to population density for OLM, SEM, UniLAG and MultiLAG models.

Fig. 2 Fitted shares and average semi-elasticities for population density

use and negative for "forest" and "other" land uses. This means that an increase in population density results in an increase of the "urban" and "other" use at the expense of "forest" and "agricultural" uses. The semi-elasticity of "other" land use remains positive until a population density around 20 households per hectare and becomes negative after that value. This indicates that in areas with low urban density, increasing population density would increase "urban" areas and "other" land uses at the expense of agriculture and forestry. For example, taking the selected point represented by a circle (agriculture dominant area) with a fitted share vector equal to $(0.006, 0.899, 0.092, 0.003)$ (in the order "urban", "agriculture", "forest" and "other") with the MultiLAG model and a vector of semi-elasticities equal to $(0.305, 0.001, -0.029, 0.131)$, an increase of one household per hectare, all other variables remaining constant, results approximately in an increase of the "urban" share by 30% and a decrease of forestry by 3% leaving the other two shares almost constant. In areas with high urban density (above 20 households per hectare), "urban" use would gain ground on all other uses including "other" land use. Note that the "other" land use may correspond to a temporary land use awaiting conversion to "urban" use either from "forest" or "agricultural" lands.

4.3.2 Direct and Indirect Impacts

Figure 3 is an illustration of the local contributions to the direct and indirect impacts caused by an increase of one unit of population density at the cell M_u for the MultiLAG model. M_u is part of the city of Marseille, the second largest city in France. Its neighborhood contains mainly cells with a dominance of "urban" or "other" uses (see Fig. 3a). Figure 3b (resp. Fig. 3c) represent the change of "urban" (resp. "other")

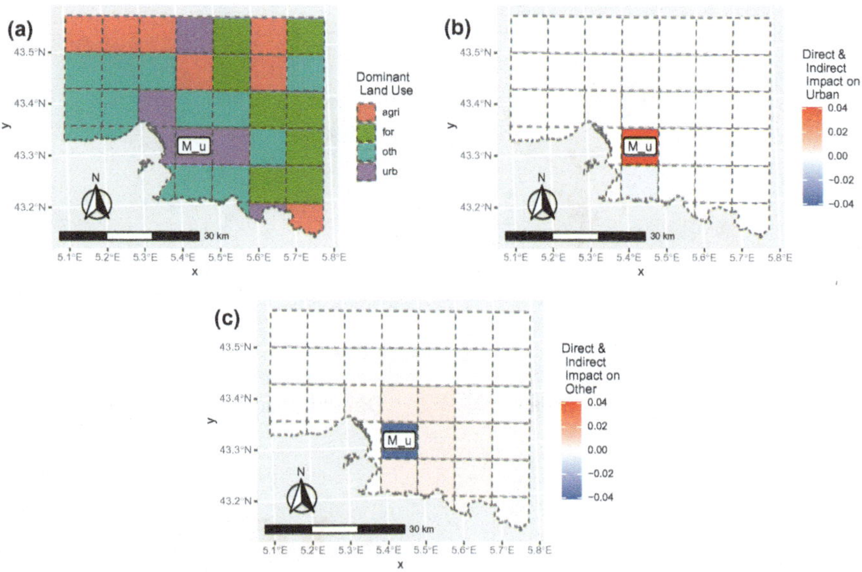

Fig. 3 Region of Marseille: **a** Dominant land uses; **b** Local impact on "urban" share; **c** Local impact on "other" share

use on M_u and its neighborhood. The local direct impact corresponds to the changes observed at M_u itself whereas the local indirect impact corresponds to the changes observed at all other cells. We observe that an increase of one unit of population density has a positive direct impact on the "urban" use at the expense of "other" use. On the contrary, we observe a positive indirect impact on the "other" use whereas the indirect impact on the "urban" use is almost zero. Moreover, the indirect impact is all the more strong as the location is closer to the point of interest (see Fig. 3c).

The maps in Fig. 4 summarize direct and indirect impacts of population density on "urban" and "agricultural" land use shares. They show that an increase in population density has a larger increase impact on "urban" land in already urbanized areas and it happens at the expense of "agricultural" land located in peri-urban areas.

The maps in Fig. 5 show direct and indirect impacts of population density on "forest" and "other" land use shares. These maps show that an increase in population density results in an increase of "other" land uses at the expense of "forest" mainly in less urbanized areas. This confirms the fact that "other" land use could be a transition use from "forest" to "urban" land use when the population density increases.

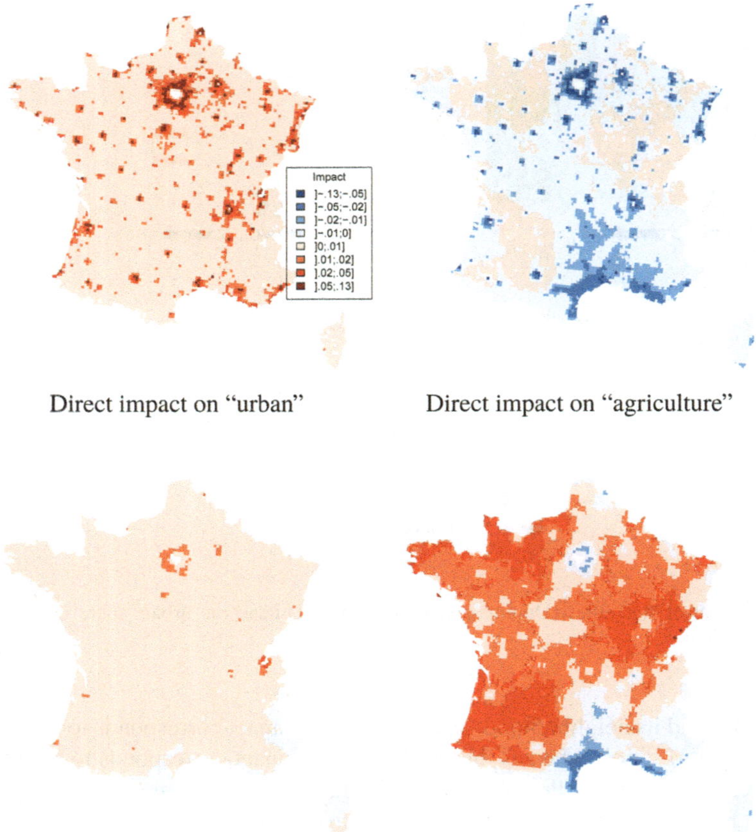

Direct impact on "urban" Direct impact on "agriculture"

Indirect impact on "urban" Indirect impact on "agriculture"

Fig. 4 Direct and indirect impact of population density on "urban" and "agricultural" shares

5 Conclusion

Through this application to land use, we have explored the differences between two families of models which we call the UniSEM and the MultiLAG. We have proved that a choice of coordinate representation does not have any impact on the parameters in the simplex as long as we do not impose further restrictions. Indeed we have seen that a restriction like the matrix of spatial autocorrelation parameters in coordinate space being diagonal is not a simplex assumption; it is dependent on a particular choice of coordinate space. Therefore a multivariate approach without any restrictions on the parameters is preferable. Finally we could have considered a MultiSEM model but did not do so due to the lack of code to fit such a multivariate model and this can be considered as a perspective for future work.

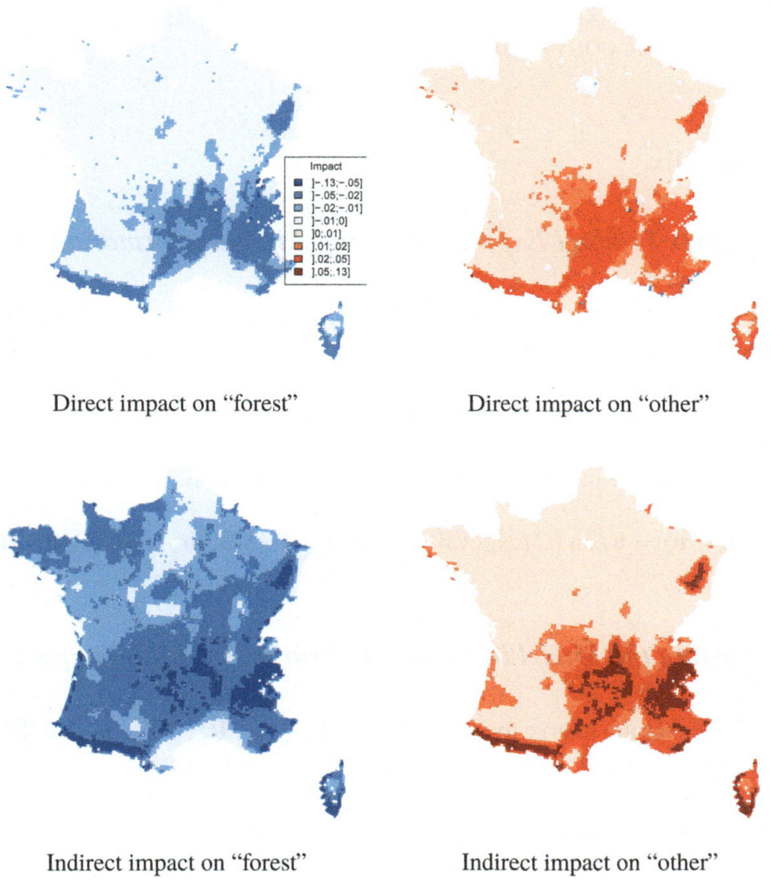

Fig. 5 Direct and indirect impact of population density on "forest" and "other" shares

We have demonstrated that approximations of the semi-elasticities can be easily computed and interpreted in the simplex. The application results yield interesting insights for understanding the drivers of land use. Several directions of improvements could be pursued. First of all, one could derive formulas for the statistical significance of the semi-elasticities. The prediction formulas which have been used in the MSE evaluation are univariate: using multivariate counterparts (which haven't been derived yet for these models to our knowledge) would be preferable.

Acknowledgements The authors are grateful to Vera Pawlowsky-Glahn for her valuable contributions to the CODA field. They also thank two anonymous referees and the editors for their helpful comments. Thibault Laurent, Anne Ruiz-Gazen and Christine Thomas-Agnan acknowledge funding from ANR under grant ANR-17-EURE-0010 (Investissements d'Avenir program). Raja Chakir and Anna Lungaska acknowledge the support of the Agence Nationale de la Recherche as part of the "Investments d'Avenir" Programme within STIMUL (Scenarios Towards integrating multiscale land use tools) flagship project (LabEx BASC; ANR-11- LABX-0034) and Cland Institut de convergence (ANR-16-CONV-0003)

6 Appendix

6.1 Matrix Product Expression Using Alr Transformation

Let us consider the alr_D transformation $\text{alr}_D(\mathbf{u}) = \mathbf{F}_D \ln(\mathbf{u})$ for a vector $\mathbf{u} \in \mathbf{S}^D$ and with $\mathbf{F}_D = \begin{bmatrix}\mathbf{I}_{D-1} & -\mathbf{j}_{D-1}\end{bmatrix}$. The result can be generalized to any alr_m transformation.

$$\text{alr}_D(\mathbf{B} \boxdot \mathbf{u}) = \mathbf{F}_D \ln(\mathbf{B} \boxdot \mathbf{u}) = \mathbf{F}_D \ln\left(\text{ilr}_V^{-1}\left(\mathbf{B}_V^* \text{ilr}_V(\mathbf{u})\right)\right),$$

where $\mathbf{VB}_V^* = \mathbf{BV}$. Since $\mathbf{VV}^T = \mathbf{I}_D - \mathbf{j}_D\mathbf{j}_D^T/D$ and $\mathbf{F}_D\mathbf{j}_D = \mathbf{0}_D$, we have

$$\text{alr}_D(\mathbf{B} \boxdot \mathbf{u}) = \mathbf{F}_D \ln\left(C\left(\exp(\mathbf{VB}_V^* \text{ilr}_V(\mathbf{u}))\right)\right) = \mathbf{F}_D\mathbf{VB}_V^* \text{ilr}_V(\mathbf{u}) = \mathbf{F}_D\mathbf{B}\ln(\mathbf{u}).$$

6.2 Writing the Model in Reduced Form in Coordinate Space

Recognizing that the matrix form of the term $\sum_{m \in S_l^{WY^*}} \mathbf{R}_{ml}^* \mathbf{WY}_{.m}^*$ is $\mathbf{WY}^*\mathbf{R}^*$, we get easily that the matrix formulation of the model can be written

$$\mathbf{Y}^* = \mathbf{WY}^*\mathbf{R}^* + \left[\mathbf{X}_{S_1^X}\boldsymbol{\beta}_{S_1^X}^* \cdots \mathbf{X}_{S_{D-1}^X}\boldsymbol{\beta}_{S_{D-1}^X}^*\right] + \boldsymbol{\epsilon}^*.$$

We are going to use the following property: if we have three matrices $\mathbf{A}$, $\mathbf{B}$ and $\mathbf{C}$ such that the number of columns of $\mathbf{A}$ is equal to the number of rows of $\mathbf{B}$ and the number of columns of $\mathbf{B}$ is equal to the number of rows of $\mathbf{C}$, then

$$\text{vec}_c(\mathbf{ABC}) = (\mathbf{C}^T \otimes \mathbf{A})\text{vec}_c(\mathbf{B}). \tag{23}$$

Using (23), we get

$$\text{vec}_c(\mathbf{WY}^*\mathbf{R}^*) = \left((\mathbf{R}^*)^T \otimes \mathbf{W}\right) \text{vec}_c(\mathbf{Y}^*).$$

Therefore the c-vectorization of the whole model is

$$\text{vec}_c\mathbf{Y}^* = \left((\mathbf{R}^*)^T \otimes \mathbf{W}\right) \text{vec}_c(\mathbf{Y}^*) + \begin{bmatrix} \mathbf{X}_{S_1}\boldsymbol{\beta}_1^* \\ \cdot \\ \mathbf{X}_{S_{D-1}}\boldsymbol{\beta}_{D-1}^* \end{bmatrix} + \text{vec}_c\boldsymbol{\epsilon}^*$$

and the reduced form of the model in vectorized form is

$$\text{vec}_c \mathbf{Y}^* = (\mathbf{I}_{n(D-1)} - ((\mathbf{R}^*)^T \otimes \mathbf{W}))^{-1} \begin{bmatrix} \mathbf{X}_{\mathbf{S}_1}\boldsymbol{\beta}_1^* \\ . \\ \mathbf{X}_{\mathbf{S}_{D-1}}\boldsymbol{\beta}_{D-1}^* \end{bmatrix} + \text{vec}_c \boldsymbol{\epsilon}^*.$$

6.3 Lemma Used in the Proof of Theorem 1

Lemma 1 *Let $\mathbf{j}_D$ (resp. $\mathbf{0}_D$) denote the D-dimensional column vector of ones (resp. zeros) and $\mathbf{I}_D$ be the $D \times D$ identity matrix. Let $\mathbf{K} = \begin{bmatrix} \mathbf{I}_{D-1} - \mathbf{j}_{D-1}\mathbf{j}_{D-1}^T/D \\ -\mathbf{j}_{D-1}^T/D \end{bmatrix}$ a $D \times (D-1)$ matrix and $\mathbf{F} = \begin{bmatrix} \mathbf{I}_{D-1} & -\mathbf{j}_{D-1} \end{bmatrix}$ a $(D-1) \times D$ matrix. Let $\mathbf{B}$ be a $D \times D$ matrix such that $\mathbf{B}\mathbf{j}_D = \mathbf{0}_D$ and $\mathbf{B}^T\mathbf{j}_D = \mathbf{0}_D$ and $\mathbf{B}^*$ be a $(D-1) \times (D-1)$ matrix. We have that $\mathbf{FB} = \mathbf{B}^*\mathbf{F}$ is equivalent to $\mathbf{B} = \mathbf{KB}^*\mathbf{F}$.*

We have

$$\mathbf{K} = \begin{bmatrix} 1-\frac{1}{D} & -\frac{1}{D} & -\frac{1}{D} & \cdots & -\frac{1}{D} \\ -\frac{1}{D} & 1-\frac{1}{D} & -\frac{1}{D} & \cdots & -\frac{1}{D} \\ -\frac{1}{D} & -\frac{1}{D} & 1-\frac{1}{D} & \cdots & -\frac{1}{D} \\ \vdots & \vdots & \vdots & \ddots & \vdots \\ -\frac{1}{D} & -\frac{1}{D} & \frac{1}{D} & \cdots & 1-\frac{1}{D} \\ -\frac{1}{D} & -\frac{1}{D} & -\frac{1}{D} & \cdots & -\frac{1}{D} \end{bmatrix} \quad \text{and} \quad \mathbf{F} = \begin{bmatrix} 1\,0\,0 \ldots 0 & -1 \\ 0\,1\,0 \ldots 0 & -1 \\ 0\,0\,1 \ldots 0 & -1 \\ \vdots\ \vdots \quad \ddots\ 0 & -1 \\ 0\,0\,0 \quad\ \ 1 & -1. \end{bmatrix}$$

Let $\mathbf{B}$ be a $D \times D$ matrix such that $\mathbf{B}\mathbf{j}_D = \mathbf{0}_D$ and $\mathbf{B}^T\mathbf{j}_D = \mathbf{0}_D$ and $\mathbf{B}^*$ be a $(D-1) \times (D-1)$ matrix. First, let us prove that if $\mathbf{B} = \mathbf{KB}^*\mathbf{F}$, then $\mathbf{FB} = \mathbf{B}^*\mathbf{F}$. It is easy to show that $\mathbf{FK} = \mathbf{I}_{D-1}$. Thus, if $\mathbf{B} = \mathbf{KB}^*\mathbf{F}$,

$$\mathbf{FB} = \mathbf{FKB}^*\mathbf{F} = \mathbf{I}_{D-1}\mathbf{B}^*\mathbf{F} = \mathbf{B}^*\mathbf{F}.$$

Let us now prove the converse. We have $\mathbf{B}\mathbf{j}_D = \mathbf{B}^T\mathbf{j}_D = \mathbf{0}_D$ and we assume $\mathbf{FB} = \mathbf{B}^*\mathbf{F}$. We write $\mathbf{B}$ in blocks as follows:

$$\mathbf{B} = \begin{bmatrix} \mathbf{B}_1 & \mathbf{b}_2 \\ \mathbf{b}_3^T & b_4, \end{bmatrix}$$

where $\mathbf{B}_1$ is a $D-1 \times D-1$ matrix, $\mathbf{b}_2$ and $\mathbf{b}_3^T \in \mathbf{R}^{D-1}$ and $\mathbf{b}_4 \in \mathbf{R}$. Using the fact that $\mathbf{B}\mathbf{j}_D = \mathbf{B}^T\mathbf{j}_D = \mathbf{0}_D$, we get

$$\begin{cases} \mathbf{b}_2 = -\mathbf{B}_1\mathbf{j}_{D-1} \\ \mathbf{b}_3 = -\mathbf{B}_1^T\mathbf{j}_{D-1} \\ \mathbf{b}_4 = \mathbf{j}_{D-1}^T\mathbf{B}_1\mathbf{j}_{D-1} \end{cases} . \tag{24}$$

To find $\mathbf{B}$, we only need to find $\mathbf{B}_1$. Using (24), we write the $(D-1) \times D$ matrix $\mathbf{FB}$ as a function of $\mathbf{B}_1$:

$$\mathbf{FB} = [(\mathbf{I}_{D-1} + \mathbf{j}_{D-1}\mathbf{j}_{D-1}^T)\mathbf{B}_1 \quad - (\mathbf{I}_{D-1} + \mathbf{j}_{D-1}\mathbf{j}_{D-1}^T)\mathbf{B}_1\mathbf{j}_{D-1}].$$

Furthermore, $\mathbf{B}^*\mathbf{F} = [\mathbf{B}^* \quad - \mathbf{B}^*\mathbf{j}_{D-1}]$. So, $\mathbf{FB} = \mathbf{B}^*\mathbf{F}$ implies

$$\left(\mathbf{I}_{D-1} + \mathbf{j}_{D-1}\mathbf{j}_{D-1}^T\right)\mathbf{B}_1 = \mathbf{B}^*.$$

The inverse matrix of $\left(\mathbf{I}_{D-1} + \mathbf{j}_{D-1}\mathbf{j}_{D-1}^T\right)$ is $\left(\mathbf{I}_{D-1} + \mathbf{j}_{D-1}\mathbf{j}_{D-1}^T/D\right)$. Thus,

$$\mathbf{B}_1 = (\mathbf{I}_{D-1} + \mathbf{j}_{D-1}\mathbf{j}_{D-1}^T/D)\mathbf{B}^*. \tag{25}$$

Using (24) and (25), it is now easy to check that $\mathbf{B} = \mathbf{KB}^*\mathbf{F}$.

References

J. Aitchison, *The Statistical Analysis of Compositional Data* (Chapman and Hall, London, 1986). Reprinted in 2003 with additional material by the Blackburn Press

R.S. Bivand, D.W.S. Wong, Comparing implementations of global and local indicators of spatial association. TEST **27**(3), 716–748 (2018)

R. Chakir, A. Lungarska, Agricultural rent in land-use models: comparison of frequently used proxies. Spat. Econ. Anal. **12**(2–3), 279–303 (2017)

R. Chakir, A.-C. Madignier, Analyse des changements d'occupation des sols en France entre 1992 et 2003. Economie Rurale 296 (2006)

J. Chen, X. Zhang, S. Li, Multiple linear regression with compositional response and covariates. J. Appl. Stat. **44**(12), 2270–2285 (2017)

A.D. Cliff, J.K. Ord, *Spatial Processes* (Pion, London, 1981)

P. Filzmoser, K. Hron, M. Templ, *Applied Compositional Data Analysis. With Worked Examples in R*. Springer Series in Statistics (Springer International Publishing, Springer Nature Switzerland AG, Cham, Switzerland, 2018)

J.A. Foley, R. DeFries, G.P. Asner, C. Barford, G. Bonan, S.R. Carpenter, F.S. Chapin, M.T. Coe, G.C. Daily, H.K. Gibbs et al., Global consequences of land use. Science **309**(5734), 570–574 (2005)

J.J. Egozcue, J. Daunis-I-Estadella, V. Pawlowsky-Glahn, K. Hron, P. Filzmoser, Simplicial regression. The normal model. J. Appl. Probab. Stat. **6**, 87–108 (2012)

M. Goulard, T. Laurent, C. Thomas-Agnan, About predictions in spatial autoregressive models: optimal and almost optimal strategies. Spat. Econ. Anal. **12**(2–3), 304–325 (2017)

H.H. Kelejian, I.R. Prucha, A generalized spatial two-stage least squares procedure for estimating a spatial autoregressive model with autoregressive disturbances. J. R. Estate Financ. Econ. **17**(1), 99–121 (1998)

H.H. Kelejian, I.R. Prucha, Estimation of simultaneous systems of spatially interrelated cross sectional equations. J. Econ. **118**(1–2), 27–50 (2004)

R. Lal, Soil carbon sequestration impacts on global climate change and food security. Science **304**(5677), 1623–1627 (2004)

T.J. Leininger, A.E. Gelfand, J.M. Allen, J.A. Silander, Spatial regression modeling for compositional data with many zeros. J. Agric. Biol. Environ. Stat. **18**(3), 314–334 (2013)

J. LeSage, R.K. Pace, *Introduction to Spatial Econometrics* (Chapman and Hall/CRC, London, 2009)

A. Lungarska, R. Chakir, Climate-induced land use change in France: impacts of agricultural adaptation and climate change mitigation. Ecol. Econ. **147**, 134–154 (2018)

J.A. Martín-Fernández, C. Barceló-Vidal, V. Pawlowsky-Glahn, Dealing with zeros and missing values in compositional data sets using nonparametric imputation. Math. Geol. **35**(3), 253–278 (2003)

J. Morais, C. Thomas-Agnan, Covariates impacts in compositional models and simplicial derivatives. Austrian J. Stat. (2020)

D.K. Munroe, D. Müller, Issues in spatially explicit statistical land-use/cover change (LUCC) models: examples from western Honduras and the central highlands of Vietnam. Land Use Policy **24**(3), 521–530 (2007)

T.H.A. Nguyen, C. Thomas-Agnan, T. Laurent, A. Ruiz-Gazen, A, A simultaneous spatial autoregressive model for compositional data. Spat. Econ. Anal. (2019). https://www.tse-fr.eu/fr/publications/simultaneous-spatial-autoregressive-model-compositional-data

V. Pawlowsky-Glahn, J.J. Egozcue, Exploring compositional data with the CoDa-dendrogram. Austrian J. Stat. **40**(1&2), 103–113 (2011)

V. Pawlowsky-Glahn, J.J. Egozcue, R. Tolosana-Delgado, *Modeling and Analysis of Compositional Data* (Wiley, Hoboken, 2015)

R.A. Pielke, Land use and climate change. Science **310**(5754), 1625–1626 (2005)

B. Pirzamanbein, J. Lindström, A. Poska, M.-J. Gaillard, Modelling spatial compositional data: reconstructions of past land cover and uncertainties. Spat. Stat. **24**, 14–31 (2018)

G. Shmueli, To explain or to predict? Stat. Sci. 289–310 (2010)

C. Thomas-Agnan, T. Laurent, A. Ruiz-Gazen, Covariates impacts in spatial autoregressive models for compositional data. TSE working paper 20-1162 (2020)

A. Veldkamp, E.F. Lambin, Predicting land-use change. Agric. Ecosyst. Environ. **85**(1), 1–6 (2001)

P.H. Verburg, K.-H. Erb, O. Mertz, G. Espindola, Land system science: between global challenges and local realities. Curr. Opin. Environ. Sustain. **5**(5), 433–437 (2013)

Applications

The Whole Versus the Parts: The Challenge of Compositional Data Analysis (CoDA) Methods for Geochemistry

Antonella Buccianti and Caterina Gozzi

Abstract In complex geochemical aqueous systems, chemical species are conceptually distinct but empirically related thanks to a large number of interactions taking place at different spatial and/or temporal scales. In this condition, common elements are shared, multiplicative interactions arise, and feedback mechanisms may be able to maintain the system far from the thermodynamical equilibrium, bearing wide fluctuations. Chemical species can have alternative stable states and transitions among them that could produce important consequences for the stability and the resilience of the solutions, also forced by climate changes and with impact on human health. Under the Compositional Data Analysis (CoDA) methodology, it is possible to appreciate the power of some tools able to take a look at the whole instead of the constituting parts, enhancing the understanding of the nature of mutual interactions. In this research work, the role of the perturbation operator governing addition/subtraction in the simplex geometry is explored as a way to trace compositional changes and investigate the system dynamics. The results of our approach on the chemistry of the Arno River waters (Central Italy) highlight the possibility to discover the resilience of chemical species under the pressure of the environmental drivers affecting the catchment. Geochemical mobility (e.g., ionic potential, ionic strength) can be associated with new tools that provide information on either the resistance to change, predisposing the system to critical shifts, or its adaptive capacity, which instead favors gradual changes. This information appears to be fundamental since river water chemistry enables to decipher processes at the boundaries among lithosphere, biosphere, hydrosphere, and atmosphere, all key reservoirs involved in the dynamics of the Earth. This knowledge will be particularly relevant if the pressure of the climatic changes on our planet will continue to increase.

A. Buccianti (✉) · C. Gozzi
Department of Earth Sciences, University of Florence, Florence, Italy
e-mail: antonella.buccianti@unifi.it

C. Gozzi
e-mail: caterina.gozzi@unifi.it

P. Filzmoser et al. (eds.), *Advances in Compositional Data Analysis*,
https://doi.org/10.1007/978-3-030-71175-7_13

1 Introduction

Compositional Data Analysis (CoDA) has been considered, at first, only as a way to open constrained data with the aim of working in the correct sample space. At the beginning, this idea has generated some misunderstanding since data that do not close to a given constant were assumed not to be compositional (Aitchison 1982). Subsequently, the idea of proportionality among the terms of a composition has gained ground and it was clear that the total to which a composition is closed is irrelevant (Aitchison 1986). What really counts in CoDA are the reciprocal relationships among the parts of the composition that are intimately linked to each other (Egozcue et al. 2011). This framework appears to well describe simultaneous chemical reactions that occur in natural processes. A geochemical system comprises, at least, an aqueous solution in which the species of many elements are dissolved, include one or more minerals, a buffering condition with a gas reservoir, and the atmosphere in the surficial environment (Bethke 2008). Water geochemists work daily with equations that describe the equilibrium of several simultaneous chemical reactions among dissolved species, minerals, and gases. Each mass action equation relates the activities of the species to the reaction equilibrium constants. How can we describe the state of such a system? A direct approach would be to write the single reactions among the species of the system, minerals, and gases. To solve equilibrium problems a set of concentrations that simultaneously satisfy the mass action equation should be written for each possible reaction. It is clear that such a system is multicomponent and due to the relationships among the parts and the presence of feedback mechanisms, its dynamics become complex. Reactions and concentrations cannot be decoupled and the reaction rates, relating sink and source terms, can induce nonlinearity (Sanchez-Vila et al. 2007). The evolution of an aqueous system is characterized by the migration of the chemical species between phases in innumerable cycles that form all together a larger, more complex interconnected whole (Kleidon 2010). The interconnections generate properties of the whole that cannot be wholly understood by examining the parts of the system in isolation. Thus, complex systems have properties that depend on the integrity of the whole. In this context, most of the geological systems are complex systems since they are dissipative and self-organizing systems, working in open and dynamic conditions and consisting of a great number of components (e.g., rocks, minerals, and elements) which interact in a nonlinear region far from the equilibrium (Shvartsev 2009). A way to take into account the simultaneous relationships among all the components of a system is to apply methods of multivariate analysis (Krzanowski 2000). If this path is followed, it is necessary to consider the appropriateness of the sample space in which data move, since it determines the variance-covariance structure. Detecting in a correct manner how variability moves is fundamental to intercept the dynamics of a complex system and to understand the response to environmental perturbations. In these terms, the CoDA approach appears to be the theoretical framework to abandon single variables and to focus our attention on the properties of the whole. The treatment of compositional data requires two possible strategies (Aitchison 1986). The first one, called "stay in the simplex" approach, adopts the Aitchison geometry to work in a constrained sample space. The second one is based on the transformation of the data to move cases out from the simplex in the real Euclidean space and it is known as "working in coordinates" (Egozcue et al. 2003; Pawlowsky-Glahn et al. 2015). In this work, our interest focused on the first approach and, in particular, on the use of the perturbation operator. It represents one of the basic tools required to give to the simplex a vector space structure and it could have a strategic role in monitoring changes, giving useful information about the dynamics of a system. An application example to the waters of the Arno (Central Italy) riverine geochemical system will be used to illustrate the procedure. The research question is whether the perturbation operator is able to provide insight about similarity among chemical elements and the coherence in their behavior within the dynamics of the system, thus offering a new geochemical tool of investigation.

2 Material and Methods

2.1 Tracing Change Through Perturbations

The vector $\mathbf{x} = (x_1, x_2, \ldots, x_D)$ indicates a D-part composition in the simplex:

$$S^D = \left[\mathbf{x} = (x_1, x_2, \ldots, x_D) : x_i > 0 (i = 1, 2, \ldots, D), \sum_{i=1}^{D} x_i = \kappa \right] \quad (1)$$

where κ is a given positive constant whose value depends on the measure unit. Two basic operations on the simplex, called perturbation and powering that are able to induce a real vector space structure on the simplex, have been introduced (Aitchison 1986). Furthermore, the introduction of an inner product, with its associated norm and distance, has been used to obtain a $D-1$-dimensional Hilbert space structure (Billheimer et al. 2001; Pawlowsky-Glahn and Egozcue 2001). In this framework, a perturbation $\mathbf{p} = (p_1, p_2, \ldots, p_D)$ is a differential scaling operator that, when applied to the composition $\mathbf{x} = (x_1, x_2, \ldots, x_D)$, yields the composition:

$$\mathbf{y} = \mathbf{p} \oplus \mathbf{x} = C(p_1 x_1, \ldots, p_D x_D) \quad (2)$$

where C is the closure operator that scales elements to remain in the simplex sample space:

$$C(\mathbf{x}) = \left[\frac{\kappa \cdot x_1}{\sum_{i=1}^{D} x_i}, \frac{\kappa \cdot x_2}{\sum_{i=1}^{D} x_i}, \ldots, \frac{\kappa \cdot x_D}{\sum_{i=1}^{D} x_i} \right]. \quad (3)$$

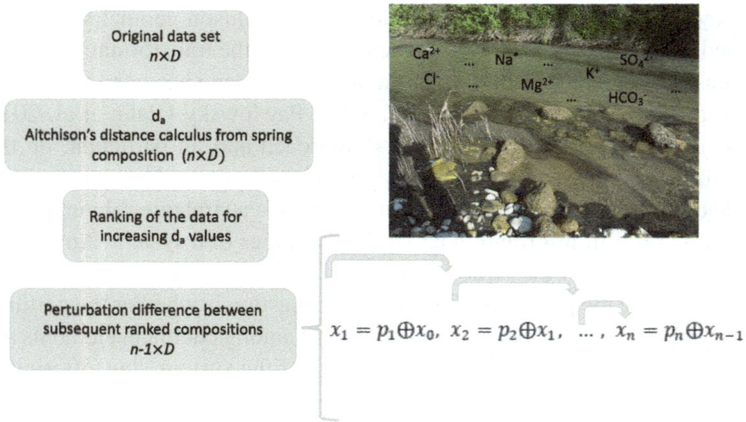

Fig. 1 Steps of the procedure to rank the cases by increasing Aichison's distance from the spring composition and to calculate the perturbation difference for subsequent compositions, $x_{i+1} - x_i$

As any composition can be expressed as a result of a perturbation on any other composition, the operator acquires a fundamental role in tracing compositional changes. This is especially true if a reference composition is available and differences with respect to this one can be calculated or when compositions are linked each other by a subsequent evolutive pattern. Thus, if $\mathbf{y} = \mathbf{p} \oplus \mathbf{x}$ corresponds to addition in the $\mathbb{R}$, the $\mathbf{y} = \mathbf{p} \ominus \mathbf{x}$ corresponds to the perturbation difference when obtained by a component-wise division of the elements of the **x** and **y** vectors (Aitchison 1986). Our approach is based on the transformation of the original dataset $n \times D$, related to the chemistry of waters collected in a river, in a perturbations matrix $n - 1 \times D$. In this matrix, each row is related to the perturbation difference among two subsequent compositions, corresponding to the difference $x_{i+1} - x_i$, after having ranked cases under some geochemical hypothesis. Thus, in order to calculate the perturbation matrix, the original dataset was first ranked by considering the increasing value of the Aitchison distance from the chemical composition of the spring, to be considered as a pristine water. The procedure is illustrated in Fig. 1. The (squared) Aitchison distance is a simplicial metric given by

$$d_a^2(\mathbf{x}, \mathbf{y}) = \sum_{i=1}^{D} \left[\ln \frac{x_i}{g_m(\mathbf{x})} - \ln \frac{y_i}{g_m(\mathbf{y})} \right]^2 \tag{4}$$

where $g_m(\cdot)$ is the geometric mean of the components, calculated considering the parts of the vectors **x** and **y** (Pawlowsky-Glahn and Egozcue 2001). The explorative map of the Aitchison distance could show where the compositional difference with respect to the spring is higher at the catchment scale. This permits the identification of homogeneous areas or spatial patterns to be related to other non-compositional environmental factors. On the other hand, the perturbation matrix informs, for each sample, about the chemical species that mostly take charge of the change offering an interesting tool to trace and interpret the behavior of the composition as a whole.

2.2 The Dataset

The catchment area of the Arno River Basin is entirely located in Tuscany, Central Italy and has a surface of 8,228 km^2 with an average elevation of 353 m. The river's headwater spring is in the Northern Apennines at 1650 m, and flows for about 242 km toward the Ligurian Sea, 10 km West of Pisa and 110 km of Florence, respectively. The drainage network follows NW-SE trending tectonic structures that form six main sub-basins from East to West: (1) Casentino, (2) Chiana Valley, (3) Sieve, (4) Upper

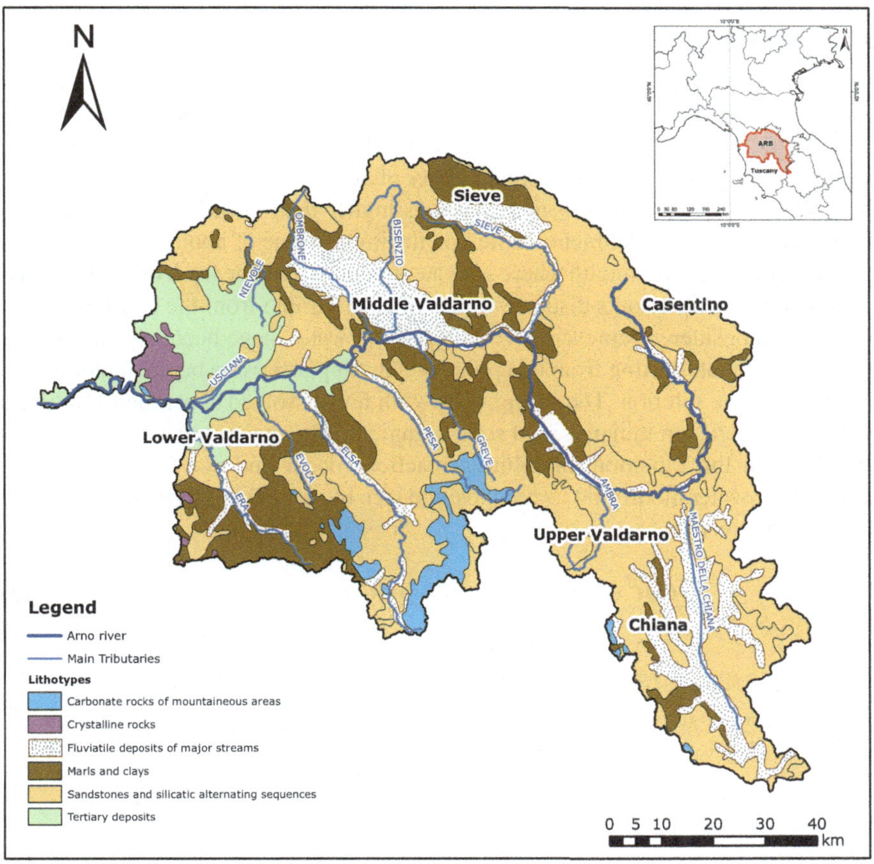

Fig. 2 Lithological map of the Arno River catchment (Tuscany, central Italy). Geological layer modified from SINAnet (ISPRA Ambiente 2019)

Valdarno, (5) Middle Valdarno, and (6) Lower Valdarno. The outcropping rocks are predominantly sedimentary folded and faulted Mesozoic and Tertiary units derived from the weathering of the Apennine Mountains. A schematic lithological map is shown in Fig. 2. The database is related to a 2-year sampling campaign (2002–2003) performed by taking into account the seasonal variability. However, no significant influence on compositional changes compared to other environmental factors (Nisi et al. 2008) appears to be attributable to seasonality. The main determined chemical species were Na^+, K^+, Ca^{2+}, Mg^{2+}, Cl^-, HCO_3^-, SO_4^{2-}, and $SiO_{2(aq)}$. The Arno River Basin, similarly to other European watersheds (Berner and Berner 1996), has suffered from the past century an increased industrial and agricultural development, so that the contribution of chemical weathering is mixed with anthropic inputs (Arrighi et al. 2018). Further details about sampling and analytical methodologies can be found in Nisi et al. (2008).

3 Results and Discussion

The analyzed dataset is given by 474 cases and 8 variables (Na^+, K^+, Ca^{2+}, Mg^{2+}, Cl^-, HCO_3^-, SO_4^{2-}, $SiO_{2(aq)}$) measured in mg/L. The chemical composition of the spring is characterized by a low conductivity of about 0.12 mS/cm and a TDS (total dissolved solids) equal to 315 mg/L. Values higher than 6.0 mS/cm are related to polluted areas of the catchment, while the maximum value of about 27.66 mS/cm is measured near the river mouth where seawater interacts with the fresh one (TDS equal to 8700 mg/L). Aitchison's distance values of each sample from the composition of the spring, considered representative of a pristine water, have been calculated. The hypothesis is that, starting from this point, weathering processes begin to modify the chemistry of the solution. The spring water with few dissolved constituents interacts with atmospheric air, minerals, and solid organic matter generating a disequilibrium condition. In this situation, dissolution reactions occur and new components are added to the water modifying its composition step by step. Typical reactions that can develop are given by

$$CaCO_3(calcite) + H^+ \rightarrow Ca^{2+} + HCO_3^- \tag{5}$$

$$\begin{aligned} &2KFe_3AlSi_3O_{10}(OH)_2(biotite) + 14H^+ + H_2O \rightarrow \\ &2K^+ + 6Fe2+ + 4H_4SiO_4^0 + Al_2Si_2O_5(OH)_4(kaolinite) \end{aligned} \tag{6}$$

$$CH_2O(organicmatter) + O_2(aq) \rightarrow CO_2(gas) + H_2O \tag{7}$$

representing the weathering of carbonate, silicate minerals, and the oxidation of the organic matter, respectively. The effectiveness of Aitchison's distance in capturing what is occurring during weathering processes is revealed by its clear positive rela-

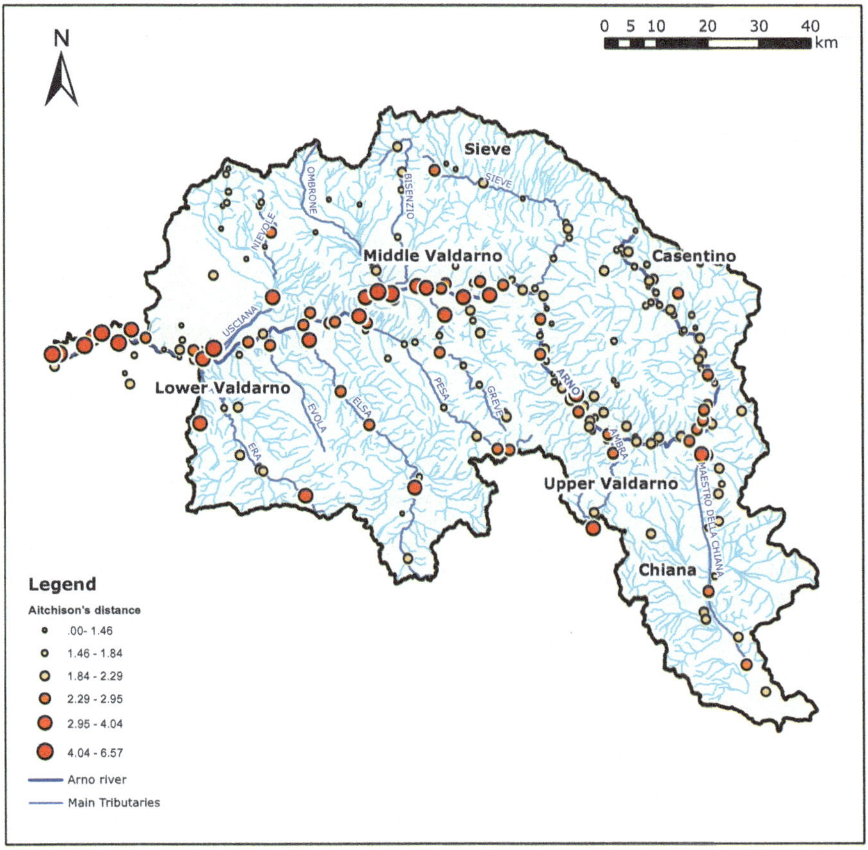

Fig. 3 Map of Aitchison's distance values of each sample from the spring water considered representative of a pristine condition

tionship with the conductivity, thus representing a possible marker of geochemical changes. The spatial distribution of the distance values is reported in Fig. 3. As we can see from the map, its values increase along the course of the main river mainly starting from the Middle Valdarno where natural processes began to mix with anthropic ones and near to the mouth due to the marine ingression and mixing processes.

After the calculus of Aitchison's distance, the data were ranked for its increasing values and the perturbation difference between subsequent compositions $x_{i+1} - x_i$ was determined. This step is important to highlight the different behavior of single chemical components inside the composition. In this way, similar Aitchison's distance values could be attributed to different processes depending on the association of variables in the perturbation. The results, as unclosed perturbation factors, are reported in Fig. 4 (Egozcue and Pawlowsky-Glahn 2011) while the histogram of Aitchison's distance values, with the associated kernel density estimation curve,

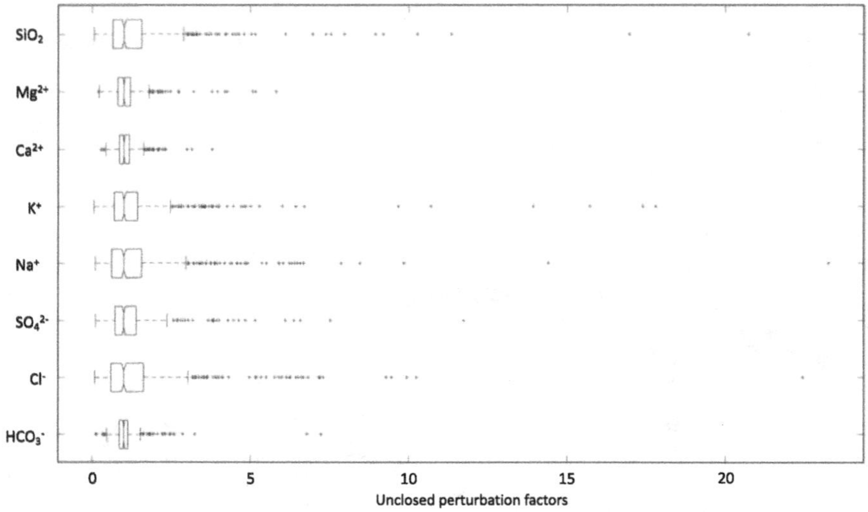

Fig. 4 Comparative box plots for the unclosed perturbation factors for each chemical species for data ranked by considering the increasing values of Aitchison's distance from the spring water. Perturbation difference represents the difference $x_{i+1} - x_i$

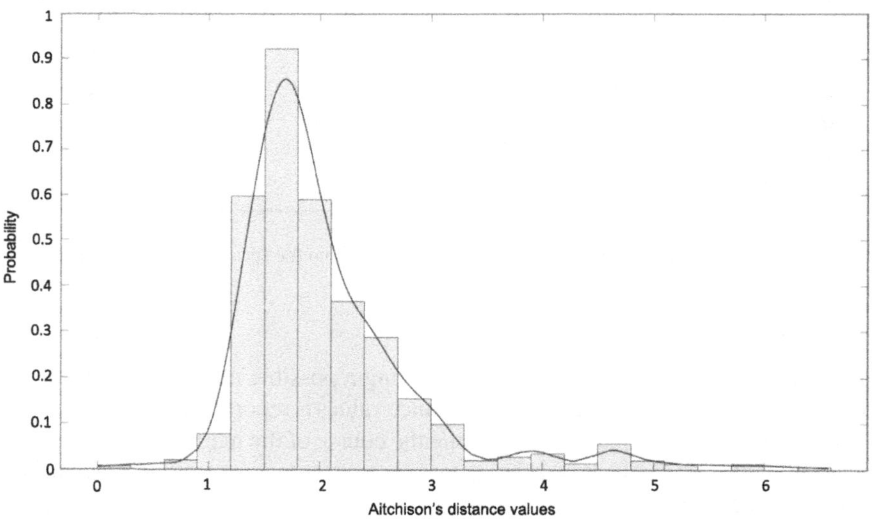

Fig. 5 Histogram and kernel density estimation of Aitchison's distance values from the spring composition considered as a pristine water

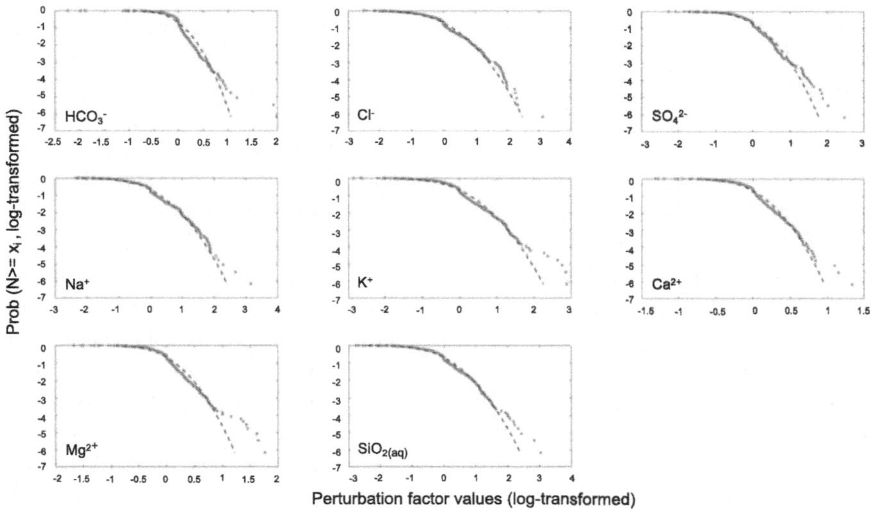

Fig. 6 Complementary cumulative distribution function for unclosed perturbation factors. The continuous line is related to the log-normal model

is shown in Fig. 5. The distance presents a positive asymmetrical distribution with rare high values in the right tail, indicating that most of the data are characterized by a distance lower than 3 with respect to the pristine condition. High values are related to the mouth (marine ingression) and to peculiar pollution conditions (e.g., Usciana and Chiana Channels) or to the presence of lenses of $CaSO_4$ and $NaCl$ in clays in the valleys of Elsa and Era Tributaries (Fig. 2). The unclosed perturbation factors of Fig. 4 provide information about the contribution of each variable to the compositional changes. $SiO_{2(aq)}$, Cl^-, Na^+, K^+, and SO_4^{2-} show a more scattered contribution while HCO_3^-, Ca^{2+}, and Mg^{2+} exhibit a more stable signal. Hence, this procedure appears to be able to divide the variables into two different groups, one more sensible to environmental changes, perhaps related to an intermittent spatial behavior, and the other more resilient. The result indicates that monitoring plans for a long temporal range, with the aim of intercepting important changes in the catchment, also related to climatic variations, should be based on the sub-composition related to the cycle of carbonates (HCO_3^-, Ca^{2+}, and Mg^{2+}). To investigate in more detail the nature of the contribution of the variables to the compositional changes, the complementary cumulative distribution function for the unclosed perturbation factors was determined and plotted (Mitzenmacher 2004; Egozcue and Pawlowsky-Glahn 2011). Results are reported in Fig. 6 with the continuous line related to the description of the patterns by the log-normal model. The comparison between experimental data and the log-normal distribution indicates that in no case it is possible to adopt it to describe the behavior of the chemical species. On the other hand, some portions of the curves are clearly linear highlighting the possible presence of more than one power law or a multifractal behavior (Seely and Macklem 2012).

The consequence is that an interaction-dominant dynamics is no longer exhaustive and the typical multiplicative processes of a log-normal model would be associated with interdependent feedback transactions covering different time or space scales (van Rooij et al. 2013). Thus, compositional changes that enrich riverine water in solutes starting from its pristine condition are apparently described by features that are typical of a complex system such as the occurrence of multifractality. This property characterizes dynamical systems in which energy dissipation can no longer be neglected. However, in this general framework, considering the unclosed perturbation factors of Fig. 4, species such as HCO_3^-, Ca^{2+}, and Mg^{2+} appear to be more efficient in stabilizing fluctuations. In fact, they reveal the presence of a sub-system characterized by homogeneity and connectivity which is particularly resistant to change but also subjected to critical transitions (Scheffer et al. 2012; Sauro Graziano et al. 2020). The cause–effect relationship able to govern this situation could be related at the catchment scale to: (i) the spatial diffusion of the lithologies from which these chemical species migrate; (ii) their homogeneity and connections; and (iii) the efficiency of geochemical processes, affecting the way in which a system, with potential local alternative states, can respond to changing conditions. The behavior of all the other chemical species, $SiO_{2(aq)}$, Cl^-, Na^+, K^+, and SO_4^{2-}, could instead indicate: (i) an heterogeneous distribution of the lithologies from which they came from; (ii) incomplete connectivity with presence of modularity; and (iii) less diffused (intermittency) or less efficient geochemical processes, all features that lead to a higher adaptive capacity and favor gradual changes. The values of the saturation indices, calculated for some fundamental minerals such as calcite, dolomite, and quartz, appear stationary when cases are ranked for increasing Aitchison's distance as well as box plots equilibrated around the median values, with the exception of some anomalous data (Zhu and Anderson 2002). Thus, chemical equilibria, able to limit the concentration of the ions in water, have a buffering effect in the compositional changes as registered by the perturbation difference from a pristine water. The result indicates this compositional tool to be very useful as an explorative aid in water geochemistry to check for different evolutive geochemical hypothesis. Therefore it is important to stress that Aitchison's distance is a compositional distance between different compositions and only the association with the perturbation difference will help us to discriminate different situations characterized by similar distance values. In our case study, the distance values are not associated with an anomalous behavior of the perturbation factors (e.g., presence of trends and pluri-modality), indicating that similar processes appear to have affected the evolutive path of the water starting from the pristine reference.

4 Conclusions and Future Developments

The research on complexity theory and nonlinear dynamics involves concepts such as dissipative structures and fractal patterning, as well as instability, resilience, adaptive cycles, and uncertainty. Such promising concepts are still developing but CoDA appears to offer the tools to investigate chemical compositions as a whole and to probe the system dynamics based on the nature of the interactions of the components under the forcing effects of different environmental drivers. In this framework, the perturbation operator seems to present powerful features able to discover the resiliency of chemical variables versus intermittency and instability. This opens important paths to reveal warning signals for relevant changes, also related to climatic variations. The results obtained for the river chemistry of the Arno catchment are very comforting. Further investigation would be performed to link perturbation differences to non-compositional variables, or environmental drivers, characterizing the catchment (e.g. slope, structure of the drainage network, erodibility, runoff, and discharge). The joint analysis would help to point out the nature of the interactions and feedback mechanisms able to govern the development of chemical reactions, thus determining the chemistry of the waters (Gozzi et al. 2019a, b). The proposed approach gives a potential basis to expand this research since Aitchison's distance and compositional changes from a pristine water synthesize the complexity of the system from the chemical point of view. Other non-compositional variables can be then called into question and interrelated with the previous ones.

Acknowledgements This work has been partially financed to A.B. by the Fondi Ateneo 2020 project (University of Firenze). A.B. wishes to thank Vera Pawlowsky-Glahn that in occasion of the IAMG (International Association for Mathematical Geosciences) meeting of Mont Tremblant, Qué bec, Canada, back in 1993, opened her mind on compositional data thus giving life to a long scientific collaboration and a deep friendship.

References

J. Aitchison, *The Statistical Analysis of Compositional Data*. Monographs on Statistics and Applied Probability (Chapman & Hall Ltd., London (UK), 1986). (Reprinted in 2003 with additional material by The Blackburn Press), 416 p

J. Aitchison, The statistical analysis of compositional data (with discussion). J. R. Stat. Soc. Ser. B (Stat. Methodol.) **44**, 139–177 (1982)

C. Arrighi, M. Masi, R. Iannelli, Flood risk assessment of environmental pollution hotspots. Environ. Model. Softw. **100**, 1–10 (2018)

E.K. Berner, R.A. Berner, *Global Environmental: Water, Air and Geochemical Cycles* (Prentice-Hall, Upper Saddle River NJ, 1996), 376 p

C.M. Bethke, *Geochemical and Biogeochemical Reaction Modeling* (Cambridge University Press, 2008), 543 p

D. Billheimer, P. Guttorp, W. Fagan, Statistical interpretation of species composition. J. Am. Stat. Assoc. **96**(456), 1205–1214 (2001)

E. Dinelli, G. Corteccci, F. Lucchini, E. Zantedeschi, Source of major and trace elements in the stream sediments of the Arno river catchment (northern Tuscany, Italy). Geochem. J. **39**, 531–545 (2005)

J.J. Egozcue, C. Barceló-Vidal, J.A. Martín-Fernández, E. Jarauta-Bragulat, J.L. Díaz-Barrero, G. Mateu-Figueras, Elements of simplicial linear algebra and geometry. In (Pawlowsky-Glahn & Buccianti, 2011), pp. 141–157, 378 p (2011)

J.J. Egozcue, V. Pawlowsky-Glahn, Basic concepts and procedure. In (Pawlowsky-Glahn & Buccianti, 2011), pp. 12–28, 378 p (2011)

J.J. Egozcue, V. Pawlowsky-Glahn, G. Mateu-Figueras, C. Barceló-Vidal, Isometric logratio transformations for compositional data analysis. Math. Geol. **35**, 279–300 (2003). https://doi.org/10.1023/A:3A1023818214614

C. Gozzi, R.S. Graziano, A. Buccianti, Statistical methods for the geochemical characterisation of surface waters: the case study of the Tiber River basin (Central Italy). Comput. Geosci. **131**, 80–88 (2019a)

C. Gozzi, R.S. Graziano, F. Frondini, A. Buccianti, Innovative monitoring tools for the complex spatial dynamics of river chemistry: case study for the Alpine region. Environ. Earth Sci. **77**(16), 579 (2019b)

R.S. Graziano, C. Gozzi, A. Buccianti, Is compositional data analysis a theory able to discover complex dynamics in aqueous geochemical systems?. J. Geochem. Explor. **211**, 106465, 1-9 (2020)

ISPRA Ambiente, Italian hydrogeological complexes [Download MAIS]. http://www.sinanet.isprambiente.it/it/sia-ispra/download-mais/complessi-idrogeologici/view (24 September, 2019) (2019)

A. Kleidon, Life, hierarchy, and the thermodynamic machinery of planet Earth. Phys. Life Rev. **7**, 424–460 (2010)

W. Krzanowski, *Principles of Multivariate Analysis (Second Edition)*. Oxford Statistical Science Series 23 (Oxford, UK, 2000), 563 p

M. Mitzenmacher, A brief history of generative models for power law and lognormal distributions. Internet Math. **1**(2), 226–251 (2004)

B. Nisi, A. Buccianti, O. Vaselli, G. Perini, F. Tassi, A. Minissale, G. Montegrossi, Hydrochemistry and strontium isotopes in the Arno River Basin (Tuscany, Italy): constraints on natural controls by statistical modeling. J. Hydrol. **1–4**, 166–183 (2008)

V. Pawlowsky-Glahn, A. Buccianti (eds.), *Compositional Data Analysis: Theory and Applications* (Wiley, 2011), 378 p

V. Pawlowsky-Glahn, J.J. Egozcue, R. Tolosana-Delgado, *Modeling and Analysis of Compositional Data*. Statistics in Practice (Wiley, Chichester, UK, 2015), 247 p

V. Pawlowsky-Glahn, J.J. Egozcue, Geometric approach to statistical analysis on the simplex. Stoch. Environ. Res. Risk Assess. **15**(5), 384–398 (2001)

X. Sanchez-Vila, M. Dentz, L.D. Donado, Transport-controlled reaction rates under local non-equilibrium conditions. Geophys. Res. Lett. **34**, 1–5 (2007)

M. Scheffer, S.R. Carpenter, T.M. Lenton, J. Bascompte, W. Brock, V. Dakos, J. van de Koppel, I.A. van de Leemput, S.A. Levin, E.H. van Nes, M. Pascual, J. Vandermeer, Anticipating critical transitions. Science **338**, 344–348 (2012)

A.J.E. Seely, P. Macklem, Fractal variability: an emergent property of complex dissipative systems. Chaos **22**, 013108-1–013108-7 (2012)

S.L. Shvartsev, Self-organizing abiogenic dissipative structures in the geologic history of the Earth. Earth Sci. Front. **16**(6), 257–275 (2009). https://doi.org/10.1007/BF02066299. https://doi.org/10.1016/S1872-5791(08)60114-1

M.M.J.W. van Rooij, B. Nash, S. Rajaraman, J.G. Holden, A fractal approach to dynamic inference and distribution analysis. Front. Physiol. **4**, 1–16 (2013)

C. Zhu, G. Anderson, *Environmental Applications of Geochemical Modeling* (Cambridge University Press, 2002), 284 p, 378 p

Groundwater Origin Determination in Historic Chemical Datasets Through Supervised Compositional Data Analysis: Brines of the Permian Basin, USA

Mark A. Engle and Julien A. Chaput

Abstract Data from historic water quality databases often lack critical measurements necessary for focused investigations, such as determining the origin of the water. The U.S. Geological Survey produced waters database contains nearly 7,000 data of good quality for the Permian Basin, the single largest oil-producing province in the United States. However, fewer than 350 of those points contain enough geochemical data (Br concentration or $\delta^{18}O$ and $\delta^{2}H$ composition) to determine whether the origin of the samples is meteoric water or paleoseawater. Three supervised methods were applied to isometric and pairwise log-ratio transformed major ion data from a subset of samples of known origin but where the Br concentration and $\delta^{18}O$ and $\delta^{2}H$ composition were excluded to predict origin: linear discriminant analysis (isometric only), support vector machines (isometric and pairwise), and random forests (pairwise only). Error rates from validation, using data of known origin (excluding Br concentration and $\delta^{18}O$ and $\delta^{2}H$ composition) that were not used in model development, found that no method performed exceptionally well. An ensemble approach of only assigning classification when all three methods provide the same classification reduced the error rate of the validation data to 11% but failed to classify 28% of the data. This latter approach was applied to the nearly 7,000 samples which only contained concentrations of major ions (Cl, Ca, HCO_3, Mg, Na, and SO_4). Spatial mapping of these newly classified data generated insight on distribution and flow of meteoric and paleoseawater across the basin.

1 Introduction

Determining the origin of water samples typically requires analysis of isotope compositions, such as oxygen and hydrogen in the water itself ($\delta^{18}O$ and $\delta^{2}H$, respectively),

M. A. Engle (✉) · J. A. Chaput
Department of Geological Sciences, University of Texas at El Paso, 500 West University Ave, El Paso, TX 79930, USA
e-mail: maengle@utep.edu

J. A. Chaput
e-mail: jachaput@utep.edu

P. Filzmoser et al. (eds.), *Advances in Compositional Data Analysis*,
https://doi.org/10.1007/978-3-030-71175-7_14

or from trace constituents whose behavior is well known. In the investigation of modern samples, this is largely a trivial issue as these analyses are routine and affordable. However, in the analysis of older data sets (e.g., pre-1990s), measurements allowing for interpretation of water origin are frequently missing. The U.S. Geological Survey Produced Waters database (Blondes et al. 2018) is the largest publicly available source of water chemistry from oil and gas wells in the United States and is dominated by data collected prior to 1980. Despite containing results for roughly 115,000 samples, fewer than 1% of the entries in the database contain corresponding water isotope data. The ability of such a large dataset to provide information on large-scale hydrogeologic flow is correspondingly diminished. Other large water chemistry databases, such as the U.S. Environmental Protection Agency's Water Quality Portal and the U.S. Geological Survey's National Water Information System, also contain vast quantities of data lacking information to determine the origin of water. One approach to handle missing geochemical data is multivariate imputation algorithms, which rely upon iterative regressions or other approaches to provide estimates based on information held in the non-missing parameters (Hron et al. 2010). However, these algorithms are not designed for sample classification or discrimination and have not been tested for this purpose.

Saline brines found in sedimentary basins present an ideal case to examine the performance of supervised methods in determining if other measurements, specifically major ion concentrations, can be used to determine water origin. In many geologic basins, such as the Permian, Williston, and Illinois basins of North America, periods of subaerial evaporation of seawater and the production of thick evaporite sequences (often halite and anhydrite) led to two dominant types of water: (1) trapped, paleoseawater from which the evaporites were generated, and (2) younger meteoric waters that intruded or displaced the older seawater, frequently gaining salinity through evaporite mineral dissolution (Rostron and Holmden 2003; Stueber et al. 1998; Walter et al. 1990). This younger meteoric water starts with a composition and salinity different than the more saline paleoseawater, but a series of geochemical reactions, including the dissolution or recycling of evaporite minerals, produce fluids from both sources that are difficult to differentiate from one another based on simple ion ratios (Engle and Blondes 2014). However, because the sources of the water are so distinct there is credence to the concept that there is enough difference in the multivariate structure that the fluids can be sourced without isotopic data. In particular, halogens, such as bromide and iodide, have successfully been used to determine the origin of brines from sedimentary basins (Engle and Brunner 2019; Walter et al. 1990). However, bromide and iodide concentration data can be quite rare in historic brine geochemistry datasets. Methods to determine the origin of water samples without the need for less common analytes (e.g., Br) have potential to provide much better information about the location and degree of meteoric inflow into sedimentary basins.

This chapter focuses on the application of supervised machine learning techniques following the mathematical principles of compositional data analysis (CoDA), referred throughout the chapter as supervised CoDA, applied to historic brine geochemistry data from the Permian Basin of west Texas and southeastern New

Mexico, USA (Fig. 1). The basin is the single most productive oil-bearing province in the United States, generating roughly 4.1 million barrels of oil per day. Many studies have examined the origin and composition of brines within the 222,000 km^2 basin, largely based on isotopic data and Na-Cl-Br systematics (Engle and Blondes 2014; Engle et al. 2016; Knauth and Beeunas 1986; Saller and Stueber 2018; Stueber et al. 1998). In particular, Engle and Blondes (2014) noted difficulty in differentiating origin of the brines in the Delaware sub-basin, due to a lack of water isotope and Br data. This lack of information prevents an accurate description of which geologic units have allowed for meteoric fluids to enter them versus those that still contain paleoseawater (Saller and Stueber 2018). Previous authors have suggested that meteoric water, driven by topographic head in the northwest, is displacing paleoseawater out of the basin, to the east (Bein and Dutton 1993). This conceptual model is generally supported by upwelling of deep paleoseawater into springs and shallow groundwater on the southeastern edge of the basin (Dutton et al. 1989), but there are few data to constrain the extent and degree of the process. Thus, methods that allow for any additional insight given a significant dearth of parameters conventionally viewed as mandatory for determining the source of data are of great benefit for constraining the behavior of groundwater flow in this important basin.

One mathematical issue to be worked through is that brine chemistry data are compositional (composed of relative parts) and exhibit some of the most obvious examples of spurious correlations between ions when conventional data analysis methods are utilized (Engle et al. 2016; Engle and Blondes 2014; Engle and Rowan 2014). Brine samples can be considered to consist of two parts: water and salt. Because the salinity of water in geologic basins ranges from fresh (<500 mg/L) to several times the salinity of the ocean (35,000 mg/L versus up to and exceeding 300,000 mg/L), water "displaces" the salt or vice versa, as salinity varies. Thus, most of the ion concentrations change in concert with one another as a function of salinity, and the corresponding relationships between the ions become spurious (Fig. 2). Application of compositional data analysis techniques, such as those developed for use with water chemistry by Buccianti and Pawlowsky-Glahn (2005), lead to results that minimize spurious relationships and provide interpretations that are internally consistent (Engle et al. 2016).

The purpose of this chapter is to combine brine chemistry data from the Permian Basin that have been properly transformed following principles of CoDA, with supervised data analysis methods to create accurate models to classify the origin of water samples based solely on their major ion composition (Ca, Cl, HCO_3, Mg, Na, and SO_4). Three supervised techniques will be applied and compared in their performance: linear discriminant analysis (LDA), random forests (RF), and support vector machines (SVMs). The techniques will be evaluated through calculation of misclassification rates during validation, using samples of known origin that were not used during model training. Finally, the models will be applied to a large dataset of previously unclassified data to create maps of water types across the basin. Results from the exercise allow for discussion of supervised CoDA methods and the final maps provide additional insight on the distribution and flow of water across the Permian Basin.

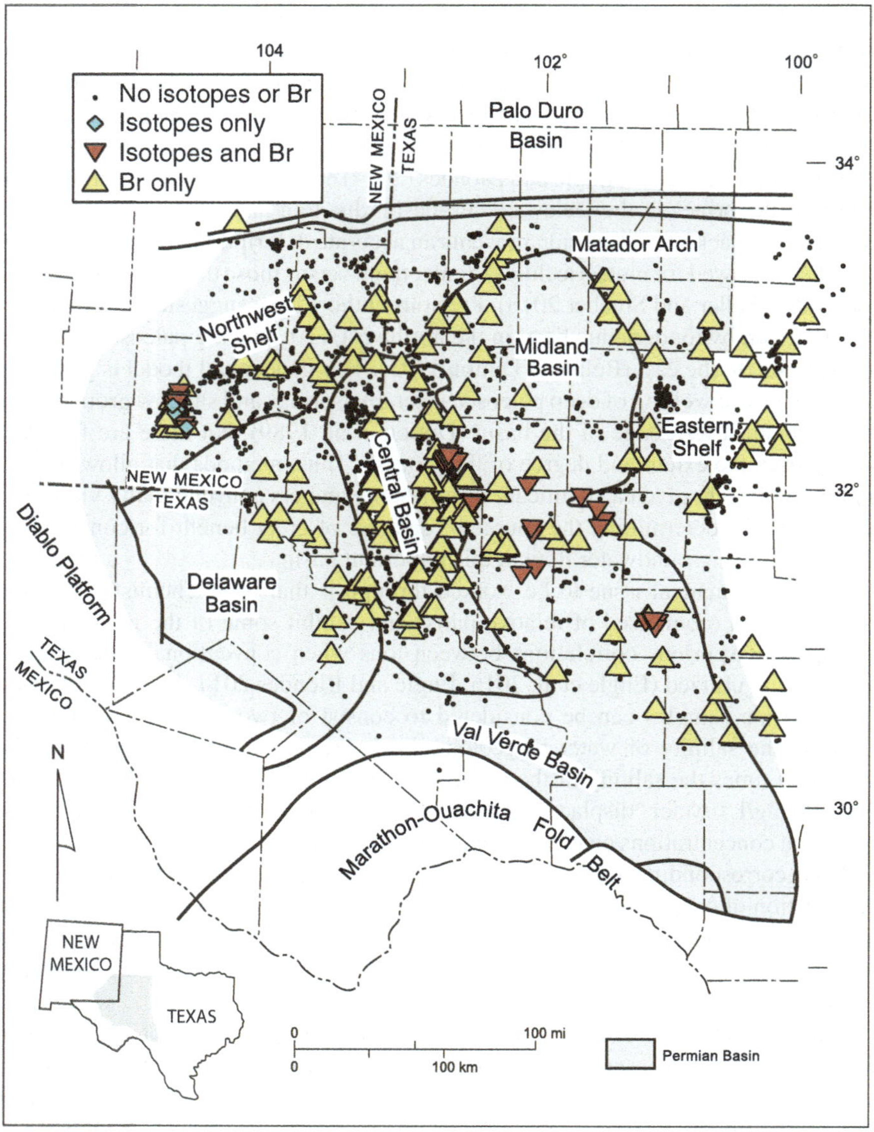

Fig. 1 Map of the study area showing major features of the Permian Basin and sample locations of brine geochemical data. All points correspond to location of data with a complete suite of major ion analyses. Symbols indicate whether corresponding data also contain data for both water isotopes ($\delta^{18}O$ and $\delta^{2}H$) and Br concentration (isotopes and Br), Br concentration but not water isotopes (Br only), water isotopes but not Br concentration (isotopes only), or neither (no isotopes or Br). Base map modified from Dutton et al. (2004)

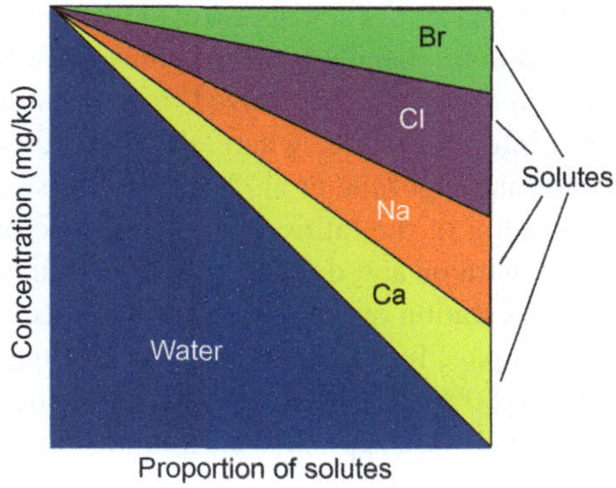

Fig. 2 Cartoon illustrating the induced co-association among solutes in water samples as a result of salinity variations. This behavior among ions can lead to spurious correlations between pairs of solutes, not because there is any active relationship, but simply because they are being co-displaced due to increase or decrease in water content

2 Methods and Materials

Water chemistry data for the Permian Basin were obtained from the U.S. Geological Survey's Produced Waters Geochemical Database, version 2.3 (Blondes et al. 2018). Roughly 55% of the nearly 16,000 data points from the Permian Basin failed the formation water culling criteria of Hitchon and Brulotte (1994) and/or were lacking a major ion (Ca, Cl, HCO_3, Mg, Na, and SO_4) and were removed from the analysis.

The subset of samples containing data for Br and/or $\delta^{18}O$ and δ^2H (see Fig. 1) had potential for classification into meteoric water versus paleoseawater using standard methods and were removed from the larger dataset to create training and validation datasets. Subsequent plotting of oxygen and hydrogen isotopic data and of the isometric log-ratio (ilr) Na-Cl-Br plot of Engle and Rowan (2013) was used to determine and assign a meteoric or paleoseawater source for a portion of these samples ($n = 246$), as will be presented in Sect. 3. This sample subset was examined in more detail using robust compositional principal component analysis on log-ratio transformed data, following the method of Filzmoser et al. (2009). Prior to supervised analysis of the 246 data points, the corresponding concentration data for the major ions (Ca, Cl, HCO_3, Mg, Na, and SO_4) were converted to ilr coordinates (Egozcue et al. 2003) and all unique pairwise log-ratios (Aitchison 1986). The ilr coordinates (z_k) were created from non-overlapping groups of parts, based geochemical principles and results from the principal component analysis, created using a sequential binary partition (SBP; Egozcue and Pawlowsky-Glahn 2005):

$$z_k = \sqrt{\frac{r_i s_i}{r_i + s_i}} \ln \frac{\left(\prod x_j\right)^{\frac{1}{r_i}}}{\left(\prod x_l\right)^{\frac{1}{s_i}}},\ k = 1, \ldots, D,$$

where r_i and s_i are the number of parts in the numerator and denominator for each coordinate in the SBP, and x_j and x_l are the concentrations in millimoles per liter of ion j (for ions those in the numerator according to the SBP) and l (for those in the denominator according to the SBP), and D is the total number of ions in the analysis

(6, in this case). As development of the SBP requires exploratory data analysis results, its creation is detailed in the following section. Once converted to ilr coordinates, the data are effectively moved from the simplex (the vector space of compositional data) into real space, where all standard numerical approaches can be completed as normal (Egozcue et al. 2003). Pairwise log-ratios were calculated by taking the natural log of the ratios of the 15 unique pairs of ion concentrations (ratios where the numerator and denominator are swapped are not considered unique). Prior to implementation of the three supervised compositional data analysis techniques, the 246 samples, for which group association was determined, were split into a training data set (80% of the data) and a validation data set (20% of the data). The classification techniques were applied to the training data set and tested using the validation data set. Classification performance of the validation dataset for each method was determined from a confusion matrix which graphically displays misclassification rates.

Linear discriminant analysis was applied only to the ilr coordinates of the training dataset, though identical results are obtained using the log-ratio pairs as distances are equal using either pairwise or isometric log-ratios. The LDA method selects a hyperplane in the multi-dimensional variable space that best separates samples from the two groups, meteoric water, and paleoseawater. It does this by maximizing the distance between the group means while minimizing the variance within each group. Because LDA performance can be strongly affected by multivariate outliers, a robust version which utilized a minimum covariance determinant (MCD) was applied. This version of the analysis, detailed by Filzmoser et al. (2011), uses the MCD estimator from a subset of the dataset ($\alpha = 0.75$, in this case) to calculate group means and variances. Because we assume that any given sample has an equal chance of either group, a prior probability of 0.5 was used for each group, though other approaches were examined (as discussed in Sect. 4).

The second technique employed to classify the data was that of Support Vector Machines (Suykens and Vandewalle 1999). Similar to LDA, SVMs work to generate a hyperplane which maximizes the distance between itself and the closest samples from either group (known as support vector points). As application of this technique to compositional data is limited, both the ilr coordinates and pairwise log-ratios were analyzed using this method and the results compared. This maximization is calculated by taking the dot product of pairs of support vector points, once the optimization problem has been formulated through a primal/dual approach derived using Lagrangian multipliers. Through the so-called kernel trick (Boser et al. 1992), lower dimensional data can be mapped into higher dimensional space by simply utilizing a kernel function that defines the necessary dot products without actually transforming the data, thus significantly saving on computation time. This can allow for better separation of the data in the typical event that class separation is not a linear hyperplane. In this case, a radial basis kernel function is typically used:

$$K(Z_1, Z_2) = e^{-\gamma(Z_1 - Z_2)^2},$$

where Z_1 and Z_2 are multivariate data and/or support vectors and γ is a scaling parameter than can be optimized for the training dataset. An additional penalty for misclassifying data during training can be specified, and generally formulated as a scaling factor C (also typically referred to as the Box Constraint). As with most classification methods, it is possible to overfit data sets by allowing γ and C to match training data too closely, and this typically results in poor classification for new data sets.

The third and final supervised method, RF, is a variation on decision trees. Decision trees, a non-linear method, utilize nodes at multiple levels to increasingly split the data using numerical divides created from the variables (i.e., $x_1 < 4$ versus $x_1 \geq 4$). The method attempts to achieve an "optimal" separation of samples from the different groups based on the logical divides. This is done by recursively splitting the data at multiple nodes until samples from only one group are present in every node. For instance, a decision tree might find that samples where $x_1 < 4$, $x_2 > 2.5$, and $x_3 < 43$ only produces samples from one group, and the same rule could be applied to classify other data. However, the decision tree method exhibits high bias and is prone to over-fitting (Dietterich 1995), particularly if very deep trees (i.e., trees with many nodes) are permitted. Random forests reduce intrinsic over-fitting of decision trees by training the trees on bootstrapped subsamples of the training dataset and randomizing subsets of variables used to create nodes (Breiman 2001). The final trained model is comprised by aggregating the bootstrapped decision trees, known as bagging, thus statistically averaging the bias inherent in individual trees. The minimum number of samples in a terminal node (leaf size) and the number of variables randomly selected to split the data at each node are other parameters that may be optimized in the course of maximizing cross-validation, using subsets of the test data. Previous work on the application of RF to compositional data by Tolosana-Delgado et al. (2019) showed that out-of-bag errors are notably much lower for pairwise log-ratios than ilr coordinates. Following from those results, only pairwise log-ratios were applied in this investigation.

Once the classification models were created and optimized using the training data set, through training and parameter tuning (as detailed in Sect. 3), they were tested through determining misclassification of the validation data set to estimate the minimum classification error. After this step, the supervised CoDA models were applied to the much larger Permian Basin sample set, which lacked results for either Br or water isotopes, to classify the origin of each sample based solely on its major ion composition. Once classified, these samples were geographically mapped to examine for patterns corresponding to known geologic features or other notable structures.

3 Exploratory Data Analysis

Origin of Permian Basin brines have been interpreted based on $\delta^{18}O$ and $\delta^{2}H$ measurements, with additional evidence from compositional data, by many previous authors (Engle et al. 2016; O'Neil et al. 1986; Saller and Stueber 2018; Stueber et al. 1998).

Only 89 samples in the dataset contain isotopic data, though there is a distinct differentiation between meteoric waters versus ancient (late Permian) seawater (Fig. 3). Samples falling along the local meteoric water line have previously been interpreted as meteoric water of various age. Engle et al. (2016) identified two meteoric aquifer systems in the deep basin brines, one above the zone of paleoseawater and one below, but they overlap in terms of $\delta^{18}O$ and $\delta^{2}H$ composition. Water samples exhibiting $\delta^{18}O > 0$ have previously been interpreted as evaporated late Permian seawater that has exchanged oxygen with carbonate minerals. Those samples falling off the local meteoric water line but with $\delta^{18}O < 0$ have been previously interpreted as late Permian seawater that has been diluted by meteoric water. As will be shown below, this subset of samples exhibits compositions that are dominated by their seawater origin. Given the primary interest interest is in identifying zones where the paleoseawater is still present, they are classified as paleoseawater in this exercise. One problem with relying completely upon data classified by origin based solely on $\delta^{18}O$ and $\delta^{2}H$ data is that they are spatially clustered and cover only a very small portion of the Permian Basin (Fig. 1).

Determination of water origin by application of Na-Cl-Br systematics divides the samples into the same groups (meteoric water and paleoseawater), based on the behavior of Br relative to Na and Cl during halite precipitation from evaporating seawater (Engle and Rowan 2013; McCaffrey et al. 1987; Walter et al. 1990). Engle and Rowan (2013) produced an isometric log-ratio plot of the relationship between the ions during seawater evaporation and halite dissolution, based on geochemical modeling pathways. The plot minimizes potential misinterpretation due to spurious correlation between ions as a function of salinity variation and can identify situations where Br is added through other sources, such as decomposition of marine kerogen (Engle et al. 2016; Engle and Brunner 2019). The x-axis of Fig. 4 can be interpreted to represent the partitioning of Br versus Na and Cl when halite is created or dissolved.

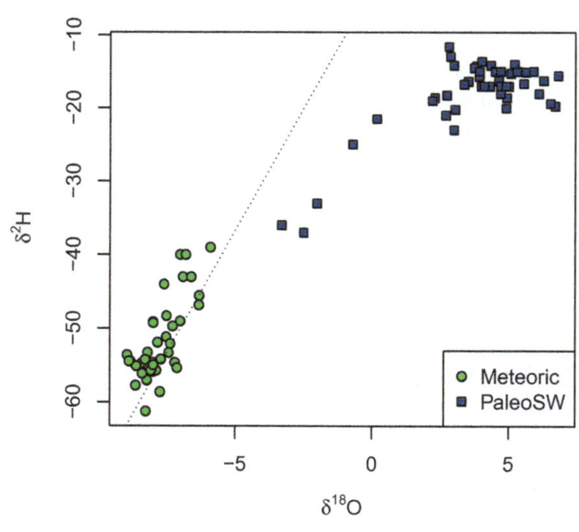

Fig. 3 Oxygen and hydrogen isotope composition of Permian Basin Brines. Dotted line represents local meteoric water line from Reyes et al. (2018). Meteoric = water previously classified as meteoric. PaleoSW = water previously classified as paleoseawater

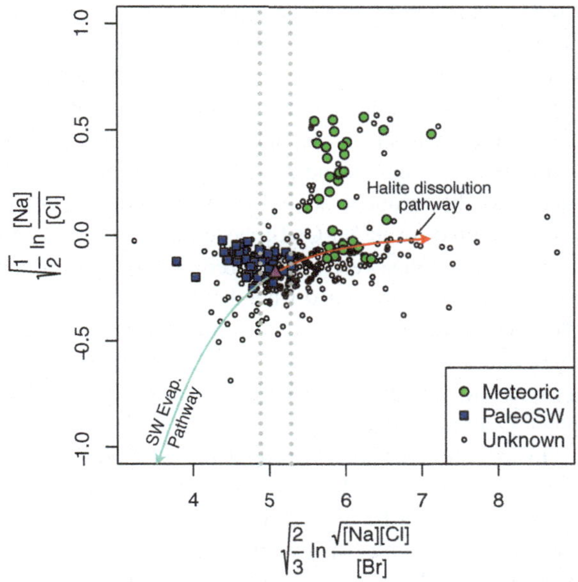

Fig. 4 Isometric log-ratio Na-Cl-Br plot of Engle and Rowan (2013) showing modeled pathways of seawater evaporation and halite dissolution projected away from seawater (pink triangle symbol). Symbols based on classification from water isotopes, where available (see Fig. 3). Vertical dotted lines used to further classify samples unknown samples (see text). PaleoSW = water previously classified as paleoseawater

If halite is created as a result of seawater evaporation, Br has little affinity for the halite structure and preferentially accumulates in the brine (McCaffrey et al. 1987), producing low values along this axis. Conversely, dissolution of halite produces brines that are very poor in Br relative to Na and Cl, leading to large values along this axis. The y-axis corresponds to the molar Na/Cl ratio, so halite dissolution trends toward values of zero (because the log of 1 is 0), while halite precipitation depletes the solution of Na relative to Cl, producing negative values (McCaffrey et al. 1987).

Examination of the data on the ilr Na-Cl-Br plot (Fig. 4) shows that samples previously classified as meteoric water or paleoseawater via interpretation of $\delta^{18}O$ and $\delta^{2}H$ data tend to plot near the modeled pathways for seawater evaporation and halite dissolution, respectively (modeling described in Engle and Rowan 2013). Data from samples classified as meteoric water that plot well above the halite dissolution trajectory have previously been interpreted as being impacted by water flooding (Engle and Blondes 2014), where surface or other water sources are injected into the reservoir to drive oil into producing wells. This interpretation is supported by the fact that total dissolved solid concentrations in these samples (4,180 to 7,680 mg/L) are much lower than those of meteoric water origin samples whose data plot closer to the halite dissolution trajectory (median $\pm$ pseudo$\sigma = 41{,}100 \pm 25{,}500$ mg/L). Similarly, a portion of the data from samples classified as paleoseawater in the $\delta^{18}O$ and $\delta^{2}H$ plot above the trajectory for seawater evaporation. Engle et al. (2016) demonstrated, through multiple lines of evidence, that this is due to excess bromide released from marine kerogen in some of the organic rich shale units, a process that is beyond the conventional interpretation of Na-Cl-Br systematics. The remaining samples which included concentration data for Br but no $\delta^{18}O$ and $\delta^{2}H$ data, labeled as unknown in

Fig. 4, tend to fall within the range of the rest of the data, suggesting that they also come from the same end-members. In order to classify these samples, we selected a tolerance along the x-axis around seawater (shown as dotted vertical lines) to account for the fact that samples whose composition plots close to this point can be of either origin. Justification for this approach comes from work in the Gulf Coast Basin that found that samples of meteoric origin can plot slightly to the left of seawater and vice versa, likely due to analytical uncertainty and error, heterogeneity, and processes not accounted for in Na-Cl-Br systematics (Engle and Brunner 2019). Data plotting beyond these lines were categorized as being part of the corresponding group, while samples found in the narrow band between the two lines were left unclassified unless their origin was previously determined through $\delta^{18}O$ and $\delta^{2}H$ analysis. This approach increased the number of samples with group assignment from 89 to 314 and provided a more uniform spatial coverage of the study area than categorization based solely on $\delta^{18}O$ and $\delta^{2}H$ data (Fig. 1).

In order for the supervised analysis to be useful, it needs to correctly assign groups with some success for larger batches of historic data from the Permian Basin that lack concentrations of Br or isotopic data. Thus, the major ion data (Ca, Cl, HCO_3, Mg, Na, and SO_4 in units of mmol/L) for the 314 samples with assigned classifications were examined using robust principal component analysis to explore classification potential and multivariate data trends. The first 3 components of the PCA accounted for 97.4% of the log-ratio variance in the dataset and are the focus here (Fig. 5). Examination of the biplots between principal component (PC) 1 versus PC 2 and PC 2 versus PC 3 shows that only one major multivariate outlier was observed, but the use of the MCD estimator should have effectively minimized its effect on the overall results. Biplots of the scores show a large degree of overlap between the two groups of samples with respect to the individual PCs. However, some separation between the two groups in scores for the PC 1 versus PC 2 biplot suggests that there is hope in methods to separate samples without use of the more classical approaches relying on Br concentration and isotopes. Examination of the loading plots for PC 1 versus PC 2 indicates that the variations which produce this modest separation between the two sample groups are the relative abundances of Ca, HCO_3, and SO_4, as noted in the corresponding loading plot. This interpretation is consistent with the fact that increasingly evaporated seawater contains increasing relative proportions of Ca, as observed in sedimentary basin brines across the U.S. (Hanor 1994), while HCO_3 and SO_4 are only found in elevated relative abundances in less saline waters, because they are quickly removed due to mineral saturation and reduction reactions, particularly when in the presence of hydrocarbons.

From this information, an SBP ideally suited to distinguish between waters of different origins in the Permian Basin was generated from the major ion data (Table 1). The first row (corresponding to z_1) balances Na and Cl with the other elements and represents the role of halite versus carbonate and sulfate mineral buffering. The second coordinate, created based on the separation of meteoric versus paleoseawater along PC1 (Fig. 5), distinguishes balances Ca and Mg with HCO_3 and SO_4. Ca is rationed against Mg in the next row (corresponding to z_3), and it can be useful indicator of dolomitization or albitization reactions which frequently occur at various

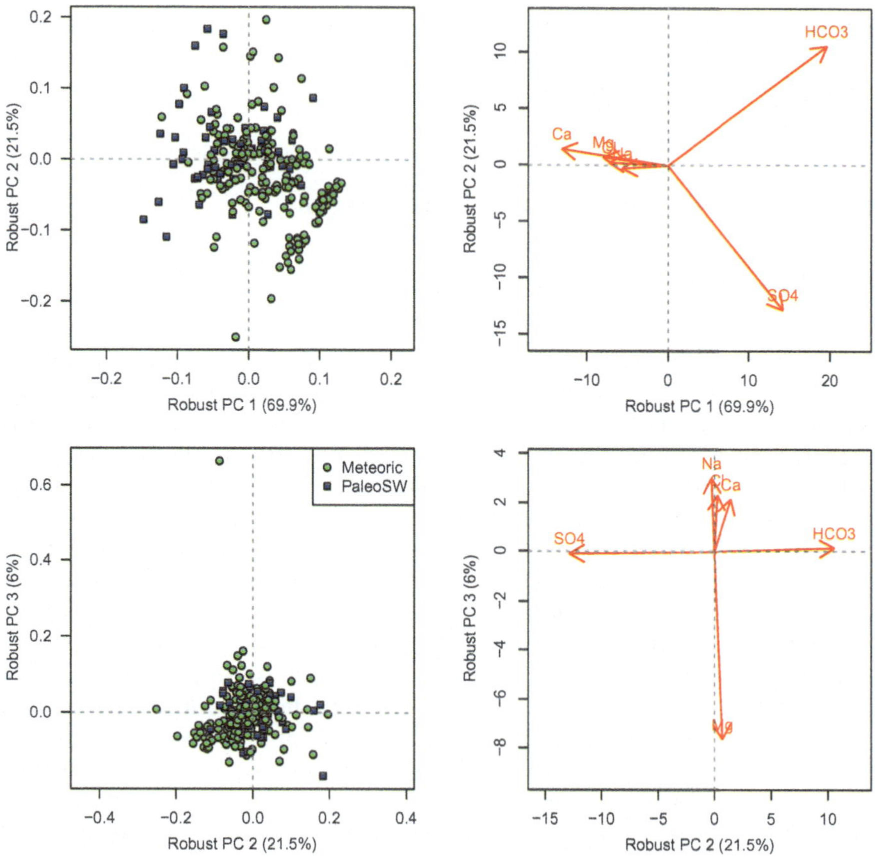

Fig. 5 PCA biplots of major ion chemistry showing loadings (right column) and scores (left column) for 314 produced water samples previously assigned groups via Na-Cl-Br systematics and $\delta^{18}O$ and $\delta^{2}H$ analysis

Table 1 Sequential binary partition of major ions showing distribution of ions in the 5 ilr coordinates (z_1–z_5). Ions labeled as + are used in the numerator, elements labeled as—are used in the numerator, and empty cells indicate that the ion is not used in that particular coordinate

Coordinate	Na	Cl	HCO_3	SO_4	Ca	Mg
z_1	+	+	−	−	−	−
z_2			−	−	+	+
z_3					+	−
z_4			+	−		
z_5	+	−				

depths. The 4th row in the SBP, corresponding to the 4th ilr coordinate, balances HCO_3 and SO_4, and is a simplified version of PC2. Finally, the last row ratios Na versus Cl, with is identical to the y-axis in the ilr Na-Cl-Br plot (Fig. 4).

4 Supervised Compositional Data Analysis

4.1 Validation and Parameter Tuning

Robust discriminant analysis was the first of three supervised techniques applied to the training dataset to examine the potential for origin of brines from historic Permian Basin data lacking Br and/or $\delta^{18}O$ and $\delta^{2}H$. Alpha (α) values for the MCD estimator, which governs that maximum proportion of data used in the group mean and covariance estimates, were varied from 0.5 to 0.99. In addition, both linear and quadratic discriminant analysis methods were applied. These variations produced negligible effects on the classification results, so discussion is focused on results for a relatively efficient, parsimonious model (LDA using an $\alpha = 0.75$). Two prior probability approaches were attempted. Under the first approach, the prior probability was set to the proportion of the two groups in the training dataset (78.6% meteoric water and 21.4% paleoseawater). The second approach was to assume equal prior probabilities. Confusion matrices (Fig. 6), which show the performance from robust LDA using the two prior probability approaches, show notably lower error from the former (total error rate of 14.0% versus 24.0%, respectively). However, results from the assumption of prior probabilities being equal to group proportion in the training set are likely biased low because the validation dataset, which was randomly pulled from the same source as the training set, exhibits a nearly identical distribution of samples (78.0% meteoric and 22.0% paleoseawater). There is no scientific reason that one would anticipate a roughly 80%/20% distribution in samples from a completely external source; conceptual models for the distribution of meteoric and paleoseawater in the basin do not support this distribution (Barnaby et al. 2004; Saller and Stueber 2018). Thus, the results assuming equal prior probabilities are probably more realistic of errors when applying this LDA model to an external dataset. Both datasets had higher misclassification rates for paleoseawater than for meteoric water, though the model using equal prior probabilities performed better (4/11 versus 7/11 misclassifications). The equal prior probability model, however, performed much worse with respect to meteoric water (8/39 versus 0/39 misclassifications, respectively). While the total error rate for the equal prior probability model (24.0%) is better than random change, it is far from ideal.

For SVMs, a broad Bayesian optimization (through Matlab's Machine Learning Toolbox) over various kernel types (linear, Gaussian, polynomial) and their hyperparameters was first performed to study variations in validation. The Radial Basis function (as described in Sect. 2) was found to perform best in general and was thus solely used thereafter. Misclassification penalty C and kernel scaling (related

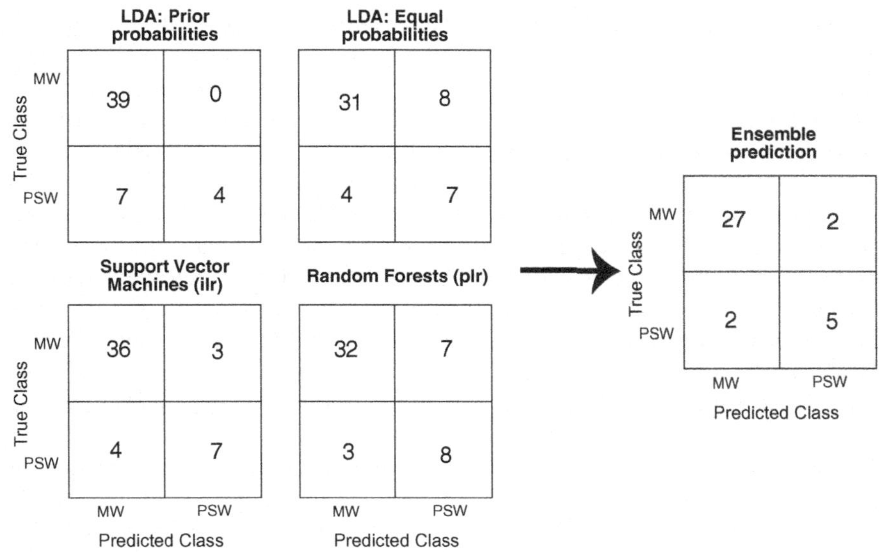

Fig. 6 Confusion matrices for the trained LDA (proportional prior probability = upper left; equal probability = upper right), SVM (lower left) and RF models (lower right), showing correct and incorrect classification of meteoric water (MW) and paleoseawater (PSW) for the test data set of 50 labeled samples not included in the training step. The rightmost confusion matrix shows results for the ensemble approach where only data classified as the same class by all three methods are shown. ilr = isometric log-ratio and plr = pairwise log-ratio

to γ) were then grid searched (from 1–10,000 to 0.1–50, respectively). A suitable model was chosen that offered a balanced prediction of both classes, rather than allowing one class to dominate the other. Given potential overlap in the classes, and thus the preference for a degree of "softness" in the decision boundary, we tested two different formulations of the RBF optimization: A 1-norm regularized optimization, and a 2-norm regularized optimization. Ultimately, the 2-norm formulation was selected given similar validation performance and superior computational speed. Given the imbalance in labeled samples in the training data set (80%/20% for meteoric water/paleoseawater samples), we further explored the impact of generating additional samples from normal distributions centered around existing paleoseawater samples, given a known data sampling error of 10–15%. Ultimately, this did not have a significant impact on validation, and a box constraint of 900 with a kernel scale of 3 was chosen. For this method, both pairwise and ilr coordinates were used in the analysis. The pairwise log-ratio performed better in the prediction meteoric water (97.4% versus 92.3% accuracy) but the ilr coordinate-based prediction was much more accurate (63.6% versus 45.5%) for the paleoseawater samples in validation analysis using the optimized SVM models. Given the accuracy trade-off and equal total error rates of 14.0%, the ilr coordinate-based analysis was chosen as the preferred model. Confusion matrix results for the ilr coordinate transformed test data set are shown in Fig. 6.

Similarly, parameters in RF methods can be chosen such that validation returns unrealistic cluster separation (i.e., by allowing for very small leaf sizes). To maintain a degree of similarity between predictions across methods, we limited the number of trees to 100 (though most averaging tapered off by roughly 30), and the leaf size to 10. RF methods can be modified to accept a penalty matrix that weighs misclassification of one class differently than another. Given the 80/20 skew in labeled classes in the training data set as mentioned above, we found that a weight ratio of 0.6 for misclassification of meteoric water/paleoseawater samples amounted to validation results that were suitably close to the SVM models (though slightly better in terms of paleoseawater classification). A compromise between smooth decision boundaries and balanced validation yielded the confusion matrix in Fig. 6 using a model with a leaf size of 8, and the model was then applied to the larger Permian Basin data set. In both cases, accurate predictions of paleoseawater resulted in very poor classification of meteoric water, owing to the overlapping nature of the data.

Given that misclassification errors of the three datasets were nearly equal at roughly 15–20% (Fig. 6), we devised a fourth and final approach of only classifying the data if all three models described here (LDA, RF, and SVM) provided the same results (i.e., all three methods classified the data for a sample as meteoric or paleoseawater). This approach assumes that the low relative accuracy for the individual methods can be overcome through ensemble prediction, which is simple form of boosting. Figure 6 shows the confusion matrix of the validation dataset using this combined approach. This generated 93.5% accuracy for the meteoric water data but only 7 of 11 paleoseawater samples were classified consistently between the three methods. Even from the remaining data, 2 of them were misclassified. Thus the approach tends to over-report the presence of meteoric water and greater confidence should be placed in samples identified as paleoseawater. The error rate was reduced to 11%, much lower than any individual method, but 28% of the validation data remained unclassified. Given the relative size of the dataset to be classified, losing a quarter or more of the results for a significant improvement in performance is reasonable.

4.2 Classification of Permian Basin Brines

Classification of the remaining Permian Basin brine geochemical data, which lack either Br or isotopes of H and O (See Fig. 1), was performed using the tuned and optimized algorithms described in Sect. 4.1, based entirely on pairwise log-ratios (SVM and RF) and the SBP-based ilr coordinates (LDA and SVM) created for the concentration of major ions (Ca, Cl, HCO_3, Mg, Na, and SO_4). Corresponding maps show locations of classified meteoric water versus paleoseawater based solely on LDA (Fig. 7) and the combined LDA, RF, and SVM approaches (Fig. 8). Samples included in Fig. 8 are a subset of those in Fig. 7, due to the aggregated nature of the agreement in classification among the three methods. This requirement left 39.3% of the data unclassified, but given the original size of the data set (n = 6,799) it still provides

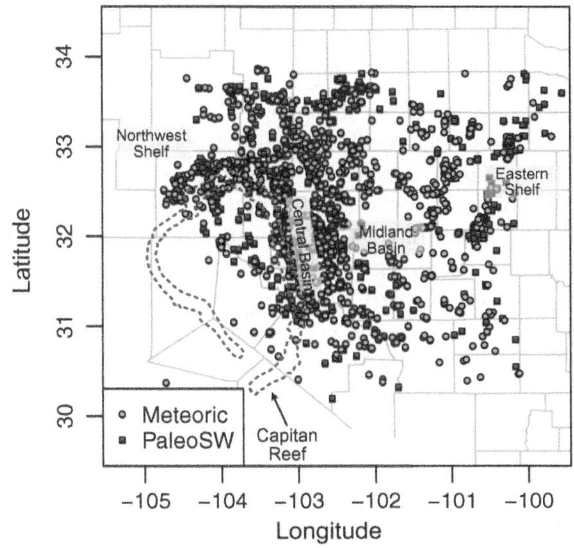

Fig. 7 Map showing assigned classification of previously unclassified brine geochemical data for the Permian Basin based on linear discriminant analysis

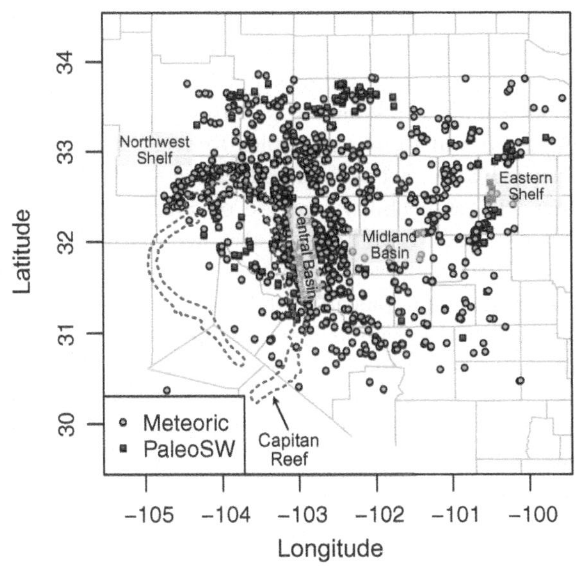

Fig. 8 Map showing assigned classification of previously unclassified brine geochemical data for the Permian Basin based on combined linear discriminant analysis, support vector machines, and random forests (see text for explanation)

a significant improvement in coverage relative to previous approaches. Despite the data loss in Fig. 8, these maps represent the most exhaustive understanding of water origin for Permian Basin brines ever released.

Comparison of the two maps (Figs. 7 and 8) highlights areas where the classification models disagree. In particular, the eastern portion of the Midland Basin and Eastern Shelf areas show notable differences in sample density between the maps, owing the largely disagreement among the classification methods in these areas. Conversely, larger proportions of the data were retained in the Delaware Basin, much of the Central Basin Platform and the Northwest Shelf, suggesting there is more confidence in the sample classification in these areas. Higher confidence in the results can be illustrated by the curvilinear groups of meteoric classified data which mirror the outline of the Permian-age Capitan Reef structure. The reef is considered a minor aquifer in the state, locally exhibits high transmissivities and yields, is hydrologically linked to regional surface recharge areas (Uliana 2001), and has been isotopically categorized as meteoric water in specific areas of the basin (O'Neil et al. 1986), supporting the hypothesis that water in the aquifer is meteoric. Conversely some regions show large proportions of data classified as paleoseawater. One region, in the northern portion of the northwest shelf, along the New Mexico-Texas border is one example. These samples were collected primarily from gas wells in Pennsylvanian-age reservoirs ranging in depth from roughly 2,800 to 3,800 m below surface, suggesting the meteoric waters have largely been excluded in this area. In contrast, Barnaby et al. (2004) demonstrated that meteoric water has entered similarly aged units in the Dagger Draw area, located to the southwest. Their work argued for deep fluid movement down vertical faults in the vicinity of local mountains ranges (i.e., Guadalupe Mountains). Steep topography is not present in the northern portion of the Northwest Shelf, indicating that topographically influenced flow driving meteoric water into these deeper reservoirs is restricted to regions closer to recharge areas in southeastern New Mexico. A second area dominated by nearly 1,000 samples classified as paleoseawater is the northern half of the Eastern Shelf. Again, the majority of samples from the dataset in this area were produced from Pennsylvanian-age reservoirs (Strawn Reef, Canyon Reef, and Strawn) suggesting that low permeability in these or overlying units has prevented displacement of paleoseawater from the aquifers since its emplacement in the Late Permian (Engle et al. 2016; Saller and Stueber 2018; Stueber et al. 1998). These interpretations put constraints on the location and extent of regional flushing of the basin by meteoric water. Previous authors have attempted to define the exact flowpaths and degree of meteoric infiltration into the basin, but have been limited in their ability to interpolate the results due to poor data distribution (Bein and Dutton 1993; Saller and Stueber 2018). Thus, more careful examination of the results produced here can be used to better constrain hydrogeologic modeling and provide a more well calibrated understanding of fluid behavior of groundwater at sub-continental scales.

5 Conclusions

This work demonstrates that despite lack of data previously considered critical for classification of water origin of brines from the Permian Basin, namely, Br concentration or $\delta^{18}O$ and $\delta^{2}H$ composition, supervised compositional data techniques were moderately successful in classifying samples using only their major ion composition. This is a major step forward in increasing the utility of historical water quality datasets, which can lack parameters seen as key for use in focused investigations. The results generated for the Permian Basin also showed that no single supervised approach (LDA, SVMs, and RF) was particularly successful in producing low classification error rates during validation. However, an aggregated approach, which only accepted the classification of a samples if all three methods provided the same classification, drastically decreased errors at the expense of leaving a portion of the data unclassified. Despite these downfalls, spatial maps showing the classification of water samples across the basin, using purely major ion data, showed many patterns not described in previous studies. In particular, meteoric water appears to be focused in reservoirs which fall in or shoulder the Capitan Reef, around the edges of the Delaware sub-basin. Similarly, the majority of data from samples along the northern Northwest Shelf and northern Eastern Shelf from Pennsylvanian-to Devonian-age reservoirs are dominated by paleoseawater. This observation is unlikely to be a result of data loss from samples identified as possibly meteoric in the aggregated approach as few wells from this reservoir range were lost during data aggregation (cf. Figs. 7, 8). Large tracts of paleoseawater in this region provide a stark contrast to west-to-east piston flow of meteoric water across the basin, including deeper reservoirs, as proposed by previous workers working with significantly fewer data.

The methods proposed and demonstrated here have broad applicability to classify historic water quality data, not just the origin of brines. However, this is a single example from a single location where the regional processes are well documented. It is less clear how universally well such an approach would work in other locations or whether the amount of error produced would be acceptable. Additionally, the scale of investigation in this case is relatively immune to small-scale perturbations or errors, given that the focus was on large-scale processes and structures. As a follow-up from this work, more localized examination and interpretation of these results will be used to future studies to examine the ability of the method to produce meaningful results at finer scales. Despite these caveats, results from this initial work are promising and suggest that insight from application of supervised CoDA techniques is of merit to assist in answering questions where other techniques have not been effective.

Acknowledgements The manuscript was assisted through discussion and feedback with geologists at numerous oil and gas operators in the Permian Basin. Funding for computing hardware provided through start-up packages for both authors and a Texas STARs grant to J. Chaput. Christian Leach provided constructive feedback on an earlier version of the chapter. M. Engle would also like to thank the encouragement, positive feedback, and guidance from Vera Pawlowsky-Glahn and Juan Jose Egozcue over the last decade.

References

J. Aitchison, The statistical analysis of compositional data. Reprinted in 2003 with additional material by Blackburn Press (Chapman & Hall, London, 1986)

R.J. Barnaby, G.C. Oetting, G. Gao, Strontium isotopic signatures of oil-field waters: applications for reservoir characterization. AAPG Bull. **88**, 1677–1704 (2004)

A. Bein, A.R. Dutton, Origin, distribution, and movement of brine in the Permian Basin (U.S.A.): a model for displacement of connate brine. Geol. Soc. Am. Bull. **105**, 695–707 (1993)

M.S. Blondes, K.D. Gans, M.A. Engle, Y.K. Kharaka, M.E. Reidy, V. Saraswathula, J.J. Thordsen, E.L. Rowan, E.A. Morrissey, U.S. Geological survey national produced waters geochemical database, version 2.3, https://eerscmap.usgs.gov/pwapp/ (2018)

B.E. Boser, I.M. Guyon, V.N. Vapnik, A training algorithm for optimal margin classifiers, in *COLT '92: Proceedings of the Fifth Annual Workshop on Computational Learning Theory* (1992), pp. 144–152

L. Breiman, Random forests. Mach. Lang. **45**, 5–32 (2001)

A. Buccianti, V. Pawlowsky-Glahn, New perspectives on water chemistry and compositional data analysis. Math. Geol. **37**, 703–727 (2005)

T. Dietterich, Overfitting and undercomputing in machine learning. ACM Comput. Surv. **27**, 326–327 (1995)

A.R. Dutton, B.C. Richter, C.W. Kreitler, Brine discharge and salinization, Concho river watershed West Texas. Ground Water **27**, 375–383 (1989)

S.P. Dutton, E.M. Kim, R.F. Broadhead, C.L. Breton, W.D. Raatz, S.C. Ruppel, C. Kerans, Play analysis and digital portfolio of major oil reservoirs in the Permian Basin: application and transfer of advanced geological and engineering technologies for incremental production opportunities. Texas Bureau of Economic Geology and New Mexico Bureau of Geology and Mineral Resources (2004)

J. Egozcue, V. Pawlowsky-Glahn, G. Mateu-Figueras, C. Barceló-Vidal, Isometric logratio transformations for compositional data analysis. Math. Geol. **35**, 279–300 (2003)

J.J. Egozcue, V. Pawlowsky-Glahn, Groups of parts and their balances in compositional data analysis. Math. Geol. **37**, 795–828 (2005)

M.A. Engle, M.S. Blondes, Linking compositional data analysis with thermodynamic geochemical modeling: oilfield brines from the Permian Basin USA. J. Geochem. Expl. **141**, 61–70 (2014)

M.A. Engle, B. Brunner, Considerations in the application of machine learning to aqueous geochemistry: origin of produced waters in the northern U.S. Gulf Coast Basin. Appl. Comput. Geosci. **3–4**, 100012 (2019)

M.A. Engle, F.R. Reyes, M.S. Varonka, W.H. Orem, L. Ma, A.J. Ianno, T.M. Schell, P. Xu, K.C. Carroll, Geochemistry of formation waters from the Wolfcamp and "Cline" shales: insights into brine origin, reservoir connectivity, and fluid flow in the Permian Basin USA. Chem. Geol. **425**, 76–92 (2016)

M.A. Engle, E.L. Rowan, Interpretation of Na-Cl-Br systematics in sedimentary basin brines: comparison of concentration, element ratio, and isometric log-ratio approaches. Math. Geosci. **45**, 87–101 (2013)

M.A. Engle, E.L. Rowan, Geochemical evolution of produced waters from hydraulic fracturing of the Marcellus Shale, northern Appalachian Basin: a multivariate compositional data analysis approach. Int. J. Coal Geol. **126**, 45–56 (2014)

P. Filzmoser, K. Hron, C. Reimann, Principal component analysis for compositional data with outliers. Environment **20**, 621–632 (2009)

P. Filzmoser, K. Hron, M. Templ, Discriminant analysis for compositional data and robust parameter estimation. Computat. Stat. **27**, 585–604 (2011)

J.S. Hanor, Origin of saline fluids in sedimentary basins, in *Geofluids: Origin, Migration and Evolution of Fluids in Sedimentary Basins*, ed. by J. Parnell (1994), pp. 151–174

B. Hitchon, M. Brulotte, Culling criteria for "standard" formation water analyses. Appl. Geochem. **9**, 637–645 (1994)

K. Hron, M. Templ, P. Filzmoser, Imputation of missing values for compositional data using classical and robust methods. Computat. Stat. Data Anal. **54**, 3095–3107 (2010)

L.P. Knauth, M.A. Beeunas, Isotope geochemistry of fluid inclusions in Permian halite with implications for the isotopic history of ocean water and the origin of saline formation waters. Geochim. Cosmochim. Acta. **50**, 419–433 (1986)

M.A. McCaffrey, B. Lazar, H.D. Holland, The evaporation path of seawater and the coprecipitation of Br- and K+ with halite. J. Sediment. Res. **57**, 928–937 (1987)

J.R. O'Neil, C.M. Johnson, L.D. White, E. Roedder, The origin of fluids in the salt beds of the Delaware Basin New Mexico and Texas. Appl. Geochem. **1**, 265–271 (1986)

F.R. Reyes, M.A. Engle, L. Jin, M.A. Jacobs, J.G. Konter, Hydrogeochemical controls on brackish groundwater and its suitability for use in hydraulic fracturing: the Dockum Aquifer, Midland Basin Texas. Environ. Geosci. **25**, 37–63 (2018)

B.J. Rostron, C. Holmden, Regional variations in oxygen isotopic compositions in the Yeomanand Duperow aquifers, Williston Basin (Canada-USA). J. Geochem. Expl. **78**, 337–341 (2003)

A.H. Saller, A.M. Stueber, Evolution of formation waters in the Permian Basin, USA: late permian evaporated seawater to Neogene meteoric water. AAPG Bull. **102**, 401–428 (2018)

A.M. Stueber, A.H. Saller, H. Ishida, Origin, migration, and mixing of brines in the Permian Basin: geochemical evidence from the eastern Central Basin platform. Texas. AAPG Bull. **82**, 1652–1672 (1998)

J.A.K. Suykens, J. Vandewalle, Least squares support vector machine classifiers. Neural Proc. Lett. **9**, 293–300 (1999)

R. Tolosana-Delgado, H. Talebi M. Khodadzadeh, K.G. van den Boogaart, On machine learning algorithms and compositional data. Presented at the CoDaWork 2017, Terrassa (2019)

M.M. Uliana, The geology and hydrogeology of the Capitan Aquifer: a brief overview, in Aquifers of West Texas, ed. by R.E. Mace, W.F. Mullican, III, E.S. Angle (Texas Water Development Board, Austin, 2001), pp. 153–166

L.M. Walter, A.M. Stueber, T.J. Huston, Br-Cl-Na systematics in Illinois basin fluids: constraints on fluid origin and evolution. Geol. **18**, 315–318 (1990)

Chronic Kidney Disease of Uncertain Aetiology and Its Relation with Waterborne Environmental Toxins: An Investigation via Compositional Balances

Jennifer M. McKinley, Ute Mueller, Peter M. Atkinson, Ulrich Ofterdinger, Siobhan F. Cox, Rory Doherty, Damian Fogarty, and J. J. Egozcue

Abstract The occurrence of environmental clusters of Chronic Kidney Disease of uncertain aetiology (CKDu), where there is no known cause for the onset of kidney dysfunction, is a concern globally. Waterborne exposure pathways in the environment may result in indirect or direct ingestion of trace elements with potential health risks. This research examines the relationship between Standardised Incidence Rates (SIRs) of CKDu and the log-ratio balances of Potentially Toxic Elements (PTEs) in regional stream water. Compositional elemental balances were created and

---

J. M. McKinley (✉) · P. M. Atkinson · U. Ofterdinger · S. F. Cox · R. Doherty
School of Natural and Built Environment, Queen's University Belfast, Belfast, Northern Ireland, UK
e-mail: j.mckinley@qub.ac.uk

P. M. Atkinson
e-mail: pma@lancaster.ac.uk

U. Ofterdinger
e-mail: u.ofterdinger@qub.ac.uk

S. F. Cox
e-mail: s.cox@qub.ac.uk

R. Doherty
e-mail: r.doherty@qub.ac.uk

U. Mueller
School of Science, Edith Cowan University, Perth, WA, Australia
e-mail: u.mueller@ecu.edu.au

P. M. Atkinson
Lancaster Environment Centre, Lancaster University, Lancaster, UK

D. Fogarty
Belfast Health Trust, Belfast, Northern Ireland, UK
e-mail: Damian.Fogarty@belfasttrust.hscni.net

J. J. Egozcue
Department of Civil and Environmental Engineering, Technical University of Catalonia (UPC), Campus Nord c/Jordi Girona 1-3, C-2, E-08034, Barcelona, Spain
e-mail: juan.jose.egozcue@upc.edu

P. Filzmoser et al. (eds.), *Advances in Compositional Data Analysis*,
https://doi.org/10.1007/978-3-030-71175-7_15

regression was used to identify the balances most associated with log-transformed SIR CKDu. At the regional scale, a statistically significant relationship was found between log (CKDu SIR) and the elemental balance Al/As which effectively delineates different geological domains across Northern Ireland. Following stratification by basalt bedrock (the dominant bedrock geology for SIR CKDu), the balance Al/Fe was identified as significantly associated with log (CKDu SIR). Superficial deposits, dissolved organic carbon and pH may act as controls on the balance of Al and Fe. With a high proportion of private water supplies registered in these areas, this research highlights the importance of considering bedrock geology and superficial deposits in understanding multi-element interactions of waterborne environmental toxins and potential links with environmental clusters of CKDu. We would like to acknowledge and thank our esteemed colleague and friend Vera Pawlowsky-Glahn who has been instrumental in the development of this research, pioneering the importance of Compositional Data Analysis (CoDA) in the study of health and the environment. For several of us, Vera provided our first introduction to CoDA and over many years has aided our shared understanding and increased awareness of the need to address the compositional nature of data such as soil and water geochemistry in our research. We are pleased to have this work on the use of compositional balances in exploring the explanatory environmental factors related to chronic kidney disease of uncertain aetiology included in this Festschrift. It represents an exciting field of innovative research where an acknowledgement of CoDA principles is vital if we are to understand fully the critical relationship between our health and the environment.

1 Introduction

The World Health Organisation (1996) divides naturally occurring elements into three groups based on their nutritional significance in humans: (1) essential elements (iodine (I), zinc (Zn), selenium (Se), copper (Cu), molybdenum (Mo) and chromium (Cr)); (2) elements which are probably essential (manganese (Mn), silicon (Si), nickel (Ni), boron (B) and vanadium (V)); and (3) potentially toxic elements (PTEs), some which may have essential functions at low levels (lead (Pb), cadmium (Cd), mercury (Hg), arsenic (As), aluminium (Al), lithium (Li) and tin (Sn)). However, individual elements or single component values may not be the most appropriate way to evaluate the risk to human health. An inherent assumption is that elemental data, such as geochemical data, represent absolute abundances. The WHO (1996) recognises that multi-element interactions can increase the bioavailability or modify the metabolism of both essential and Potentially Toxic Trace Elements (PTEs). Therefore, a multivariate approach is more suitable to investigate the relationship between human health and elements in the environment. However, because of the compositional nature of geochemical data single elements cannot be compared directly with one another. An alternative approach is presented here, using Compositional Data Analysis (CoDA). In particular, the relation between Standardised Incidence Rates (SIRs) of Chronic Kidney Disease of uncertain aetiology (CKDu) and the log-ratio balances of PTEs in regional stream water is explored.

Several factors are known to cause Chronic Kidney Disease (CKD) including age, ethnicity and pre-existing medical conditions (Lindeman et al. 1985; Couser et al. 2011; Musso and Oreopoulos 2011; Gilg et al. 2012; Lewis 2012) However, reporting of environmental clusters of CKDu has increased globally to such an extent that a WHO special task force was set up to investigate the incidence of CKDu without known risk causes (Chiu and Yang 2005; Zheng et al. 2014,2015; Jayatilake et al. 2013; Jayasumana et al. 2014; Correa-Rotter et al. 2014; Weaver et al. 2015; Lunyera et al. 2016). The UK renal registry (UKRR) reports data for nine different identifiable Primary Renal Diseases (PRD)—diabetes, glomerulonephritis, hypertension, polycystic kidney disease, pyelonephritis, renal vascular disease, other, missing and uncertain aetiology. Variation between renal centres cannot account for the significant percentage of cases of uncertain aetiology recorded (e.g., 29.4% of patients in 2017 were identified to diabetes and 14.9% to uncertain aetiology (UK Renal Registry 2019). The environmental causes of CKDu require investigation to address the global issue of geographic clusters of CKD where there is no known cause for the onset of kidney dysfunction. The causes of CKDu are most likely multifactorial (Weaver et al. 2015). Environmental factors that may cause CKDu, including the role of environmental nephrotoxins such as As, Hg and Pb (Soderland et al. 2010) and toxic organic compounds from coal (Orem and Calin 2001), require further exploration as these factors may also contribute to the heterogeneity of other causes of progressive CKD in diabetes and hypertension.

Bedrock and superficial geology have been shown previously to be important potential geogenic sources of PTEs in soils and stream waters across Northern Ireland (Barsby et al. 2012; Cox et al. 2013; McKinley et al. 2013; Palmer et al. 2015). Previous research using the UK Renal Register (UKRR) and environmental data is limited (Jackson et al. 2016; McKinley et al. 2020b,2020a). Jackson et al. (2016) used a compositional Poisson regression approach and highlighted the interaction between essential elements, PTEs and potential associations with CKDu. McKinley et al. (2020b) investigated the use of compositional balances in a study of urban soils and found geogenic and anthropogenic signatures related to CKDu, while McKinley et al. (2020a) explored the underlying causes of CKDu using social deprivation and environmental factors. In this research, compositional balances were used to explore the waterborne exposure pathways using geochemical stream water elemental data to represent potential health risks associated with indirect or direct ingestion of trace elements locally.

2 Study Area and Data

2.1 Study Area

Northern Ireland straddles three geological terrains referred to as the Central Highlands (Grampian) Terrane, Midland Valley and the Southern Uplands-Down-Longford Terrane (Fig. 1a; Cooper et al. 2013). Mid-to-late Neoproterozoic rocks of the Dalradian supergroup including clastic marine sediments, basaltic magmatism, volcaniclastic sediments and turbidites, comprise most of the Central Highlands (Grampian) Terrane. These rocks experienced regional metamorphism. The Midland Valley Terrane comprises distinct geological units bounded by faulting, including an early Palaeozoic igneous complex and Proterozoic psammites and pelitic rocks, which have experienced different folding phases and metamorphism. Continental Devonian red bed sediments and marine Carboniferous sediments overlie the Midland Valley Terrane in the south of the country. The Southern Uplands-Down-Longford Terrane is composed of Ordovician and Silurian rocks with a younger igneous intrusive complex. Permian, Triassic, Jurassic and Cretaceous rocks are overlain in the North West of the country by Palaeogene basalt lava. At least 80% of the bedrock is covered by superficial deposits including glacial till, post-glacial alluvium and peat (summarised from Cooper and Mitchell 2013; Mitchell 2004).

2.2 Chronic Kidney Disease of Uncertain Aetiology (CKDu)

The UKRR provided Standardised Incidence Rates (SIRs) for patients starting Renal Replacement Therapy (RRT) between 2006 and 2016, by Super Output Area (SOA). Data were grouped by age brackets and for uncertain aetiology (CKDu; Fig. 1b) for the 890 SOA administrative wards across Northern Ireland. The SIR for a SOA is a measure that quantifies the relationship between the actual incidence in the SOA and the expected incidence based on that of Northern Ireland as a whole. SIRs of exactly 1 indicate that the incidence of CKD for a SOA is equal to that expected based on Northern Ireland's average age-specific incidence rates. SIRs above 1 indicate that the incidence of CKD for a SOA is greater than expected and SIRs below1 indicate that the incidence for a SOA is lower than expected based on the Northern Ireland's average age-specific incidence rates.

2.3 Stream Water Geochemistry

This research used a regional geochemical survey of Northern Ireland's stream waters, completed by the Geological Survey of Northern Ireland (GSNI) and the British Geological Survey (BGS). Sampling and quality control were undertaken

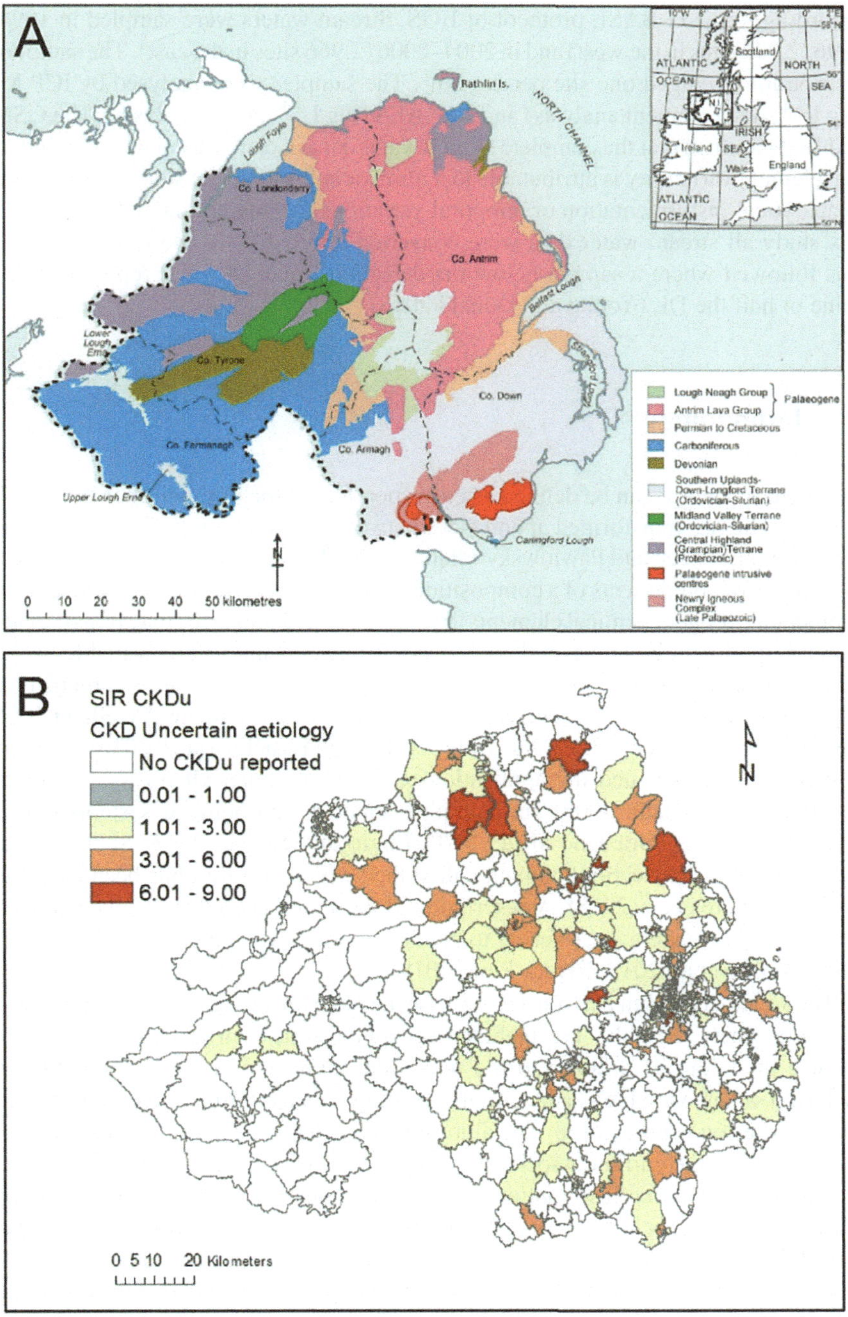

Fig.1 **a** Simplified geology of Northern Ireland (provided by Geological Survey of Northern Ireland (GSNI), Mitchell 2004). **b** Standardised Incidence Rates (SIRs) for patients starting Renal Replacement Therapy (RRT) between 2006 and 2016, by Super Output Areas (SOAs) for Chronic Kidney Disease of uncertain aetiology (CKDu)

according to the G-BASE protocol of BGS. Stream waters were sampled in 1994–1996 (2,908 sites in the west) and in 2004–2006 (2,966 sites in the east). The sampling distribution averaged one site per 2.4 km^2. The samples were analysed by ICP MS ($\mu g\ L^{-1}$; trace element analysis) and ICP-AES ($mg\ L^{-1}$; anion analysis). The GSNI and BGS confirm that the complete stream water dataset can be used with confidence that element variability is attributable to natural or anthropogenic sources as opposed to analytical instrumentation or temporal variations (Young and Donald 2013). For this study all stream water data were converted into $\mu g\ L^{-1}$. The GBase protocol was followed where a sample below the detection limit (DL) was replaced with a value of half the DL (Young and Donald 2013).

3 Methodology

In CoDA, a balance can be defined as corresponding to the normalised difference in means of the log-transformed abundances between two sub-compositions (Egozcue et al. 2003; Egozcue and Pawlowsky-Glahn 2005). The concept of balances between two groups of components of a composition provides an interpretable way to identify components (or geochemical elements) whose relative abundances may be associated with a response variable. Several approaches to select balances are available, which may be grouped as data- or knowledge-driven approaches within a compositionally compliant context (Egozcue et al. 2003; Egozcue and Pawlowsky-Glahn 2005; Aitchison and Greenacre 2002; McKinley et al. 2016; Kynčlová et al. 2016). Previous research has investigated the application of CoDA to water chemistry (Buccianti and Pawlowsky-Glahn 2005) and complex aqueous geochemical systems (Gozzi et al. 2018; Sauro Graziano et al. 2020; Engle and Rowan 2013; Shelton et al. 2018). A forward-selection method, called the *selbal* algorithm has been proposed by Rivera-Pinto et al. (2018). A description of the *selbal* procedure is given here but for full details readers are referred to studies on the development of this approach (Rivera-Pinto et al. 2018; Rivera-Pinto 2018).

The *selbal* algorithm performs a stepwise multiple linear regression using forward selection where variables are included sequentially at each step, starting with a balance including 2 variables to the maximum number of variables possible. At each step, a new variable is added to the existing balance so that the specified association criterion is maximised. The algorithm stops when none of the possible additions increases the current association. An n-fold Cross-Validation (CV) procedure provides the strength of association or discrimination value for each balance. In the CV procedure, at each iteration (20 iterations were used in this research), the data are subdivided into n disjoint subsets whose union is the entire set and each of the n subsets is used as a test set with the associated training set comprised of the remaining $(n - 1)$ folds. Then the *selbal* algorithm is applied to the training dataset and the optimal balance from the maximum number of variables is obtained.

The classification accuracy (mean squared error (MSE) in this case) of these balances is measured on the test dataset. Since there are 20 iterations and n folds for each number of variables of the balance, the mean MSE and standard error can be computed. The "optimal number" of elements to be included in the balance is defined as the least number of variables for which the mean MSE of the balance is within 1 standard error of the minimum mean MSE. As described by Rivera-Pinto (2018), this is similar to the cross-validation process in LASSO for finding the optimal penalisation parameter lambda (Hastie et al. 2009). Once the "optimal number" of elements has been determined, this defines what is termed the global balance. The CV results are used to explore the robustness of the global balance, that is, the selbal algorithm is applied to the whole dataset with the specification that the maximum number of components in the balance is the "optimal number" of elements as defined above by the MSE. All of the balances, which meet the MSE requirement, obtained in the cross-validation process are compared then with the global balance. Output tables provide the relative frequency of the different balances obtained in the CV process and the proportion of times that each element has been included into a balance.

This approach has been used previously to identify urban soil geochemical elements whose relative abundances indicate geogenic and anthropogenic sources of PTEs associated with elevated incidences of CKDu (McKinley et al. 2020b) and demonstrated for CKD data with urban soil geochemistry and social deprivation indices (McKinley et al. 2020a).

3.1 Data Preparation

The stream water geochemistry data were tagged to the SIRs of CKDu and the geometric mean was calculated for each Super Output Area (SOA). Initially, the relationship between CKDu and a subcomposition of the stream water geochemistry was explored for all SOAs with CKDu across Northern Ireland (109 SOAs). The subcomposition comprised 14 PTEs identified in the WHO report (Al, As, Cd, Cr, Cu, K, Mn, Mo, Ni, Pb, Se, U, V and Zn) and the anions (Ca, Fe, Mg, Na and Si). As the incidence of CKDu is spatially variable across the country (Fig. 1b), the stream water geochemical data were tagged to geology using a simplified geology scheme provided by GSNI. As a high proportion of SIRs of CKDu were found to overly basalt bedrock (58 SOAs), this class only was used for further analysis. The other geological domains were not explored further in this research due to limited SIRs of CKDu reported for the SOAs.

3.2 Balance Approach

Two approaches were used. The first approach was the forward-selection method using the *selbal* algorithm. A *fivefold* Cross-Validation (CV) procedure was used to identify the set of balances. An associated Generalised Linear Regression Model

(GLM) was used to calculate the mean response based on the balances identified in the parameter estimation step. The second approach used the isometric log-ratio (ilr) of the subcomposition based initially on a Sequential Binary Partition (SBP) and secondly using an ilr balanced approach, both using the R package *compositions* (Boogaart and Tolosana-Delgado 2008). GLM regression was used to examine any potential relationship between CKDu SIRs and the stream water geochemistry.

The two compositional balance approaches (log-transformed using the selbal cross-validation option; and ilr transformed using an SBP approach and ilr balanced approach) and the environmental elemental variables were used to identify the elemental balance with the greatest association with log-transformed CKDu SIR using the 2006–2016 UKRR SIR data.

3.3 Spatial Dependence

Regression models assume independence between the observations. However, this assumption may not be valid for spatial data (Pardo-Iguzquiza and Dowd 2004; Olea and Pawlowsky-Glahn 2009). To assess the nature of the impact of spatial autocorrelation in this research Moran's *I* test statistic was used to test for spatial autocorrelation in the residuals computed from the regression models for the regional geology and basalt case studies.

4 Results

4.1 Regional Geology Study

For the regional study using the *selbal* approach, the elements most frequently identified as the global balance in the CV procedure, and therefore most associated with log (CKDu SIR) were As appearing 90% and Al appearing 74% of the time (Fig. 2a, b). In addition to the global balance of Al/As, the balances of Na/As and Na/Mo were also identified in the CV procedure. Only the global balance Al/As was found to be statistically significant (Table 1, Fig. 3: significance level <0.001). The balance between As and Al was also identified using the SBP (ilr base) and the ilr balanced approaches (Table 1; both showed a significance level of 0.1). To assess the impact of spatial autocorrelation on the regression models, Moran's *I* test statistic was used to test the residuals computed from the regression models. The Moran's Index for the residuals was found to be not significantly different from random (Moran's Index −0.008782; *z*-score: 0.009158; *p*-value: 0.992693).

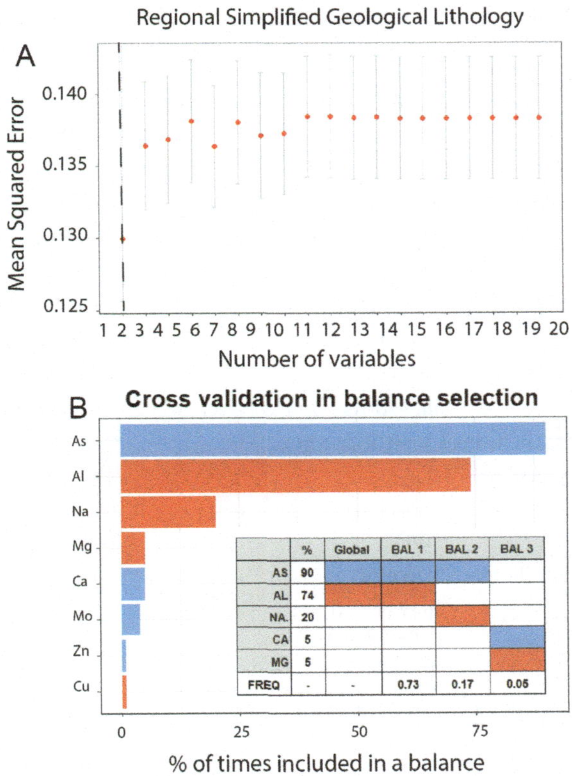

Fig. 2 Results of the forward-selection method using the *selbal* algorithm for (Super Output Areas (SOAs)) Standardised Incidence Rates (SIRs) of Chronic Kidney Disease of uncertain aetiology (CKDu), for regional study **a** Mean Squared Error (MSE) as a function of the number of components included in the balance. The optimal number of components is highlighted with a vertical dashed line; **b** The balance identified with the whole dataset is the most frequently identified in the Cross-Validation (CV) procedure and global balance and other balances identified in CV procedure

4.2 *Stratified Study for Basalt Bedrock*

Following stratification by basalt bedrock geology, a different association is indicated between log (CKDu SIR) and stream water geochemistry. The results for log (CKDu SIR) identified the balances Al/Fe (global balance with Al appearing 92% and Fe appearing 91% of the time, respectively), Pb/Fe and Na/Mo most frequently in the CV procedure (Fig. 4c, d). Regression (using GLM) suggested a correlation between log (CKDu SIR) and the global balance Al/Fe (Table 1, Fig. 5; significance level of 0.001). To assess the impact of spatial autocorrelation on the regression models, Moran's *I* test statistic was used to test the residuals computed from the regression models. The Moran's Index for the residuals was found to be not significantly different from random (Moran's Index −0.008782; *z*-score: 0.009158; *p*-value: 0.992693).

Using the SBP (ilr base) approach only a large balance which included all elements against Na was found to be significant (level 0.05) and no statistically significant relationship was found between log (CDKu SIR) and any elemental balances using the ilr balanced approach. The compositional dendrograms shown for the SBP (ilr base) and ilr balanced approach (Fig. 6a, b, respectively) show the difference between the two ilr approaches on partitioning of the elements.

Table 1 Summary of regression results for Standardised Incidence Rates (SIRs) for Chronic Kidney Disease of uncertain aetiology (CKDu) and elemental balances. only significant statistical results are shown

Dependent variable		Estimate	Std. error	t value	Pr(>|t|)	p-value
Regional study log(CKDu SIR)						
Selbal	Intercept	0.7015	0.0987	7.106	1.34E-10	<0.001
	Global balance Al/As	0.1587	0.0416	3.81	0.0002	<0.001
SPB ilr base	Intercept	−1.351	1.6997	−0.795	0.4288	
	ilr1 Al/As	−0.179	0.0958	−1.865	0.0654	0.1
ilr balanced	Intercept	−1.351	1.6997	−0.795	0.4288	
	ilr 1 Al/As	−0.179	0.0958	−1.865	0.0654	0.1
Basalt bedrock log(CKDu SIR)						
Selbal	Intercept	1.5525	0.1457	10.654	4.38E-15	<0.001
	Global balance Al/Fe	0.261	0.0847	3.081	0.0032	0.001
SPB ilr base	Intercept	−5.694	4.2771	−1.331	0.1909	
	ilr17 Na/(Al, As, Cd, Cr, Cu, K, Mn, Mo, Ni, Pb, Se, U, V, Zn, Ca, Fe, Mg)	1.4652	0.6212	2.359	0.0234	0.05
ilr balanced	ilr	No balances found to be significant				
Basalt bedrock (reordered elements) log(CKDu SIR)						
Selbal	Intercept	1.5525	0.1457	10.654	4.38E-15	<0.001
	Global balance Al/Fe	0.261	0.0847	3.081	0.0032	0.001
SPB ilr base	Intercept	−5.694	4.2771	−1.331	0.1909	
	ilr3 Na/(Al,Fe,Cr)	1.2603	0.5515	2.285	0.0278	0.01
	ilr4 Mo/(Al,Fe,Cr,Na)	−0.583	0.2511	−2.323	0.0255	0.01
ilr balanced	Intercept	−5.694	4.2771	−1.331	0.1909	
	ilr2 (Na/Cr)	1.0755	0.487	2.208	0.0332	0.05

To test the impact of the ordering of elemental variables in the analysis, the balances identified using the *selbal* approach were used (a knowledge-driven approach) to reorder the elements and all CoDA balance approaches were investigated. The elements were reordered as Al, Fe, Cr, Na, Mo, Cu, K, Mn, Ni, Pb, Se, U, Zn, V, As, Cd, Ca, Mg and Si. The results obtained using the *selbal* approach did not change with all balances and associated statistical significance remaining the same (significance level 0.001 for global balance Al/Fe). The dendrograms

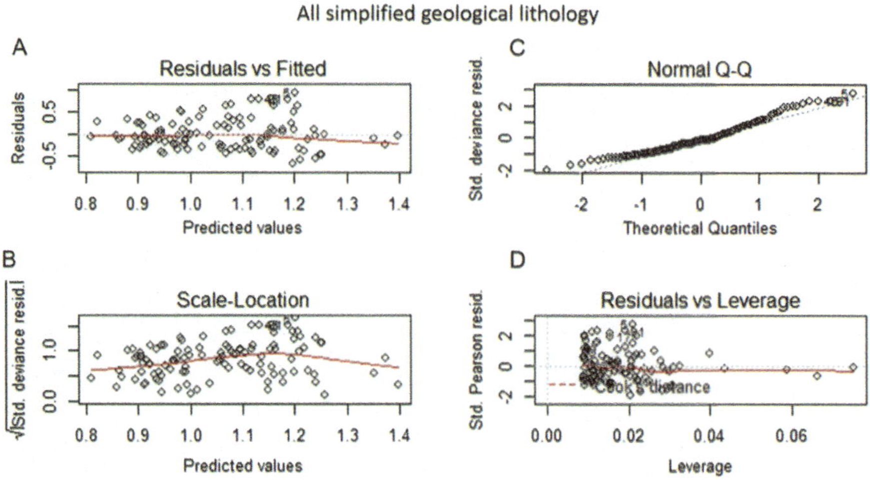

Fig. 3 Generalised Linear Model regression (GLM) summary for log (SIR_CKDu) with regional stream water data (all simplified geological lithology) for global balance Al/As; **a** Residual versus Fitted; **b** Normal Q-Q plot; **c** Scale Location and **d** Residuals versus Leverage

Fig. 4 Results of the forward-selection method using the *selbal* algorithm for (Super Output Areas (SOAs)) Standardised Incidence Rates (SIRs) of Chronic Kidney Disease of uncertain aetiology (CKDu), for basalt bedrock **a** Mean Squared Error (MSE) as a function of the number of components included in the balance. The optimal number of components is highlighted with a vertical dashed line; **b** The balance identified with the whole dataset is the most frequently identified in the Cross-Validation (CV) procedure and global balance and other balances identified in CV procedure

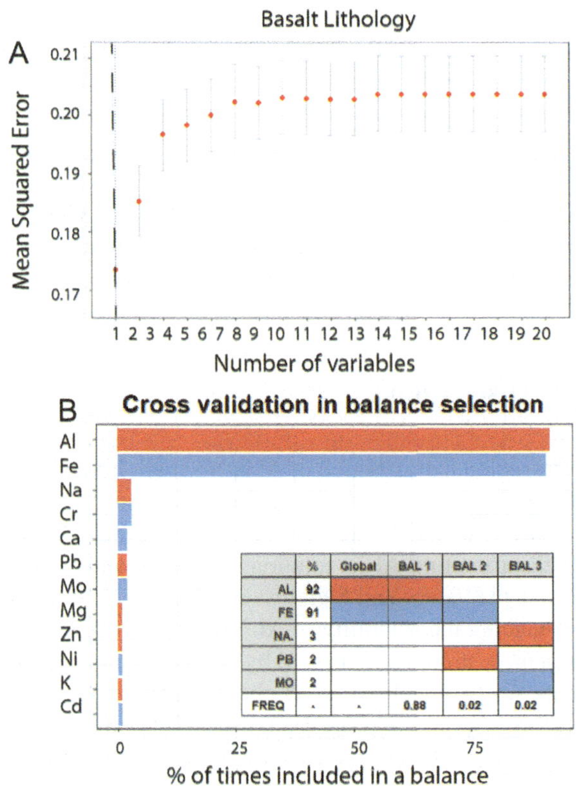

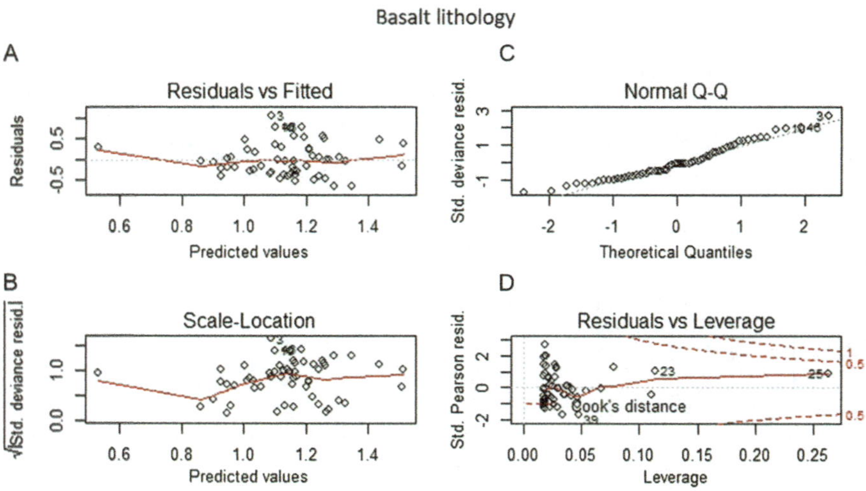

Fig. 5 Generalised Linear Model regression (GLM) summary for log (SIR_CKDu) with stream water data for basalt lithology (bedrock) for global balance Al/Fe; **a** Residual versus Fitted; **b** Normal Q–Q plot; **c** Scale Location and **d** Residuals versus Leverage

shown for the SBP (ilr base) and ilr balanced approach (Fig. 6c, d, respectively) demonstrate the impact for reodering the elements on the partitioning process. Reordering the elements using the SBP approach identified a statistically significant relationship with log (CKDu SIR) for the balances $\sqrt{\frac{3}{4}}\log\left(\frac{Na}{(Al\times Fe\times Cr)^{1/3}}\right)$ and $\sqrt{\frac{4}{5}}\log\left(\frac{Mo}{(Al\times Fe\times Cr\times Na)^{1/4}}\right)$ (Table 1 SBP ilr 3 and ilr 4; both showed a significance level of 0.05). Using the ilr balanced approach, only the balance of Na/Cr had a significant relationship with log (CKDu SIR) (significance level 0.05).

5 Discussion

The regional distribution of Al in Northern Ireland is described as directly controlled by geology and terrain (Young and Donald 2013). High relative concentrations of Al are associated with pH values >7 and high levels of dissolved organic carbon, both promoting Al in solution in stream waters. Superficial deposits of peat with lower pH values and high dissolved organic carbon are likely to be responsible for enhancing the solubility of Al. Likewise, the relative concentration of As in stream waters is strongly controlled by bedrock geology and the composition of superficial deposits (Mitchell 2004). Higher relative amounts of As are found in the Dalradian supergroup and over Devonian and Carboniferous bedrock (Fig. 1). Lower relative amounts are found associated with the Palaeogene Mourne mountain complex and

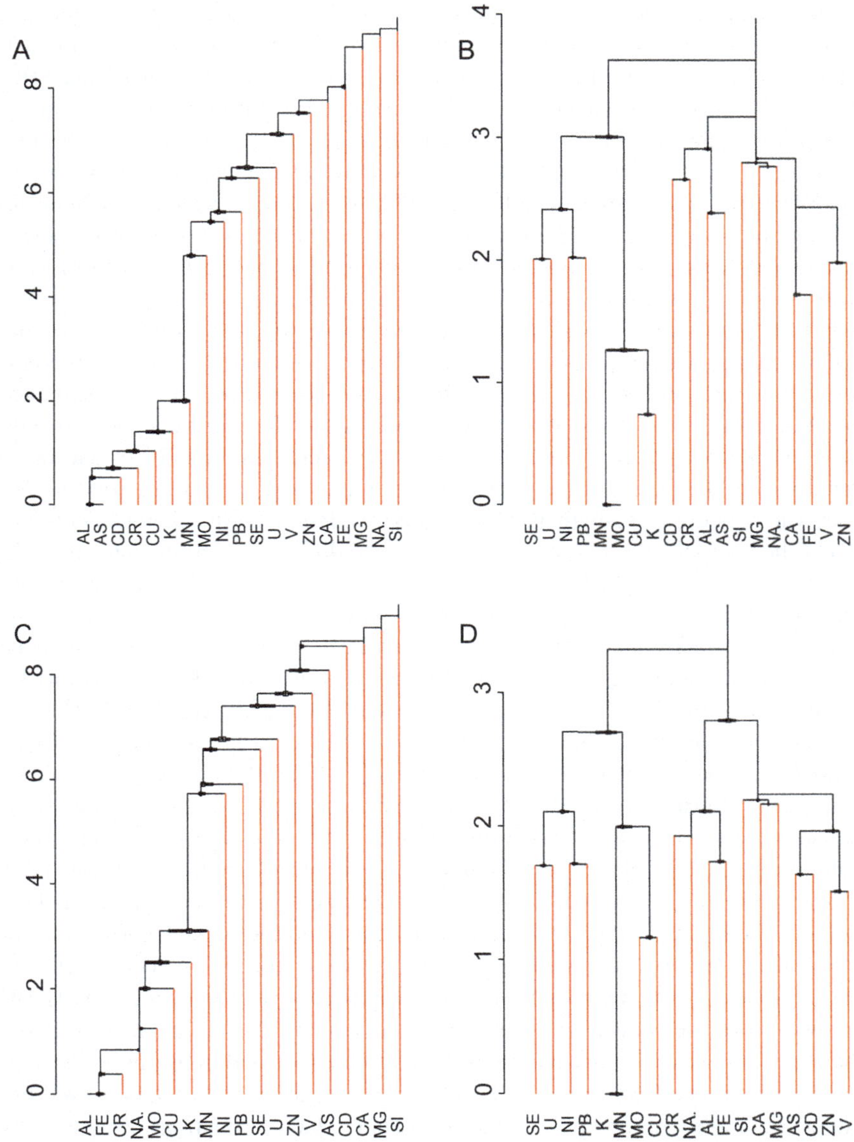

Fig. 6 Compositional dendrograms for stream water geochemistry data for basalt bedrock showing **a** SBP (ilr base), **b** balanced ilr approach; for reordered elements **c** SBP (ilr base) and **d** balanced ilr approach

Antrim Lava group. The elemental balance identified in the *selbal* approach categorizes the diversity in the geological foundation of Northern Ireland and the impact of superficial deposits. Further research into the association between CKDu SIR, bedrock geology and superficial deposits through these preliminary findings cannot be confirmed without controlling for other effects (e.g., soil PTEs and socio-economic factors; McKinley et al. 2020a). However, the significant association found between log (CKDu SIR) and the elemental balance of Al/As in stream waters may indicate that where indirect or direct ingestion of trace elements occurs through waterborne exposure pathways, there is a potential health risk in terms of CKDu. A major factor in waterborne exposure pathways is whether households have access to public water supply through the water mains network or via private supplies (NIEA 2019). The distribution of registered private water supplies as reported by NIEA (2019) coincides well with basalt bedrock. The distribution of registered private water supplies across Northern Ireland and especially in the northwest of the country suggests that in these areas there is greater potential for indirect or direct ingestion of PTEs. Groundwater flow, residence time, topography, climate and land use are all controls on stream water chemistry (Smyth and Johnson 2013). However, the chemical composition of the bedrock and superficial geology of the different environments, through which the streams pass, will impact stream waters. Therefore, it is a reasonable assumption that stream water geochemistry can provide an indication of the chemistry of private wells. An elemental balance of Al/Fe in stream waters is identified as significantly associated with log (CKDu SIR) for areas underlain by basalt bedrock. These elements are considered essential for human health at different levels (Al can be considered essential at low levels). However soluble Al is also known to be a risk for human health including patients on dialysis. The Palaeocene Antrim lava Group on the Antrim Plateau shows a wide variation in relative Al concentrations in stream waters, whereas Fe concentrations are found to be higher over the Upper Basalt Formation. Higher relative concentrations for both Al and Fe are recorded for streams draining the outcrop of Dalradian metasediments in northeast Antrim. The pH of soils and land through which the waters pass and the organic carbon of the cover are major factors in the solubility and mobility of elements (Smyth et al. 2013). Higher relative concentrations of Al on elevated areas are coincident with upland peat cover with lower pH values which enhances the solubility of Al in stream waters. Fe has relatively low mobility under most environmental conditions but is affected by redox conditions (Brewer et al. 2013). This can result in very high dissolved Fe loadings in streams draining peatland or other acid reducing situations (Smyth et al. 2013). Iron coatings on minerals can exert controls on the surface geochemistry of other elements including As and Mo (Hylander et al. 2000). Previous studies have considered the chemical factors controlling Al bioavailability and toxicity in aquatic environments and the importance of multi-element interactions (Williams 1991). The role of superficial deposits (with peat providing higher dissolved organic carbon and reducing pH and Quaternary glacial tills moderating the solubility of Al) as a control on the balance of Al and Fe and any potential link with CKDu needs to be explored further.

There has been much debate on how to interpret compositional balances (McKinley et al. 2020b, 2020b; Aitchison and Greenacre 2002; Buccianti and Pawlowsky-Glahn 2005; Kynčlová et al. 2016; Gozzi et al. 2018; Sauro Graziano et al. 2020). Several considerations on the interpretation of compositional balances have arisen from this research. It is widely recognised that while a data-driven approach for exploratory data analysis may be helpful, the ordering of elements for SBP and an ilr balanced approach does impact the results. The results from this research suggest that one should consider carefully the ordering of elements and use an informed knowledge-driven approach when using a SBP ilr base approach. Subtle changes in the dendrogram were shown in the balanced ilr approach (Fig. 6). This research found the *selbal* forward-selection approach to provide consistent and interpretable results to identify 'the most significant balances' most associated with log-transformed SIR CKDu. While it is acknowledged that further work needs to be done controlling for other factors, such as exposure pathways, spatial dependence, environment and social economic demographics, the balance approach shown provides a direction for further research. The approach provides a framework for investigating the importance of multi-element interactions of both essential and potentially toxic trace elements in understanding the environmental clusters of CKDu.

6 Conclusions

The relationship between Standardised Incidence Rates (SIRs) of CKDu and stream water geochemistry was explored through the use of log-ratio balances of Potentially Toxic Elements (PTEs). Several CoDA approaches using balances (a forward-selection procedure using a selbal algorithm, SBP and ilr balanced approaches) were used to identify the elemental balances most associated with log-transformed SIR CKDu. At the regional scale, a statistically significant relationship was found between log (CKDu SIR) and the elemental balance Al/As. This elemental balance characterises different geological domains across Northern Ireland. Following stratification by basalt bedrock (the dominant bedrock geology for SIR CKDu), the balance Al/Fe was identified as significantly associated with log (CKDu SIR). Controls on the relative amounts of Al and Fe in stream waters include bedrock geology, superficial deposits, amount of dissolved organic carbon and pH. The findings suggest that bedrock geology and superficial deposits play an important role as controls in the availability of environmental toxins with a potential health risk in terms of CKDu, where indirect or direct ingestion of trace elements occurs through waterborne exposure pathways.

Acknowledgements The study received ethics approval in March 2018, NHS National Research Ethics Committee REC reference: 15/EM/0366. We thank all the UK Renal Centres for providing data to the UK Renal Registry. The views and opinions expressed in this article are those of the authors and do not reflect the views of the UK Renal Registry or UK Renal Association. The Tellus Project was funded by the Department of Enterprise, Trade and Investment (NI) and by the 'Building Sustainable Prosperity' fund of the EU Regional Development Programme. Chemical analysis of the urban soil samples was co-financed by the Northern Ireland Environment Agency. JJE received financial support through the project METhods for COmpositional analysis of DAta (CODAMET), Ministerio de Ciencia, Innovación y Universidades, Spain (Ref: RTI2018-095518-B-C21/C22, 2019-2021). We would like to thank our two reviewers Antonella Buccianti, Ricardo Olea and Editor Peter Filzmoser for their helpful comments which have improved the research.

References

J. Aitchison, M. Greenacre, Biplots for compositional data. J. R. Stat. Soc. Ser. C (Appl. Stat.) **51**(4), 375–392 (2002)

A. Barsby, J.M. McKinley, U. Ofterdinger, M. Young, M.R. Cave, J. Wragg, Bioaccessibility of trace elements in soils in Northern Ireland. Sci. Total Environ. **433**, 398–417 (2012). https://doi.org/10.1016/j.scitotenv.2012.05.099

N. Brewer, N. Chenery, Geochemistry, in *A guide to the Tellus data*. ed. by M.E. Young, A.W. Donald (Geological Survey of Northern Ireland (GSNI), Belfast, 2013), pp. 47–54

A. Buccianti, V. Pawlowsky-Glahn, New perspectives on water chemistry and compositional data analysis. Mat. Geol. **37**(7), 703–727 (2005). https://doi.org/10.1007/s11004-005-7376

H.F. Chiu, C.Y. Yang, Decreasing trend in renal disease mortality after cessation from arsenic exposure in a previous arseniasis-endemic area in southwestern Taiwan. J. Toxicol. Environ. Health A **68**(5), 319–327 (2005)

M. Cooper, I. Mitchell, Geology, in *A guide to the Tellus data*. ed. by M.E. Young, A.W. Donald (Geological Survey of Northern Ireland (GSNI), Belfast, 2013), pp. 5–20

R. Correa-Rotter, C. Wesseling, R.J. Johnson, CKD of unknown origin in Central America: the case for a mesoamerican nephropathy. Am. J. Kidney Dis. **63**(3), 506–520 (2014)

W.G. Couser, G. Remuzzi, S. Mendis, M. Tonelli, The contribution of chronic kidney disease to the global burden of major noncommunicable diseases. Kidney Int. **80**(12), 1258–1270 (2011). https://doi.org/10.1038/ki.2011.368

S. Cox, M. Chelliah, J.M. McKinley, S. Palmer, U. Ofterdinger, M. Young, M.R. Cave, J. Wragg, The importance of solid-phase distribution on the oral bioaccessibility of Ni and Cr in soils overlying Palaeogene basalt lavas Northern Ireland. Environ. Geochem. Health **35**(5), 53–567 (2013). https://doi.org/10.1007/s10653-013-953

J.J. Egozcue, V. Pawlowsky-Glahn, Groups of parts and their balances in compositional data analysis. Math. Geol. **37**(7), 795–828 (2005)

J.J. Egozcue, V. Pawlowsky-Glahn, G. Mateu-Figueras, C. Barceló-Vidal, Isometric logratio transformations for compositional data analysis. Math. Geol. **35**(3), 279–300 (2003)

M.A. Engle, E.L. Rowan, Interpretation of Na-Cl-Br systematics in sedimentary basin brines: comparison concentration, element ratio, and isometric log-ratio approaches. Math. Geosci. **45**(1), 87–101 (2013)

R.W. Gensemer, R.C. Playle, The bioavailability and toxicity of aluminum in aquatic environments. Crit. Rev. Environ. Sci. Technol. **29**(4), 315–450 (1999). https://doi.org/10.1080/10643389991259245

J. Gilg, C. Castledine, D. Fogarty, UK Renal Registry 14th annual report: chapter 1 UK RRT incidence in 2010: National and centre-specific analyses (Southampton, Karger AG, Basel 2012)

C. Gozzi, R. Sauro Graziano, F. Frondini, A. Buccianti A, Innovative monitoring tools for the complex spatial dynamic of river chemistry: case study for the Alpine region, 579. Environ. Earth Sci. **77**(16), 1–11 (2018). https://doi.org/10.1007/s12665-018-7756-0

T. Hastie, R. Tibshirani, J. Friedman, *The Elements of Statistical Learning: Data Mining, Inference and Prediction*, 2nd edn. (Springer, Berlin, 2009)

L.D. Hylander, M. Meili, L.J. Oliveira, E.D.E. Silva, J.R.D. Guimaraes, D.M. Araujo, R.P. Neves, R. Stachiw, A.J.P. Barros, G.D. Silva, Relationship of mercury with aluminum, iron and manganese

oxy-hydroxides in sediments from the Alto Pantanal, Brazil. Sci. Total Environ. **260**, 97–107 (2000)
C.E. Jackson, J.M. McKinley, U. Ofterdinger, D. Fogarty, P.M. Atkinson, S. Palmer, Investigating relations between environmental toxins in northern Irish soils and streams and chronic kidney disease prevalence. Appl. Geochem. **75**, 236–246 (2016). https://doi.org/10.1016/j.apgeochem.2016.10.016
C. Jayasumana, R. Gajanayake, S. Siribaddana, Importance of Arsenic and pesticides in epidemic chronic kidney disease in Sri Lanka. BMC Nephrol. **15**(1), 124 (2014)
N. Jayatilake, S. Mendis, P. Maheepala, F.R. Mehta, Chronic kidney disease of uncertain aetiology: prevalence and causative factors in a developing country. BMC Nephrol. **14**, 180 (2013)
P. Kynčlová, P. Filzmoser, K. Hron, Compositional biplots including external non-compositional variables. Statistics **50**(5), 1132–1148 (2016). https://doi.org/10.1080/02331888.2015.1135155
R. Lewis, *Understanding Chronic Kidney Disease: A Guide for the Non-Specialist* (M&K Update Ltd., England, 2012), p. 157
R. Lindeman, J. Tobin, N.W. Shock, Longitudinal studies on the rate of decline in renal function with age. J. Am. Geriatr. Soc. **33**, 278–285 (1985)
J. Lunyera, D. Mohottige, M. Von Isenburg, M. Jeuland, U.D. Patel, J.W. Stanifer, CKD of uncertain etiology: a systematic review. Clin. J. Am. Soc. Nephrol. **11**(3), 379–385 (2016)
J.M. McKinley, U. Mueller, P.M. Atkinson, U. Ofterdinger, C. Jackson, S.F. Cox, R. Doherty, D. Fogarty, J.J. Egozcue, V. Pawlowsky-Glahn, Investigating the influence of environmental factors on the incidence of renal disease with compositional data analysis using balances. Appl. Comput. Geosci. **6**, 100024 (2020b). https://doi.org/10.1016/j.acags.2020.100024
J.M. McKinley, U. Mueller, P.M. Atkinson, U. Ofterdinger, C. Jackson, S.F. Cox, R. Doherty, D. Fogarty, J.J. Egozcue, V. Pawlowsky-Glahn, Chronic kidney disease of unknown origin is associated with environmental urbanisation in Belfast, UK. Environ. Geochem. Health (2020a), https://link.springer.com/article/10.1007/s10653-020-00618-y
J.M. McKinley, U. Ofterdinger, M. Young, A. Barsby, A. Gavin, Investigating local relationships between trace elements in soils and cancer data. Spat. Stat. **5**, 25–41 (2013). https://doi.org/10.1016/j.spasta.2013.05.001
J. McKinley, R. Tolosana Delgado, K. Hron, P. de Caritat, E. Grunsky, C. Reimann, P. Filzmoser, K.G. van den Boogaart, Single component map: fact or fiction? J. Geochem. Explor. **162**, 16–28 (2016). https://doi.org/10.1016/j.gexplo.2015.12.005
W. Mitchell, *The Geology of Northern Ireland,Geological Survey of Northern Ireland*, 2nd edn. (2004)
C. Musso, D. Oreopoulos, Aging and physiological changes of the kidneys including changes in glomerular filtration rate. Nephron Physiol. **119**, 1–5 (2011)
Northern Ireland Environment Agency (NIEA), Drinking Water Quality in Northern Ireland, 2018 A report by the drinking water inspectorate for Northern Ireland, ISBN 978-1-83887-020-1 (2019)
R.A. Olea, V. Pawlowsky-Glahn, Kolmogorov–Smirnov test for spatially correlated data. Stoch. Environ. Res. Risk Assess. **23**, 749–757 (2009). https://doi.org/10.1007/s00477-008-0255-1
W.H. Orem, C.A. Calin, Health effects of toxic organic compounds from coal—the case of Balkan endemic nephropathy (BEN). U.S. Geological Survey Fact Sheet 004-01 (2001), p. 4, https://pubs.usgs.gov/fs/fs004-01/
S. Palmer, R. McIlwaine, U. Ofterdinger, S.F. Cox, J.M. McKinley, R. Doherty, J. Wragg, M. Cave, The Effects of lead sources on oral bioaccessibility in soil and implications for contaminated land risk management. Environ. Pollut. **198**, 161–171 (2015). https://doi.org/10.1016/j.envpol.2015.01.004
E. Pardo-Iguzquiza, P.A. Dowd, Normality test for spatially correlated data. Math. Geol. **36**(6), 659–681 (2004)
UK Renal Registry, UK Renal Registry 21st annual report—data to 31/12/2017, Bristol, UK (2019), https://www.renalreg.org/publications-reports/. Accessed 30 Oct. 2019

J. Rivera-Pinto, Statistical methods for the analysis of microbiome compositional data in HIV infection, unpublished PhD thesis in Experimental Sciences and Technology, Universitat Central De Catalunya, Spain (2018)

J. Rivera-Pinto, J.J. Egozcue, V. Pawlowsky-Glahn, R. Paredes, M. Noguera-Julian, M.L. Calle, Balances: a new perspective for microbiome analysis. mSystems (2018). https://doi.org/10.1101/219386

R. Sauro Graziano, C. Gozzi, A. Buccianti, Is compositional data analysis a theory able to discover complex dynamics in aqueous geochemical systems? J. Geochem. Explor. **211**, 106465 (2020). https://doi.org/10.1016/j.gexplo.2020.106465

J.L. Shelton, M.A. Engle, A. Buccianti, M.S. Blondes, The isometric log-ratio (ilr)ion plot: a proposed alternative to the Piper diagram. J. Geochem. Explor. **190**, 130–141 (2018)

D. Smyth, C. Johnston, Geochemistry: Methodlogy, in *A guide to the Tel-lus data*. ed. by M.E. Young, A.W. Donald (Geological Survey of Northern Ireland (GSNI), Belfast, 2013), pp. 33–44

P. Soderland, S. Lovekar, D.E. Weiner, D.R. Brooks, J.S. Kaufman, Chronic kidney disease associated with environmental toxins and exposures. Adv. Chronic Kidney Dis. **17**(3), 254–264 (2010)

K.G. Van den Boogaart, R. Tolosana-Delgado, compositions: a unified R package to analyze compositional data. Comput. Geosci. **34**(4), 320–338 (2008). https://doi.org/10.1016/j.cageo.2006.11.017

V.M. Weaver, J.J. Fadrowsk, B.J. Jaar, Global dimensions of chronic kidney disease of unknown etiology (CKDu): a modern era environmental and/or occupational nephropathy? BMC Nephrol **16**, 145 (2015). https://doi.org/10.1186/s12882-015-0105-6; https://www.ncbi.nlm.nih.gov/pmc/articles/PMC4539684/

T.M. Williams, A sedimentary record of the deposition of heavy metals and magnetic oxides in the Loch Dee basin, Galloway, Scotland since c. AD 1500. Holocene **1**, 142–150 (1991)

World Health Organisation, Trace elements in human nutrition and health WHO Library cataloguing in publication data Geneva. ISBN 92 4 156173 4 (1996)

M.E. Young, A.W. Donald, *A Guide to the Tellus Data*. (Geological Survey of Northern Ireland (GSNI), Belfast, 2013), https://nora.nerc.ac.uk/509171/

L. Zheng, C.C. Kuo, J. Fadrowski, J. Agnew, V.M. Weaver, A. Navas-Acien, Arsenic and chronic kidney disease: a systematic review. Curr. Environ. Health Rep. **1**(3), 192–207 (2014)

L.Y. Zheng, J.G. Umans, F. Yeh, K.A. Francesconi, W. Goessler, E.K. Silbergeld et al., The association of urine arsenic with prevalent and incident chronic kidney disease: evidence from the strong heart study. Epidemiology **26**(4), 601–612 (2015)

Multivariate Classification of the Crude Oil Petroleum Systems in Southeast Texas, USA, Using Conventional and Compositional Data Analysis of Biomarkers

Ricardo A. Olea, Josep Antoni Martín-Fernández, and William H. Craddock

Abstract Chemically, petroleum is an extraordinarily complex mixture of different types of hydrocarbons that are now possible to isolate and identify because of advances in geochemistry. Here, we use biomarkers and carbon isotopes to establish genetic differences and similarities among oil samples. Conventional approaches for evaluating biomarker and carbon isotope relative abundances include statistical techniques such as principal component and cluster analysis. Considering that proportions of the different hydrocarbon molecules are relative parts of a laboratory sample, the data are compositional in nature, thus requiring the use of log-ratio approaches for adequate mathematical modeling. We apply both traditional and compositional modeling approaches to crude oil samples from an onshore area of about 50,000 square miles in southeast Texas. The data comprise 177 crude oil samples from producing oil fields that include key biomarkers, elemental, and isotopic values commonly used in source rock correlation studies. Our results indicate that compositional modeling has higher discriminating power and lower uncertainty than the traditional approach, allowing the identification of up to 16 clusters. Each cluster represents one oil family from a source rock organofacies ranging from Carboniferous to Paleogene. The families provide new insights into important petroleum systems in the Texas onshore region of the Gulf of Mexico sedimentary basin.

R. A. Olea (✉) · W. H. Craddock
U.S. Geological Survey, Geology, Energy and Minerals Science Center, Reston, VA 20192, USA
e-mail: rolea@usgs.gov

W. H. Craddock
e-mail: wcraddock@usgs.gov

J. A. Martín-Fernández
Department of Computer Science, Applied Mathematics and Statistics, University of Girona, 17003 Girona, Spain
e-mail: josepantoni.martin@udg.edu

P. Filzmoser et al. (eds.), *Advances in Compositional Data Analysis*,
https://doi.org/10.1007/978-3-030-71175-7_16

1 Introduction

Hydrocarbon biomarkers are molecules present in trace amounts in crude oil, enzymatically synthesized by organisms living millions of years ago. Biomarkers are valuable tools for interpreting paleoenvironments and establishing genetic relationships among oils and potential source rocks (He et al. 2018). Cluster analysis of biomarkers and carbon isotopes data can be used to determine the number of different oil types also called genetic oil families. Each family is supposed to have been generated from common source rock organofacies (He et al. 2014). An organofacies is a mappable rock unit comprising distinct types of organic matter independent of the mineralogy (Peters et al. 2007). The onshore part of the Western Gulf Coast is a major oil-producing region in the country. To understand the Western Gulf Coast petroleum systems, previous regional studies comprising all source rocks have used biomarker analysis, but they have not reported using cluster analysis (Kennicutt et al. 1992; Hood et al. 2002), or openly declared failure trying to obtain meaningful results applying cluster analysis (Comet et al. 1993).

This contribution has two main objectives. The first objective is to advance the knowledge of the onshore southeastern Texas petroleum systems by performing a cluster analysis of biomarker and carbon isotope data from conventional hydrocarbon reservoirs, where hydrocarbons have accumulated in a permeable formation after generation, expulsion and migration from a source rock rich in organic matter. The second objective is to use an appropriate geometry for the cluster analysis. Geochemical relative abundances are compositional data, which require an Aitchison geometry instead of Euclidean geometry for measuring the distances that need to be considered in cluster analysis. Palarea-Albaladejo et al. (2012) discussed the adequacy of different dissimilarities in the simplex, however they did not address these points in the context of biomarkers. By treating the biomarker data as compositional, our goal is to present a more accurate cluster analysis to determine the source rock organofacies for different genetic oil families in the Western Gulf Coast.

2 Geology of the Texas Gulf Coast

The Gulf of Mexico basin in the onshore part of southeastern Texas (Western Gulf Coast) is underlain by a relatively conformable succession of Mesozoic and Cenozoic strata. The strata were deposited during the continental breakup of Pangaea, roughly in the Triassic–Jurassic, and during a subsequent passive margin phase, ranging from the Late Jurassic to the present (Fig. 1). Overlying Upper Jurassic,-younger strata are more widely preserved (e.g., Salvador and Muñeton 1989; Salvador 1991; Galloway 2008). The vast quantities of petroleum produced from the Texas Gulf Coast (e.g., Nehring 1991) have been sourced from within this rift-to-rift succession (e.g., Hood et al. 2002).

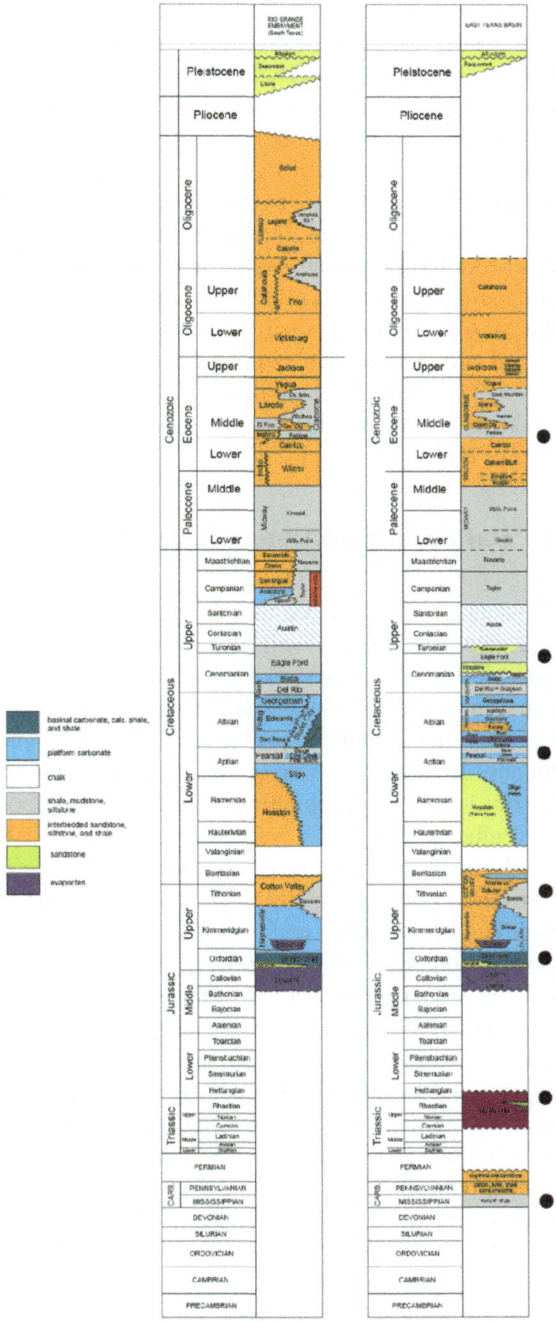

Fig. 1 Stratigraphic column (After Salvador and Muñeton 1989). The bullets denote the existence of source rocks. See Fig. 2 for the location of the Rio Grande Embayment and East Texas Basin

For the purpose of this study, a key point is that at least seven distinct petroleum source intervals have been recognized in the Gulf of Mexico region at large (Fig. 1). Hood et al. (2002) propose the following sources from oldest to youngest: Triassic, Upper Jurassic (centered on Oxfordian), Upper Jurassic (centered on Tithonian), Lower Cretaceous (centered on Aptian), Upper Cretaceous (centered on Cenomanian–Turonian), and Paleogene (centered on Eocene). These intervals generally correspond to marine transgressions (e.g., Haq et al. 1988; Hood et al. 2002), and at least in some cases to times of oceanic anoxia (e.g., Comet et al. 1993). Adjacent Paleozoic basins which overlap with the up-dip portions of the Texas Gulf Coast are also known to have Carboniferous and older source rocks. Where penetrated, the Mesozoic–Cenozoic rift-to-drift succession is mostly underlain by strata that record and/or were involved in the Paleozoic Ouachita–Marathon orogeny (Woods et al. 1991). Some limited oil and gas production occurs from strata underlying the rift-to-drift sequence near the up-dip edge of the coastal plain.

3 Data and Selection of Variables

The data initially gathered for this study included analyses on 204 crude oil samples collected from onshore wells in southeast Texas. These data, which are available by subscription from GeoMark Research, Ltd. (2019), represented all the oil analyses that were available from this region when we accessed the database in early 2019. The laboratory measurements are reported as peak heights calibrated in such a way that, for any oil sample, the ratio of the peak of one biomarker to any other peak is equal to the ratio of their concentrations (Ethan Brown (Geomark), written communication, November 2019).

The number of variables for adequately characterizing a multivariate system must bear certain proportionality to the number of objects measured, also called sample size. Too few variables and it is hard, if not impossible, to properly characterize the system. Too many variables, then redundancies abound, and one starts modeling not only the signal but also noise, rarely the objective of any study. Years of experience have produced rules for avoiding these two extremes. Some authors recommend a ratio of (sample size)/(descriptive variables) as low as 10, whereas others prefer the higher value of 30 (Harrell et al. 1996; Siddiqui 2013). The nature of the variables is also important. Multivariate analyses are highly automated. The main way for the geologist to have a significant influence in the results is through a preliminary evaluation of the geology of the study area and subsequent selection of biomarkers relevant to that geology. We posit that the nine variables below capture much of the unique information about source rock age, lithology, and depositional environmental conditions that can be discerned by biomarker and isotope analysis in southeast Texas. These variables gave a preliminary ratio (sample size)/(descriptive variables) before reviewing the data of $204/9 = 22.7$. The variables are

- Five terpane ratios:

1. C_{22}/C_{21} (C_{22}/C_{21} tricyclic terpanes, peak heights from mass-to-charge ratio (m/z) of 191)
2. Tet/C_{23} (C_{24} tetracyclic terpane/C_{23} tricyclic terpane, peak heights from m/z of 191)
3. C_{31}R/H (C_{31} homohopane/hopane)
4. C_{35}S/C_{34}S (C_{35} homohopane 22S/C_{34} homohopane 22S)
5. C_{26}T/Ts (C_{26} S + R tricyclic terpane/trisnorneohopane)

- Two sterane ratios:

1. %C_{27} steranes (%$C_{27} = 100 \cdot C_{27}/[C_{27} + C_{28} + C_{29}]$ steranes), with m/z of 218 for C_{27} (S5B), C_{28} (S10B) and C_{29} (S14B)
2. C_{28}/C_{29} (%C_{28}/%C_{29} steranes)

- Two stable carbon isotope ratios, normalized to a reference material:

1. $\delta^{13}C_{saturated}$, ($^{13}C/^{12}C$) in per mil
2. $\delta^{13}C_{aromatic}$, ($^{13}C/^{12}C$) in per mil, where

$$\delta^{13}C = 1000\left(\frac{\left(\frac{^{13}C}{^{12}C}\right)_{sample}}{\left(\frac{^{13}C}{^{12}C}\right)_{standard}} - 1\right)$$

After eliminating 27 oils with some problematic measurements, such as too high equivalent vitrinite reflectance, the size of the dataset to be used in the modeling dropped to 177. Hence, the effective (sample size)/(descriptive variables) is 177/9 = 19.7. Figure 2 shows the geologic age of the reservoir from where the crude oil samples were taken. There were 26 (1.6%) missing values, which were imputed applying log-ratio techniques implemented in the function *lrEMplus* of the R package *zCompositions* (Palarea-Albaladejo and Martín-Fernández 2015).

4 Compositional (Aitchison) Chemometrics

Relative geochemical abundances are compositional data measuring proportions of a whole, independently of the units, be them parts per million or properly calibrated chromatogram peak heights. Calculation of Euclidean distances for the purpose of making comparisons presumes that the axes are orthogonal, which is not the case for compositional data or ratios of concentrations of two biomarkers. Under that circumstance, proximity should be measured using Aitchison distances (Palarea-Albaladejo et al. 2012). This requires using a different type of ratios. While the GeoMark database fortunately provides all the data behind the calculation of the biomarker ratios, the same is not true for the carbon isotope ratios. Figure 3 shows the distributions for all the new variables: the centered log-ratio (clr) transformed variables required for the use of Aitchison distances, plus two isotope ratios. The

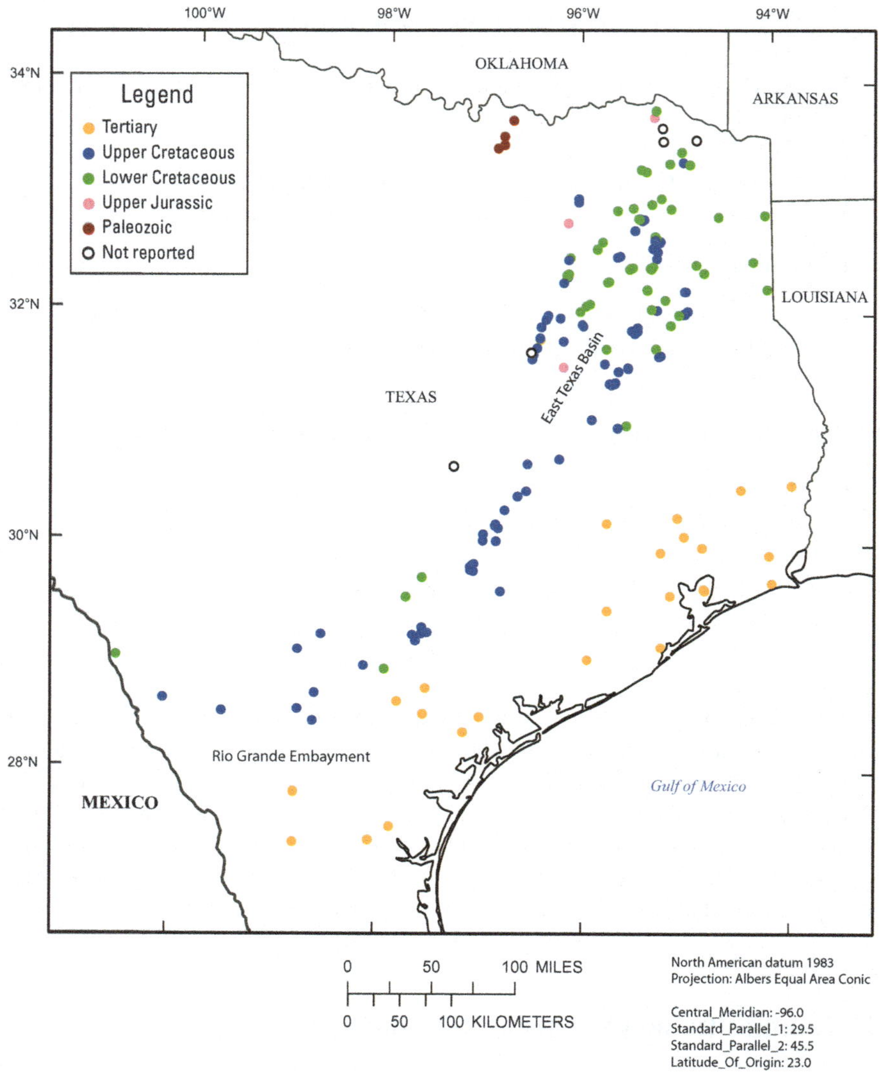

Fig. 2 Posting of the age of the reservoirs from where the 177 crude oil samples were taken. There are 26 oils from Tertiary reservoirs, 84 from upper Cretaceous, 55 from Lower Cretaceous, 3 from Upper Jurassic, 4 from Paleozoic, and 5 without reported age

number of variables increased to 15, for a (sample size)/(number of variables) of 177/15 = 11.8. A $clr_{i,j}$ transformation of the ith compositional variable of the jth object, $x_{i,j}$ is the logarithm of $x_{i,j}$ over the product of the geometric mean G_j of all the D values for that jth object (Pawlowsky-Glahn et al. 2015), 13 in our case:

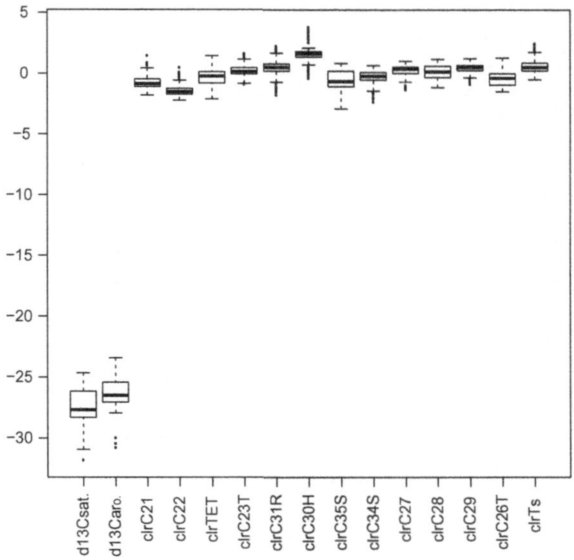

Fig. 3 Box plots before scaling of the new variables behind the ratios defined in Sect. 3, plus the isotopic ratios that remain the same

$$clr_{i,j} = \ln \frac{x_{i,j}}{G_j}.$$

In a simple example, if the first object has values (10, 20, 30), then $clr_{2,1} = \ln\left(20/(10 \cdot 20 \cdot 30)^{1/3}\right) = -0.594$. Note that the $x_{i,j}$ continue being analyzed as ratios, some more complex ratios than direct ratios of proportions of two biomarkers, but ratios after all, now involving all biomarkers for each clr.

Once the clr scores are calculated, the distances are simply the Euclidean distances for the clr scores plus all original variables not requiring or allowing the clr transformation, $\delta^{13}C_{aromatic}$ and $\delta^{13}C_{saturated}$ in our case. A scaling is required prior to starting the cluster analysis for avoiding dominance by the variables with the largest absolute values, bringing, for every variable, the means to 0 and the variances to 1.

There are two main families of methods for assessing these distances: hierarchical clustering and partitioning clustering. The most common form of reporting the results from hierarchical clustering is a dendrogram resembling an inverted tree, with the length of the branches proportional to the distances (Mirkin 2013). This graphical display of results explains the popularity of the hierarchical methods. Figure 4 is the dendrogram obtained after applying Ward's hierarchical method (Mirkin 2013) to our clr data. This contribution stays away from additionally analyzing the influence that different clustering methods would have in the modeling.

All data analyses in this contribution were done applying the R-statistical software by R Core Team (2018).

Clustering methods do not provide a direct answer in terms of the actual number of clusters in a system; any integer number between one and the sample size is an acceptable solution (Visvanathan et al. 2009). Nevertheless, there are more than 30

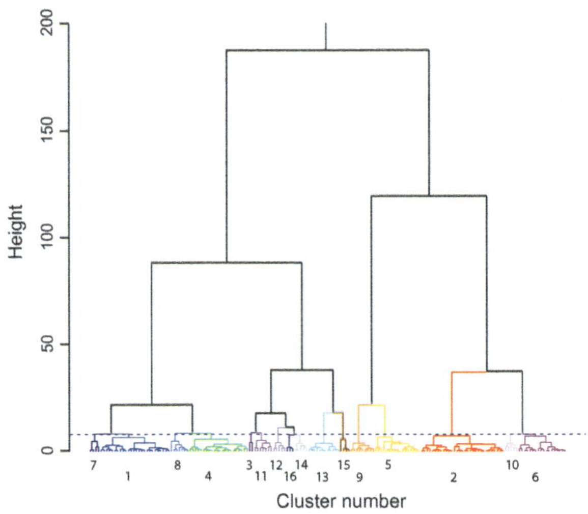

Fig. 4 Ward's hierarchical dendrogram for the variables in Fig. 3 after scaling. The similarity line in blue intends to visualize the grouping of the oil samples into 16 clusters

validation methods for mathematically postprocessing the cluster analysis results for helping to select the most adequate number of clusters (Milligan and Cooper 1985), with none appearing in the biomarker literature. The Dunn index (Brock et al. 2008), $DI(k)$, uses the minimum distance from one object to all observations in the other $k - 1$ clusters, *MD*, over the maximum separation between two objects inside the same cluster, *MS*:

$$DI(k) = \frac{MD}{MS}.$$

When a solution is the best, the index reaches a maximum value, which denotes the clusters with minimal internal distances and maximum distances among clusters.

Another validation method increasingly used in cluster analysis is the silhouette width, SW_i. For the same dataset, separate calculations and displays need to be prepared for each number of clusters. However, it has the additional advantage of rating the quality of the assignments for every observation. A width is calculated for each object *i* as the normalized difference between the average distance of the object to all other objects in the same cluster, I_i, to the minimum average distance from *i* to all other clusters, O_i (Sugar and James 2003):

$$SW_i = \frac{I_i - O_i}{\max(I_i, O_i)}.$$

The larger the index, the better the quality of the assignment. Negative values denote unreliable cluster assignments. According to both the results from the Dunn index (Fig. 5) and the average silhouette width (Fig. 6a), the quality of the clustering

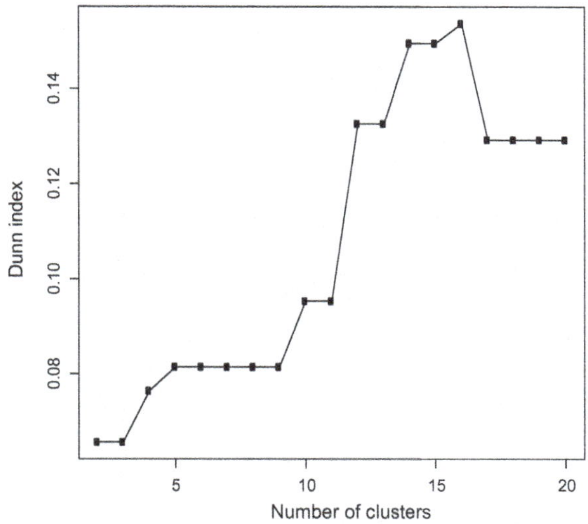

Fig. 5 Dunn index applied to the results in Fig. 4

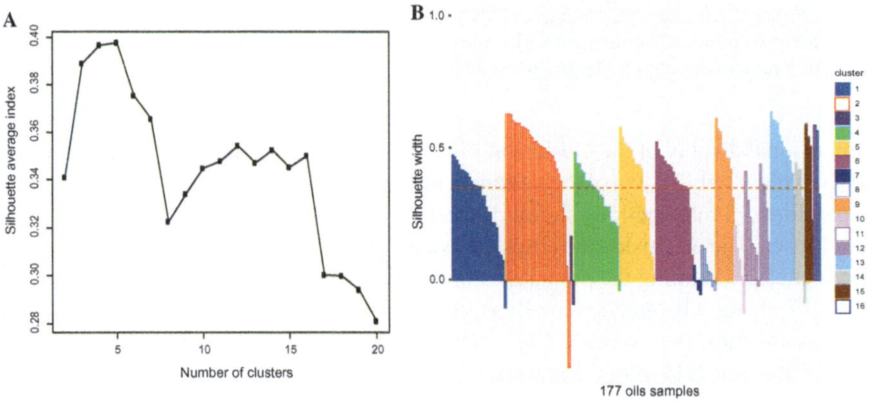

Fig. 6 Silhouette plots: **a** The average silhouette width for several number of clusters; and **b** The silhouette plot proper for the 16 clusters in Fig. 4, with the segmented line showing the average width, which is the value 0.35 in the graph above

keeps increasing up to 16 clusters, thus the breaking in Fig. 4. The quality of the assignments for the 16 clusters can be observed in Fig. 6b.

Silhouettes may be complemented with direct observation of the objects in every cluster in the multidimensional space—a 15-dimensional space in our case—which is hard to visualize beyond any dimension higher than 3. Principal component analysis (Jolliffe 2002) offers the alternative of rotating the axes and orienting them according to the style of variability of the distances. If a few components account for most of the variability in properties, then a display using two or three axes—principal components—can capture most of the geometry of the clusters. Figure 7 is a two-

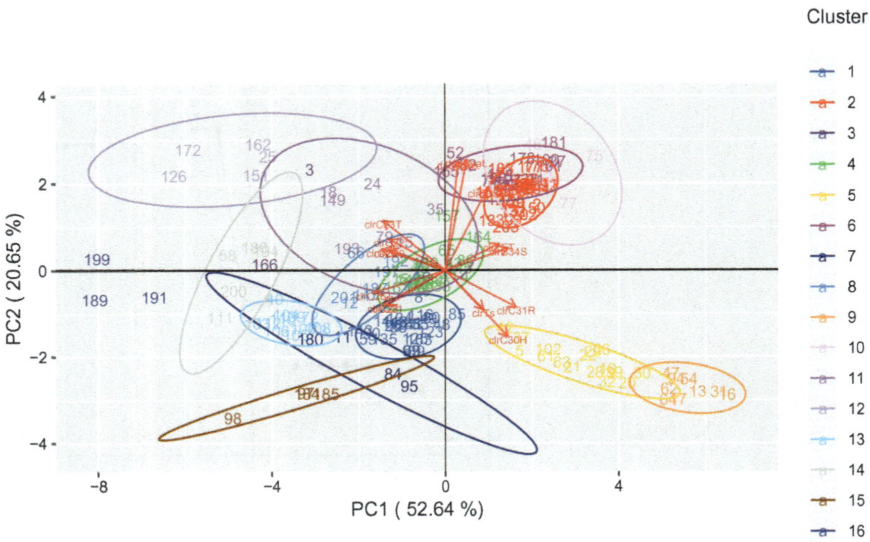

Fig. 7 Projection of the scaled values of $\delta^{13}C_{saturated}$, $\delta^{13}C_{aromatic}$ and the clr scores onto the plane of the first two principal components. The ellipses denote 95% confidence boundaries of Gaussian prediction regions for clusters larger than size 3

dimensional display of the clusters in Fig. 4. Note that the first two principal components explain only 73.3% of the variance, implying that, in the 15-dimensional space, hyperellipsoid intersections should be less numerous and prominent than those for the ellipses. The oils inside the ellipses have higher than 95% probability of belonging to the cluster assuming the points are normally distributed (Fox and Weisberg 2019). In Fig. 7, those oils outside the ellipses or at ellipse overlappings are candidates for misclassification (Rousseeuw 1987), which are likely to be some of the oils below zero in the silhouette plots (Fig. 6b).

Both the silhouettes widths (Fig. 6b) and the projection of the data in the plane of the first two principal components (Fig. 7) can be used to eliminate the most uncertain assignments from the results. The projections, despite their wide use in conventional biomarker analysis, suffer from three inconveniences: (a) it is not always possible to see the location of every object; (b) the view is partial, because fluctuations in the full space of principal component are not possible; and (c) the main references for discarding, the ellipses, are based on an assumption of normality. Instead, results from the preparation of silhouettes come with a listing of the width for every single object, which can be easily inspected and see if any object is contributing to significantly negative silhouettes. In our case, we decided to consider unacceptable any silhouette values below −0.2, thus eliminating oil sample 183 in cluster 2 from the results.

5 Source Assignments and Geographic Distribution of Oil Families

The modeling rarely ends up with the cluster analysis. Frequently, there is interest in associating the clusters, a geometric concept, with entities in the real world. While the legitimacy of the clusters is guaranteed, the assignments may be highly subjective and incorrect. In biomarker modeling, the ultimate objective is to associate the clusters with organofacies of oil source rocks. The solutions are seldom unquestionable, so some general heuristics need to be applied to help solving the conundrum in a more systematic way. In the case of southeast Texas geology, the following are some points worth to consider for limiting the number of solutions:

(a) Although extensional faults with large vertical displacements may be conduits for charging older strata, we follow the general practice of assuming that oil migrates into reservoirs that are the same age or younger than the parent source rock. A corollary is that when crude oils from the same family occur in different strata, the parent source rock should be at least as old as the oldest reservoir. For example, if the oils in one cluster have been sampled from reservoirs Upper Cretaceous, Lower Cretaceous and Upper Jurassic in age, the source rock should be Upper Jurassic or older.
(b) Although lateral migration is plausible, petroleum migration is mostly vertical (Peters et al. 2012).
(c) When a reservoir is charged from more than one source rock, the mixed oil in general has its own biomarker characteristics. Mixed oils cannot be assigned to any source rock uniquely, although an assignment can be attempted if a single source rock has a dominant contribution.
(d) In the particular case of the onshore the Gulf Coast, Comet et al. (1993) state that, with the exception of the Smackover, the oils tend to accumulate in reservoirs in close stratigraphic proximity to the source rocks, which is in agreement with the observation by He et al. (2012) that source rocks not only generate oil, but also tend to act as barrier preventing migration and, consequently, oil mixing (Levorsen and Berry 2001).

Ordinarily, some assignments are simpler than others. The location of the four Pennsylvanian–Permian reservoirs coincides exactly with the location of the crude oils making cluster 15 (Figs. 2 and 8). Based on the simplifying assumptions described above, these oils are interpreted to derive from a Paleozoic source rock, perhaps the Carboniferous Barnett Shale (Marra et al. 2015) or its equivalents.

Correlation of clusters 5 and 9 to a source rock also appears to be relatively straightforward. The clusters are located within the three tiers of counties closest to the modern coastline. Cenozoic strata are so thick beneath this area that any pre-Cenozoic source rocks should exhibit thermal maturity above the oil window (Kennicutt et al. 1992). Oils found in Paleogene–Miocene reservoirs most likely have been charged from Paleogene source rocks (Kennicutt et al. 1992). Clusters 5 and 9 are next to each other in the dendrogram (Fig. 4), with their branches joining

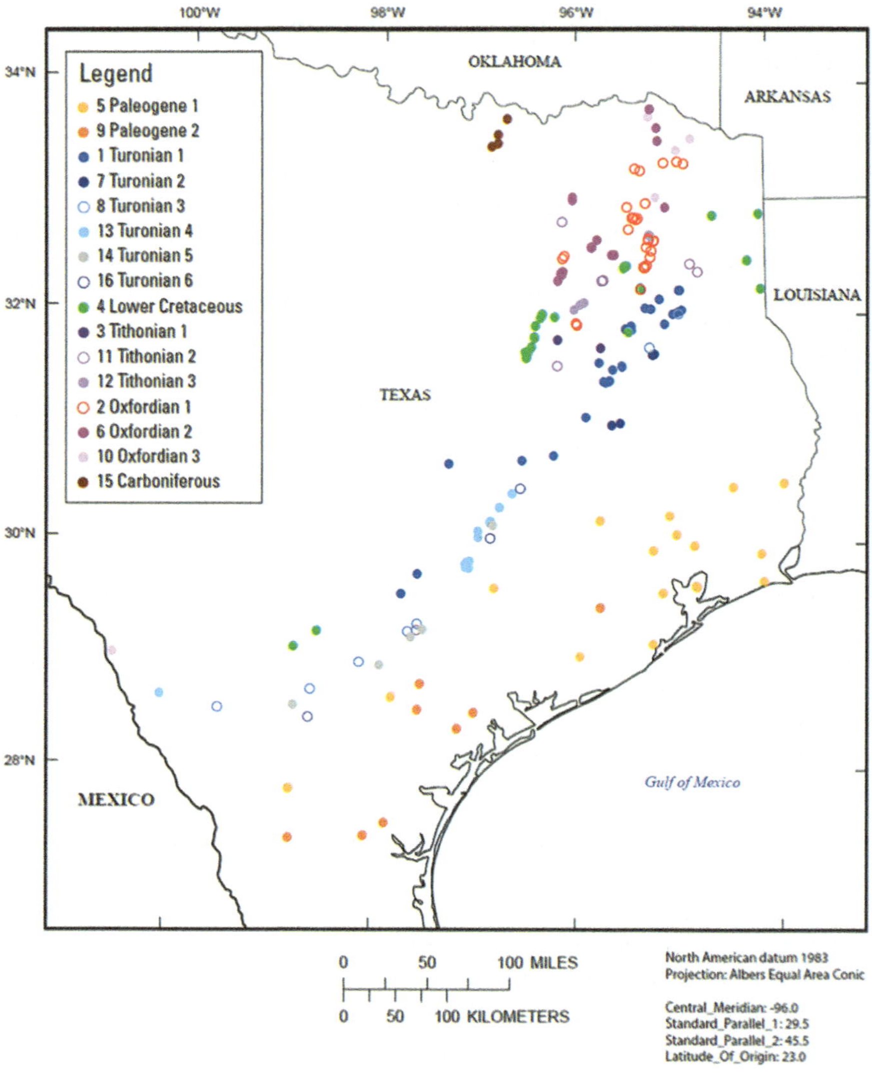

Fig. 8 Posting of 16 clusters from 176 oil samples and postulated oil families resulting from a compositional approach. In the legend, the first number after the bullet is the same cluster number in Figs. 4, 6b and 7, which is followed by the name of the correlated family

to form a larger cluster. These hierarchically related larger families are called tribes and, by the nature of the dendrogram, have similar characteristics. Cluster 5 and 9 are in the southeast quadrant of Fig. 7.

Moving to the northwest and up depositional dip from clusters 5 and 9, clusters 1, 7 and 8 also form a tribe in the dendrogram. They are located in the southwestern part of the study area, mostly in the Coniacian–Santonian Austin Chalk. Based on our simplifying assumptions, we correlate these oils to the Eagle Ford Shale. Along depositional strike to the northeast, clusters 13, 14 and 16 are located up depositional dip from the Paleogene oils and contain oils primarily from Upper Cretaceous reservoirs, mostly the Cenomanian Woodbine Formation. By our assumptions, we also correlate these oils to the Eagle Ford Shale. With six clusters, in relative terms, this is the most heterogeneous source rock. This heterogeneity may be the result of mixing, particularly for the small sized clusters: 7, 14 and 16. The likelihood of mixing is also postulated by Wenger et al. (1990) and Kennicutt et al. (1992).

Oils in the remaining seven clusters are less straightforward to correlate to a source rock. Moving up depositional dip from the Eagle Ford Shale-derived oils, cluster 4 is the third largest cluster with 22 oil samples. Oils in this cluster form three agglomerations that are not geographically contiguous. One near the Texas–Louisiana border with oils from strata as old as Neocomian–Aptian (one sample), and three of four samples are found in the Albian Glen Rose Formation. Another agglomeration is toward the southwest end of the Oxfordian oils, mostly found in Cenomanian strata. The third agglomeration is toward the southwest of the study area. Oils in cluster 4 are around the origin of the principal component graph in Fig. 7. Eight out of the 22 oil samples are from Lower Cretaceous reservoirs, the oldest reservoir age. Based on our simplifying assumptions, we correlate these oils to a Lower Cretaceous source rock (Tyler and Ewing 1986).

Clusters 2, 6 and 10 also form another tribe in Fig. 4, thus implying that they have similar properties. They are all in the northeast part of the study area (Fig. 8). They are also near the northern rim of the Gulf basin, spatially associated with major system of Cretaceous extensional faults that is thought to have been a migration path for reservoirs as young as Upper Cretaceous (Claypool and Mancini 1989; Sassen 1989; Wenger et al. 1990). In our case, 94% of reservoirs producing the oils samples in cluster 2, 6 and 10 are Cretaceous, two reservoirs have an undisclosed age, and just one of the 50 oil samples—from cluster 10—is from a Jurassic reservoir. On the basis of our simplifying assumptions we suggest that these oils were sourced from strata that are Cretaceous or older. We correlate these oils to an Oxfordian source rock.

Crude oils in clusters 3, 11 and 12 are predominantly from Lower Cretaceous reservoirs. They are in the northwest quadrant of Fig. 7. As seen in Fig. 8, they are near the southern end of the oils assigned to Oxfordian Smackover, with similar but not identical properties to the Oxfordian families. If the Tithonian source rocks charged any reservoirs, clusters 3, 11 and 12 are the prime candidates for correlating to Tithonian organofacies.

6 Conventional (Euclidean) Chemometrics

Despite violation of the orthogonality condition, a conventional cluster analysis is conducted primarily for comparison. The box plots for $\delta^{13}C_{saturated}$ and $\delta^{13}C_{aromatic}$ are the same; the one for $\%C_{27}$ fluctuates between 19.8 and 40.5%; the remainder

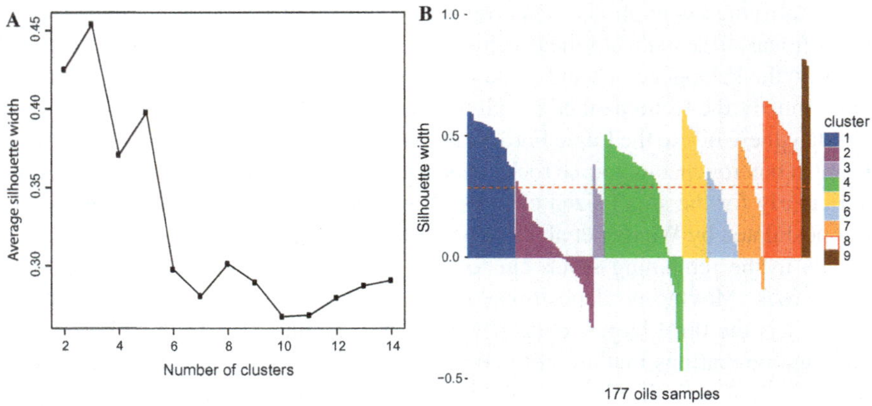

Fig. 9 Silhouettes for the Euclidean modeling: **a** Average width versus number of clusters; and **b** silhouette plot when selecting 9 clusters

of the biomarker ratios have minima close to zero and maxima between 0.5 for C_{31}R/H and 6.08 for Tet/C_{23}. These wide fluctuations make necessary scaling prior to conducting the cluster analysis. The Dunn index for the clusters resulting from the new hierarchical cluster analysis starts at 0.08 for 2 clusters, increasing to reach a maximum of 0.128 for 8 and 9 clusters, and then it starts to decrease. Based on the Dunn index values and Fig. 9, the decision was to select 9 clusters.

Oil samples with negative silhouette width values are now more numerous and reach lower values (Fig. 9b). Consistent with the previous decision of excluding from the results oil samples with silhouette widths below −0.2, we discarded oils 75 and 183 in cluster 2 and oils 85, 123 and 125 in cluster 4. Figure 10 shows the projection of the 177 ratios into the two first principal component plane that explains 78.5% of the variance.

Figure 11 is a posting of all the biomarker clusters and correlated oil families using Euclidean, rather than Aitchison distances. The labeling uses both the cluster number in Figs. 9 and 10, and the assigned age that is consistent with the choices made in Sect. 5 for the compositional chemometric analysis. Oil sample 77, the westernmost sampling location, again, like in Fig. 8, is an outlier.

7 Discussion of the Two Methods

Several authors in multiple disciplines have disregarded the special nature of parts of a whole and modeled the parts as real variables. Petroleum geochemists understood from the beginning the relative nature of biomarker data, adopting the use of ratios for the analysis, primarily one biomarker concentration over another, thus obtaining

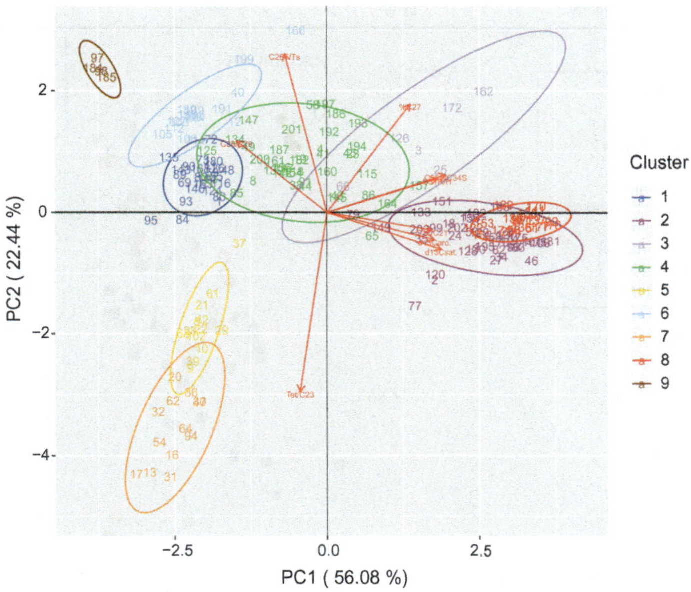

Fig. 10 Projection of all 177 scaled ratios onto the plane of the first two principal components. The ellipses denote 95% confidence boundaries of Gaussian prediction regions for clusters larger than size 3

better results than those obtained when completely ignoring the special compositional data properties. Hence, discrepancies between biomarker modeling applying Euclidean and Aitchison geometries are not as severe as in other disciplines.

In the few studies in which the modeling has been done by compositional and non-compositional approaches—none using biomarkers—practice has confirmed that the advantages are in favor of the compositional methods (e.g., Engle and Rowan 2013; Shelton et al. 2018). In our study, the advantages can be appreciated in two ways: (a) the compositional approach has a higher discriminating power, allowing the correct identification of seven more clusters, a 78% increase; and (b) the quality of the modeling is better, which is revealed by a higher average silhouette width of 0.35 versus 0.29, a 20% increase, plus just 10 negative silhouettes, only one below −0.2, against 27, with 5 below −0.2 in the traditional approach.

Overall, our results by the compositional method are comparable to previous studies that have de-emphasized statistical methods and focused more on aspects on the regional geology and oil geochemistry (e.g., Hood et al. 2002). Perhaps the main geological difference between our study and previous work is the identification of a Lower Cretaceous oil family. Whereas previous studies (e.g., Hood et al. 2002) have not formally delineated an oil family of this age within the Texas Gulf Coastal plain, they have delineated an "undifferentiated Cretaceous" oil family in the southwestern part of the study area. Moreover, the interval is a known source rock in the broader Gulf of Mexico region (e.g., South Florida basin, see Hood et al. 2002). The degree to

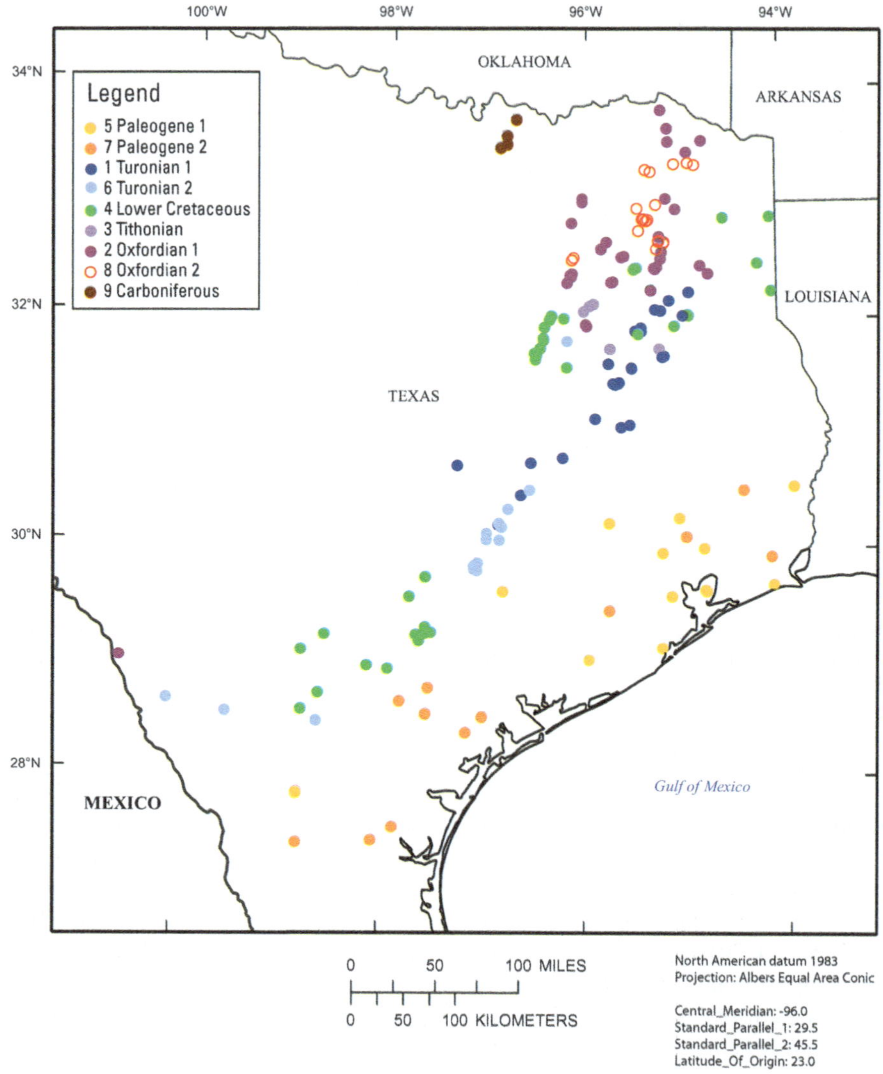

Fig. 11 Posting of clusters and oil families using Euclidean distances of 172 oil samples. In the legend, the first number after the bullet is the same cluster number in Figs. 9 and 10, which is followed by the name of the correlated family

which the Lower Cretaceous stratigraphic section in Texas hosts a robust petroleum source rock may be a topic that is worthy of additional geological and geochemical investigation.

8 Geological Interpretations

Accumulation of evidence over the years allows today for inferences to be made about the source rock organofacies based on some biomarker and carbon isotope ratios (Peters et al. 2005). Displays are customarily in the form of scatterplots or ternary diagrams. Figures 12, 13, 14, 15, 16 and 17 are six types of graphs relevant to the onshore portion of the Gulf of Mexico basin in Texas. Unless mentioned specifically, all references below relate to the compositional diagrams comprising the 16 oil families obtained with the compositional approach.

Most biomarker ratios used to analyze oils in this study provide information about source depositional environments rather than age. The scatterplot in Fig. 12 is an exception. There, C_{26}T/Ts terpane values larger than 1.0 denote a source rock older than Jurassic (Peters et al. 2007). For oils with C_{26}T/Ts terpane values less than 1.0, the C_{28}/C_{29} is interpreted to be inversely correlated with age (Peters and Moldowan 1993), provided that the sediments are free of terrigenous input (Peters et al. 2007). Overall, our correlations are consistent with these guidelines. Oils with an interpreted Paleozoic source rock exhibit the highest C_{26}T/Ts terpane values, generally larger than or equal to 1.5. Oils interpreted to derive from Mesozoic source rocks exhibit lower C_{26}T/Ts terpane values and C_{28}/C_{29} sterane values between ~0.4 and 1.4, with the older oils exhibiting the lowest values. The Paleogene oils do not conform to the regional trend in C_{28}/C_{29} values, consistent with the observation that the source rocks contain significant amounts of terrestrial organic matter (e.g., Hood et al. 2002).

Biomarker and carbon isotope ratios can also discriminate between source rock depositional settings, based on the organic matter types common to those environments. This type of discrimination is commonly performed with a variety of plots, including the Sofer (1984) diagram (Fig. 13), sterane ternary diagrams (Fig. 14), cross plots of Tet/C_{23} terpane against $C_{35}S/C_{34}S$ terpane (Fig. 15), and cross plots of C31R/H terpane against C_{24}/C_{23} terpane (Fig. 16). Figure 15 indicates a particularly clear differentiation of the Paleogene oils. They exhibit high values of Tet/C_{23} which

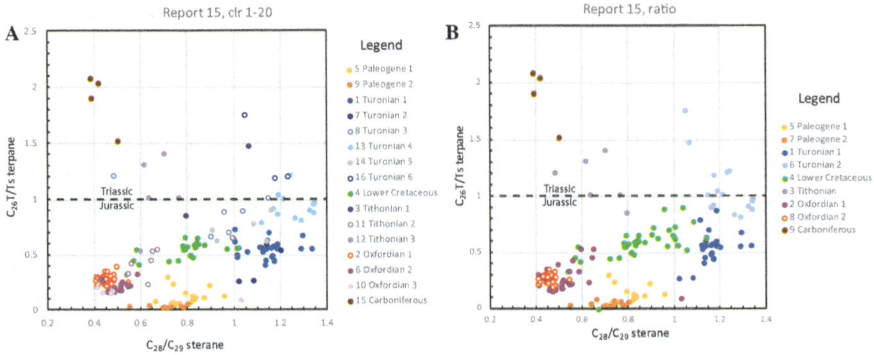

Fig. 12 Time sensitive scatterplot. The dotted line is taken from Peters et al. (2007): **a** Aitchison analysis; **b** Euclidean analysis

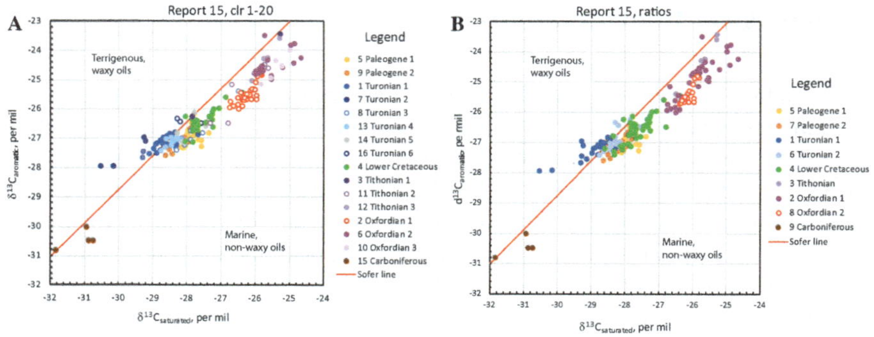

Fig. 13 Sofer (1984) diagram: **a** Aitchison analysis; **b** Euclidean analysis

indicate sourcing from organic matter deposited in shallow waters, such as marine shelves (Zumberge and Scotese 2007), in this case with partial input of terrestrial organic matter (Kennicutt et al. 1992). Figures 13 and 14 also suggest the presence of terrigenous organic matter in the Paleogene source rock, however, there is more overlap with other, more marine oils and not all samples conform to the interpretative boundaries that are applied to these diagrams. The Paleogene oils actually plot below Sofer line, suggesting that they are derived from marine strata (Fig. 13). The discrepancies among these figures are an indication that it is not possible to expect changes in relative abundance that throughout time are exactly proportional for all biomarkers and the carbon isotopic ratios. Different groups can tell different stories, and it is advisable to incorporate multiple biomarker metrics into a source rock correlation.

Cenomanian–Turonian oils have the largest number of families, and a differentiation in source rock depositional settings is also suggested by Figs. 13, 14, 15 and 16. For example, among the Turonian oil families, cluster 7 is clearly the one most distant from the Sofer line on the terrigenous side, with cluster 1 being the only other cluster completely on the terrigenous side of the Sofer line (Fig. 13). The isotopic ratios are the lowest among post-Paleozoic families and second only to the Paleogene families in terms of low $C_{35}S/C_{34}S$. The Cenomanian–Turonian families have the higher proportion of C_{28} steranes in Fig. 14, mostly on the estuarine side of the line dividing estuarine from terrestrial environments. Both in Figs. 13 and 14, family Turonian 3 is the organofacies indicating the most marine depositional environment. Our Turonian families 3 and 5 closely coincide in geographical location and $\delta^{13}C$ values in Zumberge et al.'s (2016). Turonian families 2, 3 and 7 correlated to Eagle Ford organofacies, oil samples that perhaps are the same ones in our database subset. Overall, the plots suggest a more marine influence toward the southwest and a more terrigenous influence toward the northeast. This is consistent with the gradient in depositional environments across the Texas Gulf Coast during the Cenomanian–Turonian, which ranged from ~deltaic in the northeast to shelfal in the southwest (Galloway 2008), at the southern end of the epeiric North American seaway (Zumberge et al. 2016).

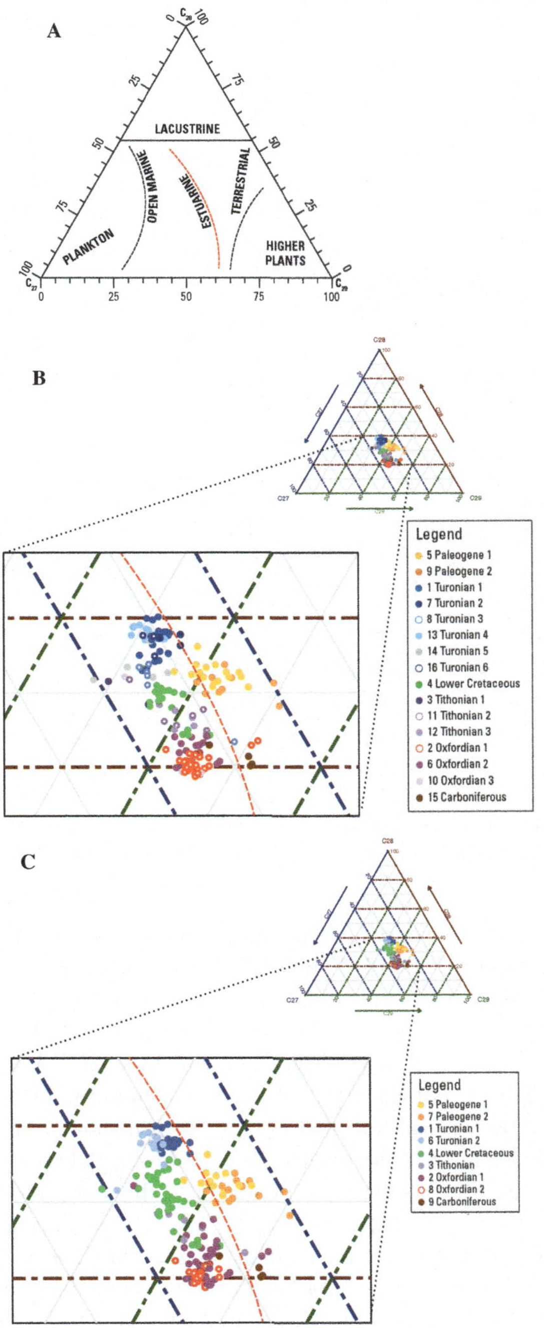

Fig. 14 Sterane ternary diagrams. The red segmented line in diagrams **b** and **c** is the same divider between estuarine and terrestrial environments in diagram **a**: **a** Postulated depositional environments (After Pedersen et al. 2006); **b** Aitchison analysis; **c** Euclidean analysis

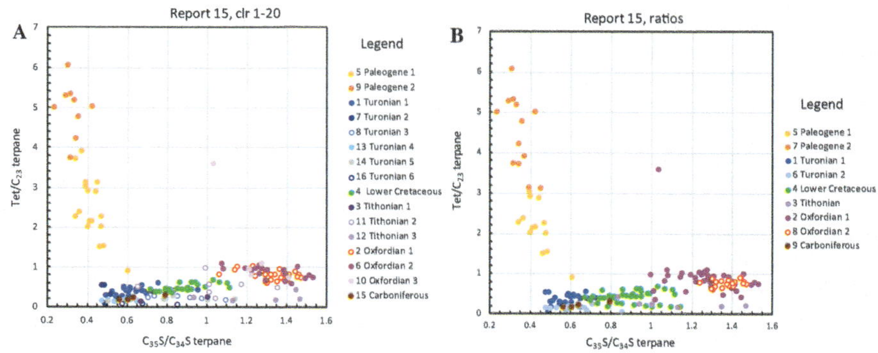

Fig. 15 Scatterplot of two terpane ratios: **a** Aitchison analysis; **b** Euclidean analysis

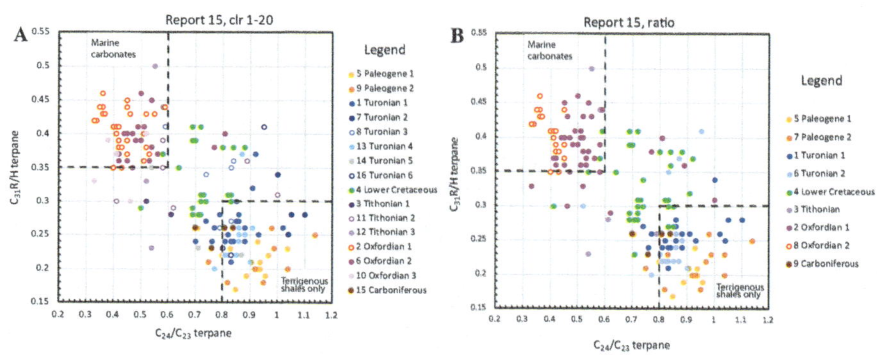

Fig. 16 Scatterplot of two terpane ratios sensitive to the depositional environment. The segmented lines are taken from Peters et al. (2007): **a** Aitchison analysis; **b** Euclidean analysis

Oils that we correlated to Upper Jurassic (Oxfordian) source rocks show extreme values in Figs. 13, 14, 15 and 16. The marine nature of these oils is confirmed by the largest distance of these oils to the Sofer line in Fig. 13, C31R/H values predominantly over 0.35 and C_{24}/C_{23} values below 0.6 in Fig. 16 (Peters et al. 2007), and the high values for $C_{35}S/C_{34}S$ in Fig. 15 (Kennicutt et al. 1992; Burton et al. 2019).

Figure 17 is a scatterplot of sulfur concentration and API gravity, two bulk physical properties of crude oils that were not used in either of the two clustering approaches, partly because of their sensitivities to biodegradation and thermal maturity (Peters et al. 2016). This diagram, incidentally, illustrates the point that it is not necessary to include a variable in the cluster analysis for later using it for preparing scatter-plots. Sulfur concentrations in excess of about 2.1 weight percent are an exclusive characteristic of the Oxfordian 1 family, coinciding with the findings by Kennicutt et al. (1992). Lack of iron favors reducing conditions that result in kerogens high in sulfur.

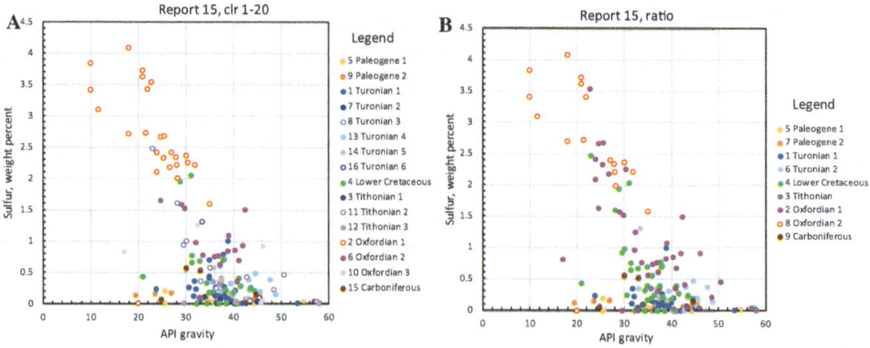

Fig. 17 Scatterplot of two bulk properties: API gravity and sulfur content. **a** Aitchison analysis; **b** Euclidean analysis

Oxfordian source rocks have been described as containing marine algae (Sassen 1989), which only exist in marine environments, and so this is consistent with our source rock correlation. Wenger et al. (1990) analyzed 11 crude oils taken from Smackover reservoirs, some close to our study area east of the triple junction of Arkansas, Louisiana and Texas. They identified two kinds of crude oils: Smackover Type I and Smackover Type II oils. Their Type I oils coincide with the values of our Oxfordian 2 family in terms of $\delta^{13}C$ values and the highest sulfur concentrations. Their Type II oils are equivalent to our Oxfordian 1 family in terms of the relatively $\delta^{13}C$ values and sulfur concentrations second only to Oxfordian 2 family. These similarities also are consistent with our Oxfordian correlation.

In summary, although the correlations in Sect. 5 could have been done in numerous other ways, the validation in this section reveals that some of the key source rock assignments do not show contradictions with the geologic properties of source rocks indicated by six types of diagrams, the known properties of the postulated source rocks, and partial findings from several other authors.

9 Conclusions

Cluster analysis and some validation methods were applied for the first time to biomarker data and carbon isotope ratios of 177 oils comprising several petroleum systems in southeast Texas (Western Gulf Coast). Cluster analysis classifies objects based on similarities and dissimilarities using distances between objects in the multivariate space of the observations. Applying Ward's method, the biomarker data were modeled using the traditional approach of calculating Euclidean distances and also considering clr transformations to conform with an Aitchison geometry. The results from the latter approach using the clr transformation were superior in two aspects:

- The discriminating power was higher. Upon optimally selecting the number of clusters on the basis of the Dunn index and the average silhouette widths, the clr cluster analysis produced 16 biomarker clusters, 7 more (78%) than the traditional approach.
- The quality of the cluster analysis was better when measured by the silhouette width: (a) the average width for the clr analysis was 0.35, 0.06 units (20%) higher than the traditional approach; (b) the number of negative silhouettes for clr analysis was 10, whereas the ones for the traditional approach was almost three times larger at 27; and (c) only one negative silhouette width for the clr analysis was below −0.2, whereas the traditional approach had 5 silhouettes less than −0.2.

Source rock assignments for each cluster were reviewed applying several biomarker tools for analyzing age and depositional environments, using attributes that were and were not used in the cluster analysis. The results are comparable between the two approaches here and the most recent published study about the area (Hood et al. 2002). The nature of the depositional environment of the Paleogene rocks is one of the main discrepancies in the geological interpretations that can be inferred from biomarkers and carbon isotopes, with the Sofer line assigning a more marine origin to the Paleogene oil families than a sterane ternary plot, finding deserving additional investigation.

The superior accuracy and discriminating power of the compositional modeling allowed the identification of seven different types of Cretaceous oils. Six of these oils appear to be derived from the Eagle Ford Shale and the differentiation of the oils appears to reflect the regional paleogeography of the Cenomanian–Turonian. A seventh oil family is located immediately up-dip of the Eagle Ford-derived oils. The age of the reservoirs implies that these are derived from a Lower Cretaceous source rock that is less well known on a regional basis than the Eagle Ford Shale.

Acknowledgements This manuscript completed a mandatory internal review following the U.S. Geological Survey (USGS) Fundamental Science Practices (https://pubs.usgs.gov/circ/1367/). We are grateful to Paul Lillis (USGS) for a thorough internal review that resulted in suggestions and comments that helped to improve the manuscript, and to Madalyn Blondes (USGS) for a critical reading of a late version of this chapter. Steven Cahan (USGS) added the geographical information to all maps and Eric Morrissey (USGS) drafted Fig. 14a. J. A. Martín-Fernández was supported by the Spanish Ministry of Science, Innovation and Universities under the project CODAMET (RTI2018-095518-B-C21, 2019-2021).

References

G. Brock, V. Pihur, S. Datta, S. Datta, clValid: an R package for cluster validation. J. Stat. Softw. **25**(4), 1–22 (2008)

Z.F.M. Burton, J.M. Moldowan, L.B. Magoon, R. Sykes, S.A. Graham, Interpretation of source rock depositional environment and age from seep oil, east coast of New Zealand. Int. J. Earth Sci. **108**(4), 1079–1091 (2019)

G.E. Claypool, E.A. Mancini, Geochemical relationships of petroleum in Mesozoic reservoirs to carbonate source rocks of Jurassic Smackover Formation, southwestern Alabama. AAPG Bull. **73**(7), 904–924 (1989)

P.A. Comet, J.K. Rafalska, J.M. Brooks, Sterane and triterpane patterns as diagnostic tools in the mapping of oils, condensates, and source rocks of the Gulf of Mexico region. Org. Geochem. **20**(8), 1265–1296 (1993)

M.A. Engle, E.L. Rowan, Interpretation of Na-Cl-Br systematics in sedimentary basin brines: comparison concentration, element ratio, and isometric log-ratio approaches. Math. Geosci. **45**(1), 87–101 (2013)

J. Fox, S. Weisberg, *An R Companion to Applied Regression*, 3rd ed. (Sage, Thousand Oaks, CA, 2019), 577 pp.

W.E. Galloway, Depositional evolution of the Gulf of Mexico sedimentary basin, in *The Sedimentary Basins of the United States and Canada*, ed. by A. Miall. Sedimentary Basins of the World, vol. 5 (Elsevier, Amsterdam, The Netherlands, 2008), pp. 505–549

GeoMark Research, Ltd., RFDBASE—Rock and fluid database (2019), https://geomarkresearch.com/database-products/

B.U. Haq, J. Hardenbol, P.R. Vail, Mesozoic and Cenozoic chronostratigraphy and cycles of sea-level change, in *Sea-Level Changes–An Integrated Approach*. SEPM Special Publication 42 (Society of Economic Paleontologists and Mineralogists, Tulsa, OK, 1988), pp. 71–108

F.E. Harrell Jr., K.L. Lee, B.D. Mark, Multivariable prognostic models: issues in developing models, evaluating assumptions and adequacy, and measuring and reducing errors. Stat. Med. **15**(4), 361–387 (1996)

M. He, S. Graham, A.H. Scheirer, K.E. Peters, A basin modeling and organic geochemistry study in the Vallecitos syncline, San Joaquin Basin, California. Mar. Pet. Geol. **49**, 15–34 (2014)

M. He, M. Moldowan, A. Nemchenko-Rovenskaya, K.E. Peters, Oil families and their inferred source rocks in the Barents Sea and northern Timan-Pechora Basin, Russia. AAPG Bull. **96**(6), 1121–1146 (2012)

M. He, J.M. Moldowan, K.E. Peters, Biomarkers: petroleum, in *Encyclopedia of Geochemistry*. ed. by W.M. White (Springer, Cham, Switzerland, 2018), pp. 136–148

K.C. Hood, O.P. Gross, L.M. Wenger, S.C. Harrison, Hydrocarbon systems analysis of the northern Gulf of Mexico: delineation of hydrocarbon migration pathways using seeps and seismic imaging, in *Surface Exploration Case Histories: Applications of Geochemistry, Magnetics, and Remote Sensing*, ed. by D. Schumacher, L.A. LeSchack. AAPG Studies in Geology No. 48 and SEG Geophysical References Series No. 11 (2002), pp. 25–40

I.T. Jolliffe, *Principal Component Analysis*, 2nd ed. (Springer, New York, 2002), 487 pp.

M.C. Kennicutt II., T.J. McDonald, P.A. Comet, G.J. Denoux, J.M. Brooks, The origins of petroleum in the northern Gulf of Mexico. Geochim. Cosmochim. Acta **56**, 1259–1280 (1992)

A.I. Levorsen, F.A.F. Berry, *Geology of Petroleum*, 2nd ed. (AAPG, 2001), 724 pp.

K.R. Marra, R.R. Charpentier, C.J. Shenk, M.D. Lewan, H.M. Leathers-Miller, T.R. Klett, S.B. Gaswirth, P.A. Le, T.J. Mercier, J.K. Pitman, M.E. Tennyson, Assessment of undiscovered shale gas and shale oil resources in the Mississippian Barnett Shale. Bend Arch-Fort Worth basin, North-Central Texas. U.S. Geological Survey Fact Sheet 2015-3078 (2015)

G.W. Milligan, M.C. Cooper, An examination of procedures for determining the number of clusters in a dataset. Psychometrika **50**(2), 159–179 (1985)

B. Mirkin, *Clustering: A Data Recovery Approach*, 2nd ed. (CRC Press, 2013), 365 pp.

R. Nehring, Oil and gas resources, in *The Gulf of Mexico Basin*, ed. by A. Salvador (Geological Society of America, Boulder, CO, 1991), pp. 445–494

J. Palarea-Albaladejo, J.A. Martín-Fernández, zCompositions—R package for multivariate imputation of left-censored data under a compositional approach. Chemom. Intell. Lab. Syst. **143**, 85–96 (2015)

J. Palarea-Albaladejo, J.A. Martín-Fernández, J.A. Soto, Dealing with distances and transformations for fuzzy C-means clustering of compositional data. J. Classif. **29**(2), 144–169 (2012)

V. Pawlowsky-Glahn, J.J. Egozcue, R. Tolosana-Delgado, *Modeling and Analysis of Compositional Data* (Wiley, Chichester, UK, 2015), p. 247

J.H. Pedersen, D.A. Karlsen, K. Backer-Owe, J.E. Lie, H. Brunstad, The geochemistry of two unusual oils from the Norwegian North Sea: implications for new source rock and play scenario. Pet. Geosci. **12**(1), 85–96 (2006)

K.E. Peters, D.J. Curry, M. Kacewicz, An overview of basin and petroleum system modelling, in *Basin Modeling: New Horizons in Research and Applications*, ed. by K.E. Peters, D. Curry, M. Kacewicz. AAPG Hedberg Series, vol. 4 (2012), pp. 1–16

K.E. Peters, J.M. Moldowan, *The Biomarker Guide: Interpreting Molecular Fossils in Petroleum and Ancient Sediments* (Prentice Hall, Englewood Cliffs, NJ, 1993), p. 363

K.E. Peters, L.S. Ramos, J.E. Zumberge, Z.C. Valin, C.R. Scotese, D.L. Gautier, Circum-Arctic petroleum systems identified using decision-tree chemometrics. AAPG Bull. **91**(6), 877–913 (2007)

K.E. Peters, C.C. Walters, J.M. Moldowan, *The Biomarker Guide*, 2nd ed. (Cambridge University Press, Cambridge, UK, 2005), p. 1155

K.E. Peters, T.L. Wright, L.S. Ramos, J.E. Zumberge, L.B. Magoon, Chemometric recognition of genetically distinct oil families in the Los Angeles basin, California. AAPG Bull. **100**(1), 115–135 (2016)

R Core Team, R: A language and environment for statistical computing. R Foundation for Statistical Computing, Vienna, Austria (2018), https://www.R-project.org

P.J. Rousseeuw, Silhouettes: a graphical aid to the interpretation and validation of cluster analysis. J. Comput. Appl. Math. **20**(C), 53–65 (1987)

A. Salvador (ed.), *The Gulf of Mexico Basin*. The Geology of North America, vol. J (The Geological Society of America, Boulder, CO, 1991), 568 pp.

A. Salvador, J.M.Q. Muñeton, compilers, Stratigraphic correlation chart: Gulf of Mexico Basin, in *The Gulf of Mexico Basin*, ed. by A. Salvador. The Geology of North America, vol. J (Geological Society of America, Boulder, plate 5, 1989)

R. Sassen, Migration of crude oil from the Smackover source rock to Jurassic and Cretaceous reservoirs of the northern Gulf rim. Org. Geochem. **14**(1), 51–60 (1989)

J.L. Shelton, M.A. Engle, A. Buccianti, M.S. Blondes, The isometric log-ratio (ilr)-ion plot: a proposed alternative to the Piper diagram. J. Geochem. Explor. **190**, 130–141 (2018)

K. Siddiqui, Heuristics for sample size determination in multivariate statistical techniques. World Appl. Sci. J. **27**(2), 285–287 (2013)

Z. Sofer, Stable carbon isotope compositions of crude oils: application to source depositional environments and petroleum alteration. AAPG Bull. **68**(1), 31–49 (1984)

C.A. Sugar, G.M. James, Finding the number of clusters in a dataset: an information-theoretic approach. J. Am. Stat. Assoc. **98**(463), 750–763 (2003)

N. Tyler, T. Ewing, Major oil plays of south and south-central Texas, in *Contributions to the Geology of South Texas* (1986), pp. 24–52

M. Visvanathan, A.B. Srinivas, G.H. Lushington, P. Smith, Cluster validation: an integrative method for cluster analysis, in *Proceedings of IEEE International Conference on Bioinformatics and Biomedicine Workshops* (2009), pp. 238–242

L.M. Wenger, R. Sassen, D. Schumacher, Molecular characteristics of Smackover, Tuscaloosa and Wilcox-reservoired oils in the Eastern Gulf Coast, in *Proceedings of the Ninth Annual Research Conference of the Gulf Coast Section SEPM Foundation* (1990), pp. 37–57

R.D. Woods, A. Salvador, A.E. Miles, Pre-Triassic, in *The Gulf of Mexico Basin*, ed. by A. Salvador. The Geology of North America, vol. J (Geological Society of America, Boulder, CO, 1991), pp. 109–129

J. Zumberge, H. Illich, L. Waite, Petroleum Geochemistry of the Cenomanian-Turonian Eagle Ford Oils of South Texas, AAPG Memoir 110, American Association of Petroleum Geologists, Tulsa, OK (2016), pp. 135–165

J. Zumberge, C.R. Scotese, Biomarkers from marine crude oils reflect modeled climatic/oceanographic conditions for the Late Cretaceous. Abstracts of reports from the 23rd International Meeting on Organic Chemistry, 9–14 September, Torquay, UK, 23, 555 (2007), https://www.researchgate.net/publication/264044691_Biomarkers_from_Marine_Crude_Oils_Reflect_Modeled_ClimaticOceanographic_Conditions_for_the_Late_Cretaceous

Finding the Centre: Compositional Asymmetry in High-Throughput Sequencing Datasets

Jia R. Wu, Jean M. Macklaim, Briana L. Genge, and Gregory B. Gloor

Abstract High-throughput sequencing datasets comprise millions of reads of genomic data and can be modelled as count compositions. These data are used for transcription profiles, microbial diversity, or relative cellular abundance in culture. The data are sparse and high dimensional. Moreover, they are often unbalanced, i.e. there is often systematic variation between groups due to presence or absence of features, and this variation is important to the biological interpretation of the data. The imbalance causes samples in the comparison groups to exhibit varying centres contributing to false positive and false negative identifications. Here, we extend the centred log-ratio transformation method used for the comparison of differential relative abundance between two groups in a Bayesian compositional context. We demonstrate the pathology in modelled and real unbalanced experimental designs to show how this causes both false negative and false positive inference. We examined four approaches to identify denominator features, and tested them with different proportions of modelled asymmetry; two were relatively robust, and recommended. We recommend the 'LVHA' transformation for asymmetric transcriptome datasets, and the 'IQLR' method for all other datasets when using the `ALDEx2` tool available on Bioconductor.

J. R. Wu
Department of Computer Science, University of Waterloo, Waterloo, Canada
e-mail: jr2wu@uwaterloo.ca

J. M. Macklaim · B. L. Genge · G. B. Gloor (✉)
Department of Biochemistry, London, ON, Canada
e-mail: ggloor@uwo.ca

J. M. Macklaim
e-mail: jean.macklaim@gmail.com

B. L. Genge
e-mail: bgenge3@gmail.com

P. Filzmoser et al. (eds.), *Advances in Compositional Data Analysis*,
https://doi.org/10.1007/978-3-030-71175-7_17

1 Background

High-throughput sequencing (HTS) technology is used to generate information regarding the relative abundance of features. In these designs, DNA or RNA is isolated, a library is made from a sample of the nucleic acid, and a random sample of the library is sequenced on an instrument. The output is a set of short sequence tags, called reads, which are mapped to reference sequences for annotation of each feature that generates a table of read counts per feature for every sample. Traditionally, samples comprise a set of features whose identity depends on the experimental design. For example, features are genes in the case of RNA-seq or metagenomic sequencing, or are operational taxonomic units (OTUs) when the objective is identifying microbial diversity.

The instruments used for HTS have an upper limit on the total number of reads delivered, for example, the same library sequenced on an Illumina MiSeq or an Illumina NextSeq will deliver approximately 20 million or 400 million reads. HTS instruments thus deliver a fixed-sum random sample of the sequences that are in the library, which itself is a random sample of what was in the environment.

Data from HTS instruments are usually analysed by count-based methods which use a negative binomial or zero-inflated Gaussian model, and assume the features are independent and identically distributed for statistical tests, and that the majority of features are invariant (Anders et al. 2013; Auer and Doerge 2010). In addition, the variance in HTS data is usually larger than the mean (overdispersed), and negative binomial and similar models are generally adequate when the underlying assumptions are approximated.

Many HTS datasets do not fit the usual statistical models. For example, in meta-transcriptomic datasets, the relative abundance of both organisms and of the genes expressed by the organisms can both change independently, while in metagenomic datasets there is no a priori reason for feature counts to follow any particular statistical model. The assumption that the majority of features are invariant is broken if there is any sort of systematic variation between groups. For example, when comparing microbial diversity between sampling sites or conditions, organisms present in one sub-site or condition may be absent from another (Macklaim et al. 2015; Hummelen et al. 2010; Gajer et al. 2012). In the case of multi-organism RNA-seq (meta-transcriptomics), organisms resident in one condition may have a different expression profile and abundance than those resident in a second condition (Macklaim et al. 2013). In the case of a single-organism RNA-seq, samples from one condition may contain more genes than samples from another condition (Lang and Johnson 2015; Peng et al. 2014; Zhao et al. 2013; Gierliński et al. 2015). These differences are represented by either zeros or low-count features that occur systematically in only one group. Meta-transcriptomic datasets are thus some of the most difficult experiments to analyse (Macklaim et al. 2013; Fernandes et al. 2013) since the datasets do not fit multiple assumptions of existing tools. We have shown previously that existing count-based tools fail to give sensible answers in meta-transcriptome and other datasets that exhibit asymmetry (Fernandes et al. 2013, 2014; Gloor et al. 2016b).

1.1 A Ratio-Based Approach to Analysing HTS

Another approach to analyse data taking into account the fixed capacity of the instrument and the random sampling of the library (and the environment) is to model the data as a multivariate probability distribution and to examine the result with the principles of compositional data analysis (Aitchison 1986; Fernandes et al. 2013). This method is instantiated in the `ALDEx2` Bioconductor package. Conceptually, the approach is similar to sequencing many different technical replicates of each of the underlying biological replicates, treating the resulting data as ratios of the features and reporting the mean value of the technical replicates. This approach has been found to be a generally useful tool that works for meta-transcriptomic datasets (Macklaim et al. 2013) and translates to many different experimental designs (Fernandes et al. 2014; McMurrough et al. 2014; Gloor et al. 2017; Macklaim and Gloor 2018; Almeida et al. 2019).

The basis for the approach used by `ALDEx2` is the centred log-ratio (CLR) transformation proposed by Aitchison (1986) and defined below. However, this approach requires that the denominator used to calculate the CLR be comparable across all samples. The ratio-based approach is similar to relative quantitative PCR (qPCR), a method in common use in molecular biology that measures relative abundance of molecules in a mixture (Thellin et al. 1999; Vandesompele et al. 2002) and is often used as the gold standard to validate HTS results. In this type of qPCR, the feature of unknown abundance is determined relative to the abundance of a feature of (presumed) known abundance. This standard can be a housekeeping gene or can be a DNA molecule of known amount, a so-called spike-in molecule, added to the mixture. It is well known that the relative abundance measure will change when a different spike-in species is used as the denominator, leading to the use of multiple (presumed invariant) species in some cases. Thus, one shortfall of any ratio-based approach is to determine the appropriate denominator.

The potential for a change in cell number and the potential for expression linkage of genes in biological systems, coupled with the inability to collect a large enough number of sequence reads, can lead to experiments with an apparent or a real asymmetry in relative abundance of many genes or features. Such an asymmetry will result in miscentring of the data when conducting differential abundance analyses, largely, but not exclusively because of the effect on the geometric mean upon which the CLR depends. The asymmetry will also affect the scale invariance of the data, since a value of 0 is not scaled when multiplied by a constant. Note that it is also entirely possible for the dataset *as a whole* to be centred, but for the particular comparison of interest to not be centred. This could arise because of a systematic experimental bias that is unknown to the investigator.

For convenience, and because the datasets are generally very complex, the analyses and discussion here are drawn from RNA-seq, or transcriptome, experiments where the data are exploring the relative abundance of features that are gene transcripts found in cells in an environment. However, the examples, results and conclusions apply without restriction to metagenomic sequencing, microbial diversity sampling (by 16S rRNA gene sequencing) or to in vitro selection experiments (Fernandes et al. 2014; McMurrough et al. 2014; Gloor et al. 2017).

1.2 Choosing the Denominator

The basic principle of compositional data analysis as developed by Aitchison is to convert the data into log-ratios between the features for each sample (Aitchison 1986). Formally, Aitchison defined a composition as a vector $\boldsymbol{x}$ of positive values $x_1 \dots x_D$ whose features sum to an arbitrary constrained constant α. Absolute values of features in a composition are uninformative, and so the only information available in compositional data are the relative magnitudes, or the ratios between the pairs of components. For example, the only knowledge available is that the gene 1:gene 2 ratio is 5, but the absolute abundance of either is unavailable. This is the case in HTS since the count observed for a feature contains no information regarding the absolute number of molecules in either the sequencing library or the environment, although the magnitude of the count contains information on the precision of the estimate.

Data collected from high-throughput sequencing are count compositions (Fernandes et al. 2014; Gloor et al. 2016b). Count-based tools do not address the compositional nature of HTS data (Gloor et al. 2016b; Fernandes et al. 2014) and assume that the features are sufficiently independent when there are enough of them, or when they fulfill certain statistical properties (Weiss et al. 2017). Much effort is placed on 'normalizing' the data to have a consistent read depth instead of treating the data as compositional (McMurdie and Holmes 2014; Sun et al. 2013).

However, since all read count totals from a machine are arbitrary we should treat the data as an equivalence class where the composition contained in vector $\mathbf{x}$ can be scaled into an identical composition $\boldsymbol{y}$ by multiplication of a constant α (Barceló-Vidal et al. 2001). Thus, in the ideal case, we can discuss any composition as being a probability vector scaled by α without loss of precision (Fernandes et al. 2013).

One way of satisfying the need to examine the ratios between features is to use the centred-log-ratio (CLR) transformation proposed by Aitchison, defined as

$$\mathbf{x}_{clr} = log\left(\frac{x_i}{G(\mathbf{x})}\right)_{i=1\dots D}, \tag{1}$$

where $G(\mathbf{x})$ = the geometric mean of the D features of $\mathbf{x}$. When interpreting CLR-transformed values, care must be taken to ensure that the analyst understands that any changes observed are always relative to the denominator of the CLR in each sample.

The CLR, and any other ratio-based method is, at least in theory, scale invariant because if the parts of $\boldsymbol{x}$ are counts with $\alpha = N$ reads, then

$$\mathbf{x}_{i,clr} = log(\frac{\alpha x_i}{G(\alpha \mathbf{x})}) = log(\frac{x_i}{G(\mathbf{x})}). \quad (2)$$

The important caveat that limits this ideal situation when dealing with high-throughput sequencing data is that the total read count, α, for each sample should be similar. The CLR is the default denominator used by `ALDEx2`. In practical terms, `ALDEx2` handles the problem of varying read depth by constructing a posterior probability distribution of the original count matrix (Fernandes et al. 2013). In this approach, features in low-count samples have broader probability distributions than features in high-count samples. Thus, the precision of the estimates varies when samples have different read depths.

Aitchison (1986) also defined the ALR, the additive log-ratio as

$$\mathbf{x}_{alr} = log(\frac{x_i}{x_D})_{i=1...D-1} \quad (3)$$

where, following from above, $\boldsymbol{x}_{alr}$ is the composition transformed by ALR. When using the ALR the denominator is the Dth feature of $\boldsymbol{x}$, which by convention is the feature chosen to be constant.

In the ALR, the log-ratio is determined by selecting one presumed invariant feature as the denominator, and so the ALR is similar to the relative qPCR approach in common use in molecular biology. The ALR and CLR can be viewed as the two limits of a continuum of incomplete knowledge about the proper internal standard, or basis, by which relative abundance should be judged. The ALR uses one presumed constant feature as the basis, while the CLR presumes that majority of features are not changed, leading to the use of the geometric mean of all features as the basis. Clearly, neither of these are satisfactory. We can, however, choose to use combinations of other features as the basis for comparison.

1.3 HTS Data Are Sparse

It is common for HTS data to be sparse, that is, for a given sample to contain features with counts of 0. One limitation of the log-ratio approach is that ratios have no meaning if the denominator contains a 0 value. The sparsity of a sample is affected by the total number of reads obtained for each sample, and by the number of features that the total reads are apportioned between. Each sample in a transcriptome contains between thousands and tens of thousands of features each of which may have a potential dynamic range of over four orders of magnitude. In many cases, a transcriptome dataset will be composed of several groups, where the expression of a feature (gene) is so low that it is below the detection limit in one group, and very high in another group. The expression of genes in biological systems is linked, and some genes control the expression of other genes, either by increasing or decreasing their relative abundance. Furthermore, the cell has a built-in control system whereby gene expression itself appears to be a composition, that is, the expression levels of all genes in a cell in a given environment are constrained by an absolute upper bound (Scott et al. 2010). In eukaryotic cells, the upper bound can be changed with surprisingly small genetic changes (Lovén et al. 2012), and changes in the upper bound are known confounder of transcriptome experiments. In the case of a meta-transcriptome, a population of cells does not necessarily exhibit total gene expression with an upper bound, since the cells themselves can change in both absolute and relative abundance in a mixture.

The assumption being made when using the CLR transformation to identify features that differ between groups is that most features are either invariant or, at worst, vary at random when comparing the two groups. However, marked asymmetry can break this assumption, and this is the problem being addressed by identifying and using alternate denominators.

1.4 Alternative Denominators

The starting point for the alternative denominators is the n samples by D features matrix of counts. The data are CLR transformed using Eq. 1 sample-wise with a uniform prior of 0.5 to give a new matrix, and the objective is to identify a subset of the features that best represent those features that are least likely to exhibit systematic change in the dataset.

The *IQLR* (interquartile log-ratio) transformation uses as the denominator for the log-ratio those features with ‘typical’ variance across all samples. The variance is calculated for each feature group-wise, and features with variance between the first and third quartile in each group are extracted. Thus, the vector of features in group A, $\mathbf{f_A}$, is chosen from the variance vector of group **A** as follows: $\mathbf{var(A)_{Q1}} > \mathbf{f_A} < \mathbf{var(A)_{Q3}}$. The interquartile-variance feature *IQVF* subset of the D features is the intersect of the group-wise typical variance features, $\mathbf{IQVF} = \mathbf{f_A} \cap \mathbf{f_B}$ if there are two groups. The geometric mean of the *IQVF* set is used in the denominator to calculate a log-ratio for each sample vector $\mathbf{x_i}$ in matrix S:

$$\mathbf{x}_{\mathbf{i},IQLR} = \mathbf{log}\left(\frac{\mathbf{x_{i,j=1...D}}}{\mathbf{G}(IQVF)}\right) \tag{4}$$

The non-zero log-ratio, *NZLR*, transformation uses as the denominator for the log-ratio those features that are non-zero in each group. The log-ratio is thus calculated using a different set of features as the denominator for each group. The indices of non-zero features in each group are retained $\mathbf{f_A} = \mathbf{min}(\mathrm{S}_{1...i_A}, \mathbf{1 \ldots D}) > \mathbf{0}; \mathbf{f_B} = \mathbf{min}(\mathrm{S}_{(i_B+1)...n}, \mathbf{1 \ldots D}) > \mathbf{0}$ and used group-wise.

Now the features in group A have their log-ratio computed using the non-zero features in the group as the denominator, and likewise for group B. Thus, the transformed vectors in group A are

$$\mathbf{x}_{\mathbf{i}\mathrm{A},NZLR} = \mathbf{log}\left(\frac{\mathbf{x}_{1\ldots\mathbf{i}_\mathrm{A},\mathbf{j}=1\ldots\mathbf{D}}}{\mathbf{G}(\mathbf{f}_\mathbf{A})}\right). \tag{5}$$

Similarly, the vectors in group B are transformed using as the denominator the $G(\mathbf{f_B})$ of the set of non-zero features in group B.

The *LVHA* (low-variance high-abundance log-ratio) transformation uses as the denominator the intersect of those features that have low variance and high relative abundance in each group. These are chosen as the features that exhibit variance below the first quartile and relative abundance above the third quartile on a per-group basis. That is, the vector of features for group A is generated as follows: $\mathbf{f_A} = (\mathbf{A} < \mathbf{var}(\mathbf{A})_{\mathbf{Q1}}) \cap \mathbf{A} > \mathbf{rab}(\mathbf{A})_{\mathbf{Q3}}$, and so on for other groups. The final set is $LVHA = \mathbf{f_A} \cap \mathbf{f_B}$ if there are two groups.

The geometric mean of the *LVHA* set is used in the denominator to calculate a log-ratio for each sample vector $\mathbf{x_i}$ in matrix S:

$$\mathbf{x}_{\mathbf{i},LVHA} = \mathbf{log}\left(\frac{\mathbf{x}_{\mathbf{i},\mathbf{j}=1\ldots\mathbf{D}}}{\mathbf{G}(LVHA)}\right) \tag{6}$$

The log median transformation uses as the denominator the median of the clr-transformed values rather than the geometric mean.

$$\mathbf{x}_{\mathbf{i},\mathbf{MED}} = \mathbf{log}(\mathbf{x}_{\mathbf{i},\mathbf{j}=1\ldots\mathbf{D}}) - \mathrm{MED}(\mathbf{log}(\mathbf{x_i})) \tag{7}$$

2 Results

Throughout, we use two plots to summarize the location of the features in multivariate datasets. The Bland–Altman (BA) plot (Altman and Bland 1983) plots the mean log-ratio abundance on the x-axis and the difference between groups on the y-axis. The BA plot is efficient at showing the relationship between (relative, mean log-ratio) abundance and difference, but contains little information on the per-feature dispersion in the data. The effect size plot (Gloor et al. 2016a) complements the BA plot by showing the relationship between a measure of dispersion (on the x-axis) and the difference between groups (on the y-axis). All plots are in log units calculated with a base of two for convenience. The ratio between the within- and between-group differences is a proxy for the effect size statistic calculated by `ALDEx2`. Difference and dispersion are calculated using methods that are indifferent to distributional assumptions and are defined in the Implementation section. Figure 1 shows that incorrect estimates of the location of the data can be achieved with seemingly minor variation within simulated data, resulting in a clear asymmetry in the data. The goal is to identify a basis that best represents each sample so features can be accurately compared even when the data contains an asymmetry.

2.1 Simulated Data

RNA-Seq data was simulated for benchmarking purposes. Assemblies from *Saccharomyces cerevisiae* uid128 and a complete reference genome of *S. cerevisiae* were drawn from GenBank. The R package `polyester v1.10.0` (Frazee et al. 2016) was used to simulate an RNA-Seq experiment with 2 groups of 10 replicates with 20x average sequencing coverage across the simulation experiment. For the base dataset, 40 genes were chosen at random to have 2–5-fold expression difference, and these were apportioned equally between the two groups. These 40 features serve as an internal control of true positives for each dataset as their fold changes are explicit and should always be displayed as differentially expressed. We used `bowtie2` (Langmead and Salzberg 2012) to align the simulated reads to the *S. cerevisiae* reference genome. Labelling each group as A and B is arbitrary and hence the first 10 samples belong to condition A, and the final 10 samples belong to condition B. There are a total of 6349 features in these simulated data, but only the first 1000 genes by order were chosen for the majority of the figures.

An additional 51 datasets derived from the base dataset were generated in order to benchmark how well the generic log-ratio (*LR) transformations centre the data in the resultant set of datasets with sparsity ranging from 0 to 51% sparse in one condition relative to the other.

2.2 Four Alternative Methods

The CLR is used for these analyses rather than the more formally correct, and recently more popular, isometric log-ratio or balance-based approaches (Rivera-Pinto et al. 2018; Silverman et al. 2017; Washburne et al. 2017) primarily because of time and space complexity or because of additional requirements. For example, the `selbal` R package (Rivera-Pinto et al. 2018) fails to complete an analysis of the largest of the datasets used here in more than 24 hours of computation time using 10 concurrent threads on an Intel i9 class processor. Additionally, tools such as `phylofactor` (Silverman et al. 2017) are tree-based methods that assume phylogeny is an important determinant of the outcome.

In its initial implementation, `ALDEx2` computes the CLR, a per-sample geometric mean using all features as the baseline for feature comparisons. The 'Symmetric dataset' panel in Fig. 1 is an 'effect plot' (Gloor et al. 2016a) and we can see that the 40 internal control features are found to be both statistically significant and to have an effect size greater than 1 between the two groups. The remainder of the features

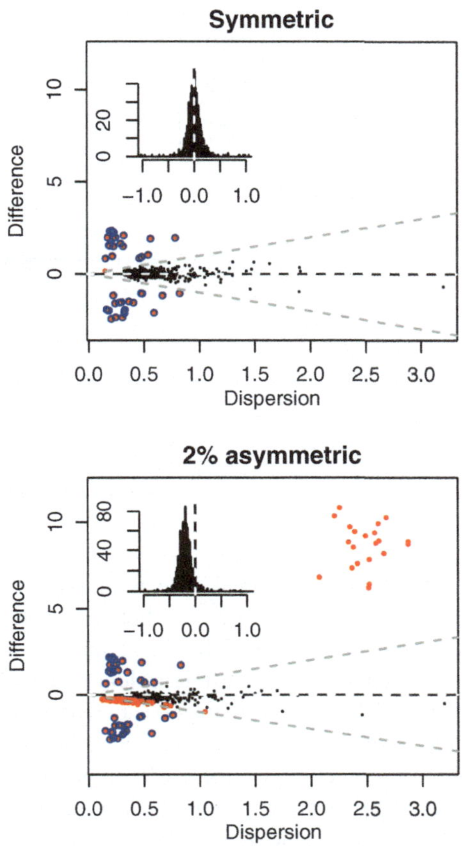

Fig. 1 Effect plots of simulated asymmetric data illustrate the problem. The effect plots show the difference between two conditions in simulated RNA-seq data with 1000 genes where 40 genes are modelled to have true difference between groups of an effect size of 1 or more. Each point is a feature (gene) and they are coloured in black if not different between groups, red if identified as being statistically significantly different between groups and red with a blue circle if they are one of the 40 genes modelled to be true positives. The red points in the top right quadrant are the genes modelled to be asymmetrically variable between groups. These are also true positive features, but are not part of the initial modelled true positives. The inset histograms show the distribution of the differences between groups as calculated by `ALDEx2`, and the vertical line shows a difference of 0. These x-axis of the histograms are truncated to show only differences near the midpoint

have very small difference and a very large dispersion, and correspondingly have an effect size much less than 1, that is, the results are as expected.

The inset histogram shows that the distribution of difference values between groups A and B is symmetric and has a location of 0. However, the introduction of small amounts of asymmetry strongly affects the results. The asymmetric 2% dataset is the base dataset modified by choosing 20 random features from Group A and setting their value to 0. It is apparent from the bottom panel of Fig. 1 that even

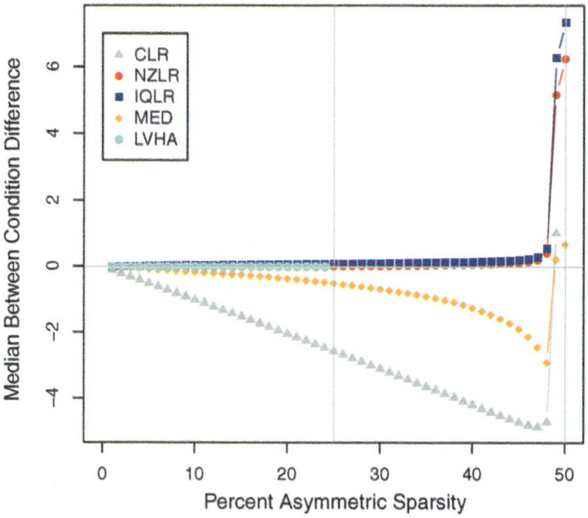

Fig. 2 The behaviour of each transform for datasets with varying asymmetry. Each point represents the median between condition difference for a given transformation in a dataset with a specified sparsity. Points closer to the location y=0 are favourable and indicate proper centring of the dataset. Abbreviations for each method for determining the denominator are the same as in the text

this low level of simulated asymmetry breaks the assumption that most features are invariant, and the location of modal no-difference between groups is no longer at the origin. This small amount of asymmetry shifts the geometric mean of the data such that the two groups are calculated using different denominators, causing bias. It is unlikely that the problem will be as easy to diagnose in real data as in simulated data.

The major determinant of the centre of a sample when using the CLR is the denominator, or basis, used to compute the CLR, thus one obvious approach to address the problem is to compute the geometric mean of a subset of features that are more representative of the central tendency of the data, and to use this value as the denominator in the equation. We examined four different approaches to identify the features to include in the denominator, and tested them with different proportions of modelled sparse asymmetry.

The transformation in Eq. 4 is termed the IQLR transformation, and relies on finding those features with 'typical' variance in the dataset. The assumption being made here is that features with typical variance are those with stochastic variance in the dataset, but that do not differ reproducibly between groups. The results of this method are shown in Fig. 2, and we see that the IQLR transformation restores the centre of the dataset to the origin even when over 40% of the modelled features are asymmetric.

The second approach uses as the denominator the set of non-zero features in each group calculated as in Eq. 5. Thus, in this case, the geometric mean of group A and group B are based on different, but potentially overlapping, sets of features, and this approach is called the non-zero log-ratio (NZLR). As shown in Fig. 2, the NZLR method also restores the centre of the data to the origin with similar efficiency in our test dataset as does the IQLR method. Note that if the asymmetry is not driven by sparsity, the NZLR method will obviously fail. This method is not recommended unless the investigator knows from prior inspection that asymmetric sparsity occurs.

The third approach was to identify the intersect between groups of those features that have variance which is in the bottom quartile and a relative abundance in the top quartile in each group. This is referred to as the low-variance high-abundance log-ratio *LVHA* method and is calculated as given in Eq. 6. In essence, the LVHA approach is an attempt to identify those features that would be similar to those chosen by an experimentalist using qPCR, being relatively abundant with low variance in all groups in the dataset. The results of this method are shown in Fig. 2. We can see that the LVHA approach perfectly centres the data up to 25% asymmetry in our test dataset.

The fourth approach replaces the geometric mean in Eq. 1 with the median of the vector as given in Eq. 7 since this should be a robust estimate of the midpoint of the data. We can see in Fig. 2 that the median is better than the geometric mean at low proportions of asymmetry, but performs worse than the previous approaches in almost every case.

2.3 *Example of a Meta-RNA-seq Dataset*

We finally introduce the example of a real meta-transcriptome dataset collected to determine the differences in gene expression of the vaginal microbial community in the healthy (H) and bacterial vaginosis (BV) states (Macklaim et al. 2013; Deng et al. 2018). The vaginal community can be dominated either by a few members of the *Lactobacillus* genus in the H state, or by a mixed group of anaerobic bacterial genera in the BV state (Ravel et al. 2011). In either state the members of the bacterial consortium from the other state are either very rare or absent. The data presented in Fig. 3 show effect size plots for the CLR and three different denominators for a count table derived from data deposited at the European Nucleotide Archive metagenomics site under the accession number PRJEB31833 using the methods in Macklaim and Gloor (2018) and annotated using the SEED subsystems (Overbeek et al. 2014). Compatible count tables for other meta-transcriptomic or metagenomic datasets may be downloaded from the EBI metagenomics website (Mitchell et al. 2018).

The values in Fig. 3:denom=all were computed using the CLR. We can see that there is a large asymmetry in distribution, the most striking of which are the functions in BV located below the midline on the y-axis. This asymmetry is driven by the greater complexity of the BV microbial community, and the generally larger and more complex genomes in the set of bacteria found in BV (Macklaim et al. 2013). The asymmetry is composed of both presence–absence (sparsity) and large differences between groups. This can be seen with the sparsity overlay colour, where functions that contain one or more zeros are coloured in a dark grey.

Note that there appears to be two clusters of functions that are just above the y-axis midline. The ones at 4,2 are composed of sparse functions expressed at low levels in BV but that are absent or very rare in H. We also see a second group composed of functions found in common between the H and BV group that is expressed at very high relative levels, centred around 1,1.5 on the plot. These are generally housekeeping

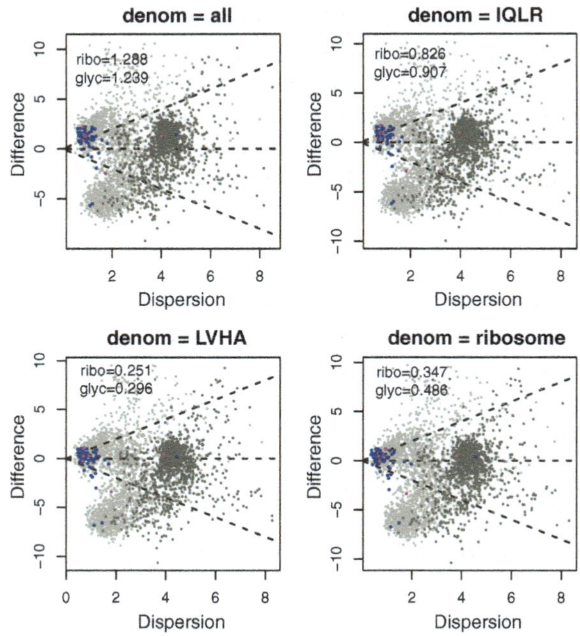

Fig. 3 Sparsity and asymmetry in a meta-transcriptome dataset of vaginal samples. The panel shows effect plots with log base 2 scaling of the data using the log-ratio values. Each point is a SEED-level 4 functional annotation category derived from the raw reads as in Macklaim and Gloor (2018) and they are coloured in light grey if no sample contains a 0 value (non-sparse), dark grey if at least one sample contains a 0 count, blue if annotated as a protein translation function or magenta if annotated as a glycolytic function. The median displacement of the functional subsets from 0 is given in the top left of the plot. The user-defined denominator was composed of functions annotated as ribosomal small or large subunit functions

functions that are central and required by all living organisms, and we can see that the two sets of housekeeping functions that are highlighted cluster here: translational protein functions in blue (rib) and glycolysis (glycolic, core sugar metabolism)-related functions in magenta. Many of the functions in these groups are often used as internal standards for comparison by qPCR as it is assumed that their expression is invariant (Scott et al. 2010), and we are attempting to minimize the displacement of these functions from a difference location of 0. The median offset of the two sets of housekeeping functions on the y-axis is given, and we can see that they are both observed to be substantially above the expected location of 0 when the CLR alone is used to determine differential relative abundance.

The other three panels in Fig. 3 show the result of applying three adjustment approaches to the asymmetric meta-RNA-seq dataset. The LVHA, IQLR, and user-defined adjustment (ribosome) methods all centred the data better than did the CLR. In the case of the user-defined functions, we used the set of translational protein functions in blue as the denominator in the denom=ribosomal panel. We can see that the bulk of the expected invariant groups are closer to the y-axis midline with two adjustments, and the offset of the translational and glycolytic functions is reduced substantially when compared to the CLR result. The LVHA adjustment brings the housekeeping functions very close to the centre of the dataset. Centring on the geometric mean of the low variance but high (relative) abundance functions provides a substantial improvement over all other methods in this dataset.

3 Discussion

Biological data derived from high-throughput sequencing is rarely ideal and exhibits many pathologies. It is currently difficult to examine complex communities using RNA-seq because of a mismatch between the assumptions of the tools and the characteristics of the data. In particular, such data can be derived from asymmetric environments, where sets of genes, operational taxonomic units, or organisms can be present or abundant in one condition and absent or rare in another. Asymmetries also often exist in metagenomic, and in vitro selection (SELEX) datasets, and the approaches described here help to recentre the data (Almeida et al. 2019; McMurrough et al. 2018; Wolfs et al. 2016). Alternatively, an asymmetry in the data can arise because of a systematic failure in experimental design, for example, through improper blocking or the presence of outlier samples. In any of these instances, the presence of an asymmetry may not be obvious.

In a biological context, it is entirely reasonable that asymmetry could occur because counts rather than simple sparsity. For example, the default gene expression condition for many genes is low-level expression, and the inclusion of a transcriptional activator could increase expression of many genes from very low expression to very high expression. In the context of 16S rRNA gene sequencing study, it is possible for samples to be dominated not only by one very abundant organism but also to contain many other taxa at low abundance. Thus, we would have a count asymmetry that is not necessarily based on sparsity, and for this reason we do not recommend the NZLR approach unless the others fail. Furthermore, sparsity is strongly affected by read depth, the same samples derived from a sequencing dataset from an Illumina NextSeq run delivering a total count of 400M reads will be substantially less sparse than those derived from an Illumina MiSeq run delivering a total count of 25M reads; however, any underlying asymmetry will be preserved.

We demonstrated that even a small number of asymmetric features can change the location of the dataset, leading to both false positive and false negative differences being identified. When the asymmetry is moderate, the IQLR correction is most appropriate. This correction makes the assumption that those features with variance between the first and third quartile of variance are a suitable proxy for the expected 'typical' variance of the data. This approach can tolerate up to at least 40% asymmetry in the data when the geometric mean of these features are used as the denominator for a log-ratio normalization. In fact, we recommend that the IQLR be used as the default when performing differential relative abundance analysis, since this normalization makes no strong assumption about the data and appears to never perform worse than the CLR normalization until failure of the approach occurs. In this case, `ALDEx2` will report an error.

More extreme asymmetry, as found in the vaginal transcriptome dataset, forces the investigator to make strong assumptions about the underlying data and use the LVHA correction, or to explicitly choose a set of housekeeping genes. The assumptions made here are thus similar to the assumptions made when performing qPCR: that there are one or more invariant features in the data, and that these will typically be relatively abundant housekeeping functions. We found that these features could usually be identified as having low variance but high relative abundance and can be used as exemplars of 'invariant' features.

4 Conclusions

We tested different methods to properly centre the data, and found that the IQLR- and LVHA-specified centring approaches were the most general purpose and thus recommended for use. All are implemented in the `ALDEx2` R package (Fernandes et al. 2014) and can be used in conjunction with the `propr` R package for compositional association (Quinn et al. 2017). We observe that not all datasets have feature sets that are compatible with the LVHA approach; in these cases, the investigator can make an even stronger assumption and choose one or more features that prior knowledge suggests would be appropriate. In any case, it must be remembered that the results of any analysis must be interpreted as *abundance relative to the chosen invariant part of the dataset*, and not as changes in absolute abundance.

5 Abbreviations

ALR: additive log-ratio transformation
BA plot: Bland–Altman plot
BV: bacterial vaginosis state
CLR: centred log-ratio transformation
H: healthy (non-BV) state
HTS: high-throughput sequencing
IQLR: the interquartile log-ratio method
LVHA: the low-variance, high-abundance method
M: million
NZLR: the non-zero log-ratio method
OTU: operational taxonomic unit
qPCR: quantitative PCR

6 Declarations

7 Availability and Requirements

The methods are included in the `ALDEx2` R package available at Bioconductor (https://www.bioconductor.org/packages/release/bioc/html/ALDEx2.html).

7.1 Ethics Approval and Consent to Participate

Not applicable

7.2 Consent for Publication

Not applicable

7.3 Availability of Data and Material

All data used in this publication are publicly available. The raw data for the yeast genome was drawn from the reference *Saccharomyces cerevisiae* genome at: https://www.ncbi.nlm.nih.gov/genome/?term=txid4932. The raw data for Fig. 3 was obtained from European Nucleotide Archive metagenomics site under the accession number PRJEB31833 using the methods in Macklaim and Gloor (2018). All R scripts for figure generation and analysis are located on a public GitHub repository at: https://github.com/JRWu/Log-Ratio-Publication.

7.4 Competing Interests

The authors declare that they have no competing interests.

7.5 Funding

This work was supported by a grant from the Natural Sciences and Engineering Research Council of Canada grant RGPIN-2015-03878 to GBG.

8 Author's Contributions

Designed the experiments GBG, JMM, JRW, BLG. Developed and applied corrections BLG, JRW, GBG, JMM. Wrote the manuscript, GBG. All authors have read and approved the manuscript.

Acknowledgements GBG first met Dr. Palwlowsky-Glahn in 2014 where she was an instructor and organizer of the CoDa summer school. At that time, our group was struggling with interpreting meta-transcriptomic datasets, and found that compositional approaches were promising. GBG thanks Vera for her everlasting patience in answering any and all questions, and for making the entry into the world of CoDa a fruitful and rewarding experience.

References

J. Aitchison, *The Statistical Analysis of Compositional Data* (Chapman & Hall, London, 1986)

A. Almeida, A.L. Mitchell, M. Boland, S.C. Forster, G.B. Gloor, A. Tarkowska, T.D. Lawley, R.D. Finn, A new genomic blueprint of the human gut microbiota. Nature **568**, 499 (2019). https://doi.org/10.1038/s41586-019-0965-1

D.G. Altman, J.M. Bland, Measurement in medicine: the analysis of method comparison studies. J. R. Stat. Soc. Ser. D (Stat.) **32**(3), 307–317 (1983), http://www.jstor.org/stable/2987937

S. Anders, D.J. McCarthy, Y. Chen, M. Okoniewski, G.K. Smyth, W. Huber, M.D. Robinson, Count-based differential expression analysis of RNA sequencing data using R and Bioconductor. Nat Protoc **8**(9), 1765–86 (2013). https://doi.org/10.1038/nprot.2013.099

P.L. Auer, R.W. Doerge, Statistical design and analysis of RNA sequencing data. Genetics **185**(2), 405–16 (2010). https://doi.org/10.1534/genetics.110.114983

C. Barceló-Vidal, J.A. Martin-Fernàndez, V. Pawlowsky-Glahn, Mathematical foundations of compositional data analysis, in *Proceedings of IAMG*, vol. 1 (Springer, 2001), pp. 1–20

Z.L. Deng, C. Gottschick, S. Bhuju, C. Masur, C. Abels, I. Wagner-Döbler, Metatranscriptome analysis of the vaginal microbiota reveals potential mechanisms for protection against metronidazole in bacterial vaginosis. mSphere **3**(3) (2018). https://doi.org/10.1128/mSphereDirect.00262-18

A.D. Fernandes, J.M. Macklaim, T.G. Linn, G. Reid, G.B. Gloor, ANOVA-like differential expression (ALDEx) analysis for mixed population RNA-seq. PLoS One **8**(7), e67019 (2013). https://doi.org/10.1371/journal.pone.0067019

A.D. Fernandes, J.N. Reid, J.M. Macklaim, T.A. McMurrough, D.R. Edgell, G.B. Gloor, Unifying the analysis of high-throughput sequencing datasets: characterizing RNA-seq, 16S rRNA gene sequencing and selective growth experiments by compositional data analysis. Microbiome **2**, 15.1–15.13 (2014). https://doi.org/10.1186/2049-2618-2-15

A.C. Frazee, A.E. Jaffe, R. Kirchner, J.T. Leek, Polyester: simulate RNA-seq reads. R package version 1.10.0 (2016)

P. Gajer, R.M. Brotman, G. Bai, J. Sakamoto, U.M.E. Schütte, X. Zhong, S.S.K. Koenig, L. Fu, Z.S. Ma, X. Zhou, Z. Abdo, L.J. Forney, J. Ravel, Temporal dynamics of the human vaginal microbiota. Sci. Transl. Med. **4**(132), 132ra52 (2012). https://doi.org/10.1126/scitranslmed.3003605

M. Gierliński, C. Cole, P. Schofield, N.J. Schurch, A. Sherstnev, V. Singh, N. Wrobel, K. Gharbi, G. Simpson, T. Owen-Hughes, M. Blaxter, G.J. Barton, Statistical models for RNA-seq data derived from a two-condition 48-replicate experiment. Bioinformatics **31**(22), 3625–3630 (2015). https://doi.org/10.1093/bioinformatics/btv425

G.B. Gloor, J.M. Macklaim, A.D. Fernandes, Displaying variation in large datasets: plotting a visual summary of effect sizes. J. Comput. Graph. Stat. **25**(3C), 971–979 (2016a). https://doi.org/10.1080/10618600.2015.1131161

G.B. Gloor, J.M. Macklaim, M. Vu, A.D. Fernandes, Compositional uncertainty should not be ignored in high-throughput sequencing data analysis. Austrian J. Stat. **45**, 73–87 (2016b). https://doi.org/10.17713/ajs.v45i4.122

G.B. Gloor, J.M. Macklaim, V. Pawlowsky-Glahn, J.J. Egozcue, Microbiome datasets are compositional: and this is not optional. Front. Microbiol. **8**, 2224 (2017). https://doi.org/10.3389/fmicb.2017.02224

R. Hummelen, A.D. Fernandes, J.M. Macklaim, R.J. Dickson, J. Changalucha, G.B. Gloor, G. Reid, Deep sequencing of the vaginal microbiota of women with HIV. PLoS One **5**(8), e12078 (2010). https://doi.org/10.1371/journal.pone.0012078

K.S. Lang, T.J. Johnson, Transcriptome modulations due to A/C2 plasmid acquisition. Plasmid **80**, 83–89 (2015). https://doi.org/10.1016/j.plasmid.2015.05.005

B. Langmead, S.L. Salzberg, Fast gapped-read alignment with Bowtie 2. Nat. Methods **9**(4), 357–359 (2012)

J. Lovén, D.A. Orlando, A.A. Sigova, C.Y. Lin, P.B. Rahl, C.B. Burge, D.L. Levens, T.I. Lee, R.A. Young, Revisiting global gene expression analysis. Cell **151**(3), 476–482 (2012). https://doi.org/10.1016/j.cell.2012.10.012

J.M. Macklaim, G.B. Gloor, From RNA-seq to biological inference: using compositional data analysis in meta-transcriptomics. Methods Mol. Biol. **1849**, 193–213 (2018). https://doi.org/10.1007/978-1-4939-8728-3_13

J.M. Macklaim, A.D. Fernandes, J.M. Di Bella, J.A. Hammond, G. Reid, G.B. Gloor, Comparative meta-RNA-seq of the vaginal microbiota and differential expression by Lactobacillus iners in health and dysbiosis. Microbiome **1**(1), 12 (2013). https://doi.org/10.1186/2049-2618-1-12

J.M. Macklaim, J.C. Clemente, R. Knight, G.B. Gloor, G. Reid, Changes in vaginal microbiota following antimicrobial and probiotic therapy. Microb. Ecol. Health Dis. **26**, 27799 (2015)

T.A. McMurrough, R.J. Dickson, S.M.F. Thibert, G.B. Gloor, D.R. Edgell, Control of catalytic efficiency by a coevolving network of catalytic and noncatalytic residues. Proc. Natl. Acad. Sci. USA **111**(23), E2376–83 (2014). https://doi.org/10.1073/pnas.1322352111

P.J. McMurdie, S. Holmes, Waste not,want not: why rarefying microbiome data is inadmissible. PLoS Comput Biol **10**(4), e1003531 (2014). https://doi.org/10.1371/journal.pcbi.1003531

T.A. McMurrough, C.M. Brown, K. Zhang, G. Hausner, M.S. Junop, G.B. Gloor, D.R. Edgell, Active site residue identity regulates cleavage preference of LAGLIDADG homing endonucleases. Nucleic Acids Res. **46**(22), 11990–12007 (2018). https://doi.org/10.1093/nar/gky976

A.L. Mitchell, M. Scheremetjew, H. Denise, S. Potter, A. Tarkowska, M. Qureshi, G.A. Salazar, S. Pesseat, M.A. Boland, F.M.I. Hunter, P. Ten Hoopen, B. Alako, C. Amid, D.J. Wilkinson, T.P. Curtis, G. Cochrane, R.D. Finn, EBI metagenomics in 2017: enriching the analysis of microbial communities, from sequence reads to assemblies. Nucleic Acids Res. **46**(D1), D726–D735 (2018). https://doi.org/10.1093/nar/gkx967

R. Overbeek, R. Olson, G.D. Pusch, G.J. Olsen, J.J. Davis, T. Disz, R.A. Edwards, S. Gerdes, B. Parrello, M. Shukla, V. Vonstein, A.R. Wattam, F. Xia, R. Stevens, The seed and the rapid annotation of microbial genomes using subsystems technology (RAST). Nucleic Acids Res. **42**(Database issue), D206–14 (2014). https://doi.org/10.1093/nar/gkt1226

J. Peng, B. Hao, L. Liu, S. Wang, B. Ma, Y. Yang, F. Xie, Y. Li, RNA-seq and microarrays analyses reveal global differential transcriptomes of Mesorhizobium huakuii 7653R between bacteroids and free-living cells. PLoS One **9**(4), e93626 (2014). https://doi.org/10.1371/journal.pone.0093626

T.P. Quinn, M.F. Richardson, D. Lovell, T.M. Crowley, propr: an R-package for identifying proportionally abundant features using compositional data analysis. Sci. Rep. **7**(1), 16252 (2017). https://doi.org/10.1038/s41598-017-16520-0

J. Ravel, P. Gajer, Z. Abdo, G.M. Schneider, S.S.K. Koenig, S.L. McCulle, S. Karlebach, R. Gorle, J. Russell, C.O. Tacket, R.M. Brotman, C.C. Davis, K. Ault, L. Peralta, L.J. Forney, Vaginal microbiome of reproductive-age women. Proc. Natl. Acad. Sci. USA **108**, 4680–4687 (2011). https://doi.org/10.1073/pnas.100611107

M. Scott, C.W. Gunderson, E.M. Mateescu, Z. Zhang, T. Hwa, Interdependence of cell growth and gene expression: origins and consequences. Science **330**(6007), 1099–1102 (2010). https://doi.org/10.1126/science.1192588

J. Rivera-Pinto, J.J. Egozcue, V. Pawlowsky-Glahn, R. Paredes, M. Noguera-Julian, M.L. Calle, Balances: a new perspective for microbiome analysis. mSystems **3**(4) (2018). https://doi.org/10.1128/mSystems.00053-18

O. Thellin, W. Zorzi, B. Lakaye, B. De Borman, B. Coumans, G. Hennen, T. Grisar, A. Igout, E. Heinen, Housekeeping genes as internal standards: use and limits. J. Biotechnol. **75**(2–3), 291–295 (1999)

J.D. Silverman, A.D. Washburne, S. Mukherjee, L.A. David, A phylogenetic transform enhances analysis of compositional microbiota data. Elife **6**, 21887 (2017). https://doi.org/10.7554/eLife.21887

J. Sun, T. Nishiyama, Shimizu,K.,Kadota,K.:TCC: an R package for comparing tag count data with robust normalization strategies. BMC Bioinformatics **14**, 219.1–219.13 (2013). https://doi.org/10.1038/10.1186/1471-2105-14-219

J. Vandesompele, K. De Preter, F. Pattyn, B. Poppe, N. Van Roy, A. De Paepe, F. Speleman, Accurate normalization of real-time quantitative RT-PCR data by geometric averaging of multiple internal control genes. Genome Biol. **3**(7), RESEARCH0034 (2002)

J.M. Wolfs, T.A. Hamilton, J.T. Lant, M. Laforet, J. Zhang, L.M. Salemi, G.B. Gloor, C. Schild-Poulter, D.R. Edgell, Biasing genome-editing events toward precise length deletions with an RNA-guided TevCas9 dual nuclease. Proc. Natl. Acad. Sci. USA (2016). https://doi.org/10.1073/pnas.1616343114

A.D. Washburne, J.D. Silverman, J.W. Leff, D.J. Bennett, J.L. Darcy, S. Mukherjee, N. Fierer, L.A. David, Phylogenetic factorization of compositional data yields lineage-level associ-ations in microbiome datasets. PeerJ **5**, e2969 (2017). https://doi.org/10.1038/10.7717/peerj.2969

Weiss, S., Zu, Z.X., Peddada, S., Amir, A., Bittinger, K., Gonzalez, A., Lozupone, C., Zaneveld, J.R., Vazquez-Baeza, Y., Birmingham, A.,, Hyde, E.R., Knight, R.: Normalization and microbial differential abundance strategies depend upon data characteristics. Microbiome **5**(1), 27 (2017). https://doi.org/10.1038/10.1186/s40168-017-0237-y

H. Zhao, C. Chen, Y. Xiong, X. Xu, R. Lan, H. Wang, X. Yao, X. Bai, X. Liu, Q. Meng, X. Zhang, H. Sun, A. Zhao, X. Bai, Y. Cheng, Q. Chen, C. Ye, J. Xu, Global transcriptional and phenotypic analyses of Escherichia coli O157:H7 strain Xuzhou21 and its pO157_Sal cured mutant. PLoS One **8**(5), e65466 (2013). https://doi.org/10.1371/journal.pone.0065466

Bayesian Balance-Regression in Microbiome Studies Using Stochastic Search

Lu Huang and Hongzhe Li

Abstract In microbiome studies, 16S rRNA sequencing is commonly used to quantify the taxonomic abundance of a microbial community. The resulting data are counts of amplicons. However, the total count is not informative because of the sampling, sample preparation, and sequencing processes. These counts are used to obtain estimates of the relative abundance of the taxa, which is compositional with a unit sum constraint. Analysis of compositional data requires special statistical treatment to account for the intrinsic dependence of the components due to this constraint. Balance, defined as the normalized log-ratio of the geometric mean of the values for the two groups of components, provides an interesting way of studying microbial community structure, where the two groups represent the beneficial and detrimental taxa, respectively. Such a balance can be used to quantify dysbiosis of the microbial community that is associated with a clinical outcome. However, identification of the outcome-associated balance is challenging. In this paper, we introduce a Bayesian balance-regression and a Markov Chain Monte Carlo (MCMC) stochastic search algorithm to identify the compositional balance that is associated with the outcome. Specifically, we propose a random walk strategy in MCMC that explores the very large space of all possible balance defined from high-dimensional compositional vector. Simulation studies suggest that the algorithm can identify the bacterial taxa that define the outcome-associated balance with a high probability. The effect of the balance on the outcome can be easily inferred from their predictive posterior distribution. We apply the proposed methods to two human microbiome studies and identify the balance of gut microbiome composition associated with body mass index and risk of inflammatory bowel disease, respectively.

L. Huang · H. Li (✉)
University of Pennsylvania, Philadelphia, PA, USA
e-mail: hongzhe@upenn.edu

L. Huang
e-mail: luhuang@pennmedicine.upenn.edu

P. Filzmoser et al. (eds.), *Advances in Compositional Data Analysis*,
https://doi.org/10.1007/978-3-030-71175-7_18

1 Introduction

Advances in sequencing technology have enabled researchers to study microbial communities without having to culture them in the laboratory. Recent large-scale microbiome studies have greatly expanded our knowledge of the role of microbiome in human health and disease. For example, dysbiosis of the gut microbiome has been linked to Crohn's disease, cancer, and cardiometabolic syndrome (Lewis et al. 2015; Halfvarson et al. 2017). Microbial community composition from a sample can be inferred from 16S rRNA sequencing, which provides data on read counts that are assigned to different taxa. However, the total count is not informative because of the sampling, sample preparation, and sequencing processes. These counts are used to obtain estimates of the relative abundance of the p taxa in the community, which is compositional with a unit sum constraint. Such data are called compositional data with p parts, which can be summarized as $\mathbf{x} = (x_1, x_2, \ldots, x_p) \in \mathbb{S}^{p-1}$, where $x_i \geq 0, i = 1, 2, \cdots, p, \sum_{i=1}^{p} x_i = 1$ and $\mathbb{S}^{p-1}$ represents $p - 1$-dimensional simplex.

Standard statistical methods cannot be applied directly to microbiome compositional data because of the dependence among the p parts (Li 2015). In addition, according to Aitchison (1982), the analysis of compositional data should satisfy two principles: *scale invariance and subcompositional coherence*. The former principle implies that the total count or sequencing read count in microbiome study should be irrelevant in statistical analysis of the microbial composition. Subcompositional coherence requires that even two microbial communities have different numbers of bacteria, and any associations found among the overlapping taxa should be not affected by other non-overlapping taxa. This principle is important, since in microbiome studies researchers rarely identify the same number of taxa in different studies.

Log-ratio analysis aims to preserve scale invariance and subcompositional coherence in compositional data analysis. Such a transformation is often performed by first choosing a component from $\mathbf{x}$ as the denominator, say x_1. Log ratios are then computed with one of the remaining components as the numerator,

$$\log \frac{x_i}{x_1} \qquad \forall i = 2, 3, \ldots, p.$$

However, the choice of the denominator is arbitrary, so this log-ratio transformation is not symmetric and may yield different results for different choices of the denominator when these log ratios are used as covariates in regression analysis. Recently, several researchers (Shi et al. 2016; Wang and Zhao 2017; Lu et al. 2018) have developed new inference approaches that alleviate the non-symmetric issue by introducing a constraint on the regression coefficients. Aitchison (1986) also proposed a symmetric log-ratio transformation, called centered log-ratio transformation and is defined as

$$\log \frac{x_i}{g(\mathbf{x})} \qquad \forall i = 1, 2, \ldots, p,$$

where $g(\mathbf{x})$ is the geometric mean of all p components. Centered log-ratio-transformed compositional data are intrinsically colinear and the resulting matrix is singular, which can lead to problems in downstream data analysis. Bates and Tibshirani (2018) proposed a penalized variable selection procedure for log ratios formed by all pairwise parts in a compositional data. The total number of log ratios is, however, very large.

All the above log-ratio transformations are developed to address the foundation of compositional data analysis that individual components have no meaning and it is their relationship to other components that has meaning. In microbiome studies, the community members may exhibit co-operative or co-exclusive relationship and the entire community maintains a homeostasis (Griffin et al. 2004; West et al. 2006). Disruptions in relative proportions of certain microbes can cause perturbed homeostasis and even pathogenesis. For example, the ratio between *Firmicutes* and *Bacteroidetes* is significantly higher in overweight and obese subjects (Castaner et al. 2018). Thus, it is critical to carry out statistical analysis based on ratios between groups of taxa, which is more informative than only considering a pair of taxa and is capable of capturing the dysbiosis of a microbial community in pathogenic status.

Log-ratio of partitions of a composition, termed balance, was first introduced in geology by Egozcue and Pawlowsky-Glahn (2005), and was recently applied to microbiome studies with meaningful discoveries (Morton et al. 2017). Balance is defined as a weighted log-ratio between two partitions of a compositional vector. Let $\mathbf{z} = \left(z_1, z_2, \ldots, z_p\right)$ be a p-dimensional indicator vector with three values, 1, -1, and 0, where $z_i = 1$ indicates that the corresponding part in the composition x_i is in the numerator of a ratio; $z_i = -1$ indicates the denominator; and $z_i = 0$ indicates the parts that are not a member of the balance. Let z_+, z_-, and z_0 be non-overlapping sets of indices of the elements in $\mathbf{z}$ whose values are 1, -1, and 0, respectively, and $z_+ \cup z_- \cup z_0 = \{1, 2, \ldots, p\}$. Given $\mathbf{z}$, the balance is defined as

$$B_z = \sqrt{\frac{m_+ m_-}{m_+ + m_-}} \log \frac{\left(\prod_{i \in z_+} x_i\right)^{1/m_+}}{\left(\prod_{i \in z_-} x_i\right)^{1/m_-}}, \tag{1}$$

where m_+, m_-, m_0 denote the number of indices in z_+, z_-, z_0, respectively. In previous applications, the value of the indicator vector $\mathbf{z}$ is chosen a priori based on a phylogenetic tree or some other biological knowledge (Morton et al. 2017; Washburne et al. 2017). However, such information may be biased or even misleading since the taxonomic positions do not always differentiate beneficial from detrimental taxa.

One important question in microbiome data analysis is to identify the vector $\mathbf{z}$ and the corresponding balance that is associated with a certain outcome variable based on the data collected. Rivera-Pinto et al. (2018) proposed a forward selection procedure to derive a balance from microbiome data. Their method adds one "best" taxon at a time into the balance based on an optimization criterion such as the mean square error and stops when a prescribed maximum number of taxa is achieved or the improvement of the optimization parameter is lower than a threshold. However, such a search only explores a small subspace of all possible balances and can lead to sub-optimal results if the optimal model is not visited in the search space. In addition, when p is large, searching for all possible 3^p balance becomes infeasible.

In this paper, we propose a Bayesian balance-regression to construct a balance and to model the association between the balance and outcome of interest using stochastic search. Unlike the search method proposed by Rivera-Pinto et al. (2018), we develop a Markov chain Monte Carlo (MCMC) stochastic search algorithm that is capable of exploring the large search space. The final structure of the balance can be inferred from the marginal posterior probability of each z_i or from the most probable configuration of $\mathbf{z}$ when the MCMC algorithm converges. The MCMC algorithm enables the posterior inference for the regression coefficients conditional on the structure of balance at each iteration of the search algorithm. Since the latent vector $\mathbf{z}$ and therefore the balance is unknown, our MCMC algorithm is very different from the standard Bayesian regression (George and McCulloch 1993) where the covariates are known. In our case, besides sampling the model parameters, we have to sample the latent indicator vector $\mathbf{z}$. In Bayesian variable selection, the covariates are observed and one samples the covariate-inclusion indicators to select the covariate that are associated with the response.

The remaining of the paper is outlined as follows. Section 2 introduces the Bayesian balance-regression for a continuous outcome and the stochastic search procedure. Extension of the methods to a binary outcome is also briefly discussed. Section 3 presents application of the methods for two microbiome studies. Section 4 presents simulation results to evaluate the proposed methods. We conclude the paper with a discussion in Sect. 5.

2 Bayesian Balance-Regression for Compositional Data

2.1 *A Bayesian Balance-Regression for Continuous Outcomes*

Consider a microbiome study with n $i.i.d.$ samples from a population. Let $\mathbf{y} = (y_1, \cdots, y_n)$ be the vector of a continuous outcome of length n, $\mathbf{X} = \{x_{ij}, i = 1, \cdots, n, j = 1 \cdots, p\}$ be the compositional matrix with dimension $n \times p$, where p is the number of taxa observed in the samples and x_{ij} is the abundance of the jth taxon in the ith sample estimated from the sequencing counts. Let $\mathbf{z}$ represent the indicator vector that determines the structure of the balance and the resulting balance B_z as defined in Eq. (1). We can calculate the balance using (1) for the entire sample when $\mathbf{z}$ is known. For a given balance configuration $\mathbf{z}$, let $\mathbf{B}_z$ be the vector of the balance values for the n individuals.

Our main interest is to identify a $\mathbf{z}$ indicator vector and the corresponding balance that is strongly associated with the outcome. Let $\mathbf{B_{1z}} = (\mathbf{1}, \mathbf{B}_z)$ be the $n \times 2$ matrix of columns of $\mathbf{1}$ and the vector $\mathbf{B}_z$. For a given balance configuration $\mathbf{z}$, we define the following Bayesian balance-regression:

$$\mathbf{y}|\mathbf{z} = \mathbf{B}_{1\mathbf{z}}\boldsymbol{\beta}_z + \boldsymbol{\epsilon}_z, \qquad \boldsymbol{\epsilon}_z \sim MVN(0, \sigma^2\mathbf{I}), \tag{2}$$

where $\boldsymbol{\beta}_z = (\beta_{0z}, \beta_{1z})^T$ is the regression coefficients that depend on $\mathbf{z}$. To complete the model, we assume the following prior distributions for all the unknown parameters, namely, $\boldsymbol{\beta}_z, \sigma^2, \mathbf{z}$, in order to perform a posterior inference. Similar to Bayesian regression, we assume a normal-inverse-gamma distribution for $\boldsymbol{\beta}_z$ and σ^2 as discussed in George and McCulloch (1993). More specifically, we assume that $\boldsymbol{\beta}_z = (\beta_{0z}, \beta_{1z})^T$ are jointly normal with mean $\mathbf{b_0} = (b_{00}, b_{01})^T$ and covariance $\sigma^2\mathbf{V}$, where $\mathbf{V}$ is a diagonal matrix with entries h and c. We also assume that σ^2 follows an inverse Gamma distribution with parameters $(v/2, v\lambda/2)$. The advantage to use a normal-inverse-gamma prior is that we can obtain a closed form of the marginal distribution $\mathbf{y}|\mathbf{z}$ after integrating out $\beta_\mathbf{z}$ and σ^2. We can therefore sample $\mathbf{z}$ with shorter chains and faster convergence, as compared with other choices of the prior distributions.

The balance configuration vector of $\mathbf{z}$ determines the search space and the prior on $\mathbf{z}$ represents our belief of the true structure of a balance. A straightforward choice is an independent multinomial distribution for each component of $\mathbf{z}$,

$$f(\mathbf{z}) = w_1^{m_+} w_2^{m_-} (1 - w_1 - w_2)^{p-m_+-m_-}, \tag{3}$$

where w_1 and w_2 are the expected number of variables in z_+ and z_- group, respectively. They affect the sizes of taxa with positive and negative effects in the posterior inference of the balance. A non-informative choice is $w_1 = w_2 = 1/3$, resulting in $f(\mathbf{z}) = (1/3)^p$. On the other hand, small values of w_1 and w_2 assume sparse structure of the balance. This prior has implicit assumptions that the sizes of the sets z_+ and z_- are approximately equal. It is possible to relax the independence specification and use a dependent prior among elements of $\mathbf{z}$. If p is large, we suggest to use the sparse prior for the balance configuration $\mathbf{z}$.

2.2 Posterior Inference Based on MCMC

Based on the Bayesian balance-regression model and the respective prior distributions of the parameters, the posterior distribution $\mathbf{z}|\mathbf{y}$ is proportional to the joint distribution of $(\mathbf{y}, \mathbf{z})$:

$$f(\mathbf{z}|\mathbf{y}) \propto |\mathbf{V}^*|^{1/2} \left(v\lambda + \mathbf{b_0}^T\mathbf{V}^{-1}\mathbf{b_0} + \mathbf{y}^T\mathbf{y} - \mathbf{u}^T{\mathbf{V}^*}^{-1}\mathbf{u}\right)^{-(v+n)/2} f(\mathbf{z}), \tag{4}$$

where $\mathbf{V}^* = (\mathbf{V}^{-1} + {\mathbf{B}_{1\mathbf{z}}}^T\mathbf{B}_{1\mathbf{z}})^{-1}$ and $\mathbf{u} = \mathbf{V}^*(\mathbf{V}^{-1}\mathbf{b_0} + \mathbf{B}_{1\mathbf{z}}^\mathrm{T}\mathbf{y})$ and $\boldsymbol{\beta}_z$ is integrated out.

We develop the following MCMC algorithm to explore high-dimensional sample space of $\mathbf{z}$ and to sample $\mathbf{z}$ from its posterior distribution given in (4). For stochastic search, three possible value changes can occur during the random exploration: (i) between 0 and 1; (ii) between 0 and -1; and (iii) between 1 and -1. Modifications between two values have three alterations: value conversions in either direction and value switch. We set equal probabilities among all possible modifications, but such probabilities can be adjusted to speed up the exploration of the sampling space. The convergence and mixing of the MCMC algorithm can be evaluated using different chains with different starting points and by examining the cumulative moving average for each component of $\mathbf{z}$. After convergence, the posterior inclusion probabilities for the sets z_+, z_-, z_0 are simply taken as the marginal proportions of $1, -1, 0$ in $\mathbf{z}$. A threshold can be set to make inference on which set of particular taxa should belong to the balance and the resulting balance definition.

Conditioning on $\mathbf{z}$, we can obtain the predictive posterior distribution of the regression coefficients $\boldsymbol{\beta}_z$. From model (2) and the normal prior distribution of $\boldsymbol{\beta}_z$, one can show that $\boldsymbol{\beta}_z|\mathbf{y}, \mathbf{z}$ has a multivariate t-density with degrees of freedom $\lambda + n$, location parameter $\mathbf{u}$, and shape matrix

$$\tilde{V} = \frac{\nu\lambda + \mathbf{b_0}^T\mathbf{V}^{-1}\mathbf{b}_0 + \mathbf{y}^T\mathbf{y} - \mathbf{u}^T\mathbf{V}^{*-1}\mathbf{u}}{\nu + n}\mathbf{V}^*. \tag{5}$$

During each iteration of MCMC, we sample $\boldsymbol{\beta}_z$ from its posterior distribution and perform posterior inference for $\boldsymbol{\beta}_z$ based on these sampling points. Finally, at the convergence, due to randomness of $\mathbf{z}$, the posterior samples of $\boldsymbol{\beta}_z$ are averaged over those $\mathbf{z}$, which gives the mean of the predictive posterior distribution of $\boldsymbol{\beta}_z$ based on model average. Algorithm 1 presents an outline of the final MCMC algorithm for fitting the balance-regression.

Algorithm 1 MCMC algorithm for the Bayesian balance-regression analysis

1 Sample $\mathbf{z}|\mathbf{y}^*$ based on its posterior probability (4) and stochastic search.
2 Sample $\boldsymbol{\beta}_z|\mathbf{y}, \mathbf{z}$ from a t-distribution with location parameter $\mathbf{u}$ and shape matrix given in (5).
3 Obtain posterior probability of $Pr(z_i|\mathbf{y}, \mathbf{X})$ for balance inference and predictive posterior distribution of $\boldsymbol{\beta}_z$.

Since the target distribution of $\mathbf{z}$ has two modes when the prior of β_{1z} has a mean value of 0. To make it identifiable, if the estimate of β_{1z} is less than 0, we flip the signs of $\mathbf{z}$ and return a positive estimate of β_{1z}. This effectively constrains β_{1z} to be positive in order to make $\mathbf{z}$ identifiable.

2.3 A Balance-Regression for Binary Outcomes

For binary outcome, let y_i be the outcome for subject i that takes only two possible values 0 and 1. Similar to model Sect. 2.1, we define $\mathbf{B_{1z}} = (\mathbf{1}, \mathbf{B}_z)$ to be the $n \times 2$ matrix with columns of $\mathbf{1}$ and the vector $\mathbf{B}_z$, where $\mathbf{z}$ is the indicator vector that defines the balance. The following model with a hidden variable y_i^* corresponds to the generalized linear model with a probit link (Holmes and Held 2006),

$$\begin{aligned} y_i &= \begin{cases} 1 & \text{if } y_i^* \geq 0 \\ 0 & \text{otherwise} \end{cases} \\ y_i^*|\mathbf{z} &= \beta_{0z} + \beta_{1z} B_{zi} + \epsilon_{iz} \\ \epsilon_{iz} &\sim N(0, 1). \end{aligned} \tag{6}$$

The prior distribution for $\boldsymbol{\beta}_z = (\beta_{0z}, \beta_{1z})$ is chosen as the multivariate normal with mean $\boldsymbol{b_0}$ and diagonal covariance matrix Σ, whose entries are v_0, v_1. After integrating out $\boldsymbol{\beta}$, the conditional distribution of $(\mathbf{y}, \mathbf{y}^*)|\mathbf{z}$ is

$$\frac{|\mathbf{\Sigma^*}|^{1/2}}{(2\pi)^{n/2}} \exp\left\{-\frac{1}{2}(\mathbf{y^*} - \mathbf{B}_{1z}\boldsymbol{b_0})^T \mathbf{\Sigma^*}(\mathbf{y^*} - \mathbf{B}_{1z}\boldsymbol{b_0})\right\} \times g(\mathbf{y}|\mathbf{y}^*) \tag{7}$$

with $\mathbf{\Sigma^*} = (\mathbf{I} + \mathbf{B_{1z}} \Sigma \mathbf{B_{1z}}^T)^{-1}$. Therefore, $\mathbf{y}^*|\mathbf{y}, \mathbf{z}$ simply follows a truncated normal distribution with truncation at zero. The posterior distribution of $\mathbf{z}$ is proportional to multiplying Eq. (7) by the prior of $\mathbf{z}$ given in Eq. (3). Due to unobserved $\mathbf{y}^*$, we need to sample both $\mathbf{z}$ and $\mathbf{y}^*$ in our algorithm. The information from the observed data $\mathbf{y}$ is used to infer $\mathbf{y}^*$, which in turn determines $\mathbf{z}$. As a result, we propose the following MCMC sampling steps (see Algorithm 2).

Algorithm 2 MCMC algorithm for the Bayesian probit balance-regression analysis

1 Sample $\mathbf{y}^*|\mathbf{y}, \mathbf{z}$ from truncated normal whose original form is Eq. (7) and truncation occurred at 0.
2 Sample $\mathbf{z}|\mathbf{y}^*$. This step is equivalent to the sampling in Sect. 2.1
3 Return to the first step until convergence.

The posterior inference of the balance structure is the same as in Sect. 2.1. Here, we propose to carry out the predictive posterior inference for $\boldsymbol{\beta}_z$ conditional on $\mathbf{y}^*, \mathbf{z}$, as the information contained in $\mathbf{y}$ is passed onto $\mathbf{y}^*$. In each MCMC iteration, given $\mathbf{z}$ and $\mathbf{y}^*$, the posterior distribution of $\boldsymbol{\beta}_z$ is a multivariate normal with mean vector $\tilde{\boldsymbol{\mu}}$ and variance covariance matrix $\tilde{\mathbf{\Sigma}}$ given as

$$\tilde{\mu} = \tilde{\Sigma}(\mathbf{B}_{1z}^T y^* + \Sigma^{-1} b_0),$$
$$\tilde{\Sigma} = (\Sigma^{-1} + \mathbf{B}_{1z}^T \mathbf{B}_{1z})^{-1}.$$

From this, we can obtain the mean of the predictive posterior distribution of $\boldsymbol{\beta}_z$ given $\mathbf{z}$.

3 Applications to Real Datasets

In this section, we apply the proposed Bayesian balance-regression to two microbiome datasets in order to understand how gut microbiome may be associated with body mass index (BMI) and the risk of inflammatory bowel disease (IBD).

3.1 Compositional Balance Associated with BMI

Our first example focuses on understanding how gut microbiome is associated with BMI. Several previous studies have reported that microbiome compositions were associated with obesity (Shi et al. 2016; Castaner et al. 2018). Wu et al. (2011) conducted a gut microbiome study using a 16*S* target sequencing to associate diet and microbial composition. In this cross-sectional study, 98 healthy volunteers were recruited and their fecal samples were collected. Sample DNA were amplified according to the V1-V2 region in the 16*S* ribosomal DNA. The Operational Taxonomic Units (OTUs) were grouped into 87 taxa, which had at least 0.2% relative abundance in at least one sample and appeared in more than 10% samples.

We aggregated the count data into genera and re-normalized the counts into relative abundance. A total of 45 genera were identified in the data. We are interested in identifying the key bacterial genera and the corresponding balance that can explain the variability of BMI. We ran the MCMC search using the proposed Bayesian balance-regression in order to identify a balance that may be associated with BMI. Specifically, the proposed MCMC approach was carried out with the hyperparameters for prior distribution of $\boldsymbol{\beta}$ chosen as $b_{00} = 0, \quad b_{01} = 0, \quad c = 10^6, \quad h = 10^6$, representing uninformative prior information. We ran five chains with different starting values to assess the convergence. Burn-in was taken to be 10^4 and the MCMC iterations were ran 5×10^4 times. We performed the analysis using sparse prior where $w_1 = w_2 = 0.1$. From the posterior probability plot (Fig. 1), *Acidaminococcus, Allisonella* had a posterior probability of 0.91 and 0.63 being in the z_+ set and *Alistipes, Clostridium* had a posterior probability of 0.64 and 0.79 being in the z_- set.

Using 0.5 as the threshold, the Bayesian balance-regression identified a balance defined by the normalized log-ratio between *Acidaminococcus* and *Allisonella*, and *Alistipes* and *Clostridium*, where *Acidaminococcus, Allisonella* and *Clostridium*

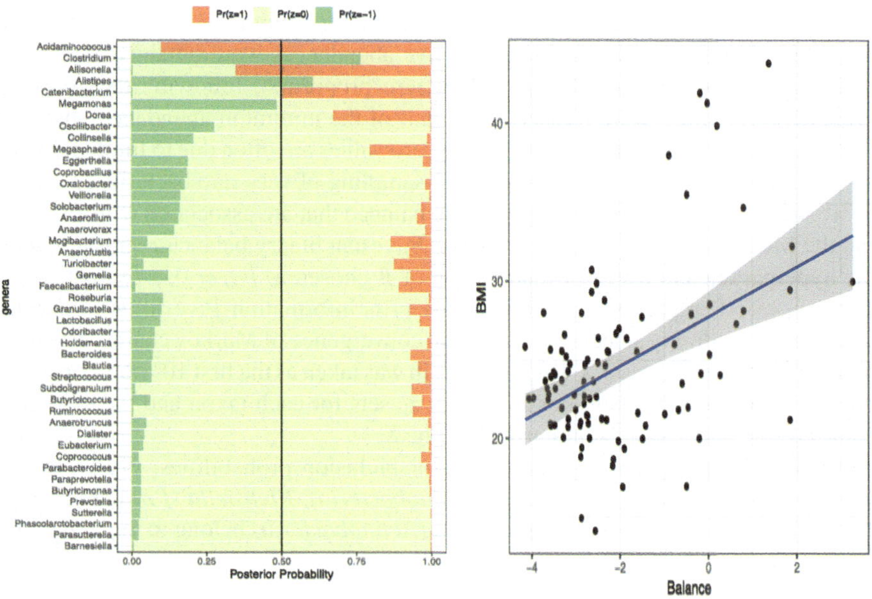

(a) Posterior probability of z_i (b) Estimated balance vs BMI

Fig. 1 Analysis of BMI and gut microbiome dataset. **a** Posterior probability of z_i for each genus i, where red, yellow, and green colors represent $Pr(z_i = 1|\mathbf{y}, \mathbf{X})$, $Pr(z_i = 0|\mathbf{y}, \mathbf{X})$, and $Pr(z_i = -1|\mathbf{y}, \mathbf{X})$, respectively. The vertical line is 0.5 cutoff line. The genera are ordered by $Pr(z_i = 1|\mathbf{y}, \mathbf{X}) + Pr(z_i = -1|\mathbf{y}, \mathbf{X})$. **b** BMI versus the estimated balance and the corresponding linear fit with 95% confidence band

belong to the *Firmicutes* phylum and *Alistipes* belongs to the *Bacteroidetes* phylum. Figure 1 shows the scatterplot of the observed BMI and value of the derived balance, showing an approximately linear relationship. The posterior mean of the regression coefficient is 2.14 with an 95% credible interval of (1.12, 3.28). These results indicate that obesity is associated with gut microbiome that include more *Acidaminococcus* or *Allisonella* genera, or fewer *Alistipes Clostridium* genera, and the balance of these four gut bacteria that is significantly associated with BMI.

3.2 *Compositional Balance Associated with Pediatric Crohn's Disease*

Crohn's disease is one major category of IBD. Dysbiosis of gut microbiome has been shown to be associated with increased risk of IBD (Lewis et al. 2015). In the study published in Lewis et al. (2015), researchers at the University of Pennsylvania collected stool samples from 85 pediatric patients with Crohn's disease and 26 healthy volunteers and performed shortgun metagenomic sequencing prior to any treatment or intervention. After quality control, 31 relatively common bacterial genera were observed in these samples and their relative abundances were obtained. Since the definition of the balance does not allow zero proportion, following the common practice, zero values were replaced with half of the minimum abundance observed (Kurtz et al. 2015; Cao et al. 2018). These zero values are often due to limited library sizes during sequencing that lead to under-sampling of very rare bacteria.

To identify the bacterial genera and the balance that are associated with IBD, we ran the MCMC search using the proposed Bayesian binary balance-regression with the hyperparameters for prior distribution of $\boldsymbol{\beta}$ chosen as $b_{00} = 0, \quad b_{01} = 0, \nu_0 = 10^6, \quad \nu_1 = 10^6$, representing uninformative prior information. Five different starting points were randomly selected to check the convergence of Markov chains. The total number of iterations was 2×10^5 and burn-in was taken as the first 10^5 iterations. The posterior inclusion probabilities in z_+, z_-, z_0 sets for each taxon and the predicted values of the balance are summarized in Fig. 2.

Using 0.5 as the threshold for posterior inclusion probabilities, Fig. 2a shows that *Veillonella (Firmicutes), Rothia (Actinobacteria), Klebsiella (Proteobacteria), Escherichia (Proteobacteria), Eggerthella (Actinobacteria)* belong to the z_+ set and

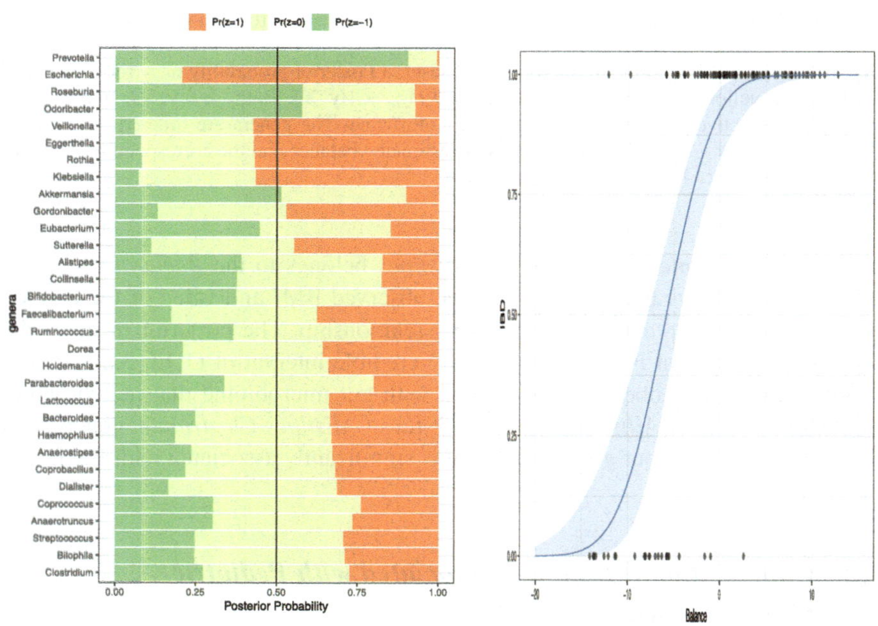

(a) Posterior probability of z_i (b) Estimated balance vs IBD status

Fig. 2 Analysis of IBD gut microbiome data. **a** Posterior probability of z_i for each genus i, where red, yellow, and green colors represent $Pr(z_i = 1|\mathbf{y}, \mathbf{X})$, $Pr(z_i = 0|\mathbf{y}, \mathbf{X})$, and $Pr(z_i = -1|\mathbf{y}, \mathbf{X})$, respectively. The vertical line is 0.5 cutoff line. The genera are ordered by $Pr(z_i = 1|\mathbf{y}, \mathbf{X}) + Pr(z_i = -1|\mathbf{y}, \mathbf{X})$. **b** Estimated balance versus IBD status and the corresponding logistic fit and pointwise 95% confidence band

Roseburia (Firmicutes), Prevotella (Bacteroidetes), Odoribacter (Bacteroidetes), Akkermansia (Verrucomicrobia) belong to the z_- set. These bacteria together define the balance of the microbial community that was associated with IBD. Our results are consistent with other reports that microbiota dysbiosis in IBD patients was characterized by an increase in *Proteobacteria* and a decrease in *Firmicutes* (Lewis et al. 2015). However, our results also suggested other bacteria especially *Actinobacteria, Bacteroidetes,* and *Verrucomicrobia* were also involved in the dysbiosis in the IBD patients, where *Actinobacteria* seems to increase the risk of IBD and *Bacteroidetes* and *Verrucomicrobia* seems to be protective of IBD.

Figure 2b shows the relationship between the outcome IBD and estimated the balance using 0.5 as the cutoff for the balance configuration probabilities. The fitted curve from generalized linear models with probit link function is also plotted for comparison, indicating that the identified balance is indeed associated with IBD. The posterior mean of the regression coefficient is 0.26 with an 95% credible interval of (0.16, 0.38).

4 Simulation Evaluations

We present simulation studies in this section to evaluate how well the proposed MCMC sampling can recover the balance indicator vector $\mathbf{z}$ and how well we can estimate the effect of the balance on the outcome. We only present results on linear balance-regression. Similar results were observed for the probit balance-regression.

4.1 Data Generation

We evaluate the performance of our proposed method using simulations. For each simulation, we set $n = 100, \quad p = 30, \quad \beta_0 = 1, \quad \sigma^2 = 1$, and β_1 taking three values 1, 0.5, and 0.1 representing strong, moderate, and small effect of the balance on the outcome. The hyperparameters for prior distribution of $\boldsymbol{\beta}$ are chosen as $b_{00} = 0, \quad b_{01} = 0, \quad c = 10^6, \quad h = 10^6$, representing uninformative prior information. For binary outcome, we set $c = 10^3, \quad h = 10^3$ due to the restrictions in Gibbs sampling of the latent vector $\boldsymbol{y}^*$ and set w_1 and w_2 to be $1/3$, representing each component being equally likely to be in the z_+, z_-, z_0 sets. The Inverse Gamma prior distribution $IG(v/2, \lambda/2)$ corresponds to a likelihood of σ^2 that comes from v independent $N(0, \lambda)$ observations. As a result, v can be treated as the prior sample size and λ as a prior estimate for σ^2. In practice, λ can be chosen as the sample variance of the outcome variable in model (2).

To generate the compositions, we first generate n independent vector of dimension p from a multivariate normal with zero mean, variance of 10, and equicorrelation of 0.2. Taxon count matrix is generated by exponentiating the above matrix and then taking the greatest integer smaller or equal to the respective numbers. Zero counts correspond to rare taxa and are replaced with 0.5. Finally, the count matrix is normalized per row and to obtain the compositional matrix $\mathbf{X}$. Across all simulations, the first 3 taxa are in the z_+ set of a balance and the next 3 taxa are in z_- set. Outcomes are generated according to model (2) and (6), respectively. The initial values for $\mathbf{z}$ are chosen randomly. The performance of mixing of the Markov chains is evaluated with several starting points.

4.2 Simulation Results

Figure 3 summarizes the posterior probabilities $Pr(z_i = 1|\mathbf{y}, \mathbf{X})$, $Pr(z_i = -1|\mathbf{y}, \mathbf{X})$, and $Pr(z_i = 0|\mathbf{y}, \mathbf{X})$ after 10^5 iteration with 10^4 burn-ins for all 30 taxa across 100 simulations. When the balance effect is strong, the posterior marginal inclusion probabilities for the first 3 taxa are above 0.75 across all simulations for the compositional taxa in the z_+ set and near 0 for those in the z_- group. Similarly, compositional taxa 4 to 6 can be correctly identified in the z_- set with high probabilities. When the balance effect becomes weaker, the elements in those two sets can sometimes be recognized with high probability. However, the posterior inclusion probabilities are pulled toward the prior, since the data has a low signal-to-noise ratio as compared to those with $\beta_{1z} = 1$. The remaining taxa, which are not part of the balance index, can sometimes enter into the z_+ or z_- set.

Figure 4 shows the posterior mean of β_{1z} over 100 simulations. These boxplots show that they are centered around the true values. All of the 95% empirical credible intervals derived from the sampling points contain the true value of β_{1z}, indicating that the MCMC algorithm provides valid estimate of the balance effect.

5 Discussion

In this paper, we have proposed a Bayesian balance-regression and an MCMC stochastic search algorithm to identify the microbial signature, termed balance, which is related to the outcome of interest, including both continuous and dichotomous outcomes. The balance signature reflects the mutualistic or repellent interactions among microbes. Balance can be regarded as the extension of the log-ratio between two groups of microbes that are highly indicative of healthy/pathological traits. The proposed method automatically selects the most relevant reference taxa in the denominator of the balance, unlike other approaches that adopt additive log-ratio or central log-ratio transformation, and at the same time also preserves the scale invariance as well as subcompositional coherence properties. Results from the analysis of the association between gut microbiome and BMI and IBD demonstrated the model and methods can indeed identify the balance feature that is closely associated with these outcomes. The balance can be used as an index for dysbiosis of the microbial community.

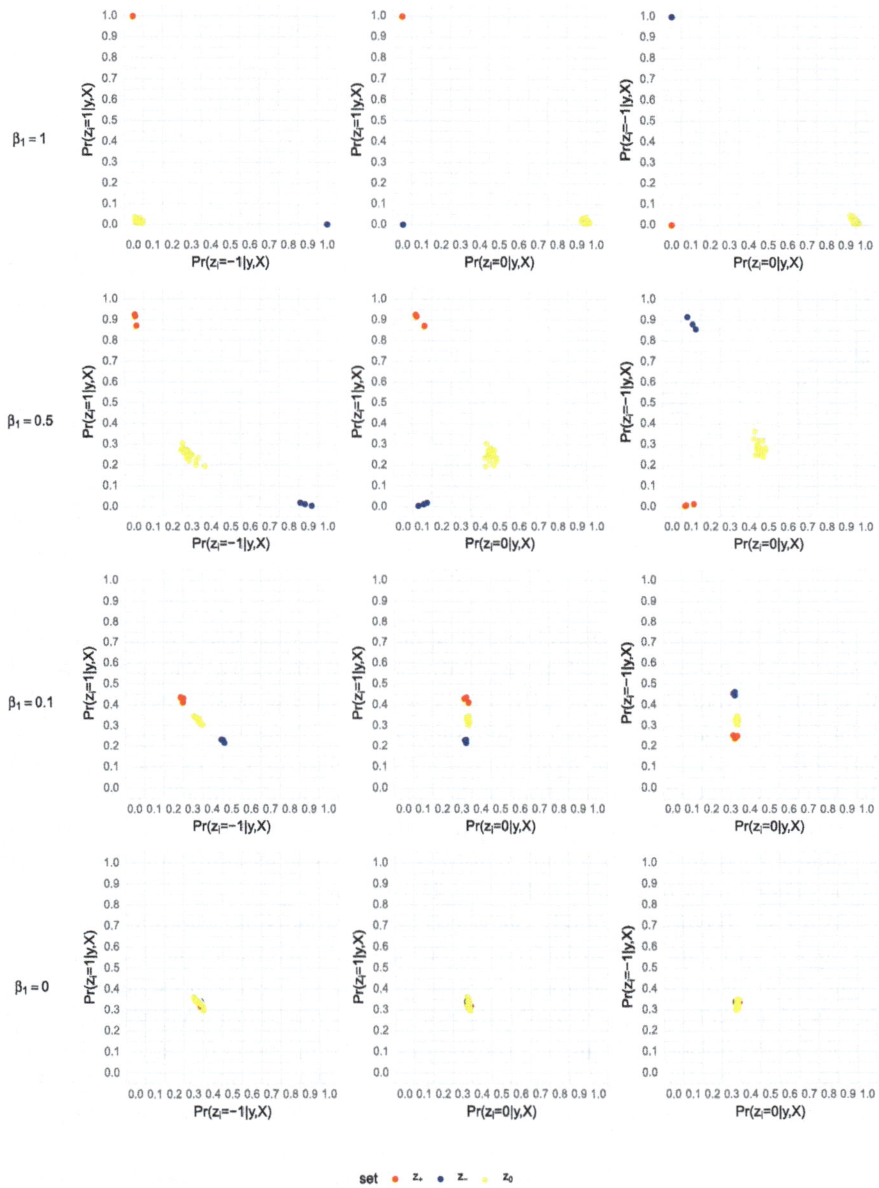

Fig. 3 Simulation results for balance-regression model with different balance effect β_1. For each of the four β_1 values, pairwise posterior probabilities (average over 100 simulations) are plotted for taxa in the positive set z_+ (red), negative set z_- (blue), and null set z_0 (yellow)

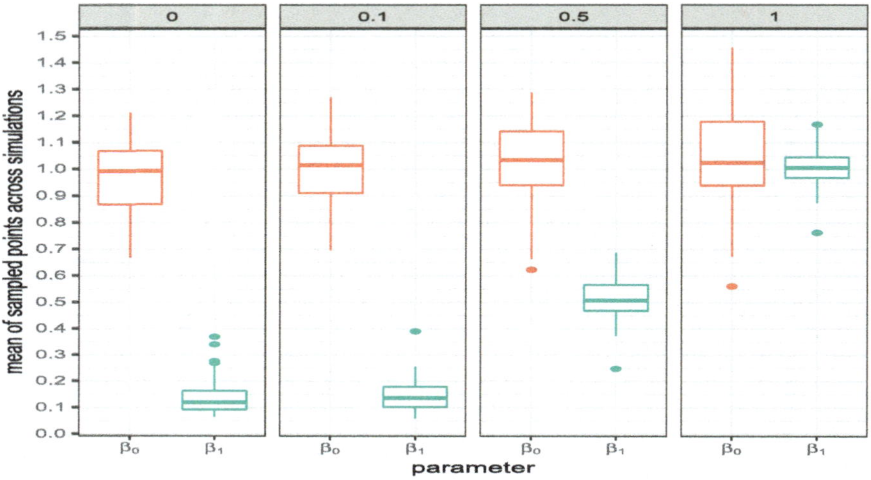

Fig. 4 Simulation results for balance-regression model with different balance effect β_1. Box plots of the posterior mean of β_{0z} and β_{1z} over 100 simulations

The MCMC stochastic search in our method allows feasible exploration of the moderate to high-dimensional space (3^p-dimensional for p taxa) whereas the search algorithm (Rivera-Pinto et al. 2018) only searches for a small space of possible balance since it only allows adding one component to the positive part or the negative part of the previous sets during the iterations. Our algorithm allows the elements in the previous sets being removed out of the balance set or switching from the positive set to the negative set or vice versa. Another advantage of using a Bayesian approach to balance-regression is that it automatically provides an assessment of uncertainty of the balance identified and its effect on the outcome.

As for any Bayesian methods, one needs to choose sensible values for the hyper-parameters in the prior distributions. In our analysis, we used non-informative or vague priors. However, in studies where sample size is small or noise level is high, the selection result can be affected by the prior parameters. During MCMC sampling, we obtain the posterior inclusion probabilities of all three sets for each of the taxa. We use the posterior modes to choose the final set of $\mathbf{z}$ and the corresponding balance. Since the conditional posterior distribution of $\boldsymbol{\beta}_z$ has a closed form, statistical inference about $\boldsymbol{\beta}_z$ can be made based on the final estimate of $\mathbf{z}$. In this case, our inference of $\boldsymbol{\beta}_z$ can be interpreted as conditioning on $\mathbf{z}$. However, in real data analysis, the posterior probabilities in the three sets might only differ slightly in magnitudes, which makes posterior inference of $\mathbf{z}$ difficult. In this case, the posterior inference about $\boldsymbol{\beta}_z$ can be performed on the sampling points obtained during each iteration of MCMC in the framework of model average.

Our proposed balance-regression model and the MCMC method are very different from the Bayesian variable selection in linear models (Mitchell and Beauchamp 1988; Holmes and Held 2006). One key difference is that the covariate (i.e., the balance)

in our model (2) or (6) is unknown and has to be identified. Our models have only one covariate, but it is not observed. In contrast, the standard Bayesian linear model selection aims to identify the relevant covariates from a set of fixed covariates. The proposed method aims to discover a particular form from a subset of taxa that act as one explanatory entity for the outcome. In this regard, we treat the regression coefficient as a nuisance parameter and integrate it out of the target distribution. Whereas in Bayesian variable selection, whether a covariate is in the final model depends on the associated regression coefficient being zero or not.

Identifying the balance that is associated with an clinical outcome is a challenging problem. Although we have demonstrated through simulations that the MCMC algorithm is effective in identifying the structure of the balance, our results are based on simulating the data from the proposed models. It would be interesting to further investigate the methods by simulating data from alternative models such as the compositional linear or generalized linear models studied in Shi et al. (2016); Lu et al. (2018) for microbiome studies. In fact, if the balance indicator vector $\mathbf{z}$ is known, the balance-regression model proposed in this paper is a special case of the model in Shi et al. (2016), where the positive coefficients of all relevant taxa are assumed to be the same and the negative coefficients of all relevant taxa are also assumed to be the same.

Acknowledgements This research was supported in part by the NIH grants GM123056 and GM129781. We thank the editors for inviting us to contribute this paper and Professor Vera Pawlowsky-Glahn for many discussions on compositional data analysis. We also thank two reviewers for their careful reading of the paper and many suggestions.

References

J. Aitchison, The statistical analysis of compositional data. J. R. Stat. Soc. Ser. B (Stat. Methodol.) **44**, 139–177 (1982)

J. Aitchison, *The Statistical Analysis of Compositional Data* (Chapman & Hall, New York, 1986)

S. Bates, R. Tibshirani, Log-ratio lasso: scalable, sparse estimation for log-ratio models. Biometrics (2018)

Y. Cao, W. Lin, H. Li, Two-sample tests of high-dimensional means for compositional data. Biometrika **105**, 115–132 (2018)

O. Castaner, A. Goday, Y.O. Park, S.H. Lee, F. Magkos, S.A. Shiow, H. Schroder, The gut microbiome profile in obesity: a systematic review. Int. J. Endocrinol. **2018**, 4095789 (2018)

J.J. Egozcue, V. Pawlowsky-Glahn, Groups of parts and their balances in compositional data analysis. Math. Geol. **7**, 795–828 (2005)

E.I. George, R.E. McCulloch, Variable selection via Gibbs sampling. J. Am. Stat. Assoc. **88**, 881–889 (1993)

A.S. Griffin, S.A. West, A. Buckling, Cooperation and competition in pathogenic bacteria. Nature **430**, 1024–1027 (2004)

J. Halfvarson, C.J. Brislawn, R. Lamendella, Y. Vazquez-Baeza, W.A. Walters, L.M. Bramer, M. D'Amato, F. Bonfiglio, D. McDonald, A. Gonzalez, E.E. McClure, M.F. Dunklebarger, R. Knight, J.K. Jansson, Dynamics of the human gut microbiome in inflammatory bowel disease. Nat. Microbiol. **2**, 17004 (2017)

C.C. Holmes, L. Held, Bayesian auxiliary variable models for binary and multinomial regression. Bayesian Anal. 145–168 (2006)

Z.D. Kurtz, C.L. Muller, E.R. Miraldi, D.R. Littman, M.J. Blaser, R.A. Bonneau, Sparse and compositionally robust inference of microbial ecological networks. PLoS Comput. Biol. **11**, e1004226 (2015)

J.D. Lewis, E.C. Chen, R.N. Baldassano, A.R. Otley, A.M. Griffiths, D. Lee, K. Bittinger, A. Bailey, E.S. Friedman, C. Hoffmann, L. Albenberg, R. Sinha, C. Compher, E. Gilroy, L. Nessel, A. Grant, C. Chehoud, H. Li, G.D. Wu, F.D. Bushman, Inflammation, antibiotics, and diet as environmental stressors of the gut microbiome in pediatric Crohn's disease. Cell Host Microbe **18**, 489–500 (2015)

H. Li, Microbiome, metagenomics, and high-dimensional compositional data analysis. Annu. Rev. Stat. Appl. **2**(1), 73–94 (2015)

J. Lu, P. Shi, H. Li, Generalized linear models with linear constraints for microbiome compositional data. Biometrics (2018)

T.J. Mitchell, J.J. Beauchamp, Bayesian variable selection in linear regression. J. Am. Stat. Assoc. **83**(404), 1023–1032 (1988)

J.T. Morton, J. Sanders, R.A. Quinn, D. McDonald, A. Gonzales, Y. Vazquez-Baeza, J.A. Navas-Molina, S.J. Song, J.L. Metcalf, E.R. Hyde, M. Lladser, P.C. Dorrestein, R. Knight, Balance trees reveal microbial niche differentiation. mSystems **2**, e00162–16 (2017)

J. Rivera-Pinto, J.J. Egozcue, V. Pawlowsky-Glahn, R. Paredes, M. Noguera-Julian, M.L. Calle, Balances: a new perspective for microbiome analysis. mSystems **3**, 4 (2018)

P. Shi, A. Zhang, H. Li, Regression analysis for microbiome compositional data. Ann. Appl. Stat. **10**, 1019–1040 (2016)

T. Wang, H. Zhao, Structured subcomposition selection in regression and its application to microbiome data analysis. Ann. Appl. Stat. **11**, 771–791 (2017)

J.D. Washburne, A.D. Silverman, J.W. Leff, D.J. Bennett, J.L. Darcy, S. Mukherjee, N. Fierer, L.A. David, Phylogenetic factorization of compositional data yields lineage-level associations in microbiome datasets. PeerJ **5**, e2969 (2017)

S.A. West, A.S. Griffin, A. Gardner, S.P. Diggle, Social evolution theory for microorganisms. Nat. Rev. Microbiol. **4**, 597–607 (2006)

G. Wu, J. Chen, C. Hoffmann, K. Bittinger, Y.Y. Chen, S.A. Keilbaugh, M. Bewtra, D. Knights, W.A. Walters, R. Knight, R. Sinha, E. Gilroy, K. Gupta, R. Baldassano, L. Nessel, H. Li, F.D. Bushman, J.D. Lewis, Linking long-term dietary patterns with gut microbial enterotypes. Science **334**, 105–108 (2011)

Compositional Data Analysis in Physical Activity and Health Research. Looking for the Right Balance

Duncan E. McGregor, Philippa M. Dall, Javier Palarea-Albaladejo, and Sebastien F. M. Chastin

Abstract The time spent in different types of sleep, sedentary behaviour and physical activity during a day represents compositional data. Previous use of compositional analysis of daily activity data through orthonormal log-ratio coordinates in the form of pivot coordinates of four components (sleep, sedentary behaviour, light physical activity, moderate to vigorous physical activity) has facilitated critical comparison with and the resolution of contradicting reports of the influence of daily behaviour on health which had arisen from the use of standard statistical analysis. For example, there has been a long-standing debate about whether the effects of time spent sedentary are independent of the time spent in moderate to vigorous activity. However, daily activity may be usefully broken down into more categories, and other compositional analysis approaches may be relevant. Here we demonstrate the use of log-ratio balances to evaluate the association between daily activity broken into six components (sleep, short periods of sedentary behaviour, long periods of sedentary behaviour, standing, slow walking and fast walking) from a nationally representative sample of adults with health markers in obesity, lipidemia and glycemia. The balanced approach to compositional analysis of daily behaviour patterns offers a way to focus the statistical assessment and modelling on variables, representing meaningful contrasts of interest between behaviours. This is fully consistent with the view that actual effects on health are associated with exchanges between some classes of daily behaviours.

D. E. McGregor · P. M. Dall · S. F. M. Chastin (✉)
School of Health and Life Sciences, Glasgow Caledonian University, Glasgow, UK
e-mail: Sebastien.Chastin@gcu.ac.uk

D. E. McGregor
e-mail: Duncan.Mcgregor@gcu.ac.uk

P. M. Dall
e-mail: Philippa.dall@gcu.ac.uk

D. E. McGregor · J. Palarea-Albaladejo
Biomathematics and Statistics Scotland, Edinburgh, UK
e-mail: javier.palarea@bioss.ac.uk

S. F. M. Chastin
Department of Movement and Sport Sciences, Ghent University, Ghent, Belgium

P. Filzmoser et al. (eds.), *Advances in Compositional Data Analysis*,
https://doi.org/10.1007/978-3-030-71175-7_19

1 Introduction

1.1 Background

Engaging in daily physical activity is associated with a very wide range of physical and mental health benefits (Warburton et al. 2010). Physical activity is a potent public health asset for the prevention of non-communicable and mental health diseases as it reduces significantly the risk of chronic diseases such as cardiovascular diseases, diabetes, stroke and some cancers, as well as the risk of premature mortality. In addition, physical activity plays a vital role in healthy ageing delaying the onset of frailty and disablement in later life (World Health Organisation 2002). Recognising the importance of physical activity to lower the burden of diseases, most countries in the world have issued guidelines and recommendations for physical activity (World Health Organisation 2010; Department of Health 2011; Piercy et al. 2018). Despite all this, physical activity levels around the globe remain very low (Guthold et al. 2018). So much so that the renowned journal The Lancet called inactivity a "pandemic" (Kohl et al. 2012) causing 5.3 million deaths annually (Lee et al. 2012) and costing in excess of $50 billion worldwide.

Physical activity research takes place in many health and medicine-related fields of science such as sports medicine, public health, health promotion, occupational health, rehabilitation sciences and behavioural and social sciences. Other domains such as environmental research, urbanism, transport and education are also interested in physical activity, as it is increasingly understood that people's inactivity is a systemic problem (Condello et al. 2016; Rutter et al. 2019). Physical activity research is concerned with studying the relationship between physical activity and health, the epidemiology of physical activity, the determinants of inactivity, and developing and evaluating interventions to promote physical activity amongst diverse populations. In this context, the synergetic approach to examine physical activity data that compositional data analysis enables is playing a crucial role in understanding the impact on health of 24-h physical behaviour and providing evidence for the new 24-h recommendations appearing around the world (Chaput et al. 2014; Tremblay et al. 2016). There is consequently a growing interest in using compositional data analysis to advance the analysis of physical activity data, to improve the accuracy of epidemiological estimates and to understand physical activity behaviour in relation to other daily physical behaviours (SMPB 2020).

1.2 Short History of Physical Activity Research

Hippocrates around 400 BC and other ancient physicians before him as far back as 2600 BC believed in physical activity as a source of health. However, modern medicine historically tended to ignore or point towards the potential negative effect of exercise. Indeed, until relatively recently, complete bed rest was the main recommendation for many ailments such as cardiovascular disease or back pain which are nowadays successfully prevented and managed with physical activity programmes. The birth of physical activity research is often attributed to Jerry Morris (Blair et al. 2010), who produced the first rigorous epidemiological studies on physical activity in 1953 (Morris et al. 1953). Since then physical activity was conceptualised as an absolute quantity, with activity and inactivity being two sides of the same coin. Interestingly, this is de-facto a two-part composition. However, in the mid-2000s new research started to appear showing the adverse effects on the health of time spent sedentary (Hu 2003; Owen et al. 2010). Sedentary behaviour is defined as any waking behaviour in a seated or reclining position with low energy expenditure (below 1.5 times the normal metabolic rate) (Tremblay et al. 2017). This launched a controversy within the field. Some researchers thought the effects associated with too much sedentary time were just due to people not being active, but other researchers claimed they could prove that effects of sedentary behaviour were in fact independent of physical activity, and thus represented a separate concept to the mere absence of physical activity. In 2015–16 two meta-analyses, which in health sciences are considered the highest level of evidence, were published in two leading medical journals. These came to diametrically opposite conclusions (Biswas et al. 2015; Ekelund et al. 2016) regarding the effect of sedentary behaviour, leaving this controversy seemingly unresolvable (Stamatakis et al. 2018). However, since then several groups of researchers have tried to find different approaches to the problem other than considering it within a traditional epidemiology framework of linear interaction between two seemingly independent variables. Part of the issue lies in how physical activity is conceptualised and considered with respect to the rest of the behaviours humans engage in during the day. Indeed, as the day consists of finite 24 h, daily exercise necessarily displaces other behaviours which might be sedentary behaviour but which might also be sleep or light physical activity (Buman et al. 2014). We recognised that daily time use is by nature compositional data and proposed that adopting a compositional approach to analysis could provide a unifying framework (Chastin et al. 2015b). It is ironic that it took so long to make this connection, since one of the examples John Aitchison worked on in his seminal monograph published in 1986 was "A statistician's time budget" (Aitchison 1986).

1.3 Compositional Analysis and Physical Activity Research, Present and Future

The use of compositional analysis in physical activity research is steadily growing (Carson et al. 2016; Foley et al. 2018; Dumuid et al. 2018; Winkler et al. 2018; McGregor et al. 2018; Gupta et al. 2018; del Pozo Cruz et al. 2020; McGregor et al. 2020). Our article (Chastin et al. 2015b) remains one of the top 10% most cited papers in PLoS One and it is in the top ten of the most cited articles about research combining sleep and physical activity (Memon et al. 2020). Health science relies heavily on univariate and bivariate inferential statistics methods but, unlike what it might appear initially, these are not exempt from the effects of the compositionality of the data (Filzmoser et al. 2009, 2010). For example, the negative bias issue of the commonly used Pearson's correlations when computed between parts of a composition has been widely reported (see e.g., Pawlowsky-Glahn et al. 2015). However, it is somehow a first hurdle for physical activity researchers that compositional methods have been more focused on multivariate exploratory data analysis which is not so frequently used in the field, probably motivated by the needs from early application areas such as geochemistry. Studies in physical activity using a compositional approach have been dominated by linear regression analysis of cross-sectional surveys (Chastin et al. 2015b; Carson et al. 2016; Dumuid et al. 2018). This has developed to look at prospective studies (McGregor et al. 2020) and at the outcome of randomised trials aimed at changing daily behaviour (Winkler et al. 2018). More modelling techniques to provide inference in longitudinal studies looking at long-term trends and evaluate interventions are in development. To date, most of the research published in physical activity has used the particular form of orthonormal log-ratio coordinates called pivot coordinates (Filzmoser et al. 2018), where the dominance of each one of the behaviours relative to an average of the remaining ones is isolated at a time with statistical inference conducted on each in an iterative process. The day has been also commonly partitioned into four parts: sleep, sedentary behaviour, light physical activity (such as walking or house chores) and moderate to vigorous activity (such as walking fast and exercising). The pivot coordinate approach has been very useful to the physical activity community. It provides the ability to look at the effect of one part of the daily behaviour composition, although it is important not to forget that this is always relative to the average of all the others, and it eases the introduction to compositional thinking as the format of the results is close to what the physical activity research community is used to.

With the introduction of body worn sensor technology, our ability to measure more precisely the various daily physical behaviours and their context is rapidly increasing. This means that we will be able to investigate daily 24-h behaviour in more detail and create compositions made of many more parts where multivariate methods will have a more prominent role. We will soon be able to look at a composition based on sleep quality, on where activities take place, with whom or what type of activity is performed. For example, we could have a daily composition made of light sleep, heavy sleep, sitting in front of a computer, sitting in front of a TV, sitting away from screens, standing, light activity indoor, light activity outdoor, commuting, occupational moderate to vigorous activity, leisure time activity, etc. The ability to understand what might be the right balance of time or energy spent across a wide range of different behaviours is probably one of the next most important questions to answer in physical activity research. Balances between behaviours could conceivably be different for different health outcomes and could lead to a different way of expressing guidelines. For example, public health recommendations could be given as "exercise x minutes for every hour you spend sitting at work". In this chapter, we look at the understanding of trade-offs between daily activity behaviours and their implications for health using the compositional approach based on the definition of meaningful balances between subsets of behaviours (Pawlowsky-Glahn et al. 2015), in contrast to looking at individual predominance relative to all the others.

2 Materials and Methods

2.1 Cohort Data and Physical Activity Measurement

Cohort

In this work, we used data from the Australian Diabetes, Obesity and Lifestyle Study (AusDiab). Full details of this cohort study can be found on the Baker IDI website (https://www.baker.edu.au/ausdiab/). Briefly, the AusDiab study is the largest Australian longitudinal population-based study examining the natural history of diabetes, pre-diabetes, heart disease and kidney disease. A baseline wave of measurement was conducted in 1999–2000 with follow up measurement taken in 2005 and 2012.

Daily Activity Measurement

In the 2012 wave, a subsample of 782 participants was recruited to measure their daily pattern of activity objectively (Healy et al. 2015). Activity was measured using a thigh worn activity monitor (activPAL3™) (Healy et al. 2015). Participants wore the sensor for seven days continually and n = 731 provided valid activity data. The data were processed using the methods described in (Winkler et al. 2016) to identify periods of sleep and awake time. During waking time, the activPAL classifies activity as sedentary behaviour (sitting/lying), standing or walking. Sedentary periods were then further sub-divided into periods of short sedentary behaviour (short SB) if the period of sedentary behaviour was <30 min and long sedentary behaviour (long SB) for periods of sedentary behaviour ≥ 30 min. Walking periods were also sub-divided into time spent in slow walking and fast walking based on a mean cadence threshold for the period. This threshold was set at 3 MET (metabolic equivalent) which corresponds to a cadence of 100 steps per minutes and delineates the difference in intensity between light and moderate to vigorous activity. This provided a six parts of the daily composition: sleep, short SB, long SB, standing, slow walk and fast walk. This composition was averaged over the seven days for each participant to obtain an estimate of the habitual daily composition of time spent in these behaviours. No zeros were found in any of the composition parts.

Health Outcomes and Covariates

In addition to daily activity measurement, blood samples, anthropomorphic data, demographic and behavioural data were also collected (Healy et al. 2015). Blood samples were processed to extract levels of triglycerides, high-density lipoprotein (HDL), low density lipoproteine (LDL), fasting blood glucose and HbA1c.

Participants' BMI was calculated from the anthropomorphic data. Health outcomes consisted of triglycerides, HDL, LDL, fasting glucose, HbA1c and BMI. Covariates were age, sex (male/female), use of blood pressure medication (yes/no), use of diabetes medication (yes/no), current smoker (yes/no), history of depression (yes/no/missing), dietary fat intake (%E) and dietary fibre intake (g/day).

2.2 *Statistical Analysis*

Log-Ratio Coordinates and Balances

Compositional data analysis (CoDA) is a methodology that explicitly considers the statistical properties of relative data by relying on three basic principles. The first and foremost is scale invariance—vectors with proportional positive parts represent the same composition and form what is known as an equivalence class. In our context, this means that changing the units in any direction between, e.g. hours, minutes or percentages should not change the relative relationships between the behaviours nor affect results and scientific conclusions. The second is that inferences from subcompositions, i.e. smaller compositions formed from subsets of behaviours (e.g. considering behaviours whilst awake), must not contradict inferences from the full day composition involving the same behaviours. The third principle states that the order of the parts of the composition must not affect the inferences. Whilst the initial work on CoDA explicitly assumed in the definition of a composition that the parts sum (are closed) to a constant total, it is nowadays well established that this closed form is only a particular representation of the data in a simplex, and equivalent non-closed compositions carry the same relative information (Barcelo-Vidal and Martín-Fernández 2016).

The use of orthonormal log-ratio coordinates allows us to consistently project the information contained in composition into the ordinary real space and facilitate statistical analysis. Importantly, the orthonormality property guarantees that distances and other measures of differences between compositions are preserved by their log-ratio counterparts. In some cases, the log-ratio representation can be regarded as a pre-processing step, so that results in log-ratio scale are transferred back by inverse transformation to be expressed in terms of the original composition (e.g. when performing clustering analysis). In this case, the family of orthonormal log-ratios used is irrelevant as they are all orthogonal rotations of each other, and hence, will provide the same results when back transformed. However, it is in many cases practically meaningful to discuss the results in terms of the log-ratio coordinates on their own, particularly in statistical modelling. In this work, we focus on the family of log-ratios known as balances (Egozcue and Pawlowsky-Glahn 2005), which produces contrasts of the relative weight of one part or group of parts with respect to another part or group of parts of the composition, with each group summarised by their respective geometric means. Formally, given a composition of D parts $\mathbf{x} = (x_1, \ldots x_D)$, a collection of $D-1$ balances b_i, with $i = 1, \ldots, D-1$, can be constructed by a sequential binary partitioning of the composition into two mutually exclusive subsets of parts, with the general expression being

$$b_i = \sqrt{\frac{r_i s_i}{r_i + s_i}} \ln\left[\frac{\left(\prod_{h=1}^{r_i} x_{i_h}^{+}\right)^{1/r_i}}{\left(\prod_{l=1}^{s_i} x_{i_l}^{-}\right)^{1/s_i}}\right], \quad (1)$$

where $x_{i_h}^{+}$ and $x_{i_l}^{-}$ refer to the subsets of r_i and s_i parts going, respectively, into the + (numerator) and − (denominator) groups. The multiplicative factor preceding the log-ratio term is a normalisation factor required to ensure comparability. This procedure offers the practitioner flexibility to define interpretable balances according to their scientific questions of interest, based on domain-specific knowledge, to be the subject of the statistical analysis and testing. In our context of application, a balance provides a measure of the relative time allocation between two groups of behaviours, e.g. active versus non-active behaviours, with a value of zero implying that average time allocations are comparable, and otherwise, the sign of the balance indicating the predominance of one subset of behaviours against the other. However, determining meaningful balances can become challenging when no relevant prior information is available or, particularly, as the number of parts of the composition increases. Thus, some data-driven strategies have been recently proposed to select optimal balances depending on the objective of the analysis (Martín-Fernández et al. 2018; Rivera-Pinto et al. 2018).

Log-Ratio Balanced Representation of Daily Activity Behaviours

The breakdown of daily behaviours in this data comprises sleep, short bouts of sedentary behaviour, long bouts of sedentary behaviour, stationary standing, slow walking and fast walking, which will be denoted as the 6-part composition (Sleep, sSB, lSB, Stand, sWalk, fWalk) where necessary to ease referencing. Table 1 outlines the sequential binary partition (SBP) used.

Table 1 Sequential binary partition (SBP) for the construction of normalised log-ratio balances ($b_i, i = 1, \ldots, 5$) between behaviours. Symbols + and – indicate whether behaviour is allocated to, respectively, numerator or denominator of balance b_i. Blanks indicate behaviour not involved. Letters r and s refer to number of behaviours into each group respectively. SB refers to sedentary behaviour

SBP	b_1	b_2	b_3	b_4	b_5
Sleep	–	+			
Short SB	–	–	–	+	
Long SB	–	–	–	–	
Stand	–	–	+		
Slow walk	+				–
Fast walk	+				+
r	2	1	1	1	1
s	4	3	2	1	1

This SBP determines a meaningful hierarchy of balances to assess common questions in physical activity research. Concretely, b_1 is a contrast between active and stationary behaviours, b_2 compares sleep to other forms of stationary behaviour, b_3 compares standing time to sitting time, b_4 is about time spent seated in short bouts versus time spent seated in long bouts, and b_5 compares walking times according to intensity (fast and slow). The individual expressions of the balances are as follows:

$$b_1 = \sqrt{\frac{4}{3}}\ln\left[\frac{(swalk \cdot fwalk)^{\frac{1}{2}}}{(Sleep \cdot sSB \cdot lSB \cdot Stand)^{\frac{1}{4}}}\right],$$
$$b_2 = \sqrt{\frac{3}{4}}\ln\left[\frac{Sleep}{(sSB \cdot lSB \cdot Stand)^{\frac{1}{3}}}\right], b_3 = \sqrt{\frac{2}{3}}\ln\left[\frac{Stand}{(sSB \cdot lSB)^{\frac{1}{2}}}\right],$$
$$b_4 = \sqrt{\frac{1}{2}}\ln\left[\frac{sSB}{lSB}\right] \text{and } b_5 = \sqrt{\frac{1}{2}}\ln\left[\frac{fwalk}{swalk}\right]. \tag{2}$$

Given these definitions, the aim is to shed light on the following questions: (1) is replacing stationary behaviour with physical activity associated with health benefits? (2) does the association between health outcomes and sleep differ from that of stationary behaviour arising whilst awake? (3) is there a qualitative difference between different forms of waking day stationary behaviour arising from posture (i.e. seated vs standing)? (4) Does the association between periods of sedentary behaviour and health outcomes differ depending on whether the periods are short or long? (5) Are different intensities of physical activity associated with different health benefits?

Linear Regression Analysis on Balances

The data set consists of three groups of variables measured on $n = 731$ individuals: daily time-use behaviours ($\mathbf{X}_{n\times D}$ matrix), health outcomes associated with obesity, lipidemia and glycemia health markers ($\mathbf{Y}_{n\times k}$ matrix), and a number of adjusting covariates ($\mathbf{C}_{n\times m}$ matrix), commonly used in epidemiology, including age, sex, and blood pressure tablets, cotinine, depression score, diabetes tablets, fat intake and fibre intake; with D, k and m being 6, 6 and 8 respectively in this case study. We investigate the influence of $\mathbf{X}$ on $\mathbf{Y}$ whilst adjusting for the covariates $\mathbf{C}$ within a multivariate linear regression analysis. For this, $\mathbf{X}$ is transformed into the matrix of log-ratio balances ($\mathbf{B}_{n\times(D-1)}$ matrix) resulting from Eq. 2. The column vector of outcomes $\mathbf{y}_k$ for the kth health outcome, $k = 1, \ldots, 6$, was then modelled as

$$\mathbf{y}_k = \mathbf{B} \cdot \beta_{k1} + \mathbf{C} \cdot \beta_{k2} + \varepsilon_k, \varepsilon_k \sim N(\mathbf{0}, \sigma_k^2\mathbf{I}) \tag{3}$$

where β_{k1} and β_{k2} are the regression coefficient vectors associated with the behaviour balances ($\mathbf{B}$) and the covariates ($\mathbf{C}$) respectively, and ε_k is the random error term assumed to be normally distributed with mean 0 and variance σ_k^2 for each outcome k. The model was fitted by the least squares method and variable selection was applied by ordinary backward stepwise elimination based on the t-test using a significance threshold of 0.2. The overall significance of the models was assessed using the ordinarily F-statistic test. The statistical significance of the individual regression coefficients was assessed by the usual Wald t-statistic at the 0.05 and 0.1 significance levels. Deviations from the assumptions of normality and homoscedasticity of the model residuals were assessed graphically using residual versus fitted and normal quantile–quantile plots (not shown). These were deemed within acceptable limits overall.

Predicting Changes in Health Outcomes

In order to further examine the practical implications of the results from the regression modelling, predicted changes in a health outcome, dy, were computed from changes dx in the composition $\mathbf{x}$ corresponding to 10, 20 and 30 min reallocations of time between each possible combination of two behaviours x_i and x_j, for $i, j = 1, \ldots, D$ and $i \neq j$. The composition $\mathbf{x}^0$ used as reference was the compositional mean $g(\mathbf{X}) = C(g_1, \ldots, g_D)$, where $g_j = (\prod_{i=1}^{n} x_{ij})^{1/n}$ for $j = 1, \ldots, D$ is the geometric mean of the jth column of $\mathbf{X}$. The closure operator C normalises the resulting vector of geometric means to be expressed in minutes. Hence, the new compositions were generically obtained as

$$\mathbf{x} = \left(\ldots, x_i^0 - dx, \ldots, x_j^0 + dx, \ldots\right), \tag{4}$$

where the times allocated to the behaviours other than x_i and x_j were not altered. The corresponding change in the outcome was obtained as

$$dy = \sum_{h=1}^{D-1} \beta_h \left(b_h - b_h^0\right), \tag{5}$$

with b_h and b_h^0 being the elements of the row vectors $\mathbf{b}$ and $\mathbf{b}_0$ of log-ratio balance coordinates corresponding to $\mathbf{x}$ and $\mathbf{x}^0$. Moreover, standard error for the expected value of dy can be obtained as

$$SE_{dy} = \sqrt{\left(\mathbf{b} - \mathbf{b}^0\right)\mathbf{V}\left(\mathbf{b} - \mathbf{b}^0\right)^T}, \tag{6}$$

where $\mathbf{V} = \sigma^2\left(\mathbf{B}^T\mathbf{B}\right)^{-1}$ and σ^2 is the residual variance from the regression model. Following ordinary results for confidence intervals of linear combinations of regression coefficients, a 95% confidence interval for the predicted changes in the outcome can be obtained as

$$dy \pm t_{n-D-m,0.975} \cdot SE_{dy} \tag{7}$$

where $t_{n-D-m,0.975}$ is the 97.5th percentiles of the Student's t distribution with $n - D - m$ degrees of freedom and SE denotes standard error. For relatively large samples this can be approximated by $dy \pm z_{0.975} \cdot SE_{dy}$, with $z_{0.975}$ being the 97.5th percentiles of the standard normal distribution.

All the data analyses and statistical modelling were conducted on the R system for statistical computing v3.6.

3 Results and Discussion

Descriptive statistics of the sample and covariates are given in Table 2. Details of the sample daily activity composition are given in Table 3. The information about daily physical activity behaviours in the studied population was represented in terms of log-ratio balances between them. This was motivated by the interest in appreciating the potential association of some functional differences in behaviour with improvement in health markers of obesity, lipidemia and glycemia. Table 4 shows a summary of the implications of the results of the regression analysis organised according to health markers and behavioural trade-offs represented by each balance. In most cases, the obtained results align with expected directions of health improvement, but some results raise questions opening potential new interpretations and insights. The use of balances allowed a different question to be asked for each balance, and the results

Table 2 Summary of covariates for sum sample of 2012 wave of the Australian diabetes, obesity and lifestyle study (AusDiab)

Covariate	Category	Count
Age group	35–44 years	71
	45–54 years	204
	55–64 years	250
	65–74 years	148
	≥75 years	58
Sex	Male	319
	Female	412
Blood pressure medication	Yes	75
	No	656
Smoker	Yes	147
	No	584
History of depression	Yes	279
	No	430
	Missing	22
Diabetes medication	Yes	36
	No	695
Covariate	**Mean**	**Standard deviation**
Dietary fibre (g)	21.24	5.62
Dietary fat (g)	34.86	5.16

Table 3 Summary statistics for 24-h time-use composition for sub sample of 2012 wave of the Australian diabetes, obesity and lifestyle study (AusDiab)

	Sleep (h)	Short sitting (h)	Long sitting (h)	Standing (h)	Short walking (h)	Long walking (h)
Compositional centre						
	8.54	4.81	3.88	4.83	0.76	1.18
Variation matrix						
Sleep	0	0.13	0.29	0.19	0.24	0.26
Short sitting	0.13	0	0.30	0.20	0.21	0.20
Long sitting	0.29	0.30	0	0.50	0.57	0.54
Standing	0.19	0.20	0.50	0	0.12	0.20
Short walking	0.24	0.21	0.57	0.12	0	0.10
Long walking	0.26	0.20	0.54	0.20	0.10	0

can be interpreted in terms of these balances to address each question individually. Additionally, the results can be interpreted from the point of view of groupings of outcomes associated with markers of health in terms of which balances are associated with them to address more general trends.

3.1 Discussion by Balance

The balance b_1 explores whether replacing stationary behaviour with physical activity was associated with health benefits. The balance is statistically significantly associated with three health outcomes (BMI, LDL cholesterol and triglycerides; $p < 0.05$), and more marginally associated with one (HDL cholesterol; $p < 0.1$). The direction of the association is as expected for BMI, with increased time being active relative to inactive helping to reduce obesity, and for two of the markers of lipidemia, with increased active versus inactive balance associated with increased HDL cholesterol and decreased triglycerides. However, increased b_1 is also unexpectedly associated with increased LDL cholesterol. It is not clear why this is the case. But given the association in the unexpected direction between the balance of sleep to stationary behaviour with HDL cholesterol, it is possible that sleep may be producing some tension between benefit and detriment compared to activity. The balance between activity and inactivity is not associated with any indicators of glycemia.

The balance b_2 explores whether the association between health outcomes and sleep differs from that of stationary behaviour (encompassing standing and sitting) whilst awake. The balance is highly associated with lipidemia, but in the opposite direction to expected. There are several potential reasons for this. The balance compares total sleep duration against other forms of inactivity. However, when it comes to healthy sleep, it is not clear whether longer always equals better, the quality of the sleep comes into this, and that is not reflected in a pure measure of duration

Table 4 Summary of associations between daily behaviour balances and health markers from multivariate regression analysis. Large plus or minus symbols indicates statistical significance at the default level ($p < 0.05$). Small plus or minus indicates marginal statistical non-significant ($p < 0.1$). (+) indicates that the coefficient is positive, and thus the marker is higher when either the numerator behaviour value is increased or the denominator decreased (i.e. if active is bigger than inactive). (–) indicates the coefficient is negative. The arrows indicate the direction for the health marker that corresponds to better health. Green (lighter) background means the association is in the expected direction, red (darker) background indicates it is not in the expected direction. Blanks indicate no statistical significance

Health marker	Outcome measure	Direction of improvement	Active versus inactive	Sleep versus stationary	Standing versus sitting	Short sitting versus long sitting	Fast walking versus slow walking
			b_1	b_2	b_3	b_4	b_5
Obesity	BMI	↓	-			-	
Lipidemia	HDL cholesterol	↑	+		+		+
	LDL cholesterol	↓	+	+	-		
	Triglycerides	↓	-	+	-	+	
Glycemia	Fasting blood glucose	↓			-		
	Hb1A1c	↓					

(Watson et al. 2015). This could potentially be addressed by sub-dividing sleep time into more categories based on sleep quality such as time spent in high quality sleep and low quality sleep, but data was not available. Additionally, there may be confounding between measurement of sleep and sedentary behaviour (the main component, by duration, of the stationary element of the balance). The activPAL accelerometer uses thigh inclination to determine when a person is in a sitting, reclining or lying position. There is thus a difficulty for a monitor worn for 24 h in distinguishing between sleep and sedentary behaviour. A common method of doing so is to introduce subjective measures and to ask individuals when they went to sleep and when they woke up. Apart from usual issues with subjective measurement, this is necessarily an estimate as no-one can really time when they went to sleep (as opposed to went to bed) and time diaries usually ask one or the other question. Some automatic methods are used, but they rely on identifying a single long period of sedentary behaviour (Lyden et al. 2016; Winkler et al. 2016). So, whilst they probably identify the period when the person was asleep, they will also include periods of being in bed before sleep and after waking, and thus accurate identification of the ends is unlikely, with some sedentary behaviour being identified as sleeping.

The balance b_3 explores whether there is a qualitative difference between different forms of waking day stationary behaviour arising from posture (i.e. sitting vs. standing). This balance is associated with most indicators of health. In all four cases, this was highly significant ($p < 0.01$) and in the expected direction, indicating that replacing sitting by standing time has a beneficial association with all three indicators of lipidemia. This balance is also the only one to have any association with outcome measures of glycemia, indicating that time standing relative to sitting is favourably association with lower fasting blood glucose. However, this balance of standing to sitting does not result in a significant association with BMI on its own. Instead, reducing obesity appears more generally associated with the trade-off between active and inactive behaviours as represented by the balance b_1.

The balance b_4 explores whether the association between periods of sedentary behaviour and health outcomes differ depending on whether the periods are short or long. This is equivocal. Increased shorter sitting over longer sitting is significantly associated with a reduction in BMI, which is the expected direction. However, the balance is also significantly associated with higher triglycerides, which is in the opposite direction to that expected. This seems to contradict the common view in observational epidemiology that shorter periods of sitting are preferable (Chastin et al. 2015a). National guidelines often recommend avoiding long periods of sitting. One possible explanation for this result is the fact that a threshold of 30 min was used in this study to differentiate between long and short periods of sitting. Thirty minutes is commonly used in intervention and behavioural measurement research, because it is a value that chimes with how people actually sit. However, many experimental studies look at the effect of breaking sitting every 20 min. More work needs to be done in this regard, but the effect of using a different threshold may change the effect on outcome measures.

Finally, the balance b_5 explores whether different intensities of physical activity are associated with different health benefits. The balance of fast against slow walking is only significantly associated with HDL cholesterol and in the expected direction. There is also debate as to the threshold used to distinguish between fast and slow walking, but the one used here is comparable to what is usually regarded to distinguish between moderate to vigorous physical activity (MVPA) and light intensity physical activity (LIPA). It is perhaps surprising that this balance is not associated with more outcomes, as the general consensus is that higher intensity of activity is more beneficial to health. This is evident, however, when looking at the changes in health outcome induced by time reallocations between behaviours. For instance, Fig. 1 shows the predicted changes in HDL cholesterol level from time reallocations between all pairs of behaviours. For each receiving behaviour (Fig. 1a–f), three points $\pm 95\%$ confidence intervals are displayed corresponding to change in health outcome from 10, 20 and 30 min reallocations (from left to right) from each giving behaviour (indicated on the x-axis). This figure illustrates that reallocating a fixed

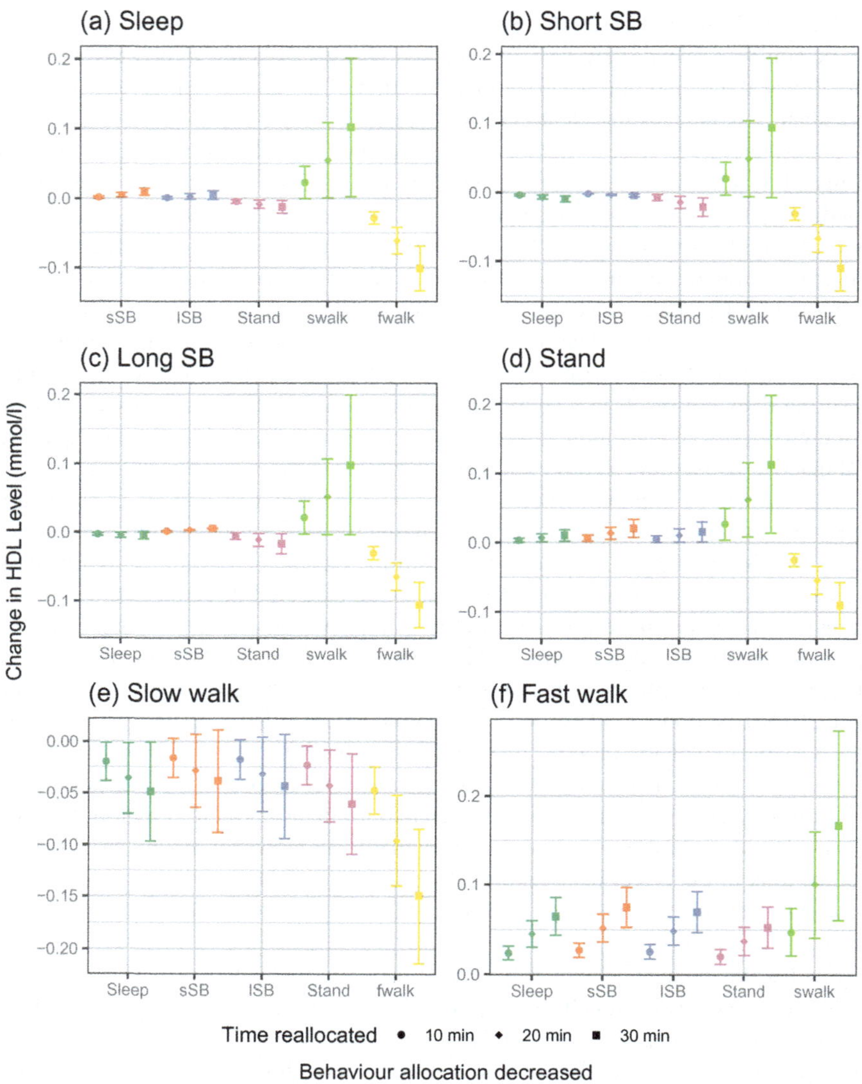

Fig. 1 Changes in HDL level predicted associated with 10, 20 and 30 min reallocations of time from the behaviours indicated on the x-axis to **a** sleep, **b** short bouts of sedentary behaviour, **c** long bouts of sedentary behaviour, **d** standing time, **e** slow walking and **f** fast walking (95% confidence intervals are presented)

amount of time from one walking speed to another result in an association with a greater effect size on HDL cholesterol than reallocating time between standing and sitting. Analogous analysis of time reallocations could be conducted for each of other health outcomes (not shown).

3.2 Discussion by Type of Outcome Measure

Looking at the association of all balances by type of outcome measures allows broader patterns of physical activity time-use across to be explored. For example, the balances are mostly associated with measures of lipidemia, although no lipidemia outcome s associated with all balances. However, all balances are associated with at least one lipidemia outcome. However, the outcomes are not always associated in the expected directions. The measure of obesity (BMI) is associated with two balances in the expected directions. More activity and longer sleep have been found to be associated with lower obesity (Cappuccio et al. 2008; Vissers et al. 2013; Watson et al. 2015). The measures of glycemia are mostly not associated with any balances. This is in contrast with some previous studies, in particular as one would expect that more activity (balance b_1) would have a beneficial association with glycemia (Colberg et al. 2016). The one exception is that a balance of standing over sitting is favourably associated with reduced fasting blood glucose. This is interesting as this is the first observational result that seems to be in line with experimental results. Numerous laboratory experiments have shown that interrupting sitting with standing is associated with better glycemia acutely, but currently, there is a dearth of observational, and in particular longitudinal, studies to ascertain if this translates into long-term benefits (Chastin et al. 2015a). With respect to lipidemia, the picture is more complex, and this is actually reflected in the physical activity literature. The results here are generally in agreement with previous research. More activity (b_1) is associated with lower triglycerides and a better balance between HDL and LDL cholesterol. Again in experimental studies, interruption of sitting with standing has sometimes been found to be associated with lower triglycerides. So, finding that the balance b_3 is associated with a decreased level of triglycerides broadly agrees with this.

4 Final Remarks

The introduction of concepts and methods from compositional data analysis into physical activity research and time-use epidemiology has enabled the unification of diverse and opposing views in the field. Further progress and adoption of compositional analysis in the field of health research would be facilitated by a collaborative and open source approach, enabling a wider audience of non-specialist to use and develop new techniques and analysis. For this reason, we have developed an open science project called www.opencoda.net. This website is a repository of resources and codes, as well as Shiny application that allow physical activity researcher to conduct compositional analysis. In the future, we will continue to develop resources, and in particular, develop more tools based on balance approach. The balance approach to conduct a compositional analysis of daily behaviour patterns offers a way to focus the statistical assessment and modelling on variables, the log-ratio balance coordinates, representing meaningful contrasts of interest between behaviours. This is fully consistent with the view that actual effects on health are associated with exchanges between some classes of daily behaviours.

The illustrative case study developed in this work uses objectively measured time budgets from a nationally representative sample. It also combines both postural and energy partitions of the daily behaviours. However, a number of the partitions are based on thresholds which are partly arbitrary and often derived from assumptions made in previous studies. This might be causing some distortion in the results. Moreover, this is a cross-sectional study, so the causal inference is limited.

Looking ahead, progress in terms of using balances involves refining the determination of thresholds between behaviour classes and investigating the sensitivity of results to those. Longitudinal models are an obvious extension, as is expanding the number of partitions in the daily time or energy budgets. Both might contribute to a better identification of meaningful balances that are relevant to all or certain dimensions of health. Note, however, that augmenting the level of granularity across the spectrum of daily behaviours also brings in new challenges. For example, in relation to the occurrence of zero time allocations for the rarer behaviours. Because compositional data analysis is based on log-ratios, a well-known issue is that these are undefined for zero values. Methods and software tools have been developed to deal with them coherently and this is an area of active research (Martín-Fernández et al. 2011; Palarea-Albaladejo and Martín-Fernández 2015). So far, typical analyses in physical activity research refer to the classic 4-part partition into sleep, sedentary behaviour, LIPA and MVPA; where the prevalence of zeros tends to be fairly moderate and concentrate in the MVPA class (e.g. when the sample includes older adults). The exploration of possible balances at this level of aggregation is not particularly involved either. However, the discussion about the treatment of zeros and the determination of balances from sensible groupings of the behaviours will need to be more careful as the number of behaviours accounted for is increased, most probably benefiting from both, a combination of expert knowledge and evidence from the data.

Acknowledgements We would like to thank the participants of the AusDiab study. We would also like to acknowledge the help of and helpful discussions with Elisabeth Winkler, Neville Owen and David Dunstan. This work was partly supported by the Scottish Government's Rural and Environment Science and Analytical Services Division [DM and JPA] and by the Spanish Ministry of Science, Innovation and Universities under the project CODAMET (RTI2018-095518-B-C21, 2019–2021) [JPA].

References

J. Aitchison, *The Statistical Analysis of Compositional Data* (Chapman and Hall, New York, 1986).

C. Barcelo-Vidal, J.-A. Martín-Fernández, The mathematics of compositional analysis. Austrian J. Stat. **45**, 57 (2016). https://doi.org/10.17713/ajs.v45i4.142

A. Biswas, P.I. Oh, G.E. Faulkner, R.R. Bajaj, M.A. Silver, M.S. Mitchell, D.A. Alter, Sedentary time and its association with risk for disease incidence, mortality, and hospitalization in adults. Ann. Intern. Med. **162**, 123 (2015). https://doi.org/10.7326/M14-1651

S.N. Blair, G. Davey Smith, I.M. Lee, K. Fox, M. Hillsdon, R.E. McKeown, W.L. Haskell, M. Marmot, A tribute to Professor Jeremiah Morris: the man who invented the field of physical activity epidemiology. Ann. Epidemiol. **20**, 651–660 (2010). https://doi.org/10.1016/j.annepidem.2010.06.001

M.P. Buman, E.A.H. Winkler, J.M. Kurka, E.B. Hekler, C.M. Baldwin, N. Owen, B.E. Ainsworth, G.N. Healy, P.A. Gardiner, Reallocating time to sleep, sedentary behaviors, or active behaviors: associations with cardiovascular disease risk biomarkers, NHANES 2005–2006. Am. J. Epidemiol. **179**, 323–334 (2014). https://doi.org/10.1093/aje/kwt292

F.P. Cappuccio, F.M. Taggart, N.-B. Kandala, A. Currie, E. Peile, S. Stranges, M.A. Miller, Meta-analysis of short sleep duration and obesity in children and adults. Sleep **31**, 619–626 (2008). https://doi.org/10.1093/sleep/31.5.619

V. Carson, M.S. Tremblay, J.-P. Chaput, S.F.M. Chastin, Associations between sleep duration, sedentary time, physical activity, and health indicators among Canadian children and youth using compositional analyses[1]. Appl. Physiol. Nutr. Metab. **41**, S294–S302 (2016). https://doi.org/10.1139/apnm-2016-0026

J.-P. Chaput, V. Carson, C.E. Gray, M.S. Tremblay, Importance of all movement behaviors in a 24 hour period for overall health. Int. J. Environ. Res. Public Health **11**, 12575–12581 (2014). https://doi.org/10.3390/ijerph111212575

S.F.M. Chastin, T. Egerton, C. Leask, E. Stamatakis, Meta-analysis of the relationship between breaks in sedentary behavior and cardiometabolic health. Obesity **23**, 1800–1810 (2015a). https://doi.org/10.1002/oby.21180

S.F.M. Chastin, J. Palarea-Albaladejo, M.L. Dontje, D.A. Skelton, Combined effects of time spent in physical activity, sedentary behaviors and sleep on obesity and cardio-metabolic health markers: a novel compositional data analysis approach. PLoS ONE (2015b). https://doi.org/10.1371/journal.pone.0139984

S.R. Colberg, R.J. Sigal, J.E. Yardley, M.C. Riddell, D.W. Dunstan, P.C. Dempsey, E.S. Horton, K. Castorino, D.F. Tate, Physical activity/exercise and diabetes: a position statement of the American diabetes association. Diabetes Care **39**, 2065–2079 (2016). https://doi.org/10.2337/dc16-1728

G. Condello, F.C.M. Ling, A. Bianco, S. Chastin, G. Cardon, D. Ciarapica, D. Conte, C. Cortis, M. De Craemer, A. Di Blasio, M. Gjaka, S. Hansen, M. Holdsworth, L. Iacoviello, P. Izzicupo, L. Jaeschke, L. Leone, L. Manoni, C. Menescardi, S. Migliaccio, J.-A. Nazare, C. Perchoux, C. Pesce, F. Pierik, T. Pischon, A. Polito, A. Puggina, A. Sannella, W. Schlicht, H. Schulz, C. Simon, A. Steinbrecher, C. MacDonncha, L. Capranica, DEDIPAC consortium, Using concept mapping in the development of the EU-PAD framework (European-physical activity determinants across the life course): a DEDIPAC-study. BMC Public Health **16**, 1145 (2016). https://doi.org/10.1186/s12889-016-3800-8

C.B. del Pozo, R.M. Alfonso-Rosa, D. McGregor, S.F. Chastin, J. Palarea-Albaladejo, C.J. del Pozo, Sedentary behaviour is associated with depression symptoms: compositional data analysis from a representative sample of 3233 US adults and older adults assessed with accelerometers. J. Affect. Disord. (2020). https://doi.org/10.1016/j.jad.2020.01.023

Department of Health, Start active, stay active: a report on physical activity from the four home countries' Chief Medical Officers (2011)

D. Dumuid, T. Olds, L.K. Lewis, J.A. Martin-Fernández, T. Barreira, S. Broyles, J.-P. Chaput, M. Fogelholm, G. Hu, R. Kuriyan, A. Kurpad, E.V. Lambert, J. Maia, V. Matsudo, V.O. Onywera, O.L. Sarmiento, M. Standage, M.S. Tremblay, C. Tudor-Locke, P. Zhao, P. Katzmarzyk, F. Gillison, C.

Maher, The adiposity of children is associated with their lifestyle behaviours: a cluster analysis of school-aged children from 12 nations. Pediatr. Obes. **13**, 111–119 (2018). https://doi.org/10.1111/ijpo.12196

J.J. Egozcue, V. Pawlowsky-Glahn, Groups of parts and their balances in compositional data analysis. Math. Geol. **37**, 795–828 (2005). https://doi.org/10.1007/s11004-005-7381-9

U. Ekelund, J. Steene-Johannessen, W.J. Brown, M.W. Fagerland, N. Owen, K.E. Powell, A. Bauman, I.-M. Lee, J. Morris, J. Heady, P. Raffle, C. Roberts, J. Parks, R. Paffenbarger, R. Hyde, A. Wing, C. Hsieh, C. Wen, J. Wai, M. Tsai, I. Lee, E. Shiroma, F. Lobelo, P. Puska, S. Blair, P. Katzmarzyk, LPASW Group, C. Matthews, K. Chen, P. Freedson, B. Hansen, E. Kolle, S. Dyrstad, I. Holme, S. Anderssen, A. Grøntved, F. Hu, J. Chau, A. Grunseit, T. Chey, A. Biswas, P. Oh, G. Faulkner, S. Carlsson, T. Andersson, A. Wolk, A. Ahlbom, K. Ensrud, T. Blackwell, J. Cauley, Group of in MS, P. Katzmarzyk, M. Inoue, H. Iso, S. Yamamoto, A. Patel, L. Bernstein, A. Deka, C. Matthews, S. George, S. Moore, H. van der Ploeg, T. Chey, R. Korda, E. Banks, A. Bauman, T. Pavey, G. Peeters, W. Brown, L. Leon-Munoz, D. Martinez-Gomez, T. Balboa-Castillo, E. Lopez-Garcia, P. Guallar-Castillon, F. Rodriguez-Artalejo, Y. Kim, L. Wilkens, S. Park, M. Goodman, K. Monroe, L. Kolonel, C. Petersen, A. Bauman, M. Gronbaek, J. Helge, L. Thygesen, J. Tolstrup, C. Matthews, S. Cohen, J. Fowke, D. Dunstan, E. Barr, G. Healy, K. Wijndaele, S. Brage, H. Besson, E. Ford, R. Seguin, D. Buchner, J. Liu, T. Jørgensen, K. Borch-Johnsen, T. Thomsen, H. Ibsen, C. Glümer, C. Pisinger, S. Krokstad, A. Langhammer, K. Hveem, B. Ainsworth, W. Haskell, S. Herrmann, L. Kmet, R. Lee, L. Cook, J. Tierney, L. Stewart, D. Ghersi, S. Burdett, M. Sydes, J. Sterne, M. Egger, WHO, H. Arem, S. Moore, A. Patel, M. Schumacher, G. Rücker, G. Schwarzer, P.S. Collaboration, F. Benatti, M. Reid-Larsen, G. Frydenlund, T. Jorgensen, U. Toft, C. Pisinger, M. Aadahl, J. Harris J. Bargh, K. Brownell, I. Kim, S. Park, J. Trombold, E. Coyle, J. van Dijk, M. Venema, W. van Mechelen, C. Stehouwer, F. Hartgens, L. van Loon, Does physical activity attenuate, or even eliminate, the detrimental association of sitting time with mortality? A harmonised meta-analysis of data from more than 1 million men and women. Lancet **388**, 1302–1310 (2016). https://doi.org/10.1016/S0140-6736(16)30370-1

P. Filzmoser, K. Hron, C. Reimann, Univariate statistical analysis of environmental (compositional) data: problems and possibilities. Sci. Total Environ. **407**, 6100–6108 (2009). https://doi.org/10.1016/j.scitotenv.2009.08.008

P. Filzmoser, K. Hron, C. Reimann, The bivariate statistical analysis of environmental (compositional) data. Sci. Total Environ. **408**, 4230–4238 (2010). https://doi.org/10.1016/j.scitotenv.2010.05.011

P. Filzmoser, M. Templ, K. Hronn, Applied compositional data analysis with worked examples in R (2018)

L. Foley, D. Dumuid, A.J. Atkin, T. Olds, D. Ogilvie, Patterns of health behaviour associated with active travel: a compositional data analysis. Int. J. Behav. Nutr. Phys. Act. **15** (2018). https://doi.org/10.1186/s12966-018-0662-8

N. Gupta, S.E. Mathiassen, G. Mateu-Figueras, M. Heiden, D.M. Hallman, M.B. Jørgensen, A. Holtermann, A comparison of standard and compositional data analysis in studies addressing group differences in sedentary behavior and physical activity. Int. J. Behav. Nutr. Phys. Act. **15**, 53 (2018). https://doi.org/10.1186/s12966-018-0685-1

R. Guthold, G.A. Stevens, L.M. Riley, F.C. Bull, Worldwide trends in insufficient physical activity from 2001 to 2016: a pooled analysis of 358 population-based surveys with 1·9 million participants. Lancet Glob. Heal. **6**, e1077–e1086 (2018). https://doi.org/10.1016/S2214-109X(18)30357-7

G.N. Healy, E.A.H. Winkler, N. Owen, S. Anuradha, D.W. Dunstan, Replacing sitting time with standing or stepping: associations with cardio-metabolic risk biomarkers. Eur. Heart. J. **36**, 2643–2649 (2015). https://doi.org/10.1093/eurheartj/ehv308

F.B. Hu, Sedentary lifestyle and risk of obesity and type 2 diabetes. Lipids **38**, 103–108 (2003)

H.W. Kohl, C.L. Craig, E.V. Lambert, S. Inoue, J.R. Alkandari, G. Leetongin, S. Kahlmeier, The pandemic of physical inactivity: global action for public health. Lancet **380**, 294–305 (2012)

I.-M. Lee, E.J. Shiroma, F. Lobelo, P. Puska, S.N. Blair, P.T. Katzmarzyk, Effect of physical inactivity on major non-communicable diseases worldwide: an analysis of burden of disease and life expectancy. Lancet **380**, 219–229 (2012). https://doi.org/10.1016/S0140-6736(12)61031-9

K. Lyden, D. John, P. Dall, M.H. Granat, Differentiating sitting and lying using a thigh-worn accelerometer. Med. Sci. Sport Exerc. **48**, 742–747 (2016). https://doi.org/10.1249/MSS.0000000000000804

J.A. Martín-Fernández, J. Palarea-Albaladejo, R. Olea, Dealing with zeros, in *Compositional Data Analysis: Theory and Applications*, ed. by V. Pawlowsky-Glahn, A. Buccianti (Wiley, Chichester, 2011), pp. 43–58

J.A. Martín-Fernández, V. Pawlowsky-Glahn, J.J. Egozcue, R. Tolosona-Delgado, Advances in principal balances for compositional data. Math. Geosci. (2018). https://doi.org/10.1007/s11004-017-9712-z

D. McGregor, V. Carson, J. Palarea-Albaladejo, P. Dall, M. Tremblay, S. Chastin, D.E. McGregor, V. Carson, J. Palarea-Albaladejo, P.M. Dall, M.S. Tremblay, S.F.M. Chastin, Compositional analysis of the associations between 24-h movement behaviours and health indicators among adults and older adults from the Canadian health measure survey. Int. J. Environ. Res. Public Health **15**, 1779 (2018). https://doi.org/10.3390/ijerph15081779

D. McGregor, J. Palarea-Albaladejo, P. Dall, K. Hron, S. Chastin, Cox regression survival analysis with compositional covariates: application to modelling mortality risk from 24-h physical activity patterns. Stat. Methods Med. Res. **29**, 1447–1465 (2020). https://doi.org/10.1177/0962280219864125

A.R. Memon, C. Vandelanotte, T. Olds, M.J. Duncan, G.E. Vincent, Research combining physical activity and sleep: a bibliometric analysis. Percept. Mot. Skills **127**, 154–181 (2020). https://doi.org/10.1177/0031512519889780

J.N. Morris, J.A. Heady, P.A.B. Raffle, C.G. Roberts, J.W. Parks, Coronary heart-disease and physical activity of work. Lancet **262**, 1111–1120 (1953). https://doi.org/10.1016/S0140-6736(53)91495-0

N. Owen, G.N. Healy, C.E. Matthews, D.W. Dunstan, Too much sitting: the population health science of sedentary behavior. Exerc. Sport Sci. Rev. **38**, 105–113 (2010). https://doi.org/10.1097/JES.0b013e3181e373a2

J. Palarea-Albaladejo, J.A. Martín-Fernández, zCompositions—R package for multivariate imputation of left-censored data under a compositional approach. Chemom. Intell. Lab. Syst. **143**, 85–96 (2015). https://doi.org/10.1016/j.chemolab.2015.02.019

V. Pawlowsky-Glahn, J. J. Egozcue, R. Tolosana-Delgado, *Modeling and Analysis of Compositional Data* (2015)

K.L. Piercy, R.P. Troiano, R.M. Ballard, S.A. Carlson, J.E. Fulton, D.A. Galuska, S.M. George, R.D. Olson, The physical activity guidelines for Americans. JAMA **320**, 2020–2028 (2018). https://doi.org/10.1001/jama.2018.14854

J. Rivera-Pinto, J.J. Egozcue, V. Pawlowsky-Glahn, R. Paredes, M. Noguera-Julian, M.L. Calle, Balances: a new perspective for microbiome analysis. mSystems (2018). https://doi.org/10.1128/msystems.00053-18

H. Rutter, N. Cavill, A. Bauman, F. Bull, Systems approaches to global and national physical activity plans. Bull. World Health Organ. **97**, 162–165 (2019)

SMPB, International Society for the measurement of physical behaviour (2020), https://ismpb.org/

E. Stamatakis, U. Ekelund, D. Ding, M. Hamer, A.E. Bauman, I.M. Lee, Is the time right for quantitative public health guidelines on sitting? A narrative review of sedentary behaviour research paradigms and findings. Br. J. Sports Med. **53**, 377–382 (2018). https://doi.org/10.1136/bjsports-2018-099131

M. Tremblay, S. Aubert, J. Barnes, T. Saunders, V. Carson, A.E.C. Latimer-Cheung, S, Altenburg T, Chinapaw M, , Sedentary behavior research network (SBRN)–terminology consensus project process and outcome. Int. J. Behav. Nutr. Phys. Act. **14**, 75 (2017)

M.S. Tremblay, V. Carson, J.-P. Chaput, S. Connor Gorber, T. Dinh, M. Duggan, G. Faulkner, C.E. Gray, R. Gruber, K. Janson, I. Janssen, P.T. Katzmarzyk, M.E. Kho, A.E. Latimer-Cheung, C.

LeBlanc, A.D. Okely, T. Olds, R.R. Pate, A. Phillips, V.J. Poitras, S. Rodenburg, M. Sampson, T.J. Saunders, J.A. Stone, G. Stratton, S.K. Weiss, L. Zehr, Canadian 24-hour movement guidelines for children and youth: an integration of physical activity, sedentary behaviour, and sleep[1]. Appl. Physiol. Nutr. Metab. **41**, S311–S327 (2016). https://doi.org/10.1139/apnm-2016-0151

D. Vissers, W. Hens, J. Taeymans, J.P. Baeyens, J. Poortmans, L. Van Gaal, The effect of exercise on visceral adipose tissue in overweight adults: a systematic review and meta-analysis. PLoS ONE **8** (2013). https://doi.org/10.1371/journal.pone.0056415

D.E.R. Warburton, S. Charlesworth, A. Ivey, L. Nettlefold, S.S.D. Bredin, A systematic review of the evidence for Canada's physical activity guidelines for adults. Int. J. Behav. Nutr. Phys. Act. **7**, 39 (2010). https://doi.org/10.1186/1479-5868-7-39

N.F. Watson, M.S. Badr, G. Belenky, D.L. Bliwise, O.M. Buxton, D. Buysse, D.F. Dinges, J. Gangwisch, M.A. Grandner, C. Kushida, R.K. Malhotra, J.L. Martin, S.R. Patel, S.F. Quan, E. Tasali, E. Tasali, Recommended amount of sleep for a healthy adult: a joint consensus statement of the American academy of sleep medicine and sleep research society. Sleep **38**, 843–844 (2015). https://doi.org/10.5665/sleep.4716

E.A.H. Winkler, D.H. Bodicoat, G.N. Healy, K. Bakrania, T. Yates, N. Owen, D.W. Dunstan, C.L. Edwardson, Identifying adults' valid waking wear time by automated estimation in activPAL data collected with a 24 h wear protocol. Physiol. Meas. **37**, 1653–1668 (2016). https://doi.org/10.1088/0967-3334/37/10/1653

E.A.H. Winkler, S. Chastin, E.G. Eakin, N. Owen, A.D. Lamontagne, M. Moodie, P.C. Dempsey, B.A. Kingwell, D.W. Dunstan, G.N. Healy, Cardiometabolic impact of changing sitting, standing, and stepping in the workplace. Med. Sci. Sports Exerc. **50**, 516–524 (2018). https://doi.org/10.1249/MSS.0000000000001453

World Health Organisation, Active ageing: a policy framework (2002)

World Health Organisation, Global recommendations on physical activity for health. Geneva (2010)

Compositional Data Analysis in Time-Use Epidemiology

Dorothea Dumuid, Željko Pedišić, Javier Palarea-Albaladejo, Josep Antoni Martín-Fernández, Karel Hron, and Timothy Olds

Abstract How we allocate time to activities impacts our health. Daily times spent in activities are interrelated because they compete for time-shares within a finite 24 h window. If more time is spent in one activity, time must be taken from one or more of the remaining activities to maintain the fixed total of 24 h. Thus, time-use data have a relative nature and can be analysed accordingly using compositional data analysis. In this chapter, we demonstrate exploratory and cross-sectional inferential analyses of an eight-part time-use composition using data from the Longitudinal Study of Australian Children ($n = 2224$, 52% boys, mean age = 34 months, standard deviation = 3). For inferential analyses, time-use compositions are expressed as a specific choice of balance coordinates to separate between types of activities. Considering the balance coordinates as explanatory variables, we explore the relationship between

D. Dumuid (✉)
Allied Health & Human Performance, Alliance for Research in Exercise, Nutrition and Activity (ARENA), University of South Australia, Adelaide 5001, Australia
e-mail: dot.dumuid@unisa.edu.au

Ž. Pedišić
Institute for Health and Sport, Victoria University, Melbourne 3000, Australia
e-mail: zeljko.pedisic@vu.edu.au

J. Palarea-Albaladejo
Biomathematics and Statistics Scotland, Edinburgh EH9 3FD, Scotland, UK
e-mail: javier.palarea@bioss.ac.uk

J. A. Martín-Fernández
Department of Computer Science, Applied Mathematics and Statistics, University of Girona, 17003 Girona, Spain
e-mail: josepantoni.martin@udg.edu

K. Hron
Department of Mathematical Analysis and Applications of Mathematics, Palacký University, 77146 Olomouc, Czech Republic
e-mail: hronk@seznam.cz

T. Olds
Allied Health & Human Performance, Alliance for Research in Exercise, Nutrition and Activity (ARENA), University of South Australia, Adelaide 5001, Australia
e-mail: tim.olds@unisa.edu.au

P. Filzmoser et al. (eds.), *Advances in Compositional Data Analysis*,
https://doi.org/10.1007/978-3-030-71175-7_20

children's time-use composition and their socio-emotional health. Subsequently, we consider the balance coordinates as dependent variables and explore the relationship between parental perception of neighbourhood liveability and their child's time-use composition.

1 Introduction to Time-Use Epidemiology

How people spend their time has important implications on their health. Epidemiologists have explored the relationships between the time spent in various daily activities and health outcomes for many decades. Some common examples or research topics include the relationship between daily physical activity and obesity (Poitras et al. 2016), nocturnal sleep duration and cardiometabolic risk (Yin et al. 2017), and screen time and depression (Wang et al. 2019). A common type of statistical analysis used in early studies was linear regression of one activity against one health outcome, often adjusted for sociodemographic, lifestyle and health factors. Increasingly, statistical models included adjustment for one or more other time-use components by adding them as covariates to the same model. The issue with this approach was that all daily activities could not be included due to the perfect multi-collinearity between them (Pedišić 2014). Given that all daily activities always sum to 24 h for every observation, it is not possible to alter one activity without a compensatory change in the remaining activities (Pedišić et al. 2017). In addition, the interpretation of individual regression coefficients is not logical—it does not make sense to consider the independent effect of one activity when the activities are in reality not independent but co-dependent on each other (Dumuid et al. 2020).

Between the early 2000s and mid-2010s, researchers began to account for the co-dependency of activity behaviour data by using compositional data analysis (Zhu et al. 2002; Zhang 2013; Taylor et al. 2013; Williams et al. 2014). Rather than looking for independent effects of individual activities, researchers started to consider the relationships between time use and health (Pedišić 2014; Pedišić et al. 2017; Chastin et al. 2015; Dumuid et al. 2018a; Chaput et al. 2014). This new area of investigation was termed "time-use epidemiology" in the late 2010s and defined as "the study of determinants, incidence, distributions and effects of health-related time-use patterns in populations and of methods for preventing unhealthy time-use patterns and achieving the optimal distribution of time for population health" (Pedišić et al. 2017).

The research area has developed rapidly in recent years, with notable papers in a variety of areas, such as occupational health (Stevens et al. 2020; Gupta et al. 2018a, 2018b, 2020; Rasmussen et al. 2018, 2019), active commuting (Foley et al. 2018, 2019), measurement of time use (Coenen et al. 2020; Hallman et al. 2019), health outcomes of sleep, sedentary behaviour and physical activity (Rosen et al. 2020; Pelclová et al. 2020; Carson et al. 2019, 2016; Dumuid et al. 2018b, 2018c; McGregor et al. 2019a, 2019b; Pozo Cruz et al. 2020; Fairclough et al. 2017a; Rodríguez-Gómez et al. 2019; Taylor et al. 2020; Biddle et al. 2018; Talarico and

Janssen 2018; Gába et al. 2020), methodology of sleep research (Matricciani et al. 2018), development of 24 h movement guidelines (Tremblay 2020; Tremblay et al. 2016) and compositional data analysis methods for time-use epidemiology (Chastin et al. 2015; Dumuid et al. 2018a, 2019; Štefelová et al. 2018; Zhao et al. 2019; McGregor et al. 2019c). Overall, in the last few years, more than 70 time-use studies have applied and/or further developed methods based on compositional data analysis.

In this chapter, we use data from the Longitudinal Study of Australian Children (LSAC) to demonstrate descriptive analyses of time-use compositions, regression with time-use composition as the exposure (independent variable) and regression with the time-use composition as the outcome (dependent variable). Issues and challenges are discussed, as well as potential avenues for further development. The second section introduces the LSAC dataset and briefly describes how the time-use data were collected. Section 3 describes the eight-part time-use composition and explores differences in time use by sex, age and socioeconomic status. Section 4 shows an application of multiple linear regression models to explore the relationship between time-use composition and children's socio-emotional health. In Sect. 5, multivariate linear regression models are used to explore how parental perceptions of the neighbourhood environment may influence children's time-use composition. Finally, Sect. 6 summarises the current applications of compositional data analysis within time-use epidemiology and discusses limitations and future possibilities.

2 Time-Use Compositions from the Longitudinal Study of Australian Children

The time-use compositions explored in this chapter were derived from 24 h time-use diaries from LSAC (Gray and Sanson 2005). The study is funded by the Australian Government and conducted in partnership with the Australian Government Department of Social Services, the Australian Institute of Family Studies and the Australian Bureau of Statistics, with advice provided by a consortium of leading researchers. Data are available under licence at https://www.dss.gov.au/national-centre-for-longitudinal-data-ncld/access-to-dss-longitudinal-datasets. LSAC has a longitudinal design, with biennial waves. It commenced with a nationally-representative sample in 2003 and has two cohorts: a birth (B) cohort and a child (K) cohort. The following analyses use data from Wave 2 of the B cohort, collected in 2005–6, when the children were 2–3 years old.

Time-use data were collected via 24 h time-use diaries (Corey et al. 2014). A parent was asked to complete a paper-based diary for two days. The diary contained a list of 26 possible activities alongside a grid representing every 15 min interval of the 24 h day. Parents marked the grid to indicate their child's participation in the respective activity at each time slot.

Before analyses, the time-use diary data were treated as follows. Some parents had marked more than one activity per time slot. Where this occurred, the 15 min were shared equally by the competing activities, unless the multi-tasked activity was sleep, because it was considered impossible to simultaneously sleep and perform other activities. If sleep was marked together with other activities, the other activities were ignored, and the full 15 min were attributed to sleep. The 26 activities were collapsed into eight time-use components for manageability, and to be consistent with other time-use surveys/recalls (Ridley et al. 2006) (Table 1).

There were 5112 child participants in Wave 1 of the B cohort. Of these, 4606 were retained in Wave 2. At least one time-use diary was available for 3423 children. For the ensuing analyses, we included diaries with complete data (i.e. all 1440 min accounted for), and plausible values (sleep between 500 and 1000 min/d; self-care (including eating) between 30 and 350 min/d; domestic/social activities <550 min/d and education-related activities <250 min/d). The final time-use dataset included 2224 children and 2975 diary days (activity compositions)

The activity compositions were checked for zero values. Zero values must be dealt with because the logratio methodology (i.e. expressing the input compositional data in logratio coordinates before proceeding with further statistical analysis (Aitchison

Table 1 Derivation of eight-part time-use compositions from 26 LSAC diary activities

Activity	Time-use component	Activity	Time-use component
Travel in pusher or on a bicycle seat	Passive transport	Being reprimanded, corrected	Domestic/social
Travel in car	Passive transport	Being held, cuddled, comforted, soothed	Domestic/social
Travel on public transport	Passive transport	Active free play (e.g. running, climbing, ball game)	Physical activity
Listening to tapes, CDs, radio, music	Quiet time	Organised lessons/activities	Education
Doing nothing, bored/restless	Quiet time	Drawing, colouring, looking at a book, educational game	Education
Awake in bed	Quiet time	Being taught to do chores, read, etc	Education
Being read to, told a story, or sung to	Quiet time	Eating and drinking, being fed	Self-care
Being taken places with adult (e.g. shopping)	Domestic/social	Bathing, dressing, hair care, health care	Self-care
Quiet free play (e.g. jigsaw, craft, dress-ups)	Domestic/social	Watching TV, video, DVD, movie	Screen time
Visiting people, special event, outing	Domestic/social	Using computer/computer game	Screen time
Crying, upset, tantrum	Domestic/social	Sleeping/napping	Sleep
Destroying things, creating a mess	Domestic/social	Ride bicycle, trike, etc. (for travel or fun)	Physical activity
Arguing, fighting	Domestic/social	Walking (for travel or fun)	Physical activity

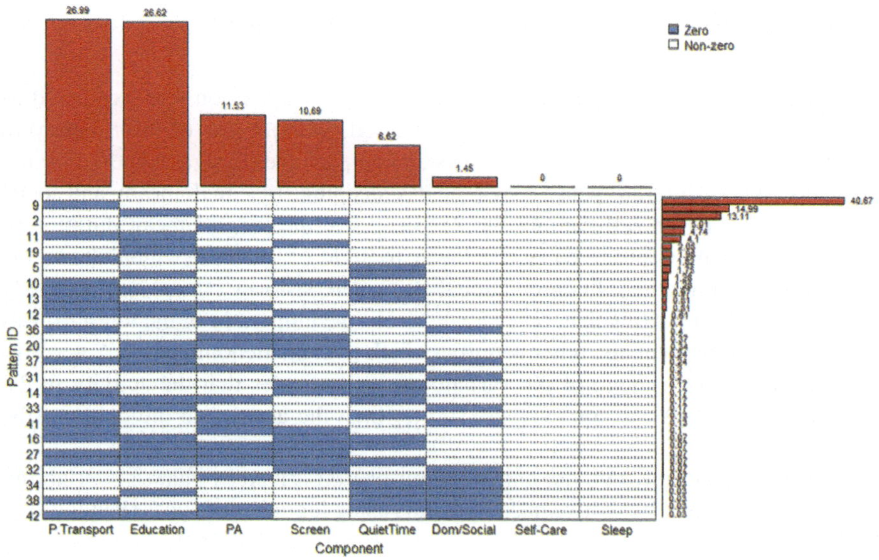

Fig. 1 Zero patterns in Wave 2 LSAC time-use diaries (B cohort). P. Transport = passive transport; PA = physical activity; Dom/Social = domestic/social. Dataset has 10.5% of zero entries

1982) is not applicable if zeros are present (Martín-Fernández et al. 2011). In this case, it was expected that if children were monitored over a long enough period, they would accumulate some time in all eight components. Thus, it seemed reasonable to assume that the zeros in the dataset were *rounded* zeros (generically left-censored values) and replace them with sensible small values below a given observation threshold (Palarea-Albaladejo and Martín-Fernández 2015). For included participants, there were zeros in all time-use components except for sleep and self-care (Fig. 1). Zero patterns were plotted using *zPatterns()* from the *zCompositions* package (Palarea-Albaladejo and Martín-Fernández 2015). Overall, zeros accounted for 10.5% of all values. In 41% of diaries, there were no zero values in any of the time-use components. Of all diaries, 27% had zeroes in passive transport (top axis), and of these, 15% had zeros *only* in passive transport (right-hand axis). This was closely followed by education-related activities (27% of diaries had zeros, 13% had zeros in education only). The zeros were replaced using a robust model-based iterative algorithm with the *lrEM()* function from zCompositions, considering 15 min as the observation threshold for all components as this was the granularity of the time-use instrument. If children had two time-use diaries, the compositional mean of the two days was calculated so that each child had one "average" daily time-use composition for analysis.

3 Descriptive Analyses of Time-Use Compositions

The *centre* or *compositional mean* of the time-use composition was expressed as a vector of geometric means of its parts, rescaled (without loss of information) to sum to 24 h (or 1440 min), using the *compositions* package (Boogaart and Tolosana-Delgado 2008). Children spent 56% of the day sleeping, 14% in domestic and social activities and 11% in self-care (Table 2). Sleep had the longest mean duration of all activities. This is observed in the lower triangle of Table 2, which shows the mean pairwise logratios of compositional parts. The last row (sleep) contains only positive values, indicating that, on average, sleep (the numerator of the logratio) was of longer duration than any other activity (which is in the denominator). A negative value in the lower triangle of the variation array indicates that the activity in the row (the numerator) was, on average, shorter than the activity in the column (the denominator).

The *variation* of the time-use compositions was described by pairwise variability (i.e. (dis)proportionality) between components using logratio variances (Pawlowsky-Glahn et al. 2015) in the variation array (the upper triangle of Table 2). Low values were observed in logratios involving sleep and self-care. This suggests that the duration of these two activities have relatively low variability in this sample. The highest variability was observed between education and passive transport, and education and physical activities, indicating low co-dependences in terms of proportionality or the existence of subgroups with homogenous time-use patterns.

A composition with eight components (seven dimensions) cannot be exactly represented in two-dimensional space. An option is to use a reduced-dimension representation of the data. This is commonly achieved by producing a biplot based on the first two dimensions derived from a principal components analysis (PCA) of a logratio representation of the compositional data, where the points refer to the individuals and the rays represent the variables (Aitchison and Greenacre 2002). However, how well the data are represented depends on how much of the variation is explained by the first two principal components. We performed a PCA of centred logratio coefficients (Filzmoser et al. 2018) with a robust method (MCD [Minimum Covariance Determinant] estimator of location and scale) (Filzmoser et al. 2009) using the *pcaCoDa()* function in the *robCompositions* package (Templ et al. 2011), and plotted the first two principal components with the *ggbiplot* package (Vu 2011). The first two components explained 52% of the total variance of the time-use dataset.

The compositional biplot was used to explore time-use distribution by sex (Fig. 2). In general, boys (blue) appeared to accumulate more sleep, physical activity, self-care and screen time (each relative to the remaining activities), while girls (orange) accumulated more education-related activities and quiet time (relative to the remaining activities). However, these observations are treated with caution as the biplot only explained 52% of variance in the time-use data.

Differences between subgroups defined by sex, age and socioeconomic status captured by a LSAC composite score derived from parental reports of their education, employment and income (Blakemore et al. 2009) were further explored in logratio

Table 2 Descriptive summary of time-use compositions: variation array

	Domestic /Social	Physical activity	Passive transport	Quiet time	Education	Screen time	Self-care	Sleep	Centre (%)	Centre (min/d)
Domestic/Social		1.61	1.15	1.10	1.66	1.34	0.61	0.45	13.9	200.5
Physical activity	−0.85		2.00	1.56	2.03	1.83	0.99	0.89	6.0	86.1
Passive transport	−2.00	−1.16		1.58	2.17	1.98	1.10	0.99	1.9	27.0
Quiet time	−1.19	−0.34	0.82		1.39	1.44	0.70	0.61	4.2	61.2
Education	−1.82	−0.97	0.18	−0.63		1.75	1.09	1.05	2.3	32.5
Screen time	−1.08	−0.23	0.93	0.11	0.75		0.88	0.77	4.7	68.4
Self-care	−0.23	0.62	1.77	0.96	1.59	0.85		0.13	11.1	159.3
Sleep	1.39	2.24	3.39	2.58	3.21	2.47	1.62		55.9	805.1

The upper triangle contains the variances of the logratio, for example, 1.61 is the variance of ln(DomesticSocial/Physical activity). Note, 1.61 is also the variance of the reciprocal logratio ln(Physical activity/DomesticSocial) which is not shown, i.e. the logratio variances are symmetrical. The lower triangle contains pairwise logratio expectations (the mean of the pairwise logratio, e.g. the mean of ln(Physical activity/DomesticSocial) = −0.85). The mean of the reciprocal logratio (not shown) is the opposite (e.g. mean of ln(DomesticSocial/Physical activity) = 0.85). Compositional centre of the dataset is shown in the final two columns

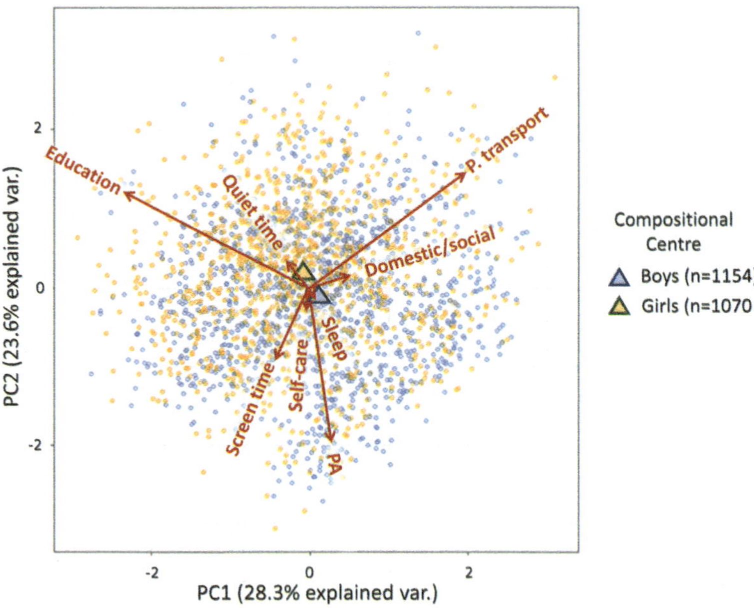

Fig. 2 Compositional biplot of eight-part time-use compositions. Triangles indicate compositional centres of subgroups (boys and girls). PC = principal component; var = variance; P. transport = passive transport; PA = physical activity

difference plots (Martín-Fernández et al. 2015) comparing the centres (geometric means) of their time-use compositions to the overall sample's time-use compositional centre. If the mean duration of an activity is identical in the subgroup (e.g. girls) as it is in the overall population, the logratio equals zero. If the activity has a higher mean duration in the subgroup, the logratio is positive, and if the activity has a lower mean in the subgroup, the logratio is negative.

The logratio difference plot for sex (Fig. 3a) supports some of the differences in time use observed on the compositional biplot. It appears that some of the high variation in education and physical activities (these components have relatively long rays in the compositional biplot (Fig. 2) may be explained by sex differences. Figure 3a shows that girls spend relatively more time in education-related activities and less time in physical activities compared to boys. Passive transport also had a long ray on the compositional biplot (Fig. 2), but in Fig. 3, little variation in passive transport was observed by sex or age, suggesting socioeconomic status or a different factor (perhaps remoteness of residence) may be driving the variation in passive transport observed on the biplot. Sleep duration appeared to have almost no variation in the overall sample and across subgroups. However, when considering raw values (min/d), sleep differences seemed relevant. For example, the youngest age group (<35 months) had an average sleep duration of 809 min/d, while the oldest age group (>40 months) had 774 min/d. Although 35 min may appear substantial, it is a small fraction of the average sleep duration in the overall sample (4% of 805 min/d). In comparison,

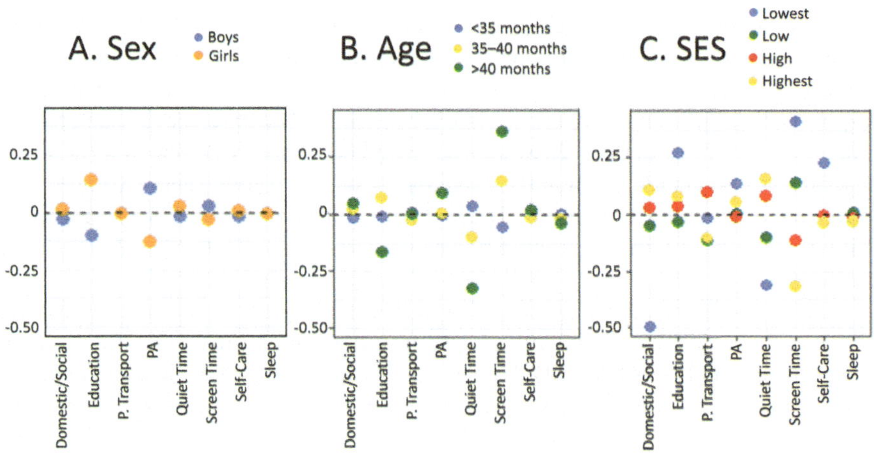

Fig. 3 Logratio differences in mean time-use composition across population subgroups. SES = socioeconomic status; P. Transport = passive transport; PA = physical activity

35 min is a large proportion of average daily screen time (51% of 68 min/d). The relative scale is clearly reflected by the respective logratios. Figure 3c suggests there are socioeconomic gradients in time use, notably, children tend to accumulate more screen time and less domestic/social and quiet time as their socioeconomic status decreases.

This section demonstrated tools for exploratory analysis of time-use compositions with more than three components. The compositional centre, variation array, and dimension-reduction methods were used to descriptively analyse and explore patterns in an eight-part time-use composition. The next section explores regression analysis with explanatory time-use compositions.

4 Regression with an Explanatory Time-Use Composition

Most commonly, analyses in time-use epidemiology explore whether time-use compositions (considered as explanatory variables) are associated with health outcomes. This section demonstrates regression of the eight-part time-use composition against a measure of the child's socio-emotional health. The aim of the analysis was to explore the relationship between children's daily time-use composition (particularly the amounts of time spent in sleep, sedentary behaviours and physically active behaviours) and their socio-emotional health. The time-use composition was expressed as a set of isometric logratio (ilr) coordinates (Mateu-Figueras et al. 2011) using a specific type of partitioning to create balances of non-overlapping groups of parts that are relevant to the research aim.

The outcome measure is child socio-emotional health, captured by the Brief Infant–Toddler Social and Emotional Assessment (BITSEA) Problem Scale (Briggs-Gowan and Carter 2002), a summary measure derived from the sum of parental responses to 23 questions about the child's behaviour. The parent was asked to tick a box: (1) Not true/Rarely; (2) Somewhat true/Sometimes; (3) Very true/Often to describe the child in the last month for statements such as "Often gets very upset" and "Has trouble adjusting to changes". The problem score ranged from 21 to 51. For analysis, the score was reversed, so that higher values indicated better socio-emotional health.

The set of explanatory variables was the eight-part time-use composition (domestic/social, physical activity, passive transport, quiet time, education, screen time, self-care and sleep) described in Sect. 3. The key aspect of the logratio methodology is to express the time-use data as a set of logratio coordinates capturing all information conveyed by the composition (Aitchison 1982). The logratio coordinates form vectors in the ordinary real Euclidean space, which can then be used to represent the time-use data in regression models. Until now, most time-use epidemiology studies have used pivot coordinates as a specific choice of ilr balance coordinates, which isolate (in one of the coordinates) the contribution of a specific part of the time-use composition relative to the remaining parts (Chastin et al. 2015; Hron et al. 2017, 2012). A few studies have used general balances of groups of parts as an alternate way to explore relationships between time-use compositions and outcomes (Dumuid et al. 2018a; McGregor et al. 2019c). We demonstrate the balance approach using the Wave 2 LSAC data.

A sequential binary partitioning (SBP) process was used to first divide the time-use composition vector into two groups of parts, followed by sequentially dividing each of the groups from the previous step until only one part remained in each group. The SBP was designed to produce balances of groups of parts that were meaningful to the research question. Because we aimed to explore the health associations of time-use composition in terms of sleep, sedentary behaviours and physically active behaviours, the compositional parts were grouped accordingly (Table 3). The signs in Table 3 indicate whether a time-use component will be in the numerator (+), denominator (–) or not included (blank) in the resulting ilr coordinate (hereafter termed *balance coordinates*, or *balances*). More details on the construction of balances through SBP can be found elsewhere (Mateu-Figueras et al. 2011; Egozcue and Pawlowsky-Glahn 2005; Egozcue et al. 2003).

The linear regression model included the set of balance coordinates created by the SBP in Table 3 as explanatory variables. Robust MM-type estimators were used. Sex (binary variable), age and socioeconomic status (numeric variables) were also included in the model as covariates. Children with complete time-use, socio-emotional and covariate data were included in the analysis. The analytical sample had 2197 participants, of which 52% were boys. The average age was 34 months (standard deviation [SD] = 3) and socioeconomic z-score was slightly higher than in the initial LSAC sample (mean = 0.19, SD = 0.96, vs mean = 0, SD = 1. Blakemore et al. 2009). The outcome (reversed BITSEA problem score) was log-transformed to

Table 3 Sequential binary partition for meaningful time-use balances: sign matrix

		Domestic /Social	Physical activity	Passive transport	Quiet time	Education	Screen time	Self-care	Sleep
B_1	Sleep vs wake	−	−	−	−	−	−	−	+
B_2	Sedentary vs active	−	−	+	+	+	+	−	
B_3	Moderate-to-vigorous vs light activities	−	+					−	
B_4	Screen vs non-screen sedentary			−	−	−	+		
B_5	Education vs other non-screen sedentary			−	−	+			
B_6	Passive transport vs quiet time			+	−				
B_7	Domestic/social vs self-care	+						−	

B = balance coordinate, vs = versus

Table 4 Relationship between time-use composition and socio-emotional health: regression coefficients

	Balance	Beta	SE	*p*
B_1	Sleep vs wake	0.032	0.009	<0.001
B_2	Sedentary vs active	−0.001	0.003	0.830
B_3	Moderate-to-vigorous vs light activities	0.005	0.003	0.128
B_4	Screen vs non-screen sedentary	−0.007	0.002	0.001
B_5	Education vs other non-screen sedentary	0.005	0.002	0.013
B_6	Passive transport vs quiet time	0.005	0.002	0.040
B_7	Domestic/social vs self-care	0.000	0.004	0.942

Note: Outcome is log-transformed BITSEA Problems Score (reversed so higher value is better). Model adjusted for child's sex, age and household socioeconomic status. B = balance coordinate, vs = versus

honour its relative scale and to better accommodate the normally distributed residuals assumption. Regression coefficients for the balance coordinates are presented in Table 4.

The individual regression coefficients can be interpreted as the expected change in the log-transformed outcome variable when the activity(ies) in the numerator of the respective logratio increase at the expense of the activity(ies) in the logratio's denominator to increase the logratio by 1, while all remaining logratios are kept constant. Table 4 shows that longer sleep, relative to waking activities, was associated with better socio-emotional health ($B_1, p = <0.001$). More time spent in education-related activities, relative to other sedentary activities (excluding screen time) was also associated with better socio-emotional health ($B_5, p = 0.013$). Conversely, more screen time, relative to non-screen sedentary activities, was associated with worse socio-emotional health ($B_4, p = 0.001$).

While the estimates in Table 4 tell us if the association was positive or negative, they may not allow for a direct interpretation of effect sizes that would be relevant in the context of time-use epidemiology. Health researchers, policymakers and research translation specialists may require a straightforward interpretation of the expected change in the outcome for a change in the allocation of time across time-use components. For example, by *how many units* is the socio-emotional score expected to improve if a child decreases screen time by *30* min and replaces it with non-screen sedentary activities? Such questions can be answered through a method called "compositional isotemporal substitution". Generalisable expressions underpinning this method can be found in Dumuid et al. (2019). A regression model is used to estimate the outcome for the mean time-use composition, and for a "new", theoretical time-use composition, where time has been reallocated between activities according to the research question. The "new" time-use composition could have 30 min less screen time than the mean composition, and these 30 min can be equally redistributed to the non-screen sedentary activities. In our data example, the difference between the two expected scores (estimated from the mean composition and

Table 5 Expected differences in the outcome (socio-emotional health) for isotemporal substitution of time (30 min) between activities (screen and non-screen sedentary activities)

	Predicted outcome (log-transformed)	Socio-emotional score (exp(prediction))
Mean composition	3.809	45.11
New composition	3.814	44.32
Difference (new – mean)	+ 0.01	+ 0.21

Note: The new composition has 30 min less screen time compared to the mean composition. The time is proportionally redistributed to non-screen sedentary activities (quiet time, passive transport and education-related activities)

from the "new" composition) represents *how many units* the log-transformed socio-emotional score is estimated to improve by when a child decreases screen time by *30* min and replaces it with non-screen sedentary activities. Note, when the compositional regression model is used in this way for prediction purposes, the choice of SBP is inconsequential. All the relative information regarding the composition is contained in the set of log ratios, regardless of SBP.

Table 5 shows that the estimated improvement in socio-emotional score for the reallocation of 30 min from screen time to non-screen sedentary activities was quite small relative to the standard deviation observed in the sample (SD of log(outcome) = 0.10). A standardised effect size of 0.1 is small, and perhaps not very clinically meaningful if considered at the individual level. The estimated improvement associated with reducing screen time may have been somewhat diluted because the 30 min were reallocated to three activities (quiet time, passive transport and education-related activities) each of which may be differentially associated with the outcome. For example, parameters in Table 4 suggest that passive transport and quiet time were differentially associated with socio-emotional health (see B_6). It may be interesting to create new compositions where time is reallocated between two activities only, i.e. *one-for-one* reallocations (McGregor et al. 2019a, 2019b, 2019c; Dumuid et al. 2019; Fairclough et al. 2017b) or create new compositions which reflect the way children commonly reallocate time in real life (i.e. *empirical* reallocations).

This section demonstrated compositional linear regression with explanatory time-use compositions. The compositions were expressed as balance coordinates following a specifically defined SBP, guided by the research aim. While a meaningful SBP is crucial to interpreting the model, model-based predictions are invariant to the choice of SBP. We showed how model-based predictions allow a direct interpretation of effect size through compositional isotemporal substitution. The following section explores regression with a time-use composition as the outcome, also using the LSAC time-use dataset.

5 Regression with a Dependent Time-Use Composition

Researchers in time-use epidemiology are often interested in determinants of a time-use composition (Rasmussen et al. 2018, 2019; Foley et al. 2018, 2019; Hunt et al. 2018). In this section, we used multivariate linear regression with robust MM-type estimators to estimate the association between the child's neighbourhood and their time-use composition. The parent's perception of the child's environment was obtained from the Neighbourhood Liveability scale (Stone and Hughes 2002) composed of two items in a questionnaire asking how strongly they agreed or disagreed with the following two statements: "This is a safe neighbourhood" and "There are good parks, playgrounds and play spaces in this neighbourhood". Possible responses were (1) Strongly agree; (2) Agree; (3) Disagree; (4) Strongly disagree. A neighbourhood liveability score was computed as the average between the two items. We reversed the values so that a higher score indicated a more liveable neighbourhood. The score was treated as a continuous explanatory variable for simplicity and log-transformed to correct for skewness. Regression models were adjusted for children's sex (binary variable), age and household socioeconomic status (numeric variables).

The eight-part time-use composition was expressed as balance coordinates, using the SBP described in Sect. 4. This meant that there were seven balances which were each considered as the dependent variable in separate regression models, following the procedures in Egozcue et al. (2011). The children with complete data ($n = 2189$) for all variables were included in the analytical models. Table 6 summarises the regression coefficients for the neighbourhood liveability score (explanatory variable) against each balance (outcome variable).

Table 6 Relationship between parents' perceptions of neighbourhood liveability (log-transformed) and children's time-use composition: results of multivariate regression analysis

	Balance	Beta	SE	p
B_1	Sleep vs wake	0.013	0.033	0.691
B_2	Sedentary vs active	0.019	0.086	0.828
B_3	Moderate-to-vigorous vs light activities	0.044	0.097	0.650
B_4	Screen vs non-screen sedentary	−0.257	0.097	0.008
B_5	Education vs other non-screen sedentary	−0.184	0.119	0.120
B_6	Passive transport vs quiet time	0.219	0.107	0.040
B_7	Domestic/social vs self-care	0.049	0.059	0.412

The regression coefficient (Beta) represents the expected change in the respective balance for a one-unit increase in the neighbourhood liveability score. All regression models were adjusted for child's age, sex and household socioeconomic status. B = balance coordinate, vs = versus

The coefficients in Table 6 indicate that as the neighbourhood environment was perceived to be more liveable, the amount of screen time decreased relative to non-screen sedentary behaviours (B_4). It could be that, as parents perceived their neighbourhood to be less liveable, they may have kept their children in the home, entertained on screens, perhaps due to fears of safety and lack of facilities such as parks, playgrounds and public transport. With better neighbourhood liveability, time spent in passive transport increased relative to time spent in quiet time activities (B_6). This may be explained by the negative relationship between reported liveability and remoteness (Fig. 4), that is, those living in remote communities have less access to, and therefore spend less time in, public transport.

Another way to interpret results from regression with compositional dependent variables is to use the models to predict logratio coordinates for different values of the explanatory variable. As noted in the previous section, while the specifically chosen SBP enables meaningful interpretation of the model, model-based predictions are invariant to the choice of SBP. We predicted each of the seven balances for incrementally increasing neighbourhood liveability score. The predicted balance coordinates were then re-expressed in the original raw units of min/d. This enabled the association between neighbourhood liveability score and time used to be interpreted in directly meaningful, absolute terms. Predicted time use across neighbourhood liveability scores is presented in Table 7. As neighbourhood liveability score increased from

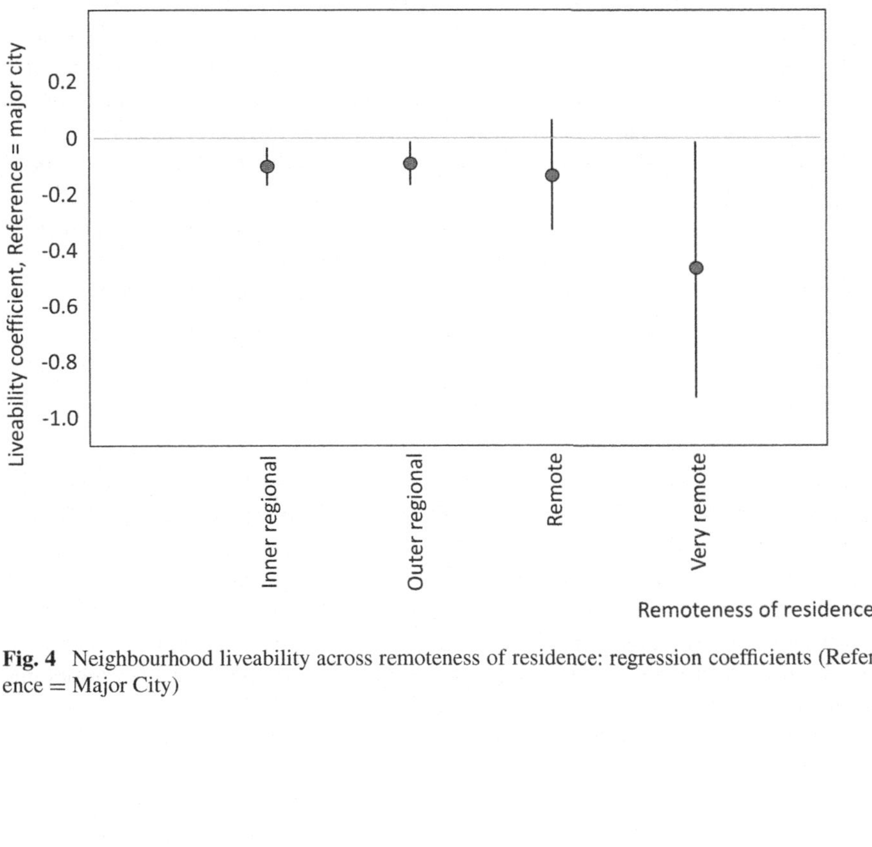

Fig. 4 Neighbourhood liveability across remoteness of residence: regression coefficients (Reference = Major City)

Table 7 Predicted eight-part time-use compositions (min/d) across neighbourhood liveability scores

Liveability	Domestic/ Social	Physical Activity	Passive Transport	Quiet Time	Education	Screen Time	Self-Care	Sleep
1	190	98	19	61	31	91	165	787
1.5	191	99	21	61	30	84	161	794
2	191	100	23	61	29	79	159	798
2.5	192	101	25	61	29	75	157	801
3	192	101	26	61	29	72	155	804
3.5	193	102	28	61	28	70	154	806
4	193	102	29	61	28	68	152	807

Note: higher value represents more liveable neighbourhood. Rows do not all add to 1440 min/d due to rounding. All analyses were adjusted for child's age, sex and household socioeconomic status. Green = component increased with better liveability, blue = component decreased with better liveability.

the lowest to the highest level, estimated decreases in screen time (23 min/d) and self-care (13 min/d), were largely compensated by increases in sleep (20 min/d), passive transport (10 min/d). These results suggest that neighbourhood factors (perceptions of safety, availability of facilities such as parks) should be considered as potential targets for interventions to improve children's time use (e.g. to decrease screen time).

This section demonstrated regression with a time-use composition as the outcome. Once the time-use data are expressed as ilr coordinates (here, specific balances were used), they can be included in standard analytical models. Interpretation of regression coefficients requires an understanding of how the ilr coordinates have been constructed. The model can be used to predict the balance coordinates (which can be re-expressed as raw time-use compositions) for various values of the explanatory variable(s). This method provides an indication of how raw values of the time-use composition (min/d) are estimated to change in relation to the predictor(s).

6 Summary

Compositional data analysis has been adopted relatively quickly by physical activity, sedentary behaviour, sleep and time-use researchers. The ability to include all daily activities into a single analytical model, and to draw inferences about the whole day rather than about individual activities in isolation has brought about a new area of research called time-use epidemiology (Pedišić et al. 2017; Chaput et al. 2014; Rosenberger et al. 2019). Alongside these developments, government bodies around the world and the World Health Organization have revised their public health guidelines for physical activity, sedentary behaviour and sleep, and have released harmonised, 24 h guidelines incorporating all of these activities together (Tremblay et al. 2016,

Australian Government Department of Health 2019, Centre and of Excellence in Human Development and Laureus "Sport for good" 2018, Jurakic and Pedišić 2019, New Zealand Ministry of Health 2017, World Health Organization 2019.

Time-use epidemiology now has its own international body, called The International Network of Time-Use Epidemiologists (INTUE 2020) which aims to draw together researchers and other stakeholders, to strengthen collaborative research on time use and to disseminate findings effectively. The network currently has >80 members from 26 countries and six continents.

This chapter demonstrated some applications of the logratio methodology for time-use compositions, involving descriptive analyses and linear regression models. In the context of time-use epidemiology, compositional data analysis can be, and has been, applied together with other statistical methods, such as cluster analysis, partial least squares and non-linear models. Future studies may combine machine learning, such as decision trees to find best fitting models, and optimisation theory, including response surface methodology (Hejazi et al. 2012; Myers et al. 2016) with compositional data analysis, to determine the optimal mix of daily activity behaviours for good health.

There are limitations and challenges to consider when using the logratio methodology for time-use compositions. For logratio coordinates, there cannot be zero values in any part of the composition. Zeros frequently appear in time-use datasets, but usually there are not many of them, and theoretically, it makes sense to replace them with small values, as they are likely to be rounded zeros. However, where zeros are essential (e.g. time spent at work for a retiree), replacing zeros with small values may not be the best solution (Martín-Fernández et al. 2011). Other options might be to explore subgroups defined by zero patterns (the package zCompositions (Palarea-Albaladejo and Martín-Fernández 2015) includes specific tools to investigate the influence of subgroups in logratios), or using approaches that deal indirectly with the zeros. When subcompositions are of interest (or 24 h data are not available), the data no longer add to a constant meaningful total time (i.e. 24 h or 1440 min), meaning it is not straightforward to express compositions in min/day. However, a basic feature of the logratio approach is that the total is regarded as irrelevant, subcompositions are still compositional data conveying relative information, and there are increasing examples of compositional data analysis for subcompositions in time-use epidemiology (Rasmussen et al. 2018; Pelclová et al. 2020; Fairclough et al. 2018).

The interpretation of findings from compositional studies can be challenging as we are used to thinking about activities in a univariate manner, and in raw units of time (minutes, or hours). With compositional data analysis, the unit of analysis is relative (reflected by the use of logratios which honour the scale invariance of compositions), thus, the interpretation is in relative rather than in absolute terms. The parts of the composition are always considered relative to one or more other parts. Variability in the composition is considered in relative terms, so a 30 min change in a typically long activity (e.g. sleep) is much smaller than a 30 min change in a typically short activity (e.g. exercise). The relative scale of time-use data makes intuitive sense, and it is increasingly being acknowledged in research and translation landscapes. Public health messages such as "Move More, Sit Less, Sleep Well" (New Zealand Ministry of Health 2017) and the Australian "Swap It" (O'Hara et al. 2016) campaign are some examples.

Compositional data analysis has enabled the integrated analysis of the daily time-use composition, potentially one of the most important determinants of human health. Regression models with ilr coordinates are currently the most commonly used application of compositional data analysis within time-use epidemiology. It is inspiring to consider how future applications of compositional data analysis to time-use data may contribute to improved health outcomes in the population.

Acknowledgements This chapter uses unit record data from Growing Up in Australia: The Longitudinal Study of Australian Children (LSAC). The LSAC study was conducted in partnership with the Department of Social Services (DSS), the Australian Institute of Family Studies (AIFS) and the Australian Bureau of Statistics (ABS). The findings and views reported in the paper are those of the authors and should not be attributed to DSS, AIFS or the ABS.

Funding D.D. was supported by the National Health and Medical Research Council Early Career Fellowship (1162166) and the National Heart Foundation of Australia Postgraduate Fellowship (102084). J.P.A. and J.A.M.F. were supported by the Spanish Ministry of Science, Innovation and Universities under the project CODAMET (RTI2018-095518-B-C21, 2019-2021). J.P.A. was partly supported by the Scottish Government's Rural and Environment Science and Analytical Services Division. K.H. was funded by a research grant from the Czech Science Foundation no. 18-09188S.

References

J. Aitchison, The statistical analysis of compositional data. J. Roy. Stat. Soc. B **44**(2), 139–160 (1982)

J. Aitchison, M. Greenacre, Biplots of compositional data. J. Roy. Stat. Soc.: Ser. C (Appl. Stat.) **51**, 375–392 (2002)

Australian Government Department of Health, Australian 24-hour movement guidelines for children and young people (5–17 years) – an integration of physical activity, sedentary behaviour and sleep (2019), https://www1.health.gov.au/internet/main/publishing.nsf/Content/health-24-hours-phys-act-guidelines

G.J.H. Biddle, C.L. Edwardson, J. Henson, M.J. Davies, K. Khunti, A.V. Rowlands, T. Yates, Associations of physical behaviours and behavioural reallocations with markers of metabolic health: a compositional data analysis. Int. J. Environ. Res. Public Health **15**(10) (2018)

T. Blakemore, L. Strazdins, J. Gibbings, Measuring family socioeconomic position. Aust. Soc. Policy **8**, 121–168 (2009)

M.J.Briggs-Gowan, A.S. Carter, Brief infant-toddler social and emotional assessment (BITSEA) manual, version 2.0 (Yale University, New Haven, 2002)

V. Carson, M.S. Tremblay, J.P. Chaput, S.F.M. Chastin, Associations between sleep duration, sedentary time, physical activity, and health indicators among Canadian children and youth using compositional analyses. Appl. Physiol. Nutr. Metab. **41**(6), S294–S302 (2016)

V. Carson, M.S. Tremblay, J.P. Chaput, D. McGregor, S. Chastin, Compositional analyses of the associations between sedentary time, different intensities of physical activity, and cardiometabolic biomarkers among children and youth from the United States. PLOS One **14**(7) (2019)

DST-NRF Centre of Excellence in Human Development and Laureus "Sport for good", South African 24-hour movement guidelines for birth to five years: an integration of physical activity, sitting behaviour, screen time and sleep (Cape Town, SA, 2018)

J.-P. Chaput, V. Carson, C.E. Gray, M.S. Tremblay, Importance of all movement behaviors in a 24 hour period for overall health. Int. J. Environ. Res. Public Health **11**(12), 12575–12581 (2014)

S.F.M. Chastin, J. Palarea-Albaladejo, M.L. Dontje, D.A. Skelton, Combined effects of time spent in physical activity, sedentary behaviors and sleep on obesity and cardio-metabolic health markers: a novel compositional data analysis approach. PLoS ONE **10**(10), e0139984 (2015)

P. Coenen, S.E. Mathiassen, A.J. van der Beek, D.M. Hallman, Correction of bias in self-reported sitting time among office workers – a study based on compositional data analysis. Scand. J. Work Environ. Health **46**(1), 32–42 (2020)

J.Corey, J. Gallagher, E. Davis, M. Marquardt, The times of their lives: collecting time use data from children in the longitudinal study of Australian children (LSAC). Technical Paper 13. In: Growing up in Australia. The longitudinal study of australian children (LSAC) (Australian Bureau of Statistics, 2014)

B. del Pozo Cruz, R.M. Alfonso-Rosa, D. McGregor, S.F.M. Chastin, J. Palarea-Albaladejo, J. del Pozo Cruz, Sedentary behaviour is associated with depression symptoms: compositional data analysis from a representative sample of 3233 US adults and older adults assessed with accelerometers. J. Affect. Disord. **265**, 59–62 (2020)

D. Dumuid, L.K. Lewis, T.S. Olds, C. Maher, C. Bondarenko, L. Norton, Relationships between older adults' use of time and cardio-respiratory fitness, obesity and cardio-metabolic risk: a compositional isotemporal substitution analysis. Maturitas **110**, 104–110 (2018)

D. Dumuid, T.E. Stanford, J.A. Martin-Fernández, Ž Pedišić, C.A. Maher, L.K. Lewis, K. Hron, P.T. Katzmarzyk, J.P. Chaput, M. Fogelholm, G. Hu, E.V. Lambert, J. Maia, O.L. Sarmiento, M. Standage, T.V. Barreira, S.T. Broyles, C. Tudor-Locke, M.S. Tremblay, T. Olds, Compositional data analysis for physical activity, sedentary time and sleep research. Stat. Methods Med. Res. **27**(12), 3726–3738 (2018a)

D. Dumuid, T.E. Stanford, Ž. Pedišić, C. Maher, L.K. Lewis, J.A. Martín-Fernández, P.T. Katzmarzyk, J.P. Chaput, M. Fogelholm, M. Standage, M.S. Tremblay, T. Olds, Adiposity and the isotemporal substitution of physical activity, sedentary time and sleep among school-aged children: a compositional data analysis approach. BMC Public Health **18**(311), 1–10 (2018)

D. Dumuid, Ž Pedišić, T.E. Stanford, J.A. Martín-Fernández, K. Hron, C.A. Maher, L.K. Lewis, T. Olds, The compositional isotemporal substitution model: a method for estimating changes in a health outcome for reallocation of time between sleep, physical activity and sedentary behaviour. Stat. Methods Med. Res. **28**(3), 846–857 (2019)

D. Dumuid, Ž Pedišić, J. Palarea-Albaladejo, J.A. Martín-Fernández, K. Hron, T. Olds, Compositional data analysis in time-use epidemiology: what why, how. Int. J. Environ. Res. Public Health **17**(7), 2220 (2020)

J.J. Egozcue, V. Pawlowsky-Glahn, Groups of parts and their balances in compositional data analysis. Math. Geol. **37**(7), 795–828 (2005)

J.J. Egozcue, V. Pawlowsky-Glahn, G. Mateu-Figueras, C. Barcelo-Vidal, Isometric logratio transformations for compositional data analysis. Math. Geol. **35**(3), 279–300 (2003)

J.J. Egozcue, J. Daunis-i-Estadella, V. Pawlowsky-Glahn, K. Hron, P. Filzmoser, Simplicial regression. The normal model. J. Appl. Prob. Stat. **6**, 87–108 (2011)

S.J. Fairclough, D. Dumuid, S. Taylor, W. Curry, B. McGrane, G. Stratton, C. Maher, T. Olds, Fitness, fatness and the reallocation of time between children's daily movement behaviours: an analysis of compositional data. Int. J. Behav. Nutr. Phys. Act. **14**(64), 1–12 (2017)

S.J. Fairclough, D. Dumuid, S. Taylor, W. Curry, B. McGrane, G. Stratton, C. Maher, T. Olds, Fitness, fatness and the reallocation of time between children's daily movement behaviours: an analysis of compositional data. Int. J. Behav. Nutr. Phys. Act. **14**(1), 64 (2017)

S.J. Fairclough, D. Dumuid, K.A. Mackintosh, G. Stone, R. Dagger, G. Stratton, I. Davies, L.M. Boddy, Adiposity, fitness, health-related quality of life and the reallocation of time between children's school day activity behaviours: a compositional data analysis. Prev. Med. Rep. **11**, 254–261 (2018)

P. Filzmoser, K. Hron, C. Reimann, Principal component analysis for compositional data with outliers. Environmetrics **20**, 621–632 (2009)

P. Filzmoser, K. Hron, M. Templ, *Applied Compositional Data Analysis: With Worked Examples in R* Springer Series in Statistics. (Springer Nature Switzerland AG, Cham, 2018).
L. Foley, D. Dumuid, A.J. Atkin, T. Olds, D. Ogilvie, Patterns of health behaviour associated with active travel: a compositional data analysis. Int. J. Behav. Nutr. Phys. Act. **15**(1), 26 (2018)
L. Foley, D. Dumuid, A.J. Atkin, K. Wijndaele, D. Ogilvie, T. Olds, Cross-sectional and longitudinal associations between active commuting and patterns of movement behaviour during discretionary time: a compositional data analysis. PLoS ONE **14**(8), e0216650 (2019)
A. Gába, Ž. Pedišić, N. Štefelová, J. Dygrýn, K. Hron, D. Dumuid, M. Tremblay, Sedentary behavior patterns and adiposity in children: a study based on compositional data analysis. BMC Pediatr. **20**(1) (2020)
M. Gray, A. Sanson, Growing up in Australia: the longitudinal study of Australian children. Fam. Matters **72**, 4–9 (2005)
N. Gupta, S.E. Mathiassen, G. Mateu-Figueras, M. Heiden, D.M. Hallman, M.B. Jørgensen, A. Holtermann, A comparison of standard and compositional data analysis in studies addressing group differences in sedentary behavior and physical activity. Int. J. Behav. Nutr. Phys. Act. **15**(1), 53 (2018)
N. Gupta, D. Dumuid, M. KorshØj, M.B. JØrgensen, K. SØgaard, A. Holtermann, Is daily composition of movement behaviors related to blood pressure in working adults? Med. Sci. Sport. Exerc. **50**(10), 2150–2155 (2018)
N. Gupta, D.M. Hallman, D. Dumuid, A. Vij, C.L. Rasmussen, M.B. Jørgensen, A. Holtermann, Movement behavior profiles and obesity: a latent profile analysis of 24-h time-use composition among Danish workers. Int. J. Obes. **44**(2), 409–417 (2020)
D.M. Hallman, S.E. Mathiassen, A.J. van der Beek, J.A. Jackson, P. Coenen, Calibration of self-reported time spent sitting, standing and walking among office workers: a compositional data analysis. Int. J. Environ. Res. Public Health **16**(17) (2019)
T.H. Hejazi, M. Bashiri, J.A. Dı, K. Noghondarian, Optimization of probabilistic multiple response surfaces. Appl. Math. Model. **36**(3), 1275–1285 (2012)
K. Hron, P. Filzmoser, K. Thompson, Linear regression with compositional explanatory variables. J. Appl. Stat. **39**(5), 1115–1128 (2012)
K. Hron, P. Filzmoser, P. de Caritat, E. Fišerová, A. Gardlo, Weighted pivot coordinates for compositional data and their application to geochemical mapping. Math. Geosci. **49**(6), 797–814 (2017)
T. Hunt, M. Williams, T. Olds, D. Dumuid, Patterns of time use across the chronic obstructive pulmonary disease severity spectrum. Int. J. Environ. Res. Public Health **15**(3), 533 (2018)
INTUE: International Network of Time-Use Epidemiologists (2020), https://www.intue.org/
D. Jurakic, Ž Pedišić, Croatian 24-hour guidelines for physical activity, sedentary behaviour, and sleep: a proposal based on a systematic review of literature. Medicus **28**(2), 143–153 (2019)
J.A. Martín-Fernández, J. Palarea-Albaladejo, R.A. Olea, Dealing with zeros, in *Compositional Data Analysis: Theory and Applications*, ed. by V. Pawlowsky-Glahn, A. Buccianti (Wiley, Chichester, 2011)
J.A. Martín-Fernández, J. Daunis-i-Estadella, G. Mateu-Figueras, On the interpretation of differences between groups for compositional data. SORT-Stat. Oper. Res. Trans. **39**(2), 231–252 (2015)
G. Mateu-Figueras, V. Pawlowsky-Glahn, J.J. Egozcue, The principle of working on coordinates, in *Compositional Data Analysis*, ed. by V. Pawlowsky-Glahn, A. Buccianti (Wiley, Chichester, 2011), pp. 29-42
L. Matricciani, Y.S. Bin, T. Lallukka, E. Kronholm, M. Wake, C. Paquet, D. Dumuid, T. Olds, Rethinking the sleep-health link. Sleep Health **4**(4), 339–348 (2018)
D.E. McGregor, J. Palarea-Albaladejo, P.M. Dall, E. Stamatakis, S.F.M. Chastin, Differences in physical activity time-use composition associated with cardiometabolic risks. Prev. Med. Rep. **13**, 23–29 (2019)

D.E. McGregor, J. Palarea-Albaladejo, P.M. Dall, B. del Pozo Cruz, S.F.M. Chastin, Compositional analysis of the association between mortality and 24-hour movement behaviour from NHANES. Eur. J. Prev. Cardiol. (2019)

D.E. McGregor, J. Palarea-Albaladejo, P.M. Dall, K. Hron, S.F.M. Chastin, Cox regression survival analysis with compositional covariates: application to modelling mortality risk from 24-h physical activity patterns. Stat. Methods Med. Res. (2019)

R.H. Myers, D.C. Montgomery, C.M. Anderson-Cook, *Response Surface Methodology: Process and Product Optimization Using Designed Experiments* (Wiley, Hoboken, 2016)

New Zealand Ministry of Health, Sit less, move more, sleep well: physical activity guidelines for children and young people (2017), https://www.health.govt.nz/system/files/documents/pages/physical-activity-guidelines-for-children-and-young-people-may17.pdf. Accessed 26 Jan 2020

B.J. O'Hara, A. Grunseit, P. Phongsavan, W. Bellew, M. Briggs, A.E. Bauman, Impact of the swap it, don't stop it Australian national mass media campaign on promoting small changes to lifestyle behaviors. J. Health Commun. **21**(12), 1276–1285 (2016)

J. Palarea-Albaladejo, J.A. Martín-Fernández, zCompositions—R package for multivariate imputation of left-censored data under a compositional approach. Chemometr. Intell. Lab. Syst. **143**, 85–96 (2015)

V. Pawlowsky-Glahn, J.J. Egozcue, R. Tolosana-Delgado, *Modeling and Analysis of Compositional Data* (Wiley, Hoboken, 2015).

Ž. Pedišić, Measurement issues and poor adjustments for physical activity and sleep undermine sedentary behaviour research—the focus should shift to the balance between sleep, sedentary behaviour, standing and activity. Kinesiology **46**(1), 135–146 (2014)

Ž. Pedišić, D. Dumuid, T. Olds, Integrating sleep, sedentary behaviour, and physical activity research in the emerging field of time-use epidemiology: definitions, concepts, statistical methods, theoretical framework, and future directions. Kinesiology **49**(2), 252–269 (2017)

J. Pelclová, N. Štefelová, D. Dumuid, Ž Pedišić, K. Hron, A. Gába, T. Olds, J. Pechová, I. Zając-Gawlak, L. Tlučáková, Are longitudinal reallocations of time between movement behaviours associated with adiposity among elderly women? A compositional isotemporal substitution analysis. Int. J. Obes. **44**(4), 857–864 (2020)

V.J. Poitras, C.E. Gray, M.M. Borghese, V. Carson, J.-P. Chaput, I. Janssen, P.T. Katzmarzyk, R.R. Pate, S. Connor Gorber, M.E. Kho, Systematic review of the relationships between objectively measured physical activity and health indicators in school-aged children and youth. Appl. Physiol. Nutr. Metab. **41**(6), S197–S239 (2016)

C.L. Rasmussen, J. Palarea-Albaladejo, A. Bauman, N. Gupta, K. Nabe-Nielsen, M.B. Jørgensen, A. Holtermann, Does physically demanding work hinder a physically active lifestyle in low socioeconomic workers? A compositional data analysis based on accelerometer data. Int. J. Environ. Res. Public Health **15**(7) (2018)

C.L. Rasmussen, J. Palarea-Albaladejo, M. Korshøj, N. Gupta, K. Nabe-Nielsen, A. Holtermann, M.B. Jørgensen, Is high aerobic workload at work associated with leisure time physical activity and sedentary behaviour among blue-collar workers? A compositional data analysis based on accelerometer data. PLOS One **14**(6) (2019)

K. Ridley, T.S. Olds, A. Hill, The multimedia activity recall for children and adolescents (MARCA): development and evaluation. Int. J. Behav. Nutr. Phys. Act. **3**(1), 10 (2006)

I. Rodríguez-Gómez, A. Mañas, J. Losa-Reyna, L. Rodríguez-Mañas, S.F.M. Chastin, L.M. Alegre, F.J. García-García, I. Ara, The impact of movement behaviors on bone health in elderly with adequate nutritional status: compositional data analysis depending on the frailty status. Nutrients **11**(3) (2019)

M.E. Rosenberger, J.E. Fulton, M.P. Buman, R.P. Troiano, M.A. Grandner, D.M. Buchner, W. Haskell, The 24-hour activity cycle: a new paradigm for physical activity. Med. Sci. Sports. Exerc. **51**(3), 454–464 (2019)

N. Štefelová, J. Dygrýn, K. Hron, A. Gába, L. Rubín, J. Palarea-Albaladejo, Robust compositional analysis of physical activity and sedentary behaviour data. Int. J. Environ. Res. Public Health **15**(10), 2248 (2018)

M.L. Stevens, P. Crowley, C.L. Rasmussen, D.M. Hallman, O.S. Mortensen, C.H. Nygård, A. Holtermann, Accelerometer-measured physical activity at work and need for recovery: a compositional analysis of cross-sectional data. Ann. Work Expo. Health **64**(2), 138–151 (2020)

W. Stone, J. Hughes, Measuring social capital: toward a standardised approach. Paper presented at the Citeseer (2002).

R. Talarico, I. Janssen, Compositional associations of time spent in sleep, sedentary behavior and physical activity with obesity measures in children. Int. J. Obes. **42**(8), 1508–1514 (2018)

R.W. Taylor, S.M. Williams, V.L. Farmer, B.J. Taylor, Changes in physical activity over time in young children: a longitudinal study using accelerometers. PLoS ONE **8**(11), e81567 (2013)

R.W. Taylor, J.J. Haszard, V.L. Farmer, R. Richards, L. Te Morenga, K. Meredith-Jones, J.I. Mann, Do differences in compositional time use explain ethnic variation in the prevalence of obesity in children? Analyses using 24-hour accelerometry. Int. J. Obes. **44**(1), 94–103 (2020)

M. Templ, K. Hron, P. Filzmoser, robCompositions: an R-package for robust statistical analysis of compositional data, in *Compositional Data Analysis: Theory and Applications*, ed. by V. Pawlowsky-Glahn, A. Buccianti (Wiley, Chichester, 2011)

M.S. Tremblay, Introducing 24-hour movement guidelines for the early years: a new paradigm gaining momentum. J. Phys. Act. Health **17**(1), 92–95 (2020)

M.S. Tremblay, V. Carson, J.P. Chaput, S. Connor Gorber, T. Dinh, M. Duggan, G. Faulkner, C.E. Gray, R. Grube, K. Janson, I. Janssen, P.T. Katzmarzyk, M.E. Kho, A.E. Latimer-Cheung, C. LeBlanc, A.D. Okely, T. Olds, R.R. Pate, A. Phillips, V.J. Poitras, S. Rodenburg, M. Sampson, T.J. Saunders, J.A. Stone, G. Stratton, S.K. Weiss, L. Zehr, Canadian 24-hour movement guidelines for children and youth: an integration of physical activity, sedentary behaviour, and sleep. Appl. Physiol. Nutr. Metab. **41**(6), S311–S327 (2016)

K.G. van den Boogaart, R. Tolosana-Delgado, "Compositions": a unified R package to analyze compositional data. Comput. Geosci. **34**(4), 320–338 (2008)

P. von Rosen, I.M. Dohrn, M. Hagströmer, Association between physical activity and all-cause mortality: a 15-year follow-up using a compositional data analysis. Scand. J. Work Environ. Health **30**(1), 100–107 (2020)

Vu, V.: ggbiplot: a ggplot2 based biplot. R package version 0.55 (2011), https://github.com/vqv/ggbiplot

X. Wang, Y. Li, H. Fan, The associations between screen time-based sedentary behavior and depression: a systematic review and meta-analysis. BMC Public Health **19**, 1524 (2019)

S.M. Williams, V.L. Farmer, B.J. Taylor, R.W. Taylor, Do more active children sleep more? A repeated cross-sectional analysis using accelerometry. PLoS ONE **9**(4), e93117 (2014)

World Health Organization, Guidelines on physical activity, sedentary behaviour and sleep for children under 5 years of age (2019), https://apps.who.int/iris/handle/10665/311664

J. Yin, X. Jin, Z. Shan, S. Li, H. Huang, P. Li, X. Peng, Z. Peng, K. Yu, W. Bao, Relationship of sleep duration with all-cause mortality and cardiovascular events: a systematic review and dose-response meta-analysis of prospective cohort studies. J. Am. Heart Assoc. **6**(9), e005947 (2017)

B. Zhang, On compositional data modeling and its biomedical applications. Columbia University (2013)

J. Zhao, L. Mackay, K. Chang, S. Mavoa, T. Stewart, E. Ikeda, N. Donnellan, M. Smith, Visualising combined time use patterns of children's activities and their association with weight status and neighbourhood context. Int. J. Environ. Res. Public Health **16**(5) (2019)

W. Zhu, B. Ainsworth, Y. Liu, A comparison of urban black and white women's physical activity patterns, in *Paper Presented at the 2002 Annual Convention of the American Alliance for Health, Physical Education, Recreation and Dance*, San Diego, California, USA, April 9–13 (2002)

The manufacturer's authorised representative in the EU is Springer Nature Customer Service Centre GmbH, Europaplatz 3, 69115 Heidelberg, Germany. If you have any concerns regarding our products, please contact ProductSafety@springernature.com

Printed and bound by CPI Group (UK) Ltd, Croydon, CR0 4YY
30/09/2026
02225293-0013